Modern Power Converters for Renewable Energy Applications

As the world transitions toward sustainable energy solutions, power converters have become indispensable in enabling the efficient integration and operation of renewable energy systems. *Modern Power Converters for Renewable Energy Applications: Modeling, Analysis, Design, and Control* offers a comprehensive guide to the modeling, analysis, design, and control of these critical technologies, tailored for solar photo voltaic, wind energy, and energy storage applications.

This book delves into the unique challenges and requirements of power converters, with detailed coverage of DC–DC, DC–AC, and multilevel converter technologies. Readers will gain insights into advanced control strategies for ensuring system stability and reliability under varying conditions.

Bridging theory and practice, this book is packed with case studies, simulation examples, and design methodologies to help readers transition from conceptual understanding to practical implementation. Using industry-standard tools, readers can analyze converter performance, optimize designs, and address real-world challenges in renewable energy systems.

Key topics include grid synchronization, power quality improvement, and compliance with international standards, equipping readers to handle the complexities of modern power grids. This book also explores the integration of energy storage systems, emphasizing their role in stabilizing renewable outputs and enhancing system flexibility.

Ideal for researchers, engineers, and students, this book provides the expertise needed to excel in power electronics for renewables. Whether advancing research, driving innovation, or solving practical challenges, *Modern Power Converters for Renewable Energy Applications: Modeling, Analysis, Design, and Control* is the definitive resource for mastering the technologies shaping the future of sustainable energy.

Modern Power Converters for Renewable Energy Applications

Modeling, Analysis, Design, and Control

Edited by

Natarajan Balasubramanian Muthu Selvan, Venkatraman Thiyagarajan, and Cheng Siong Chin

CRC Press
Taylor & Francis Group
Boca Raton London New York

CRC Press is an imprint of the
Taylor & Francis Group, an **informa** business

Designed cover image: Shutterstock

First edition published 2026
by CRC Press
2385 NW Executive Center Drive, Suite 320, Boca Raton FL 33431

and by CRC Press
4 Park Square, Milton Park, Abingdon, Oxon, OX14 4RN

CRC Press is an imprint of Taylor & Francis Group, LLC

ISBN: 9781032804514 (hbk)
ISBN: 9781032804651 (pbk)
ISBN: 9781003496946 (ebk)

DOI: 10.1201/9781003496946

Typeset in Times
by codeMantra

Contents

Preface

The integration of renewable energy sources into modern power systems has introduced new challenges and opportunities in the realm of power conversion. *Modern Power Converters for Renewable Energy Applications: Modeling, Analysis, Design, and Control* addresses these complexities by focusing on the essential role of power converters in renewable energy systems. This book provides an in-depth exploration of advanced converter technologies, offering insights into the modeling, analysis, design, and control strategies crucial for efficient and reliable energy conversion.

With a rapid shift toward sustainable energy, the need for effective power converters in applications like wind, solar, and hybrid energy systems is more pressing than ever. This book serves as a comprehensive resource, presenting state-of-the-art methodologies and practical approaches that cater to both researchers and practitioners in the field. The content spans from foundational principles to advanced techniques, including discussions on grid integration, system stability, and fault tolerance.

By presenting a blend of theoretical analysis and real-world case studies, this book aims to equip readers with the knowledge needed to develop resilient, efficient, and adaptable power converter solutions in renewable energy applications.

Editors

Natarajan Balasubramanian Muthu Selvan, Ph.D., is an associate professor in the Department of Electrical and Electronics, Sri Sivasubramaniya Nadar College of Engineering, Chennai,. He has 20 years of teaching experience which includes 10 years of research experience in the field of power system optimization. Prior to joining SSN, he was with Karunya Institute of Technology and Sciences, Coimbatore, during the year 2004–2005. He received his B.E. (EEE – Distinction) degree (2000) and M.B.A. (Finance) degree (2002) from Madurai Kamaraj University. He obtained an M.E. degree (2004) and a Ph.D. degree (2013) in the field of power system engineering from Anna University, Chennai, India. His areas of interest include power system optimization, application of power electronics controllers to power system, and renewable energy systems.

During his Ph.D., he developed an enhanced Particle Swarm Optimization algorithm by the combined application of Gaussian and Cauchy distribution for solving various power system optimization problems. His Ph.D. thesis, "Application of Gaussian and Cauchy Inspired PSO Algorithms for Power System Optimization Problems", was well commended by the examiners and Eminent Professors from IISc Bangalore and IIT Madras. He has published 20 research papers in refereed international journals and more than 25 research papers in proceedings of international and national conferences. He is a recognized supervisor to guide Ph.D. degree programs offered by Anna University. He has successfully guided four Ph.D. scholars in the field of power electronics and power systems. He has received a research grant of Rs. 5.5 lakhs from SSN Trust for the project titled "Realization of Swarm Intelligence Using Real-Time Deep Learning Algorithm".

He is a senior member of The Institute of Electrical and Electronics Engineers and life member of Indian Society for Technical Education, and a fellow of The Institution of Engineers, India. He is a recognized chartered engineer from The Institution of Engineers (India). He is the recipient of the Professional Award titled "Innovative Technological Scientific Research & Dedicated Professor Award" from the Society of Engineers and Technicians Kuala Lumpur, Malaysia, in association with the *Journal of Engineering Technological Research* in 2017. He has organized several international and national conferences, workshops, and short term training program courses along with Indian society for technical education at SSN. He has also successfully conducted 6-day induction/refresher programmes sponsored (Rs. 3 lakhs) by AICTE-ISTE titled "Renewable Energy Sources Integrated Smart Power Grid Scenarios and Control Aspects Using HVDC and FACTS Controllers" in July 2018. He has delivered invited lectures, keynote addresses, and chaired technical sessions at various international conferences. He is also an active reviewer for various international journals and technical book publications.

Venkatraman Thiyagarajan, Ph.D., is an associate professor in the Department of Electrical and Electronics Engineering, Sri Sivasubramaniya Nadar College of

Engineering, Chennai. He has about 8 years of teaching experience. He received his B.E. degree in Electrical and Electronics Engineering with distinction and M.E. degree in Power Systems Engineering with distinction from Anna University, Chennai, India. He received his Ph.D. degree from the Division of Power System Engineering, College of Engineering Guindy (CEG), Anna University, Chennai. He has published 30 research papers in refereed international journals and 75 research papers in the proceedings of various international conferences. He has coordinated and organized various national-level workshops and international conferences. He has received several best paper awards for his research papers at various international conferences. He is a recognized supervisor for Anna University. His areas of research include power converters, stability, and control of electric power grids, renewable energy systems, and microgrid. He is the reviewer for many international journals. He is a senior member of The Institute of Electrical and Electronics Engineers and a member of The Institution of Engineers, India.

Cheng Siong Chin, Ph.D., is a chair professor in Intelligent Systems Modelling and Simulation at Newcastle University and an Adjunct Professor at Chongqing University, School of Automotive Engineering. He received his Ph.D. in Applied Control Engineering at Research Robotics Centre at Nanyang Technological University (NTU) in 2008 and his M.Sc. (Distinction) in Advanced Control and Systems Engineering from The University of Manchester in 2001. Before moving into academia, he worked in the consumer electronics industry for a few years. He obtained six Economic Development Board (EDB)-Industrial Postgraduate Programme (IPP) and two Singapore Maritime Institute (SMI) grants from the Singapore government in intelligent systems design, simulation, and predictive analytics as a PI and Co-PI. He has made more than 170 publications including 5 authored books, and has 2 Singapore Patent Applications and 3 US Patents. He also received DCASE2019 Judges' Award in IEEE AASP Challenge on DCASE2019. He is an associate editor for *IEEE Access Journal*, *Applied Artificial Intelligence*, and *IEEE Transportation Electrification Community (TEC) eNewsletter*, and is an editorial board member of *Electronics*. He has served as general chair, keynote speaker, session chair, and technical committee member at numerous international conferences.

Contributors

S. Allirani
Electrical and Electronics Engineering
Sri Ramakrishna Engineering College
Coimbatore, India

Angalaeswari Sendraya Perumal
School of Electrical Engineering
Vellore Institute of Technology
Chennai, India

Anitha Perumalsamy
Electrical and Electronics Engineering
University VOC college of Engineering
Thoothukudi, India

I. Arul Doss Adaikalam
Electrical and Electronics Engineering
Easwari Engineering College
Chennai, India

G. Arun Sampaul Thomas
Electrical and Electronics Engineering
Velammal Institute of Technology
Chennai, India

P. Aruna
Electrical and Electronics Engineering
Anand Institute of Higher Technology
Chennai, India

Booma Jayapalan
Electrical and Electronics Engineering
PSNA College of Engineering and Technology
Dindigul, India

V. S. Chandrika
Electrical and Electronics Engineering
KPR Institute of Engineering and Technology
Coimbatore, India

Divya Thangaraj
Electrical and Electronics Engineering
Sri Sivasubramaniya Nadar College of Engineering
Chennai, India

J. Divya Navamani
Electrical and Electronics Engineering
SRM Institute of Science and Technology
Chennai, India

S. Edward Rajan
Electrical and Electronics Engineering
Mepco Schlenk Engineering College
Sivakasi, India

Gnanavadivel Jothimani
Electrical and Electronics Engineering
Mepco Schelnk Engineering College
Sivakasi, India

Gunasekari Rethinam
Electrical and Electronics Engineering
Sri Sairam College of Engineering
Bangalore, India

B. Hemanth Kumar
Electrical and Electronics Engineering
Mohan Babu University
Tirupati, India

M. Jawahar
Electrical and Electronics Engineering
Mepco Schlenk Engineering College
Sivakasi, India

Jayanthi Kathiresan
Electrical and Electronics Engineering
Mepco Schlenk Engineering College
Sivakasi, India

Kamaraja Arunavathi Soundarapandian
Electrical and Electronics Engineering
Kalasalingam Academy of Research and Education
Srivilliputur, India

Kanimozhi Gunasekaran
Centre for Smart Grid Technologies
Vellore Institute of Technology
Chennai, India

Kanimozhi Kannabiran
Electrical and Electronics Engineering
NPR College of Engineering and Technology
Natham, India

Karthik Kumar Kalirajan
Electrical and Electronics Engineering
National Engineering College
Kovilpatti, India

M. Karthikeyan
Centre for Advanced Wireless Integrated Technology
Chennai Institute of Technology
Chennai, India

B. Kaviya
Electrical and Electronics Engineering
Sri Ramakrishna Engineering College
Coimbatore, India

S. Kiruthika
Electrical and Electronics Engineering
M. Kumarasamy College of Engineering
Karur, India

Krishnaveni Subramani
Electrical and Electronics Engineering
Sri Sivasubramaniya Nadar College of Engineering
Chennai, India

A. Lavanya
Electrical and Electronics Engineering
SRM Institute of Science and Technology
Chennai, India

D. Logesh
Electrical and Electronics Engineering
Sri Ramakrishna Engineering College
Coimbatore, India

J. Madhan Kumar
Electrical and Electronics Engineering
Sri Ramakrishna Engineering College
Coimbatore, India

Karthikeyan Madhu
Mechatronics Engineering
Velammal Institute of Technology
Chennai, India

Manimegalai Damodaran
Electrical and Electronics Engineering
Rajalakshmi Engineering College
Chennai, India

Jawahar Marimuthu
Department of Electrical and Electronics Engineering
Mepco Schlenk Engineering College
Sivakasi, India

Muralikrishnan Gopalakrishnan
Electrical and Electronics Engineering
S.A. Engineering College
Chennai, India

S. Muthukaruppasamy
Electrical and Electronics Engineering
Velammal Institute of Technology
Chennai, India

J. Nandha Gopal
Electrical and Electronics Engineering
Velammal Institute of Technology
Chennai, India

Nazrin Salma Sheriff
Electrical and Electronics Engineering
Thamirabharani Engineering College
Tirunelveli, India

Niyas Ahamed Allavudeen
Electronics and Communication Engineering
Thamirabharani Engineering College
Tirunelveli, India

E. Parimalasundar
Electrical and Electronics Engineering
Mohan Babu University
Tirupati, India

S. Prakash
School of Electrical Engineering
Vellore Institute of Technology
Chennai, India

Preetha Krishnasamy
Electrical and Electronics Engineering
Sengunthar Engineering College
Erode, India

S. Priyadarshini
School of Electrical Engineering
Vellore Institute of Technology
Chennai, India

R. Raghul Kumar
Electrical and Electronics Engineering
Amrita School of Engineering
Coimbatore, India

Ramaprabha Ramabadran
Electrical and Electronics Engineering
Sri Sivasubramaniya Nadar College of Engineering
Chennai, India

P. Sakthi
Electrical and Electronics Engineering
M. Kumarasamy College of Engineering
Karur, India

Edward Rajan Samuel Nadar
Department of Electrical and Electronics Engineering
Mepco Schlenk Engineering College
Sivakasi, India

S. Sathya Agila
Electrical and Electronics Engineering
Sri Sivasubramaniya Nadar College of Engineering
Chennai, India

Senthil Kumar Natarajan
Electrical and Electronics Engineering
Mepco Schelnk Engineering College
Sivakasi, India

K. Shobika
Electrical and Electronics Engineering
Sri Ramakrishna Engineering College
Coimbatore, India

M. S. Sivagama Sundari
Electrical and Electronics Engineering
Amrita School of Engineering
Coimbatore, India

S. Sridharan
Electrical and Electronics Engineering
St. Joseph's College of Engineering
Chennai, India

L. Sridutt
School of Electrical Engineering
Vellore Institute of Technology
Chennai, India

Srinivasan Gopalan
Electrical and Electronics Engineering
MVJ College of Engineering
Bangalore, India

N. Subramani
Electrical and Electronics Engineering
Sri Ramakrishna Engineering College
Coimbatore, India

K. Suresh
Electrical and Electronics Engineering
Christ University
Bangalore, India

Venkatraman Thiyagarajan
Electrical and Electronics Engineering
Sri Sivasubramaniya Nadar College of Engineering
Chennai, India

Titus Sigamani
Electrical and Electronics Engineering
K. Ramakrishnan College of Engineering
Trichy, India

V. Vasan Prabhu
Tech-Manager- EV
Great Learning
Chennai, India

P. Velmurugan
Electrical and Electronics Engineering
St. Joseph's College of Engineering
Chennai, India

1 Power Converters for Renewable Energy Systems
An Introduction

Jothiperumal Nandha Gopal,
Subburaj Muthukaruppasamy,
Irudaya Maria Arulappan Arul Doss Adaikalam,
and Balasubramanian Madhu

1.1 INTRODUCTION

Because of their negative impacts, it is vital to switch from using traditional fossil fuels to produce power using clean, RESs, such as geothermal, hydropower, wind, and solar energy. An estimated hourly energy input from the sun is equal to the annual energy intake of humans, underscoring the possibility of using this plentiful a tool to reduce reliance on fossil fuels and mitigate global warming for a better and cleaner world. In rural and hilly areas, using distributed energy resources that comprise RESs can eliminate the need for expensive distribution and transmission infrastructure [1]. Delivering power to remote locations at a reasonable cost is made possible by using distributed energy resources as independent sources of renewable energy. Power generation in the near future will be dominated by wind farms, fuel cells (FC), and photovoltaic (PV) systems. The low DC voltage of PV systems makes them unsuitable for direct microgrid integration, which poses a hurdle. PV modules are often linked in series to achieve higher voltages, which requires a large amount of space and multiple components. To optimize the utilization of renewable energy, this voltage must be raised via efficient DC–DC converters. These converters have to be affordable and lightweight, should have a low switching voltage, and should have a high power density [2]. As a result, for increased power conversion efficiency, research interest in AC–DC and DC–DC conversion methods has increased significantly. The use of several electrical power converters in a system of renewable energy situations is shown in Figure 1.1.

A number of academics have focused on various types of DC–DC converters intended for energy storage and renewable energy applications in an effort to increase efficiency and address converter shortcomings. Based on how they operate, these converters are often divided into three main categories, as shown in Figure 1.2: three modes of switching: mild, hard, and linear. Linear mode converters' benefits

DOI: 10.1201/9781003496946-1

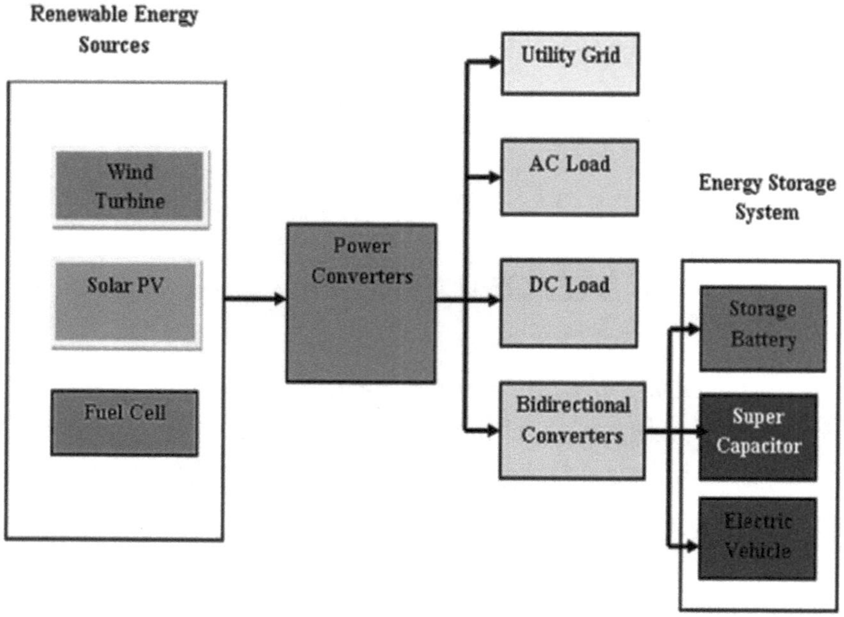

FIGURE 1.1 A renewable energy system's power converter.

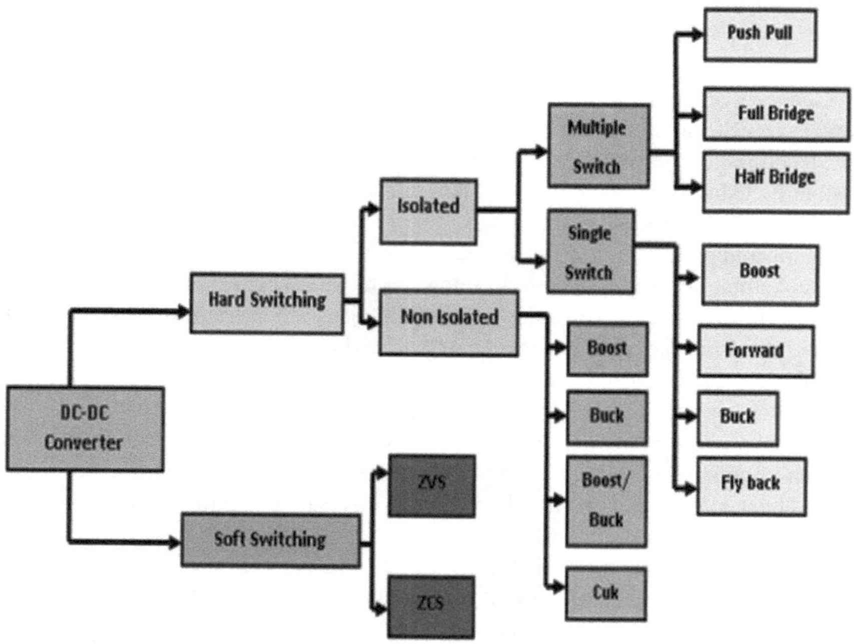

FIGURE 1.2 Types of bidirectional converters

include their simplicity, efficient control, low noise, and quick response. However, power losses under various operating situations reduce their efficiency. Depending on whether galvanic isolation is present, in hard switching mode, there are two further categories for converters: isolated and non-isolated [3]. Chopper circuits, which include buck, boost, buck-boost, and Ćuk converters, are examples of non-isolated power converter topologies. Galvanic isolation, on the other hand, which is usually accomplished via transformers, is required when the converters are connected to the power grid, for reasons of safety. When used in conjunction with the right control strategies, power converters are essential for controlling node voltages in microgrids that support a variety of load types, including constant power, inductive, resistive, and loads that are not linear. However, constant power loads are sometimes difficult to control using conventional control approaches and can lead to instability in the voltage output from power converters [4].

The isolated converter is another subtype of the switching mode that uses a variety of switching techniques, including full-bridge, fly-back, push-pull, dual half-bridge, and half-bridge converters. The achievable switching frequency of hard switching mode converters may be limited by their huge size and weight, strong electromagnetic interference (EMI), and significant switching losses. Soft switching, sometimes known as resonant converters, is the third category of DC–DC converters. Several research studies, especially those pertaining to their industrial uses, have demonstrated that the converters were developed in response to the shortcomings of hard switching mode converters. Two subtypes of soft switching are zero-voltage switching (ZVS) and zero current switching (ZCS). ZVS and ZCS converters are more efficient than linear regulators and are able to function at greater switching frequencies, allowing for smaller ferrite transformer cores to be used. In addition, as opposed to linear regulators, they may accept a wider range of DC input voltages [5].

In the academic and industrial domains, an enormous amount of interest has been shown in bidirectional DC–DC converters. These are mostly used to maintain system reliability and function as a bridge between batteries and supercapacitors as storage devices. These converters are becoming more and more common in numerous situations where bidirectional power flow is required, including electric vehicles (EVs), energy storage devices, renewable energy systems, and uninterruptible power supply (UPS) [6]. The following part elaborates on the many types and provides detailed information on bidirectional converters as illustrated in Figure 1.2.

DC–DC converters with resonance are currently the most widely used choice for power conversion in situations with both high and low voltages. Features of switching losses at output rectifier diodes and inverter switches are decreased by resonant DC–DC converters, allowing for higher switching frequencies and increased efficiency. This allows for the development of smaller converters. There are several different topologies for series, parallel, and series-parallel resonant DC–DC converters, in addition to parallel resonant DC–DC converters (PRCs), which are examples of resonant DC–DC converters. The usage of resonant converters is common and has been thoroughly studied because of their ease of use and popularity [7].

To accommodate AC loads, DC–AC conversion is also necessary in addition to DC–DC conversion and enables distributed energy resource integration into the grid. Among power converter topologies investigated in home and commercial settings, multilevel inverters are particularly noteworthy. Because of their many benefits, they have grown in popularity in high-power and medium-voltage applications in recent years [8]. Multilevel inverters have several benefits over traditional two-level inverters, including less harmonic distortion, low EMI, and reduced voltage stress on semiconductor components. They do, nevertheless, need more semiconductor components. Multilevel inverters are superior at reducing harmonic components in voltage and current waveforms in contrast to ideas for two-level inverters with equivalent power ratings [9]. Different types of level-by-level topologies are distinguished, including H-bridge inverters that are cascaded, diode-clamped, and capacitor-clamped. They are extensively utilized in clean energy sources to connect high-power loads to PV modules and other RESs, and they are essential in high-power converter applications [10].

Multilevel inverters are used in many different applications, such as devices for the Flexible Alternative Current Transmission System, power converters, and reactive power correction systems for high-power AC motors. In addition, topology hybridization has been more popular lately. For example, bidirectional operation of resonant converters allows for both directions of power transfer. Bidirectional resonant converters are frequently utilized in high-voltage power supply applications, EVs, battery chargers, supercapacitors, and renewable energy systems because they are relatively simple to integrate with other components [11].

Multilevel inverters are connected to DC–DC converters to facilitate compact step-up DC–AC conversion. Topologies like multilayer bidirectional converters and multilevel bidirectional resonant converters have been proposed as a result. The following are the main goals and objectives of multilayer bidirectional converters:

- Significant Reduction in Switching Losses: One of the primary goals is to minimize switching losses, thereby enhancing overall efficiency.
- Reduction in Harmonic Distortion: Another key target is to mitigate harmonic distortion in the output waveform, ensuring cleaner power delivery to loads.
- High Efficiency with Low Component Count: Achieving high efficiency while keeping the component count low is crucial for providing efficient and compact solutions.

By addressing these targets, multilevel bidirectional converters aim to deliver efficient and conveniently sized solutions for DC–AC conversion, meeting the demands of various applications effectively. The paper explores both resonant and bidirectional DC–DC converter topologies in addition to multilevel inverter topologies. It also looks at various converter topologies' hybrid configurations. Every topology's characteristics, contrasts, benefits, drawbacks, and uses are evaluated closely. Additionally, the difficulties and potential of the newest converter topologies are explained [12]. The remainder of this chapter is organized like this:

- In Section 1.2, converters that operate both ways are reviewed, along with an analysis of separated and un-separated converters and their classification.

- Many types of resonant converter structures are described in Section 1.3.
- In Section 1.4, Converters' topologies such as cascaded, diode-clamped, and capacitor-clamped with an emphasis on multilevel inverter topologies are discussed.
- Section 1.5 delineates the utilization of these converters in energy storage systems and grid connectivity.
- Section 1.6 provides an overview of the difficulties and potential outcomes.
- Section 1.7 outlines the study's conclusion.

1.2 DC–DC BIDIRECTIONAL CONVERTERS

Battery charge and discharge management has to be made simpler as energy storage devices like batteries and supercapacitors become more common. In order to do both charging and discharging tasks with a single power train, bidirectional DC/DC converters are essential. Renesas Electronics demonstrates in this paper how it has incorporated bidirectional control into its most recent controller, making it possible to perform reverse-direction power flow and control on-the-fly with ease. This development makes energy storage system administration easier while maintaining battery health and effectively increasing runtime [13].

As we embrace RESs and lessen our dependency on foreign oil, the globe is changing dramatically. The advancement of energy storage technology and its widespread application of high-density devices such as supercapacitors and lithium-ion batteries are important to this change. These energy storage devices are essential to RESs, such as wind and solar electricity, because they gather and store energy for a steady supply to the grid and different end users, both residential and commercial. Moreover, the proliferation of battery-operated portable devices, including cell phones, drones, robots, and EVs, underscores the growing importance of energy storage devices in our daily lives. Additionally, the demand for uninterrupted operation of information systems, such as data centers and telecom networks, in the fast-growing data economy necessitates reliable backup power solutions, further driving the need for energy storage devices. In managing energy storage devices, precise control of the charge and discharge processes is essential to ensure safety, reliability, and longevity. While conventional approaches often involve separate power trains for charging and discharging to meet different control targets, certain applications require fast transitions between charge and discharge operations. For instance, seamless power delivery in a data center server's DC battery backup system necessitates a rapid transition from charge to discharge; on the other hand, a motor drive system used for braking activities needs to switch from discharge to charge quickly [14]. To address these demands, a single power train on-the-fly bidirectional charge and discharge converter is necessary. Such converters lower system costs and allow smooth, quick transitions by combining both functions into a small design. For energy storage device control applications, the ISL81601 buck-boost controller provides a dependable on-the-spot bidirectional DC/DC power conversion solution. Customers benefit from tremendous economic value and technical confidence because of its unique architecture and control algorithm.

1.2.1 DC/DC Converter Topologies in Both Directions

With converters that use a synchronous rectifier, bidirectional operation is possible for both isolated and non-isolated topologies. For the sake of simplicity, this study only considers non-isolated topologies, yet its conclusions can be applied to isolated topologies by adding an isolation transformer. Figure 1.3 illustrates some instances of non-isolated bidirectional converters, including the buck, boost, and four-switch buck-boost designs. Of particular note are the facts that a buck converter can also act as a boost converter when reversed, and vice versa. Like a single buck or the four-switch buck-boost converter, boost converter can smoothly switch to reverse operation while operating in either two-switch buck or two-switch boost mode [15].

Two examples of common bidirectional operation systems are shown in Figure 1.4:

- Battery Charge/Discharge Bidirectional Operation System: This system makes use of a buck-boost controller with four switches that can be adjusted to enable bidirectional operation on-the-fly. If the output voltage (V_{out}) of the bidirectional DC/DC converter is near the battery operation voltage, it can function as a four-switch buck-boost converter during battery charge or discharge. If V_{out} continuously exceeds the battery voltage, it can also

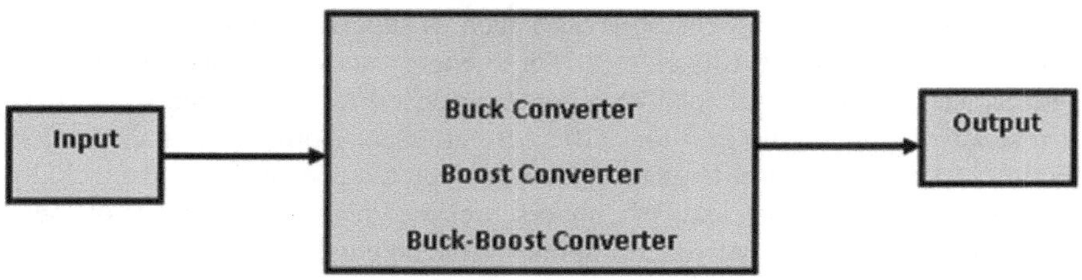

FIGURE 1.3 Bidirectional DC/DC converter topologies.

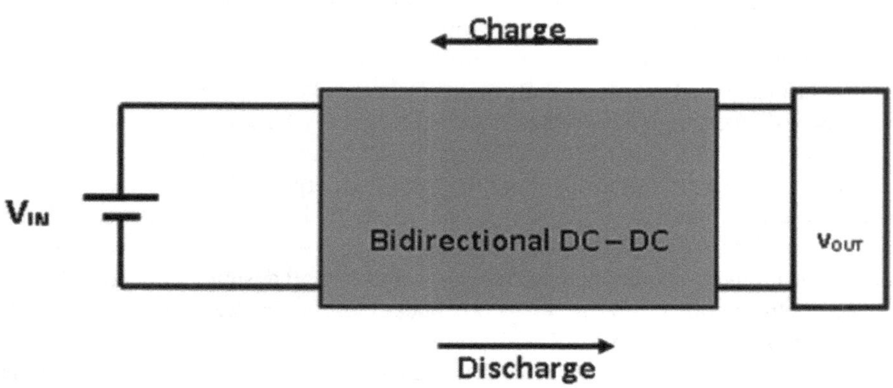

FIGURE 1.4 DC/DC converter with two channels for charging and discharging.

function as a boost converter; conversely, if V_{out} stays below the battery voltage, it can function as a buck converter. The buck-boost converter with four switches operates at maximum efficiency by using power devices and operating currents with lower voltage ratings. Furthermore, it offers complete over-current and short-circuit protection throughout both charge and discharge operations, guaranteeing safe battery operation.
- A buck-boost is recommended to utilize a bidirectional DC/DC converter in the super capacitor voltage backup system. This decision was made because, in order to use all of the capacitor's capacity, it must discharge to a very low voltage. A buck-boost converter facilitates discharge to very low voltages, allowing the super capacitor's capacity to be used efficiently.

These bidirectional operation systems highlight the versatility and efficiency benefits offered by different converter configurations, tailored to the specific requirements of battery charge/discharge and super capacitor backup applications.

1.2.2 Divide DC–DC Bidirectional Converters into Different Categories

The two types of bidirectional DC–DC converters discussed in the preceding section are Isolated Bidirectional DC–DC Converters (IBDC) and Non-Isolated Bidirectional DC–DC Converters (NIBDC), as illustrated in Figure 1.5 [16]. Each kind has unique benefits and uses, which are listed below.

1.2.2.1 DC to DC Bidirectional Non-Separated Converter

High-frequency transformers are typically employed in designs that require electrical isolation between the input and the output in order to operate NIBDC output because of safety concerns. NIBDCs are especially useful in low-power applications because of their lightweight construction and ease of control. Numerous parameters, including low EMI, high step-up/down ratio, low complexity of circuit layouts, and ease of operation, have been thoroughly examined by researchers when evaluating NIBDCs [17]. These evaluations have highlighted the effectiveness and suitability of NIBDCs for a wide range of applications, particularly in scenarios where electrical isolation is not required but efficiency, simplicity, and ease of control are paramount.

- Operating in parallel but directed in opposite directions, the Bidirectional Buck-Boost Converter is essentially a hybrid arrangement that combines both the buck and boost converter topologies. Similar to how a buck converter works, this converter allows power to move down to the low-voltage (Va) side from the high-voltage (Vb) side. However, it can also operate in reverse, transferring power from the high-voltage (Vb) side to the low voltage (Va) through a boost converter. Because of its ability to transfer energy across two voltage domains effectively, its bidirectional capacity makes it suitable for a wide range of applications that require bidirectional power flow.
- The SEPIC-ZETA DC to DC Bidirectional Converter boasts a flexible design featuring two unique power flow modes: backward and positive modes.

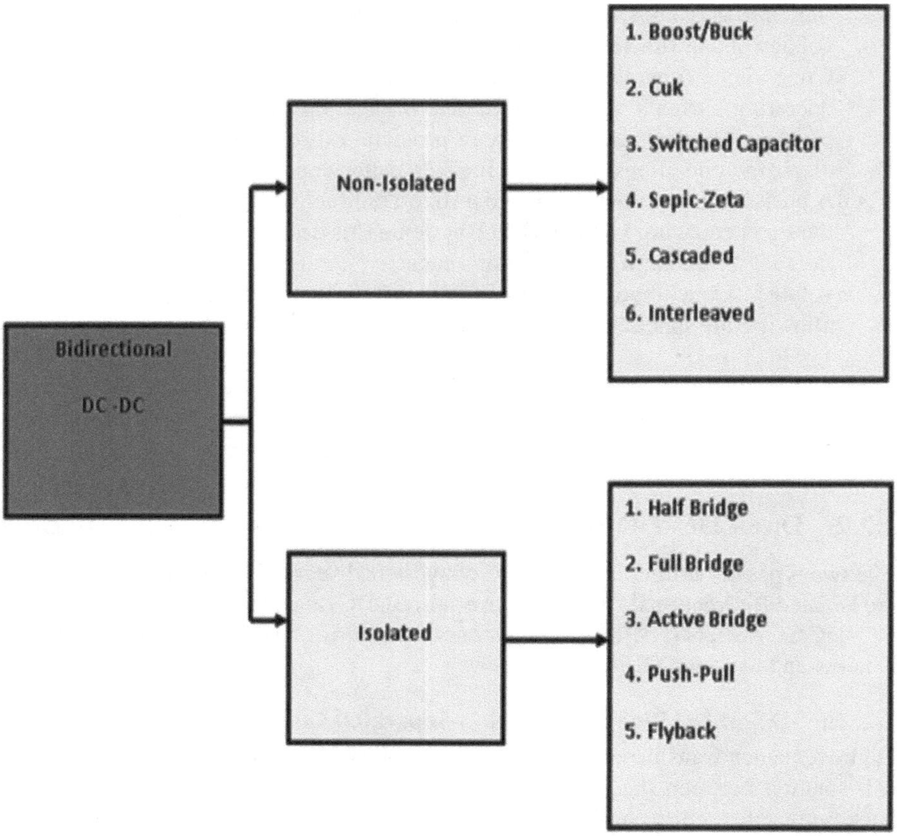

FIGURE 1.5 Classification of bidirectional DC/DC converters.

1. Forward Power Transfer**:** The converter operates as a Single-Ended Primary Inductor Converter in this mode. Because it may have the ability to convert voltage both up and down, the single-ended primary inductor converter architecture is an effective option for a variety of applications that need flexible voltage regulation.
2. Reverse Power Flow**:** The converter functions as a ZETA converter when it is in the negative mode. Bidirectional power flow, which permits energy to be transported in both step-up and step-down directions, is a feature of the ZETA topology. This mode increases the overall adaptability of the converter by facilitating effective energy transfer in applications where reverse power flow is required.

- Compared to other configurations like bidirectional and cascaded buck-boost converters, the bidirectional Cuk converter has the advantage of producing fewer output ripples, making it an ideal option for wiring batteries and supercapacitors together in circuits. The following are the benefits of this converter's use of Cuk topology:

1. Reduced Output Ripples: The Cuk converter inherently minimizes output ripples, resulting in smoother voltage and current waveforms. This characteristic is crucial for applications requiring stable and reliable power delivery.
2. Efficient Energy Transfer: The Cuk converter efficiently transfers energy between input and output, making it suitable for bidirectional power flow applications where energy storage devices like supercapacitors and batteries need to be interfaced with circuits.
3. Compact Design: Cuk converters can be designed with compact and lightweight components, making them ideal for space-constrained applications such as portable electronics and automotive systems.

   Utilizing these benefits, the bidirectional Cuk converter provides increased dependability and performance, making it the recommended option for connecting batteries and supercapacitors in circuits where lower output ripples are crucial.

- Connecting two buck-boost converters results in the Sequential Bidirectional Buck-Boost Converter architecture. This configuration allows for functioning in all four quadrants and allows for both boost and buck functionality to be used in both directions. The adaptability of this converter is highlighted by the flexibility it offers in all quadrants. It does have certain disadvantages, though. For example, higher switching losses are caused by more switches. In addition, inverse diode recovery results in additional operational losses and the need for complicated control methods.
- Within the domain of bidirectional, the Switched Capacitor DC–DC Converter and other DC–DC converters utilize integrated circuits within converter circuits to accomplish capacitor exchange. The lack of magnetic components in DC–DC converters that are not separated facilitates this integration of integrated circuits. But these converters have a lot of output ripple because of all the passive parts that are there to cause EMI. Implementing control measures, such as voltage and current management approaches, can help to reduce this difficulty, but doing so may increase system complexity and related costs.
- Interleaved DC–DC converters in non-isolated configurations employ a setup in which two or more converters operate simultaneously, each with a phase shift of 360 degrees divided by "n", where "n" is the number of interleaved phases. Many benefits come with this configuration, such as reduced output current ripples, higher thermal capacity, increased power density, improved system efficiency, and current division (I/n). Fewer switches are needed, and conduction losses are minimized by dividing the current along parallel pathways. By reducing current and voltage ripple, the interleaving method increases overall system efficiency by decreasing switching losses at the DC–DC converter input. The major characteristic of this architecture is a two-phase interleaved NIBDC with two output stages that operate 180 degrees out of phase.

1.2.2.2 DC–DC Bidirectional Isolated Converters

Applications involving NIBDCs are often limited due to the lack of secure galvanic isolation standards, which are typically required in NIBDC systems. IBDCs use high-frequency transformers to produce galvanic isolation, which is essential for many applications to manage voltage differences between environments, maintain safety, and decrease noise. The DC–AC converter performs high-frequency switching in two stages; when there are two sources, galvanic isolation is maintained by a high-frequency transformer and synchronous voltage across stages for the best possible system performance and design. It is suggested that IBDCs be used as a flexible power processing interface between various components and energy storage systems. They include benefits such as gentle switching without the requirement for a basic transformer topology, and active or passive components make design and maintenance work easier [18]. On the other hand, switch current strains provide problems for both converter components. For a quicker dynamic reaction, this method usually uses peak current mode control or average current control techniques. IBDCs have disadvantages despite their advantages. Under light-load circumstances, they could lose their capacity to switch softly, and they might become sensitive to even little changes in flux, particularly at high bus voltages. High ripple content in DC bus currents is another problem that requires intricate filtering circuits to solve. There are several different topologies for isolated converters, like dual half-bridge, Cuk, push-pull, forward, fly-back, and dual active full-bridge IBDC. Among these, the full-bridge and half-bridge topologies are favored due to their efficacy in many applications.

- Because of its power density, gentle switching qualities, and controllability, the IBDC with dual half-bridge is highly appreciated and used, especially for EV applications. The efficiency of these converters is usually between 92% and 94%. The dual half-bridge converter works by having two sides: a low-voltage side (also known as the "B side" or boost half-bridge) that receives modified current and a high-voltage side (also known as the "A side") where the half-bridge converter receives its voltage. The boost half-bridge is often found on the lower side and is coupled to a battery or capacitor to provide DC energy, preferably with minimal current ripple. For applications like EVs, this arrangement facilitates effective bidirectional power transfer between the high- and low-voltage sides.
- A popular topology that makes use of back-to-back bidirectional approaches separated by a high-frequency transformer is the bidirectional DC–DC Converter with Separated Active Full-Bridge technology. Back-to-back converters in this format can be configured as half-bridges or full-bridges, and they can be supplied either voltage or current. This flexibility in configuration allows for efficient bidirectional power transfer and isolation, making separated active full-bridge converters suitable for various applications requiring high efficiency and robust power handling capabilities.
- A separate, two-way DC–DC converter was presented using a hybrid configuration of half-bridge and full-bridge topologies in the context of UPS architecture. In this converter, the secondary side of the transformer has

a voltage-fed full-bridge topology, while the primary side has voltage-fed half-bridge architecture. Using this hybrid topology offers several benefits, one of which is reduced control complexity; a Dual Active Bridge (DAB) converter is utilized. This configuration makes it possible to implement a full UPS topology and is particularly useful for adding a two-switch buck-boost converter to the half-bridge side. Regarding UPS applications, this method maintains robustness and adaptability while optimizing efficiency and control.

- To accomplish magnetic isolation, a transformer is used in place of the buck-boost converter's inductor to create the bidirectional fly-back converter. In order to attain gain during forward power flow, the converter is subjected to both voltage-second and charge-second balancing, which appropriately reflects the voltage gain ratio of the fly-back converter. Voltage clamp snubbers and transformer design techniques must be used in order to regulate the fly-back transformer's leakage current.
- The bidirectional push-pull DC–DC converter, which builds upon the concepts of the unidirectional push-pull converter, allows power flow in both directions. Like their unidirectional cousins, bidirectional push-pull converters transfer power using a multiwinding transformer. This method is also supported by a three-phase bidirectional push-pull converter that has been created for high-power applications.

1.3 RESONANT CONVERTERS

Resonant converters are gadgets that use a system of inductors and capacitors that are adjusted to resonate at specific frequencies in order to offer switching between zero voltage and zero current. These inter-changeable re-incorporated into electric circuits to reduce switching losses in components like metal-oxide-semiconductor field-effect transistors (MOSFETs) or insulated gate bipolar transistors (IGBTs). They function as electrical inverters based on resonant current oscillation. Switching devices used in circuits can lead to high rates of change of current (di/dt) and voltage (dv/dt), resulting in increased power losses and voltage stresses. This issue becomes more pronounced with smaller component sizes, which allow for higher switching frequencies and increase losses and stress. Additionally, high di/dt and dv/dt can cause EMI. To mitigate these drawbacks, resonant converters are designed to turn switching devices on and off when the voltage and current across the device are zero, minimizing losses and stresses. The necessary resonance for these converters is achieved through the inductors (L) and capacitors (C) in the circuit [19].

1.3.1 Resonant Converter Structure

The switch network, resonant tank, and rectifier network are the three primary components of a typical resonant converter (see Figure 1.6). From a DC voltage or current source, the switch network produces pulsing voltage and current. The rectifier network receives this output from the switch network after that. The pulsing DC signal must be transformed into a constant DC voltage or current via the rectifier network.

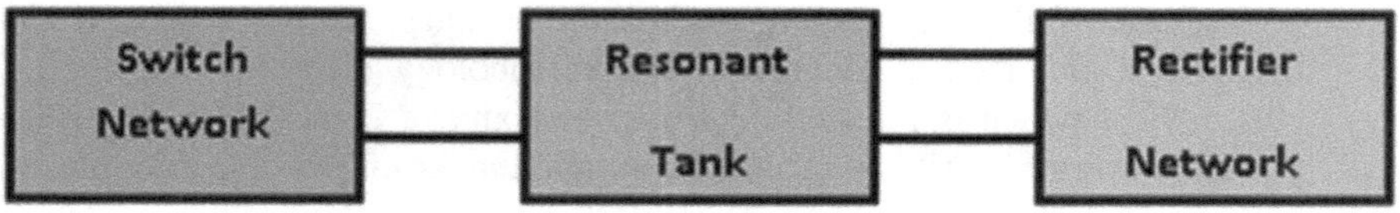

FIGURE 1.6 Resonant converter structure.

To accomplish this rectification, it processes the pulsing voltage or current that it receives from the switch network. By altering the frequency of the pulsating signal, the inductor and capacitor that make up the resonant tank are essential in changing the amplitude gain of the signal. This helps optimize the efficiency and performance of the converter. In some designs, a DC blocking capacitor (Cb) is incorporated between the switch network and rectifier network to prevent transformer saturation and ensure proper operation of the converter. This capacitor helps control the flow of DC current while allowing AC components to pass through effectively [20].

- The purpose of the Switching Network is to transform DC electricity into AC power. Full- and half-bridge inverters are common switching network types that are selected based on power requirements. While the half-bridge inverter only gives 50% of the input voltage, the full-bridge inverter is recommended for high-power applications. The half-bridge is appropriate for situations with high input voltages because of its reduced voltage transitions. Both resonant power converters and traditional DC–DC converters can achieve high switching frequencies and gentle switching. This similarity results from the fact that inverters may convert DC to AC. To send DC power to the load, voltage can then be stepped up or down utilizing electromagnetic components and rectifier networks. Usually, full- or half-bridge inverters are paired with either a full-bridge or center-tap rectifier when resonant converters are used.
- The most important part of a resonant converter, or resonant circuit, is the resonant tank. An inductive and capacitive (LC) circuit, which functions as a reactive element storing electrical energy that oscillates at the frequency of the resonant circuit, is part of this network. Resonance is reached in the LC circuit through energy oscillations between the inductor and capacitor. Electromagnetic frequencies produced by this oscillation are helpful in fields such as telecommunications. Choose the right values for the reactive components to alter the tank's resonance frequency. Three main parameters are used to categorize the different types of rectifier tank networks.
- In the architecture of a resonant network converter, the rectifier network is essential, especially for the diode rectifier with capacitive low-pass filter. After the resonant tank produces sinusoidal current and voltage waveforms, this network is utilized to correct and filter the AC waveform in order to produce the necessary DC output waveform. Center-tapped or full-bridge rectifier diode rectifier with capacitive low-pass filter solutions are frequently taken into consideration in resonant power converter research. Nevertheless, because of the high voltage stress on diodes, center-tapped

rectifiers are not as feasible; low-pass filtering with two instances of capacitance or inductance is recommended instead.

1.3.2 Resonant Converter Classification

Figure 1.7 illustrates the three primary types of resonant converters: multiresonant, quasi-conventional, and quasi-resonant converters. Load resonant converters and phase-shift modulated converters are two further classifications for conventional resonant converters. In an effort to replace conventional power switches with resonant switches, quasi-resonant converters integrate features of resonant and Pulse Width Modulation converters. Research offers succinct explanations of quasi-resonant converters and other kinds of multiresonant converters made to overcome quasi-resonant converter shortcomings [21]. To accomplish ZVS, multiresonant converters use a multiresonant network made up of two resonant capacitors and resonant inductance. Different varieties of traditional resonant converters, such as those with series, parallel, and series-parallel loads, are covered in the following sections.

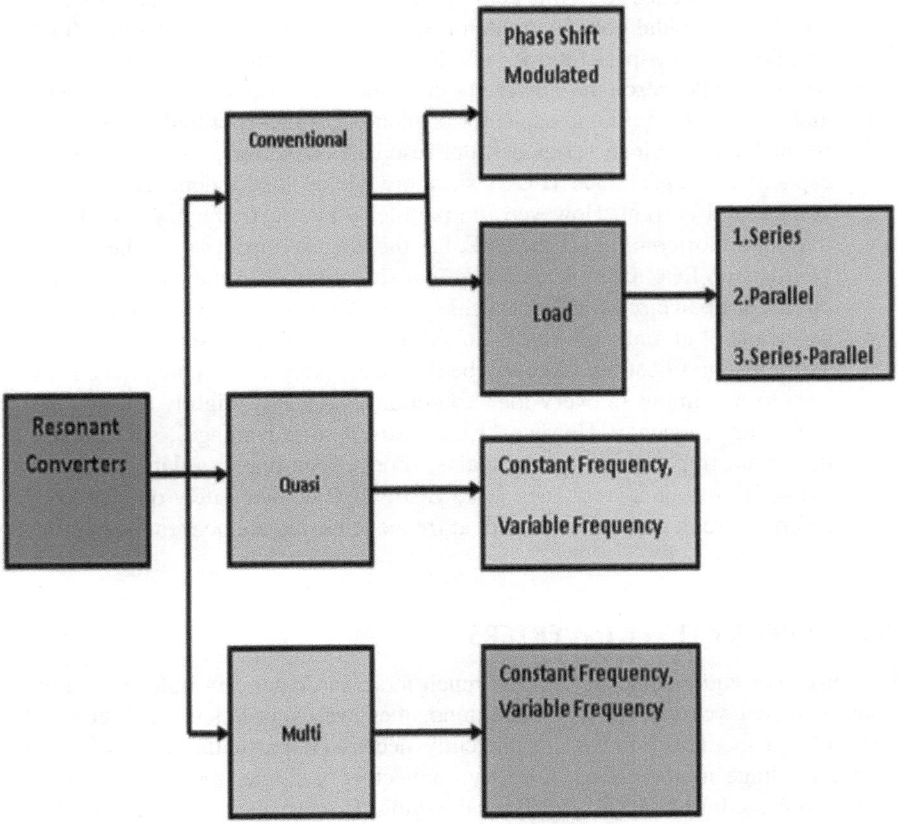

FIGURE 1.7 Resonance DC–DC converter classification.

- High-voltage applications might benefit from the unique benefits that load resonant converters, especially Series Resonant Converters (SRCs), provide over traditional power converters. High-frequency operation is made possible by SRCs, which also minimize switch stress and reduce equipment space while preserving power conversion efficiency. The load network and rectifier are connected in series with the tank circuit in a SRC. The voltage division in the circuit determines the voltages generated by the load and resonant tank. Modifying the resonant frequency applied to the tank is necessary to adjust the resonant tank impedance. Loads in SRCs are wired in series with the circuit of the resonant tank, which is usually constructed using component Lr (inductance) and Cr (capacitance).
- Tank converters consisting of two components fall under the category of PRCs. The load and the diode rectifier network are connected in parallel with the resonant capacitor C_r in this arrangement. The resonant current becomes load independent when the effective load resistance R is much greater than the reactance of the resonant capacitor (Cr). Additionally, the voltage across the resonant capacitor and reducing the load in parallel might increase resistance R. In this configuration, the Rectifier Tank Network produces a sinusoidal voltage regulator, which necessitates an inductive diode rectifier with low-pass filter for efficient voltage filtering.
- Series-parallel resonant converters combine the advantages of both PRCs and SRCs. An additional capacitor or inductor is incorporated into the resonant tank circuit in series-parallel resonant converters. The inductance-capacitance-capacitance (LCC) structure yields load-independent output voltage and current. However, the parallel resonant frequency, which has the same shortcomings as the PRC, has the biggest impact on it. These safeguards stop LCC from functioning in a safe manner in the case of a short circuit or open circuit. In contrast, LLC (LLC) achieves no-load regulation by means of an inductor that is linked in parallel to the resonant capacitor. With this modification, LLC will be able to regulate the output voltage from zero to maximum in every load condition while only slightly altering the switching frequency. However, LLC also has disadvantages, such as difficulty starting and protection against short circuit due to a flat gain above the series resonant frequency. Like LCC, LLC cannot safely operate both an open circuit and a short circuit at frequencies that are near to the parallel frequency.

1.4 LEVEL-BY-LEVEL INVERTERS

Higher-power equipment has become much more in demand in industrial applications in recent years. Due to this demand, megawatt power level converters that operate at a medium voltage are currently necessary, particularly for utility and medium-voltage motor drives. Directly connecting a single power semiconductor switch to a medium-voltage grid is challenging. Consequently, in high-power and medium-voltage situations, multiple power converter systems have emerged as a backup option. These requirements can be met by a variety of multilayer converter topologies, of which the most popular ones are the subject of this study. There is a lot of discussion about the benefits and drawbacks of these multilevel inverter topologies [22]. Multilevel converters connect photovoltaic modules on one side and high-power loads on the other; they are frequently used in high-power converters as well as RESs like FCs and wind. Multilayer Inverters, or MLIs, were modeled after the three-level inverter. The main distinction between a multilayer converter and a two-level voltage source inverter is their voltage levels. An MLI is capable of producing an endless variety of voltage levels. Usually, at least three, whereas a two-level voltage source inverter can only produce two voltage levels. Waveforms for voltage and current are more consistent as a result of this capacity. In order to seamlessly transition between voltage levels, minimize dv/dt and harmonic distortion, and create smoother waveforms, MLIs employ a mechanism that varies the waveform. On the other hand, creating multilayer converters with several voltage levels can be expensive and difficult. More parts are needed for this complexity, and advanced inverter controllers are needed. Despite these difficulties, multilevel converters are recommended due to their low EMI, ability to function at high voltages efficiently, minimum switching losses, and use of both high-frequency and fundamental Pulse Width Modulation. In conclusion, multilevel converters have difficulties in terms of complexity and expense because of the increased number of components and control needs, even if they have several benefits, such as high-voltage operation and efficiency [23].

1.4.1 Conventional and Multilevel Inverter Comparison

Table 1.1 presents a comparison between the traditional and multilevel inverters.

This comparison shows that MLIs have more complexity and a larger initial investment, but they also have benefits like lower switching losses, less harmonic distortion, and increased efficiency. Conventional and multilevel inverters can be

TABLE 1.1
Multilevel Inverter and Traditional Inverter Comparison

S.No.	Aspect	Traditional Inverters	Level-by-Level Inverters
1	Voltage levels	Typically two-level (high and low)	Multiple levels (e.g., three-level, five-level)
2	Switching losses	Higher due to higher voltage stress on switches	Lower due to distribution of voltage across multiple devices
3	Efficiency	Moderate to high depending on design	Higher efficiency, especially at medium to high-power levels
4	Complexity and cost	Relatively simple and lower cost	More complex and higher cost due to additional components and control requirements
5	Harmonic distortion	Higher harmonic distortion in the output waveform	Lower harmonic distortion, resulting in a smoother output waveform
6	Applications	Ideal for applications requiring low to moderate power	Perfect for medium- to high-power uses, including EVs, motor drives, and RESs.

selected in accordance with the specific requirements of the application, accounting for elements such as power level, waveform quality, and cost [24].

1.4.2 Categorization of Multilevel Inverters

Due to their necessity for medium- and high-voltage applications, MLIs can synthesize sinusoidal voltages at different DC levels, leading to improved output waveform quality and reduced harmonic distortion. There are several MLI topologies, which are commonly classified into three main types, as shown in Figure 1.8.

- Diode-Clamped Multilevel Inverter: This topology produces several voltage levels by adding more diodes and capacitors. Multiple voltage levels can be generated above and below the neutral DC source's midway point, which is connected to the inverter output.

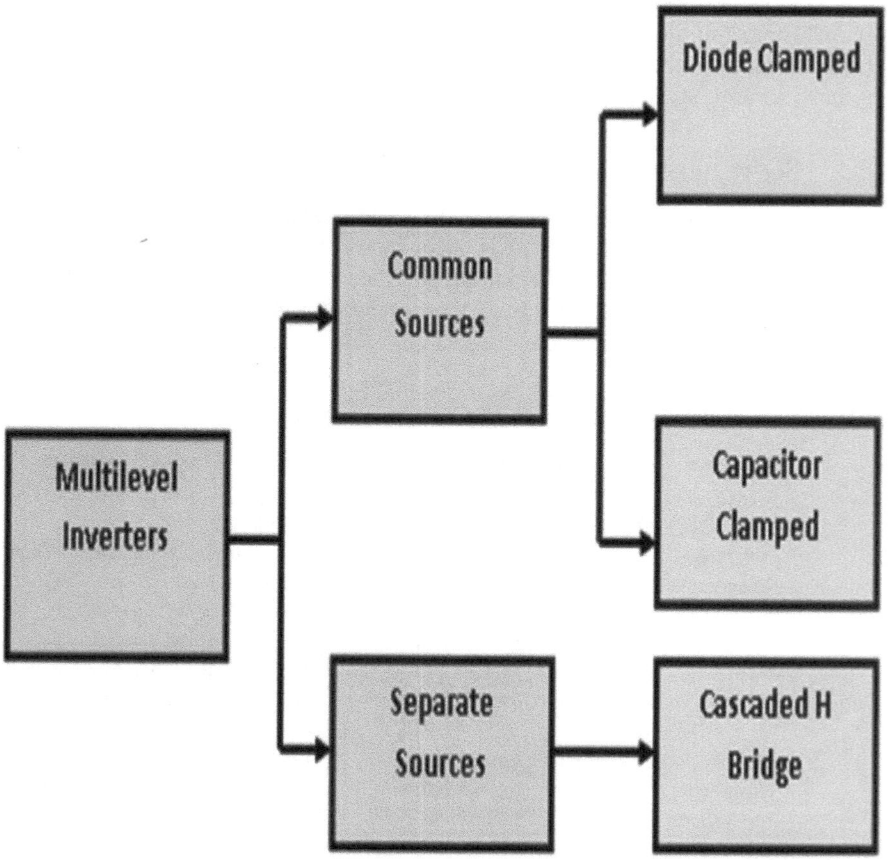

FIGURE 1.8 Multilevel inverter classification.

- Capacitor-**Clamped**: Multilevel flying capacitor inverters use capacitors that "fly" or switch between different voltage levels to generate the desired output waveform. This topology requires careful control of capacitor voltages to maintain balance and stability.
- The cascaded H-bridge architecture is made up of several power cells, or modules each connected in series. Each cell can generate a voltage level, and by stacking these cells, multiple voltage levels can be achieved. The cascaded topology offers scalability and flexibility in terms of voltage levels.

1.5 APPLICATIONS

It is imperative to utilize sustainable energy resources like wind and solar energy, to minimize emissions caused by the production of electricity. However, electricity quality and grid stability may be impacted by these sources' sporadic nature. Energy storage devices are essential for mitigating these issues by reducing oscillations in the output of renewable energy. It is feasible to provide a more reliable and continuous supply of clean electricity while lowering dependency on traditional power sources that release CO2 by combining energy storage with RESs. Effective advanced grid technology is needed for the grid to integrate renewable energy technologies, such as load shifting, power quality enhancement, frequency and voltage control, grid stabilization, and reliability enhancements. These technological advancements guarantee that RESs can provide a consistent and effective contribution to the total energy supply. Restructured power markets also make it possible to take advantage of energy arbitrage, which is the practice of buying cheap electricity during off-peak hours and selling it for more during peak demand. In addition to maximizing cash generation, this can help optimize the application of green energy [25]. An overview of the different types of DC–DC converters that are bidirectional and their uses is provided below:

1. Bidirectional DC–DC Converter for Microgrid Energy Storage:
   - Suitable for energy storage applications using microgrids.
   - Includes various topologies like Buck-Boost, Full-Bridge, and Interleaved designs.
   - Requires appropriate control approaches for microgrid resilience and steadiness.
2. Boost DC–DC Bidirectional Converter for Photovoltaic Uses:
   - Combines bidirectional boost converters and buck-boost converters.
   - Designed for PV energy harnessing and storage.
   - Offers higher conversion gain and adjustable power flow direction for energy storage.
3. UPS Bidirectional DC–DC Converter in Three Phases:
   - Used to interface batteries with DC bus in UPS.
   - Ideal for higher-power applications with reduced ripple current for battery life preservation.
4. DC–DC Resonant LLC Converter for Battery Provisioning:
   - Specifically designed for lead-acid battery charging with high efficiency.

    - Uses LLC topology and proportional integral controller for voltage regulation.
5. LC Resonant DC–DC Converter for Grid-connected RESs:
    - Utilized in grid-linked RESs for efficient power delivery.
    - Features zero-voltage and active switches with ZCS.
6. Resonant DC–DC Converter LCC-T for Photovoltaic and Fuel Cell Uses:
    - Soft-switching current-fed converter for fuel cell and solar applications.
    - Attains ZVS and ZCS to improve efficiency.
7. Five-level Cascaded Multilevel Inverter for RESs:
    - A medium-voltage converter is used to directly integrate renewable energy into the grid.
    - Minimizes problems with common mode and voltage imbalance.
8. Fuel cell energy cascaded H-bridge inverter having five, seven, and fifteen levels:
    - Multilevel inverter designed for fuel cell-based systems.
    - Offers various output voltage levels by utilizing H-bridge devices that are coupled in series.
9. The LLC Resonant Converter for Light EVs is a Full-Bridge Bidirectional device.
    - Fast charger prototype for light EVs with lithium-ion batteries.
    - Feature includes an LLC resonant converter for stress-free switching.
10. Full-Bridge Inverter with bidirectional functionality for on-board electric car battery chargers:
    - Designed to work with smart grid-installed EV battery chargers.
    - Utilizes a bidirectional multilevel topology for enhanced power factor correction.

Every type of converter is customized to meet the needs of a particular application, providing effective power transfer and control for a range of energy storage and renewable energy integration requirements. These designs are essential for improving dependability and energy efficiency in a variety of industries [26].

1.6 OBSTACLES AND PROSPECTS FOR THE FUTURE

There are several technical, financial, and design obstacles in the way of the development and broad use of converters between DC and AC. Interest has been generated by one specific type of bidirectional converter: the DAB-IBDC. Its benefits include symmetric design, soft-switching implementation, and bidirectional power transfer capabilities. It is also modular in construction. The goal of this field's future research and development will be to maximize the performance and design of DAB-IBDCs by utilizing cutting-edge silicon carbide (SiC) and gallium nitride (GaN) are examples of power semiconductor materials [27]. Some key challenges and future directions for DAB-IBDCs and similar converters include:

- Developing techniques to optimize the architecture and electrical characteristics in order to effectively capitalize on the control techniques of

DAB-IBDCs based on SiC/GaN power devices' advantages, which include low power loss, high temperature operation, and high-frequency capability. This is known as electrical optimization design.

- Mechanical Optimization Design: Utilizing SiC/GaN power devices can improve the power density, modularity, efficiency, and reliability of High-Frequency Link power conversion systems, hence optimizing the mechanical design of DAB-IBDCs.
- Creating modular, intelligent, and versatile high-frequency link power conversion systems with high power density and efficiency by designing them with DAB-IBDCs as the central circuit. To improve system performance and adaptability, this entails creating sophisticated control strategies and incorporating intelligent features.

Future developments aim to make DAB-IBDCs more efficient, compact, and suitable for several high-power uses, including renewable energy systems, electric cars, and industrial power sources. This will be accomplished by addressing these challenges and concentrating on optimizing DAB-IBDCs with advanced semiconductor technologies [28].

Present-day power converters encounter the difficulty of providing greater power while preserving excellent efficiency over a wide variety of loads. Resonant converter topologies such as the LLC and SRCs have emerged to address these challenges. Particularly helpful for front-end DC–DC conversion applications are these topologies. The LLC resonant converter offers several benefits, including:

- ZVS and ZCS: LLC resonant converters enable ZVS for main side MOSFETs or IGBTs and ZCS for secondary side power devices, which reduce switching losses and boost efficiency.
- The LLC resonance converter functions within a restricted switching frequency band, which makes control easier and allows for rapid transient response.
- Cost-Effectiveness: Reduces component count and total cost by using leakage inductance of the transformer acting as the resonant inductor.

Similar to this, switched-mode power supply applications have seen a rise in the use of SRCs because of:

- High Switching Frequency: Increasing power density and lowering switching losses are two factors that lead to improved efficiency while operating at high switching frequencies.
- Historical Significance: A family of resonant converters with high-density DC/DC mode that are extensively employed in real-world applications is the result of the SRC's lengthy technological development.

Despite the advantages of resonant converters, challenges remain that require attention in future research:

- The optimization of resonant converter performance is contingent upon the appropriate assignment and selection of parameters, including the inductance ratio, load values, voltage gain, and frequency ratio.
- Control Optimization: To further improve efficiency, dependability, and performance, future research should concentrate on refining the control techniques of resonant converter topologies.

All things considered, the increasing interest in resonant power converters suggests that there is a lot of room for more research and development. Resonant converters might continue to be a vital component in raising the effectiveness and performance of contemporary power conversion systems by tackling the issues that have been identified and enhancing control techniques. The discussion on bidirectional DC–DC converters and their applications encompasses various important aspects related to power flow control and galvanic isolation [29]. Here are key points from the provided text:

- Bidirectional power flow is made possible by bidirectional converters, which allow voltage levels to be stepped up or down in response to control requirements. Applications like energy storage systems and electric cars that need two-way energy transfer must have this capacity.

1.6.1 Bidirectional Converter Types

- Bidirectional DC–DC converter that is not isolated: These converters have good efficiency for low-power applications and are easy to manage because they do not employ transformers with high frequencies used for electrical separation.
- Galvanic separation of an IBDC's input and output results in source security, noise reduction, and voltage matching, among other advantages. Applications needing dependability and safety are better served by isolated converters.

1.6.2 Resonant Network Types with Isolated Bidirectional Converters

To achieve effective power flow management and performance, a number of research studies have combined separate, two-way DC–DC converters with various resonance network types. The symmetrical features of CLLC and CLLLC (Capacitor – L – L – C and Capacitor – L – L – L – C) resonant IBDCs are emphasized both in the directions of forward and backward power flow. However, problems with charging efficiency degradation with fluctuating battery voltage and significant frequency variation during discharging operations occur.

1.6.3 Bidirectional Use of LLC Resonant Converters

This converter, known for its tight gain range and improved performance, is extensively used in reverse power transfer applications. Challenges such as achieving high efficiency over a wide gain range, particularly with designs using twelve switches, must be addressed for optimal converter performance. In summary, bidirectional DC–DC converters are necessary in applications where control over power flow must be flexible, and different types of converters (isolated vs. non-isolated) offer unique advantages and challenges. Future research focuses on optimizing resonance network combinations and addressing efficiency and control challenges to enhance the performance of bidirectional converters in various practical applications.

MLIs offer significant advantages over traditional two-level converters, particularly in terms of producing sinusoidal output voltage and achieving low total harmonic distortion in output current. These benefits make MLIs attractive for various industrial applications, including drives, solar inverters, UPS, EVs, and static synchronous compensators. Key advantages of MLIs include:

- Reduced Device Voltage Rating and Switching Frequency: MLIs can achieve the same output voltage as conventional converters with lower device voltage ratings and switching frequencies. This reduction in voltage and frequency leads to decreased switching losses and improved overall efficiency.
- Low-Harmonic Waveforms: MLIs generate low-harmonic waveforms, which are crucial for industrial applications requiring high-quality power and reduced interference.

However, there are challenges associated with MLIs that must be addressed for widespread commercial adoption:

- Adding additional switches and auxiliary capacitors to MLI structures may result in higher production costs. In addition, the control circuit becomes more complex with the addition of sophisticated switching algorithms, novel gate drivers, and voltage balancing strategies, which increases expenses and design difficulties.
- Comparing Complexity and Performance Trade-off: Industries try to strike a compromise between MLIs' operational simplicity and design and the quantity of voltage levels (complexity). Although higher voltage step levels can lead to better performance, they also come at a higher cost and complexity.

To drive commercial adoption, ongoing improvements in MLI topology are essential:

- Size Reduction of Passive Components: Efforts should focus on reducing the size and weight of auxiliary components, such as capacitors, to lower overall system costs.
- Reduced Switching Frequency and Losses: Innovations in switching technologies and control algorithms can help reduce switching losses and optimize efficiency.
- Enhanced Reliability and Efficiency: Continued research and development are needed to enhance the reliability and efficiency of MLIs, making them more attractive for mass production and industrial applications.

To sum up, sectors are concentrating more and more on streamlining MLI designs without sacrificing effectiveness or performance. Continuous improvements in component technology, architecture, and control schemes are necessary to propel MLIs into broad commercialization across multiple industrial domains.

Because power system applications require separate DC sources and fewer switches to efficiently create different voltage levels, choosing the right MLI layout can be difficult. Multiple-DC-source topologies have lost ground to single-source MLI topologies in a number of applications, such as electrified transport and systems for converting renewable energy. A continuous problem for industry is to ensure acceptable power quality, minimize switching frequency, and increase power rating while using a single-DC-source MLI. There are particular difficulties with integrating MLIs into bidirectional converters, like battery storage systems (BSSs). These include choosing the best topology of MLI and managing the battery state of charge. MLIs are used in vehicle-to-grid systems, EV operations, and large-scale renewable energy integration. This highlights the need for effective and affordable MLI topologies [30].

Key challenges and advancements in MLI-based BSSs include:

- Battery Pack Imbalances: Because MLI-based BSSs use second-life batteries and have varying electrochemical properties, imbalances in state of charge and voltage between battery packs could be dangerous. Addressing these imbalances requires advanced control techniques and battery management strategies.
- Advancements in Battery Technology: Control methods, battery topologies, and battery technologies have advanced as a result of the use of MLIs in BSSs. MLI-based BSSs make considerable use of lithium-ion batteries, which are renowned for their extended lifespan and excellent power density. Because metal–air batteries are inexpensive and have a high energy density, research into this technology seems promising for EV applications.
- Superior Performance of MLIs: Compared to conventional two- or three-level converter schemes, MLIs perform better because of their small size, low harmonic distortion, and decreased dependency on magnetic circuits. These advantages make MLIs attractive for a wide range of power system applications.

Overall, selecting the most suitable MLI topology and overcoming challenges related to battery pack imbalances and technology integration are critical for advancing the efficiency and reliability of MLI-based power systems, including battery storage and electrified transportation applications. Ongoing research and development in this field aim to optimize MLI designs and control strategies to meet the evolving demands of modern power systems.

1.7 CONCLUSIONS

The overview of bidirectional and resonant DC–DC converters, as well as MLIs, highlights how important it is to leverage these technologies for grid-connected and energy storage applications. Based on their classifications, benefits, drawbacks, and functions in enhancing energy conversion in contemporary energy systems, each form of converter was examined. In order to reduce switching losses, the study stresses the use of soft-switching approaches and suggests galvanic separation in bidirectional DC–DC converters for hybrid energy systems. To assess the effectiveness of different types of converters for grid integration and energy storage, including multilevel and conventional inverters, comparisons were made. The difficulties in creating effective multilayer resonant converters with soft-switching, bidirectional functionality, and high-voltage operation were also covered. It is anticipated that integrated bidirectional resonant multilevel converters, with advantages like lower costs, smaller size, low harmonic distortion, and high switching frequencies, will satisfy the needs of a range of industrial applications, such as medium-voltage drives, vehicle-to-grid, and EV charging.

REFERENCES

1. Sinha, S.; Chandel, S. Review of recent trends in optimization techniques for solar photovoltaic–wind based hybrid energy systems. *Renew. Sustain. Energy Rev.* 2015, 50, 755–769.
2. Khare, V.; Nema, S.; Baredar, P. Status of solar wind renewable energy in India. *Renew. Sustain. Energy Rev.* 2013, 27, 1–10.
3. Viswanathan, B. Solar energy: Fundamentals. In Viswanathan, B. (ed.) *Energy Sources*; Elsevier: Amsterdam, the Netherlands, 2017; pp. 139–147.
4. Nandha Gopal, J.; Muthuselvan, N.B.; Muthukaruppasamy, S. Model predictive controller–based quadratic boost converter for WECS applications. *Int. Trans. Electr. Energy Syst.* 2021, 31(12), 1–17.
5. Eccher, M.; Salemi, A.; Turrini, S.; Brusa, R. Measurements of power transfer efficiency in CPV cell-array models using individual DC–DC converters. *Appl. Energy* 2015, 142, 396–406.
6. Duran, E.; Andujar, J.; Segura, F.; Barragan, A. A high-flexibility DC load for fuel cell and solar arrays power sources based on DC–DC converters. *Appl. Energy* 2011, 88, 1690–1702.
7. Saravanan, S.; Babu, N.R. Analysis and implementation of high step-up DC-DC converter for PV based grid application. *Appl. Energy* 2017, 190, 64–72.
8. Nandha Gopal, J.; Muthuselvan, N.B. Current mode fractional order PID control of wind-based quadratic boost converter inverter system with enhanced time response. *Circuit World* 2020, 47(4), 368–381.
9. Outeiro, M.T.; Visintini, R.; Buja, G. Considerations in designing power supplies for particle accelerators. In *Proceedings of the IECON 2013–39th Annual Conference of the IEEE Industrial Electronics Society*, Vienna, Austria, 10–13 November 2013; pp. 7076–7081.
10. Outeiro, M.T.; Buja, G.; Carvalho, A. Resonant converters for electric equipment power supply. In *Proceedings of the IECON 2014–40th Annual Conference of the IEEE Industrial Electronics Society*, Dallas, TX, USA, 29 October–1 November 2014; pp. 5065–5071.
11. Richelli, A.; Salem, M.; Colalongo, L. A review of fully integrated and embedded power converters for IoT. *Energies* 2021, 14, 5419.
12. Monsalve-Rueda, M.; Candelo-Becerra, J.E.; Hoyos, F.E. Dynamic behavior of a sliding-mode control based on a washout filter with constant impedance and nonlinear constant power loads. *Appl. Sci.* 2019, 9, 4548.

13. Nandha Gopal, J.; Muthuselvan, N.B. Educational tool for analysis of PI and fractional order PI controlled quadratic boost boost converter system using MATLAB/Simulink. *Int. J. Electr. Eng. Educ.* https://doi.org/10.1177/00207209211013435, 2021.
14. Hrigua, S.; Costa, F.; Gautier, C.; Revol, B. New method of EMI analysis in power electronics based on semiconductors transient models: Application to SiC MOSFET/ Schottky diode. In *Proceedings of the IECON 2012–38th Annual Conference on IEEE Industrial Electronics Society*, Montreal, QC, Canada, 25–28 October 2012; pp. 590–595.
15. Britto, K.A.; Dhanasekaran, R.; Vimala, R.; Saranya, B. EMI analysis and evaluation of an improved flyback converter. In *Proceedings of the 2012 International Conference on Computer Communication and Informatics*, Coimbatore, India, 10–12 January 2012; pp. 1–7.
16. Salem, M.; Jusoh, A.; Idris, N.R.N.; Das, H.S.; Alhamrouni, I. Resonant power converters with respect to passive storage (LC) elements and control techniques–An overview. *Renew. Sustain. Energy Rev.* 2018, 91, 504–520.
17. Ravi, D.; Reddy, B.M.; Shimi, S.; Samuel, P. Bidirectional dc to dc converters: An overview of various topologies, switching schemes and control techniques. *Int. J. Eng. Technol.* 2018, 7, 360–365.
18. Gorji, S.A.; Sahebi, H.G.; Ektesabi, M.; Rad, A.B. Topologies and control schemes of bidirectional DC–DC power converters: An overview. *IEEE Access* 2019, 7, 117997–118019.
19. Vivek, P.; Muthuselvan, N.B.; Nanadhagopal, J. Modeling of solar PV system for DC-DC converter with improved voltage stability using hybrid-optimization techniques. In *International Conference for Phoenixes on Emerging Current Trends in Engineering and Management*, Chennai, India; Atlantis Press, 2018; pp. 187–192.
20. Salem, M.; Jusoh, A.; Idris, N.N.; Sutikno, T.; Buswig, Y.M. Phase-shifted series resonant converter with zero voltage switching turn-on and variable frequency control. *Int. J. Power Electron. Drive Syst. (IJPEDS)* 2017, 8, 1184–1192.
21. Li, Z.; Xue, B.; Wang, H. An interleaved secondary-side modulated LLC resonant converter for wide output range applications. *IEEE Trans. Ind. Electron.* 2019, 67, 1124–1135.
22. Alhamrouni, I.; Hamzah, M.B.; Salem, M.; Jusoh, A.; Khairuddin, A.B.; Sutikno, T. A bidirectional resonant converter based on wide input range and high efficiency for photovoltaic application. *Int. J. Power Electron. Drive Syst.* 2019, 10, 1469.
23. Rashid, M.H. *Power Electronics Handbook*; Butterworth-Heinemann: Oxford, UK, 2018.
24. Kabalci, E. *Multilevel Inverters: Control Methods and Advanced Power Electronic Applications*; Elsevier: Amsterdam, the Netherlands, 2021.
25. Dokić, B.L.; Blanuša, B. (eds). Introduction to multilevel converters. In *Power Electronics*; Springer: Berlin/Heidelberg, Germany, 2015; pp. 559–592.
26. Salem, M.; Yahya, K. (eds). Resonant power converters. In *Electric Power Conversion*; IntechOpen: London, UK, 2019; pp. 87–102.
27. Pan, L.; Zhang, C. An integrated multifunctional bidirectional AC/DC and DC/DC converter for electric vehicles applications. *Energies* 2016, 9, 493.
28. Rajeswari, K.; Vivek, P.; Nandhagopal, J. Swing up and stabilization of rotational inverted pendulum by fuzzy sliding mode controller. In Hemanth, D.; Kumar, V.; Malathi, S.; Castillo, O.; Patrut, B. (eds) *Emerging Trends in Computing and Expert Technology. COMET 2019. Lecture Notes on Data Engineering and Communications Technologies*, vol 35; Springer: Cham, 2020. https://doi.org/10.1007/978-3-030-32150-5_40.

29. Oliveira, F.J.D.; Ribeiro, E.K.D.B.; Luiz, A.-S.A.; Stopa, M.M. Study and implementation of a high gain bidirectional DC-DC converter for photovoltaic on-grid systems. In *Proceedings of the 2017 IEEE 8th International Symposium on Power Electronics for Distributed Generation Systems (PEDG)*, Florianopolis, Brazil, 17–20 April 2017; pp. 1–8.
30. Costa, L.; Buticchi, G.; Liserre, M. Bidirectional series-resonant DC-DC converter with fault-tolerance capability for smart transformer. In *Proceedings of the 2016 IEEE Energy Conversion Congress and Exposition (ECCE)*, Milwaukee, WI, USA, 18–22 September 2016; pp. 1–7.

2 Energy Conversion for Renewable Power Systems

Srinivasan Gopalan, Niyas Ahamed Allavudeen, and Nazrin Salma Sheriff

2.1 OVERVIEW OF RENEWABLE ENERGY SOURCES

Renewable energy sources are derived from natural processes that are replenished constantly, making them sustainable alternatives to traditional fossil fuels. These sources produce energy without depleting natural resources or causing significant harm to the environment. Solar energy is harnessed from the sun's radiation and can be converted into electricity through photovoltaic cells or concentrated solar power systems. Wind energy, on the other hand, is generated by capturing the kinetic energy of wind using wind turbines, which then convert it into electrical power. Hydropower is produced by harnessing the energy of flowing or falling water, commonly through the use of dams or water turbines. Biomass energy is derived from organic materials such as wood, agricultural residues, and waste, which can be burned to produce heat or converted into biofuels. Geothermal energy originates from the heat within the Earth's crust and is accessed by tapping into hot water or steam reservoirs underground. Lastly, tidal and wave energy are captured from the movement of tides and waves using specialized turbines or generators, further diversifying the renewable energy mix.

2.2 IMPORTANCE OF POWER CONVERTERS IN RENEWABLE ENERGY SYSTEMS

Power converters are essential components in renewable energy systems, playing a crucial role in the efficient conversion, control, and management of electrical power. One of their primary functions is voltage conversion, where they transform the variable voltage output from renewable sources into a stable voltage suitable for the grid or storage systems. Additionally, converters contribute to improving power quality, ensuring that the electricity generated meets the required standards for seamless grid integration. Their role in optimizing energy conversion processes enhances efficiency by minimizing energy losses during transmission and distribution. Furthermore, power converters facilitate the integration of renewable energy systems with existing electrical grids, supporting grid stability and reliability. In systems equipped with energy storage solutions like batteries, converters also manage the charging

DOI: 10.1201/9781003496946-2

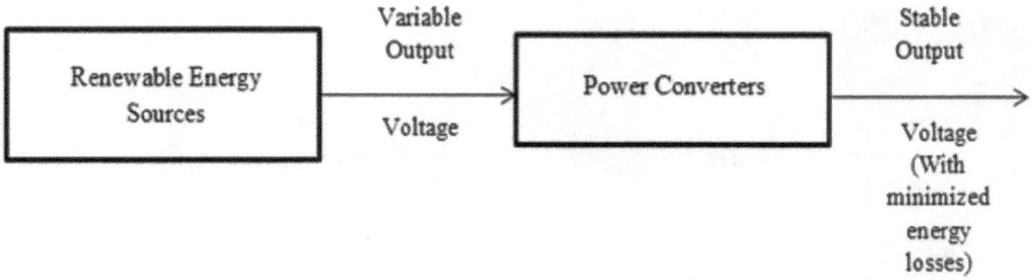

FIGURE 2.1 Power converters in enhancing renewable energy systems.

and discharging processes, maximizing storage efficiency and overall system performance [7–10]. The power converters in enhancing renewable energy sources are shown in Figure 2.1.

2.3 BRIEF HISTORY OF POWER CONVERTER TECHNOLOGY

The evolution of power converter technology has seen significant advancements over the years, spurred by innovations in semiconductor devices, control algorithms, and system integration. The journey began in the early 20th century with the development of electromechanical converters, such as rotary converters, designed to convert alternating current (AC) to direct current (DC) and vice versa for electrical distribution systems. The landscape shifted dramatically in the 1950s and 1960s with the emergence of semiconductor devices like diodes and thyristors, paving the way for static converters that offered enhanced efficiency and compactness in power conversion. The subsequent decades, particularly the 1970s and 1980s, witnessed a revolution in power converter design with the introduction of pulse-width modulation (PWM) techniques and insulated gate bipolar transistors, which further elevated efficiency and control capabilities. Fast forward to the 1990s and the present day, ongoing advancements in semiconductor technology, digital control systems, and power electronics have culminated in the development of high-performance converters adept at handling the demands of renewable energy sources such as solar and wind power. Evolution of power converter technology is shown in Figure 2.2.

Power converters have become indispensable components in renewable energy systems, bridging the gap between renewable energy sources and electrical grids or storage systems. Their evolutionary trajectory has played a pivotal role in driving the adoption and seamless integration of renewable energy technologies on a global scale.

2.4 FUNDAMENTALS OF POWER CONVERTERS

Power converters serve as the backbone of modern electrical systems, facilitating the efficient transformation and control of electrical energy. Understanding the fundamentals of power converters involves delving into the basic principles of power electronics, exploring the different types of power converters, and examining the various switching techniques and topologies employed in their design.

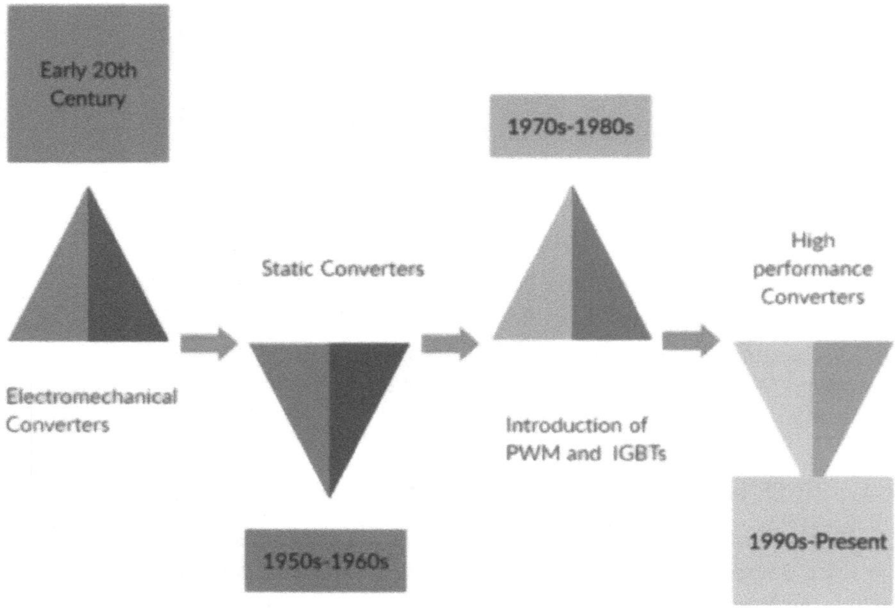

FIGURE 2.2 Evolution of power converter technology.

2.5 BASIC PRINCIPLES OF POWER ELECTRONICS

Power electronics revolves around the manipulation and control of electrical power using electronic devices. The key principles governing power electronics include:

Switching: Power electronic devices, such as transistors and diodes, operate by switching between ON and OFF states to control the flow of electrical current.

Conversion: Power converters transform electrical energy from one form to another, such as AC to DC or vice versa, while maintaining efficient energy transfer.

Control: Advanced control techniques, including modulation and feedback systems, are employed to regulate the output voltage and current of power converters, ensuring optimal performance and stability.

2.6 TYPES OF POWER CONVERTERS

Power converters can be categorized based on their input and output voltage levels, conversion topology, and application. Some common types of power converters include:

AC/DC Converters (Rectifiers): Convert AC to DC, commonly used in power supplies and battery chargers.

DC/DC Converters: Convert one DC voltage level to another, crucial for voltage regulation in various electronic systems and renewable energy applications.

DC/AC Converters (Inverters): Convert DC to AC, essential for interfacing renewable energy systems with the grid and powering AC loads.

AC/AC Converters: Convert AC voltage levels or frequency, used for applications like speed control of AC motors and voltage regulation.

2.7 SWITCHING TECHNIQUES AND TOPOLOGIES

The efficiency and performance of power converters are greatly influenced by the switching techniques and circuit topologies employed. Common switching techniques include:

Pulse-Width Modulation (PWM): Modulates the width of the switching pulses to control the output voltage or current, offering precise control and reduced harmonic distortion.

Resonant Switching: Utilizes resonant circuits to achieve soft switching, minimizing switching losses and enhancing efficiency.

Different circuit topologies, such as buck, boost, and buck-boost, determine the specific functionality and application of the power converter. These topologies define how the switching devices, passive components, and control circuits are interconnected to achieve the desired conversion characteristics.

2.8 MATHEMATICAL MODELING OF POWER CONVERTERS

Mathematical modeling lies at the heart of understanding and optimizing the performance of power converters, pivotal in converting electrical power from one form to another with specific voltage and current levels. Engineers employ various techniques for this purpose. Circuit analysis deconstructs converters into their components, applying Kirchhoff's laws and component equations to derive relationships between variables. State-space modeling describes the system's dynamic behavior in terms of state variables and inputs, accommodating nonlinearities and control actions [1–6]. Average and small-signal modeling simplify steady-state analysis and linearize equations around operating points, respectively. Switching function modeling captures the dynamics of switches, crucial in rapid on-off operations. Control strategies, including proportional-integral-derivative and model predictive control, rely on accurate models for design and simulation. Engineers validate models using simulation software, such as Simulation Program with Integrated Circuit Emphasis or MATLAB/Simulink, refining them through parameter estimation techniques to account for real-world variability [11–15]. By honing these models, engineers can design efficient and reliable power converters for diverse applications, spanning renewable energy systems, electric vehicles, and consumer electronics. The flow chart for the mathematical modeling in power converter design is shown in Figure 2.3.

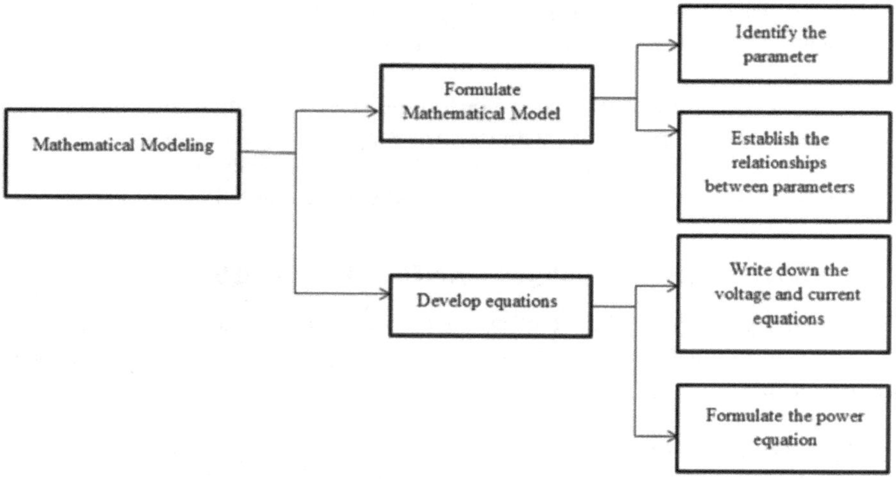

FIGURE 2.3 Flowchart of mathematical modeling in power converter design.

2.9 SIMULATION TOOLS AND METHODOLOGIES

Engineers rely on a variety of simulation tools and methodologies to develop, analyze, and optimize power converters efficiently. Among these, Simulation Program with Integrated Circuit Emphasis) stands out as a widely utilized tool for its ability to accurately model electronic circuits, including power converters. Its support for transient and AC analyses enables thorough performance evaluation by considering components like switches, diodes, capacitors, and inductors. MATLAB/Simulink is also prominent, offering a graphical interface for modeling dynamic systems through block diagrams, complemented by MATLAB's computational capabilities for advanced analysis and optimization tasks.

Specialized software such as PSIM caters specifically to power electronics, providing a user-friendly environment with pre-built component libraries and support for various simulation types. Piecewise Linear Electrical Circuit Simulation adopts a piecewise linear approximation approach for rapid simulation, seamlessly integrating with MATLAB/Simulink for system-level modeling. Additionally, hardware-in-the-loop simulation merges physical hardware with software simulation, enabling real-time validation of control algorithms and system behavior, crucial for safety-critical applications.

Simulation tools often incorporate features like parameter sweeping and optimization, automating the exploration of design parameters and facilitating the identification of optimal configurations. Moreover, co-simulation with circuit and thermal analysis tools allows engineers to assess multi-domain interactions comprehensively, considering electrical, thermal, and mechanical aspects [16]. Table 2.1 lists the simulation tool/methodology used in the design of a power converter. By leveraging these simulation tools and methodologies, engineers can streamline the design process, assess design alternatives, and ensure the reliability and performance of power converters across a broad spectrum of applications.

TABLE 2.1
Simulation Tool/Methodology Used in the Design of a Power Converter

Simulation Tool/ Methodology	Description
Simulation Program with Integrated Circuit Emphasis	Widely used tool for accurately modeling electronic circuits, including power converters. Supports transient and AC analyses for thorough performance evaluation.
MATLAB/Simulink	Graphical interface for modeling dynamic systems through block diagrams. Integrates with MATLAB for advanced analysis and optimization tasks.
PSIM	Specialized software catering specifically to power electronics. Provides a user-friendly environment, pre-built component libraries, and support for various simulation types.
PECS	Adopts a piecewise linear approximation approach for rapid simulation. Seamlessly integrates with MATLAB/Simulink for system-level modeling.
Hardware-In-the-Loop Simulation	Merges physical hardware with software simulation for real-time validation of control algorithms and system behavior. Crucial for safety-critical applications.

2.10 CHALLENGES IN MODELING NONLINEAR RENEWABLE ENERGY SYSTEMS

Modeling nonlinear renewable energy sources presents several intricate challenges stemming from the inherent complexity and variability of these systems. First and foremost is grappling with their nonlinear behavior, where the relationship between input conditions (e.g., solar irradiance, wind speed) and output power isn't linear. This complexity makes accurately capturing dependencies between inputs and outputs difficult. Furthermore, the variability and uncertainty inherent in renewable sources due to factors like weather conditions and time of day add another layer of complexity. Accounting for stochastic variations and uncertainties becomes paramount, impacting system performance and stability [17].

Another hurdle lies in modeling the dynamic response of these systems accurately. Renewable energy systems exhibit dynamic behavior in response to changes in environmental conditions and load demand. Capturing these dynamic responses is crucial for predicting transient behavior, stability, and control performance, but doing so accurately can be computationally demanding and require sophisticated modeling techniques.

Moreover, the interface of renewable sources with power electronic converters introduces additional nonlinearities due to switching operations and control strategies. Modeling these interactions accurately necessitates considering effects such as harmonics, voltage spikes, and control loop dynamics [18].

Parameter variations and degradation further complicate modeling efforts. Parameters like Photo Voltaic panel efficiency or wind turbine performance can change over time due to aging, environmental factors, and maintenance issues. Integrating these variations into models requires accounting for long-term trends and statistical variations, which adds complexity. Additionally, renewable energy systems often involve multiple physical domains such as electrical, mechanical, and thermal, requiring accurate modeling of their interactions. However, integrating multi-physics models introduces additional computational challenges.

Lastly, validating and verifying models of nonlinear renewable energy sources is challenging due to limited experimental data and the complexity of real-world systems. Ensuring that models accurately represent real system behavior under various conditions demands careful validation against empirical data.

Overcoming these challenges demands interdisciplinary approaches that blend expertise in renewable energy systems, control theory, numerical methods, and data analysis. Advanced modeling techniques, such as nonlinear system identification and probabilistic methods, are essential for developing accurate and reliable models. Continuous research and development efforts are necessary to enhance model fidelity, computational efficiency, and validation methodologies in this rapidly evolving field. Table 2.2 gives details of the challenges faced in the design of a power converter [19].

TABLE 2.2
Challenges Faced in the Design of a Power Converter

Component/ Challenge	Description
Nonlinear behavior	Nonlinear relationship between input conditions (e.g., solar irradiance, wind speed) and output power.
Variability and uncertainty	Variability and uncertainty in renewable sources due to factors like weather conditions and time of day.
Dynamic response	Dynamic behavior in response to changes in environmental conditions and load demand.
Power electronic converter interactions	Nonlinearities introduced by the interface of renewable sources with power electronic converters due to switching operations and control strategies.
Parameter variations and degradation	Changes in parameters like PV panel efficiency or wind turbine performance over time due to aging, environmental factors, and maintenance issues.
Multi-physics integration	Interaction of multiple physical domains such as electrical, mechanical, and thermal.
Model validation and verification	Validation against empirical data to ensure accurate representation of real system behavior under various conditions.
Interdisciplinary approaches	Blending expertise in renewable energy systems, control theory, numerical methods, and data analysis to overcome challenges.
Advanced modeling techniques	Techniques like nonlinear system identification and probabilistic methods used for developing accurate and reliable models.
Continuous research and development	Ongoing efforts to enhance model fidelity, computational efficiency, and validation methodologies in the field of renewable energy.

2.11 PERFORMANCE METRICS

Voltage and current regulation are vital aspects of power converters, ensuring stable output despite fluctuations in input voltage, load conditions, and other factors. These metrics assess the deviation of output voltage and current from their desired values, reflecting the converter's ability to maintain steady operation. Transient response measures how swiftly and accurately a power converter adapts to changes in load or input conditions. This capability is crucial for recovering from sudden load variations or disturbances like voltage spikes or dips, ensuring system stability and reliability.

Ripple and noise analysis evaluates small variations in the output voltage or current waveform, typically caused by switching actions within the converter. Excessive ripple can compromise system performance and reliability, making it essential to keep it within acceptable limits. Analyzing ripple and noise levels is integral to assessing power quality.

Switching frequency, the rate at which power converters switch on and off, influences efficiency, size, and electromagnetic interference (EMI). Higher frequencies can reduce the size of passive components but might elevate switching losses and EMI. Metrics associated with switching frequency help optimize trade-offs between size, efficiency, and EMI, guiding design choices for optimal converter performance.

2.12 EFFICIENCY ANALYSIS

Conversion efficiency stands as a pivotal metric in power converter performance assessment, indicating the ratio of output power to input power, usually expressed as a percentage. A higher efficiency value denotes fewer losses incurred during the conversion process. Analyzing efficiency entails scrutinizing losses across various components like switches, diodes, inductors, and capacitors, encompassing both conduction and switching losses along with losses attributed to magnetic cores.

Delving deeper into loss breakdown is imperative for optimization endeavors. Power losses stem from multiple sources, including conduction (resistive losses in switches and diodes), switching (energy dissipation during transition periods), and magnetic domains (hysteresis and eddy current losses in inductors and transformers). Efficiency analysis involves quantifying these losses meticulously and devising strategies to minimize them through judicious component selection, refined control schemes, and effective thermal management techniques [20].

Moreover, load-dependent efficiency acknowledges that power converters often exhibit varying efficiency levels contingent on the load condition. Efficiency curves delineate how efficiency fluctuates across different load levels, furnishing invaluable insights for designers to fine-tune efficiency optimization strategies tailored to typical operational scenarios.

2.13 POWER QUALITY CONSIDERATIONS

Harmonic distortion, a concern with power converters, refers to the introduction of unwanted harmonic frequencies into the electrical grid. This distortion can lead to issues such as voltage fluctuations and increased losses in connected equipment.

Power quality analysis is crucial in evaluating the harmonic content in the output waveform and ensuring compliance with relevant standards and regulations to mitigate adverse effects.

Low power factor is another critical issue that can arise, leading to increased energy consumption, reduced efficiency, and potential penalties from utilities. Power factor correction (PFC) techniques, such as active or passive PFC, are employed to address this issue by adjusting the phase relationship between voltage and current waveforms. Analyzing power factor and implementing PFC techniques can significantly enhance system efficiency and reduce harmonic distortion.

EMI generated by power converters can disrupt nearby electronic devices and degrade overall system performance. EMI analysis involves identifying sources of interference, minimizing parasitic capacitances and inductances, and implementing shielding and filtering techniques to mitigate EMI, ensuring electromagnetic compatibility with surrounding systems [21–23].

Furthermore, power converters connected to the electrical grid must adhere to grid codes and standards to ensure stable and reliable operation. Power quality considerations encompass aspects such as voltage and frequency regulation, fault ride-through capabilities, and grid synchronization techniques. Ensuring compliance with grid requirements is essential for seamless integration with the grid and maintaining system reliability.

2.14 DESIGN OBJECTIVES AND SPECIFICATIONS

2.14.1 Performance Requirements

1. **Output Voltage and Current Levels:** Defining the required levels of output voltage and current ensures that the power converter delivers the necessary electrical power to the load or system it is intended to supply.
2. **Efficiency:** Establishing efficiency targets ensures that the power converter operates with minimal power losses during the conversion process, maximizing energy utilization and minimizing waste heat generation.
3. **Transient Response:** Specifying the desired transient response characteristics ensures that the power converter can quickly and accurately adapt to changes in load conditions or input voltage, maintaining stable operation under dynamic conditions.
4. **Power Factor:** Setting power factor requirements helps optimize the power quality of the electrical system by minimizing reactive power consumption and improving overall system efficiency.

2.15 ENVIRONMENTAL CONDITIONS

1. **Temperature:** Assessing temperature variations helps determine the thermal management requirements of the power converter to ensure reliable operation over a range of operating temperatures.

2. **Humidity:** Considering humidity levels is important for selecting appropriate materials and coatings that can withstand moisture and prevent corrosion, ensuring long-term reliability.
3. **Altitude:** Accounting for altitude variations helps determine the impact on component performance and cooling requirements due to changes in atmospheric pressure.
4. **Vibration Levels:** Evaluating vibration levels helps design the power converter to withstand mechanical stresses and prevent component failures caused by vibration-induced fatigue.

2.16 REGULATORY COMPLIANCE

1. **Power Quality Standards:** Compliance with power quality standards ensures that the power converter's output meets acceptable levels of voltage regulation, harmonic distortion, and other parameters, minimizing adverse effects on the electrical grid and connected equipment.
2. **Electromagnetic Compatibility Regulations:** Adherence to electromagnetic compatibility regulations helps prevent interference with other electronic devices and ensures reliable operation in electromagnetic environments.
3. **Safety Standards:** Compliance with safety standards ensures that the power converter meets requirements for electrical insulation, protection against electrical shock, and reliability under fault conditions, ensuring user safety and preventing accidents [24].
4. **Energy Efficiency Regulations:** Adhering to energy efficiency regulations helps minimize energy consumption and reduce environmental impact, ensuring that the power converter meets efficiency targets set by regulatory authorities.

By defining performance requirements, considering environmental conditions, and ensuring regulatory compliance, designers can develop power converters that meet the needs of their intended applications while ensuring reliability, safety, and compliance with industry standards and regulations.

2.17 COMPONENT SELECTION AN SIZING

Semiconductor Devices: Selecting appropriate semiconductor devices such as MOSFETs, insulated gate bipolar transistors, diodes, and thyristors is crucial for the design of power converters. Factors to consider include:

1. **Voltage and Current Ratings:** Choose devices with voltage and current ratings that exceed the maximum expected operating conditions to ensure reliability and safety margins.
2. **Switching Speed:** Consider the switching speed of semiconductor devices, as faster switching speeds can reduce switching losses and improve efficiency. However, higher switching speeds may also lead to increased EMI and require more sophisticated control techniques.

3. **Conduction Losses:** Evaluate conduction losses, which contribute to overall power dissipation in the device. Select devices with low conduction losses to minimize power loss and improve efficiency.
4. **Voltage and Current Margins:** Choose devices with suitable voltage and current margins to accommodate variations in operating conditions, such as fluctuations in input voltage or load current. This ensures stable operation and prevents device overstress under transient conditions.

Passive Components: Selecting passive components such as inductors, capacitors, and resistors involves considering various factors:

1. **Ratings:** Choose components with voltage and current ratings that exceed the maximum expected operating conditions to ensure reliability and prevent component failure.
2. **Tolerance:** Select components with tight tolerance specifications to ensure consistency and accuracy in the performance of the power converter.
3. **Temperature Stability:** Consider the temperature stability of passive components, as variations in temperature can affect their electrical properties. Choose components with stable performance over the expected temperature range of the application [25].
4. **Frequency Response:** Evaluate the frequency response of passive components, especially inductors and capacitors, as it affects their ability to filter out noise and maintain stability in the converter circuit.

Control and Protection Circuitry: Designing control circuitry, including feedback loops, PWM controllers, and protection circuits, is essential for ensuring proper operation and protection of the power converter. Key considerations include:

1. **Regulation:** Implement feedback loops and control algorithms to regulate output voltage and current within specified tolerances, ensuring stable and accurate operation.
2. **Pulse-Width Modulation (PWM):** Use PWM controllers to modulate the duty cycle of the converter's switching signals, allowing precise control of output voltage and current levels.
3. **Protection:** Design protection circuits to safeguard the converter against overvoltage, overcurrent, and over temperature conditions, preventing damage to components and ensuring system reliability.

By carefully selecting semiconductor devices, passive components, and designing control and protection circuitry, engineers can develop power converters that meet performance requirements, operate reliably under varying conditions, and provide efficient and stable power conversion.

2.18 THERMAL MANAGEMENT TECHNIQUES

Heat Sink Design: Efficient heat sink design is crucial for dissipating heat generated by power semiconductors and passive components. Key considerations include:

1. **Optimized Size:** Determine the appropriate size of the heat sink to accommodate the heat dissipation requirements of the power converter while considering space constraints and airflow considerations.
2. **Material Selection:** Choose heat sink materials with high thermal conductivity, such as aluminum or copper, to facilitate efficient heat transfer from the components to the surrounding environment.
3. **Fin Geometry:** Optimize the fin geometry of the heat sink to maximize surface area for heat dissipation while minimizing airflow resistance and pressure drop.
4. **Thermal Resistance:** Minimize thermal resistance between the components and the heat sink by ensuring proper mounting and thermal interface materials.

Cooling Systems: Implementing effective cooling systems is essential for removing heat from critical components and maintaining acceptable operating temperatures. Considerations include:

1. **Fan Cooling:** Utilize fans to enhance airflow over the heat sink and components, increasing convective heat transfer and improving overall cooling performance.
2. **Heat Pipes:** Incorporate heat pipes to efficiently transfer heat from hotspots to areas with better airflow, reducing temperature gradients and improving thermal management.
3. **Liquid Cooling:** Implement liquid cooling systems, such as heat exchangers and coolant loops, to remove heat from components more efficiently, especially in high-power applications or confined spaces.
4. **Airflow Considerations:** Consider factors such as airflow direction, velocity, and distribution to ensure adequate cooling of critical components and minimize temperature gradients within the system.

Thermal Interface Materials: Using high-performance thermal interface materials improves heat transfer between components and heat sinks. Considerations include:

1. **Material Selection:** Choose thermal interface materials with high thermal conductivity, such as thermal pads, greases, or phase-change materials, to minimize thermal resistance and improve heat transfer efficiency.
2. **Proper Application:** Ensure proper application and mating of thermal interface materials between components and heat sinks to minimize air gaps and maximize thermal contact.
3. **Thermal Resistance:** Minimize thermal resistance across the interface by selecting appropriate thickness and applying sufficient pressure during installation.

By carefully designing heat sinks, implementing effective cooling systems, and using high-performance thermal interface materials, engineers can optimize thermal management in power converters, ensuring reliable operation and prolonging the lifespan of critical components.

2.19 OVERVIEW OF CONTROL TECHNIQUES

1. **Open-Loop Control:** In open-loop control, the system's output is controlled based on pre-defined inputs without feedback. While simple, it lacks the ability to respond to changes or disturbances in the system.
2. **Closed-Loop Control (Feedback Control):** Closed-loop control utilizes feedback from the system's output to adjust the control inputs dynamically. This approach enables precise regulation and robustness against disturbances.
3. **Proportional-Integral-Derivative Control:** Proportional-integral-derivative control is a widely used closed-loop control technique that adjusts control inputs based on proportional, integral, and derivative terms of the error between the desired and actual outputs. It provides stable and responsive control in many applications.
4. **Model Predictive Control:** Model predictive control uses a dynamic model of the system to predict future behavior and optimize control inputs over a finite time horizon. It can handle complex systems with constraints and uncertainties, making it suitable for advanced control applications.
5. **Sliding Mode Control:** Sliding mode control aims to drive the system's state onto a pre-defined sliding surface, where it exhibits robustness against disturbances and uncertainties. It achieves this by applying discontinuous control actions.

2.20 GRID-CONNECTED CONTROL STRATEGIES

1. **Voltage and Frequency Regulation:** Grid-connected power converters must regulate output voltage and frequency to match grid requirements. Voltage-source converters and current-source converters employ control techniques such as PWM modulation and grid synchronization algorithms to maintain grid compatibility [26].
2. **Active Power Control:** Grid-connected converters may be required to regulate active power flow into or out of the grid. Control strategies such as droop control, voltage control, and power factor control adjust converter operation to meet grid demand and support grid stability.
3. **Grid Fault Ride-Through:** Grid-connected converters must withstand and ride-through grid faults without disconnecting from the grid. Control strategies such as fault detection, current limitation, and fault ride-through control ensure grid stability and system resilience during fault conditions [27].

2.21 ISLANDING AND MICROGRID CONTROL

1. **Islanding Detection:** In islanded operation, power converters must detect grid disconnection and transition smoothly to standalone operation. Control techniques such as frequency and voltage-based detection methods enable timely detection of islanding events.

2. **Microgrid Control:** Microgrid control strategies coordinate multiple power sources, loads, and energy storage systems within a microgrid to optimize energy management, ensure stability, and maintain grid synchronization. Control techniques include hierarchical control, distributed control, and centralized control approaches.
3. **Energy Management:** Microgrid controllers optimize energy dispatch and storage based on factors such as energy prices, demand response, and renewable energy availability. Control strategies balance supply and demand, prioritize renewable energy integration, and minimize operating costs.

2.22 EMERGING TECHNOLOGIES AND ADVANCEMENTS

2.22.1 Wide Band Gap Semiconductor Devices

Wide-bandgap (WBG) semiconductor materials like silicon carbide (SiC) and gallium nitride (GaN) represent a significant advancement in power electronics technology, offering numerous advantages over traditional silicon-based devices. WBG devices exhibit lower switching losses and reduced conduction losses compared to silicon devices, leading to higher overall efficiency in power conversion applications. Additionally, WBG devices can operate at higher switching frequencies, enabling the design of smaller, lighter, and more compact power converters with increased power density. Furthermore, WBG devices can withstand higher operating temperatures than silicon devices, enhancing their robustness and reliability in harsh environmental conditions. These advantages make WBG semiconductor materials particularly well-suited for various applications, including renewable energy systems, electric vehicles, and aerospace electronics [28] (Figure 2.4).

2.23 MULTI-LEVEL AND MODULAR CONVERTERS

Multi-level and modular converters represent a significant advancement in power electronics technology, leveraging multiple voltage levels to achieve higher voltage and power ratings while mitigating harmonic distortion and switching stress. These converters offer several key features, including improved power quality through the generation of output waveforms with reduced harmonic content, resulting in

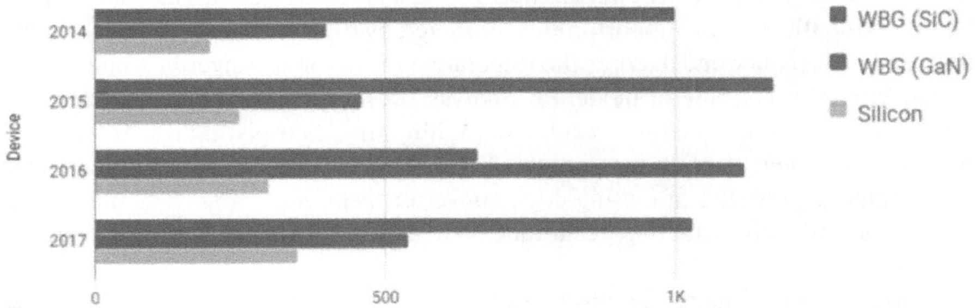

FIGURE 2.4 Switching loss characteristics of WBG (SiC, GaN) and silicon devices.

enhanced power quality and minimized EMI. Furthermore, modular converters, comprising multiple power modules connected in parallel or series, offer scalability and redundancy, enhancing system flexibility, reliability, and ease of maintenance. Additionally, multi-level converters can achieve high voltage conversion ratios without necessitating high-voltage semiconductor devices, thereby reducing cost and complexity in power conversion systems. These attributes make multi-level and modular converters well-suited for various applications, including renewable energy systems, electric vehicles, and industrial power supplies.

2.24 ADVANCED CONTROL ALGORITHMS AND ARTIFICIAL INTELLIGENCE APPLICATIONS

Advanced control algorithms and artificial intelligence techniques are transforming power electronics, offering enhanced performance, efficiency, and flexibility. Model predictive control stands out by utilizing dynamic models of the power system to predict future behavior and optimize control actions over a finite time horizon, resulting in superior transient response, stability, and robustness. Machine learning and artificial intelligence techniques are increasingly prevalent in power electronics, facilitating tasks such as fault detection, condition monitoring, optimization, and predictive maintenance. These data-driven approaches harness insights to enhance efficiency, reliability, and overall performance. Additionally, adaptive control techniques dynamically adjust controller parameters based on system dynamics and operating conditions, allowing for adaptive responses to uncertainties and disturbances. Nonlinear control techniques address inherent nonlinearities in power electronic systems, improving stability and performance further. Collectively, these advanced control methods and artificial intelligence applications are revolutionizing power electronics, offering unprecedented capabilities and enabling the development of more efficient, reliable, and adaptable systems for various applications.

2.25 CONCLUSION

Power converters stand as the linchpin in the seamless integration of renewable energy sources into modern power systems. This chapter has delved deeply into the multifaceted role of these devices, dissecting their design, operation, and application within the dynamic renewable energy landscape. By meticulously examining the unique challenges and opportunities presented by diverse renewable energy technologies, this chapter underscores the imperative for tailored converter solutions.

The intricate interplay of modeling, analysis, design methodologies, and control strategies has been meticulously explored, highlighting their pivotal role in optimizing system efficiency, reliability, and flexibility. This chapter further illuminates the transformative potential of cutting-edge converter technologies, such as multi-level converters and soft-switching techniques, in elevating system performance and sustainability.

Recognizing the inherently interdisciplinary nature of renewable energy systems, this chapter emphasizes the synergistic integration of power electronics, control theory, and renewable energy engineering principles. This holistic approach is essential for driving innovation and advancing the field. By providing a comprehensive foundation in fundamental concepts, advanced techniques, and emerging trends, this chapter empowers researchers, engineers, and practitioners to develop and implement next-generation power converter solutions. Ultimately, these advancements will accelerate the widespread adoption of renewable energy technologies, contributing to a cleaner, more sustainable, and resilient energy future [29–31].

This chapter serves as a catalyst for fostering innovation and propelling the industry forward. By equipping the reader with a deep understanding of power converter fundamentals and their application in renewable energy systems, it inspires the development of novel solutions to address the evolving challenges and opportunities in this dynamic field[32–51].

REFERENCES

1. Agorreta, J.L. et al. (2009) 'Fuzzy switching technique applied to PWM boost converter operating in mixed conduction mode for PV systems', *IEEE Transactions on Industrial Electronics*, 56(11), pp. 4363–4373. Available at: https://doi.org/10.1109/TIE.2009.2019567.
2. Babul Reddy, K. and Santhi Mary Antony, A. (2015) 'A single sensor based PFC SEPIC converter fed BLDC motor drive for fan applications', *International Journal of Applied Engineering Research*, 10(5), pp. 12187–12196.
3. Bai, B., Mi, C.C., and Gargies, S. (2008) 'The short-time-scale transient processes in high-voltage and high-power isolated bidirectional DC-DC converters', *IEEE Transactions on Power Electronics*, 23(6), pp. 2648–2656. Available at: https://doi.org/10.1109/TPEL.2008.2005106.
4. Kalantar, M. and Mousavi G., S.M. (2010) 'Posicast control within feedback structure for a DC-DC single ended primary inductor converter in renewable energy applications', *Applied Energy*, 87(10), pp. 3110–3114. Available at: https://doi.org/10.1016/j.apenergy.2010.04.012.
5. Kamaraj, V. and Nallaperumal, C. (2020), 'Modified multiport Luo converter integrated with renewable energy sources for electric vehicle applications', *Circuit World*, 46(2) pp. 125–135. Available at: https://doi.org/10.1108/cw-08-2019-0104
6. Kim, J.H., Sul, S.K., and Enjeti, P.N. (2008) 'A carrier-based PWM method with optimal switching sequence for a multilevel four-leg voltage-source inverter', *IEEE Transactions on Industry Applications*, 44(4), pp. 1239–1248. Available at: https://doi.org/10.1109/TIA.2008.926201.
7. Kim, S.Y. et al. (2008) 'Fault diagnosis of a ZVS DC-DC converter based on DC-link current pulse shapes', *IEEE Transactions on Industrial Electronics*, 55(3), pp. 1491–1494. Available at: https://doi.org/10.1109/TIE.2007.910627.
8. Kobayashi, K., Matsuo, H., and Sekine, Y. (2006) 'An excellent operating point tracker of the solar-cell power supply system', *IEEE Transactions on Industrial Electronics*, 53(2), pp. 495–499. Available at: https://doi.org/10.1109/TIE.2006.870669.
9. Kolluri, S. and Narasamma, N.L. (2015) 'A new isolated auxiliary current pump module for load transient mitigation of isolated/nonisolated step-up/step-down DC-DC converter', *IEEE Transactions on Power Electronics*, 30(10), pp. 5991–6000. Available at: https://doi.org/10.1109/TPEL.2014.2387695.
10. Li, P., Sun, W., and Shen, J.X. (2019) 'Flux observer model for sensorless control of PM BLDC motor with a damper cage', *IEEE Transactions on Industry Applications*, 55(2), pp. 1272–1279. Available at: https://doi.org/10.1109/TIA.2018.2873529.

11. Li, R., Wang, W., and Xia, M. (2017) 'Cooperative planning of active distribution system with renewable energy sources and energy storage systems', *IEEE Access*, 6, pp. 5916–5926. Available at: https://doi.org/10.1109/ACCESS.2017.2785263.
12. Lim, S.H. et al. (2008) 'Power quality improving operations of a flux-lock type SFCL integrated with voltage or current controlled-voltage source inverter', *IEEE Transactions on Applied Superconductivity*, 18(2), pp. 644–647. Available at: https://doi.org/10.1109/TASC.2008.921317.
13. Lu, D.D.C. and Agelidis, V.G. (2009) 'Photovoltaic-battery-powered DC bus system for common portable electronic devices', *IEEE Transactions on Power Electronics*, 24(3), pp. 849–855. Available at: https://doi.org/10.1109/TPEL.2008.2011131.
14. Mahmoud, J.A. et al. (2014) 'Grid connected photo voltaic powerplant controlled by using FLC and CR with DC-DC boost converter', *International Journal of Scientific Research Engineering and Technology*, 3(6), pp. 946–952.
15. Marouani, K. et al. (2012) Independent control of a two-motor drive system fed by a single voltage source inverter, *IFAC Proceedings Volumes (IFAC-PapersOnline)*. IFAC. Available at: https://doi.org/10.3182/20120902-4-fr-2032.00097.
16. Martinez-Salamero, L. et al. (2011) Sliding-mode control of DC-DC switching converters, *IFAC Proceedings Volumes (IFAC-PapersOnline)*. IFAC. Available at: https://doi.org/10.3182/20110828-6-IT-1002.00557.
17. Mezaroba, M., Martins, D.C., and Barbi, I. (2007) 'A ZVS PWM half-bridge voltage source inverter with active clamping', *IEEE Transactions on Industrial Electronics*, 54(5), pp. 2665–2672. Available at: https://doi.org/10.1109/TIE.2007.901361.
18. Magdy, M. et al. (2021) 'Design control systems for Buck-boost and Cuk converters for solar power applications', *International Journal of Scientific and Engineering Research*, 12(2), pp. 556–563.
19. Mohamed, Y.A.R.I. and El-Saadany, E.F. (2009) 'A control method of grid-connected PWM voltage source inverters to mitigate fast voltage disturbances', *IEEE Transactions on Power Systems*, 24(1), pp. 489–491. Available at: https://doi.org/10.1109/TPWRS.2008.2006996.
20. Mukerjee, A.K. and Dasgupta, N. (2008) 'Power factor controller used as DC-DC converter for photovoltaic sources', *Renewable Energy*, 33(6), pp. 1374–1378. Available at: https://doi.org/10.1016/j.renene.2007.07.005.
21. Narimani, M. and Moschopoulos, G. (2014) 'An investigation on the novel use of high-power three-level converter topologies to improve light-load efficiency in low power DC/DC full-bridge converters', *IEEE Transactions on Industrial Electronics*, 61(10), pp. 5690–5692. Available at: https://doi.org/10.1109/TIE.2014.2300063.
22. Navamani, J.D., Geetha, A., Almakhles, D., Lavanya, A., and Ali, J.S.M. (2021) 'Modified LUO high-gain DC-DC converter with minimal capacitor stress for electric vehicle application', *IEEE Access*, pp. 122335–122350. Available at: https://doi.org/10.1109/ACCESS.2021.3109273.
23. Priyadarshi, N. (2022) 'Performance evaluation of solar PV based non isolated switched inductor and switched capacitor high step up cuk converter', *Electronics*, 11(9), pp. 1–17. Available at: https://doi.org/10.3390/electronics11091381.
24. Ofoli, A.R. and Rubaai, A. (2006) 'Real-time implementation of a fuzzy logic controller for switch-mode power-stage DC-DC converters', *IEEE Transactions on Industry Applications*, 42(6), pp. 1367–1374. Available at: https://doi.org/10.1109/TIA.2006.882669.
25. Ogunrinde, O., Shittu, E., and Dhanda, K.K. (2018) 'Investing in renewable energy: Reconciling regional policy with renewable energy growth', *IEEE Engineering Management Review*, 46(4), pp. 103–111. Available at: https://doi.org/10.1109/EMR.2018.2880445.

26. Park, K. and Chen, Z. (2015) 'A double uneven power converter-based DC-DC converter for high-power DC grid systems', *IEEE Transactions on Industrial Electronics*, 62(12), pp. 7599–7608. Available at: https://doi.org/10.1109/TIE.2015.2458301.
27. Peng, F.Z., Shen, M., and Qian, Z. (2005) 'Maximum boost control of the Z-source inverter', *IEEE Transactions on Power Electronics*, 20(4), pp. 833–838. Available at: https://doi.org/10.1109/TPEL.2005.850927.
28. Saleh, E. et al. (2018) 'Overview of boost converters for photovoltaic systems', *Journal of Power and Energy Engineering*, 6(4), pp. 16–31. Available at: https://doi.org/10.4236/jpee.2018.64002.
29. Samosir, A.S. and Yatim, A.H.M. (2010) 'Dynamic evolution control for synchronous buck DC-DC converter: Theory, model and simulation', *Simulation Modelling Practice and Theory*, 18(5), pp. 663–676. Available at: https://doi.org/10.1016/j.simpat.2010.01.010.
30. Sant, A.V., Rajagopal, K.R., and Sheth, N.K. (2011) 'Permanent magnet synchronous motor drive using hybrid PI speed controller with inherent and noninherent switching functions', *IEEE Transactions on Magnetics*, 47(10), pp. 4088–4091. Available at: https://doi.org/10.1109/TMAG.2011.2159831.
31. Seok, J.K., Lee, J.K., and Lee, D.C. (2006) 'Sensorless speed control of nonsalient permanent-magnet synchronous motor using rotor-position-tracking PI controller', *IEEE Transactions on Industrial Electronics*, 53(2), pp. 399–405. Available at: https://doi.org/10.1109/TIE.2006.870728.
32. Seol, K.S. et al. (2009) 'A synchronous multioutput step-up/down DC-DC converter with return current control', *IEEE Transactions on Circuits and Systems II: Express Briefs*, 56(3), pp. 210–214. Available at: https://doi.org/10.1109/TCSII.2009.2015372.
33. Sharifian, A., Sasansara, S.F., and Balgori, A.A. (2016) 'A new control method based on type-2 fuzzy neural PI controller to improve dynamic performance of a half-bridge DC–DC converter', *Neurocomputing*, 214, pp. 718–728. Available at: https://doi.org/10.1016/j.neucom.2016.07.001.
34. Shireen, W. and Arefeen, M.S. (1996) 'An utility interactive power electronics interface for alternate/renewable energy systems', *IEEE Transactions on Energy Conversion*, 11(3), pp. 643–649. Available at: https://doi.org/10.1109/60.537037.
35. Shrivastava, Y. et al. (2000) 'Harmonic analysis of nondeterministic switching methods for DC-DC power converters', *IEEE Transactions on Circuits and Systems I: Fundamental Theory and Applications*, 47(6), pp. 868–884. Available at: https://doi.org/10.1109/81.852940.
36. Singh, B. and Bist, V. (2015) 'A BL-CSC converter-fed BLDC motor drive with power factor correction', *IEEE Transactions on Industrial Electronics*, 62(1), pp. 172–183. Available at: https://doi.org/10.1109/TIE.2014.2327551.
37. Singh, G.K., Pant, V., and Singh, Y.P. (2003) 'Voltage source inverter driven multi-phase induction machine', *Computers and Electrical Engineering*, 29(8), pp. 813–834. Available at: https://doi.org/10.1016/S0045-7906(03)00036-3.
38. Taher, S.A. and Mansouri, S. (2014) 'Optimal PI controller design for active power in grid-connected SOFC DG system', *International Journal of Electrical Power and Energy Systems*, 60, pp. 268–274. Available at: https://doi.org/10.1016/j.ijepes.2014.02.010.
39. Tan, B. et al. (2017) 'A new approach of minimizing commutation torque ripple for BLDCM', *Energies*, 10(11), pp. 3483–3490. Available at: https://doi.org/10.3390/en10111735.
40. Tan, L. et al. (2016) 'Comprehensive DC power balance management in high-power three-level DC-DC converter for electric vehicle fast charging', *IEEE Transactions on Power Electronics*, 31(1), pp. 89–100. Available at: https://doi.org/10.1109/TPEL.2015.2397453.

41. Vázquez, M.J.V., Márquez, J.M.A., and Manzano, F.S. (2008) 'A methodology for optimizing stand-alone PV-system size using parallel-connected DC/DC converters', *IEEE Transactions on Industrial Electronics*, 55(7), pp. 2664–2673. Available at: https://doi.org/10.1109/TIE.2008.924034.
42. Wai, R.J. and Wang, W.H. (2008) 'Grid-connected photovoltaic generation system', *IEEE Transactions on Circuits and Systems I: Regular Papers*, 55(3), pp. 953–964. Available at: https://doi.org/10.1109/TCSI.2008.919744.
43. Walker, G.R. and Sernia, P.C. (2004) 'Cascaded DC-DC converter connection of photovoltaic modules', *IEEE Transactions on Power Electronics*, 19(4), pp. 1130–1139. Available at: https://doi.org/10.1109/TPEL.2004.830090.
44. Wang, Y. et al. (2018) 'A precise voltage distortion compensation strategy for voltage source inverters', *IEEE Transactions on Industrial Electronics*, 65(1), pp. 59–66. Available at: https://doi.org/10.1109/TIE.2017.2716866.
45. Wang, Y., Choi, S., and Lee, E. (2009) 'Efficient and ripple-mitigating DC-DC converter for residential fuel cell system', *International Journal of Electrical Power and Energy Systems*, 31(1), pp. 43–49. Available at: https://doi.org/10.1016/j.ijepes.2008.10.005.
46. Wu, T., Member, S., and Chen, Y. (1998) 'Modeling PWM DC/DC converters', *IEEE Transaction on Power Electronics*, 13(5), pp. 870–881.
47. Xin, Z. et al. (2018) 'Mitigation of grid-current distortion for LCL-filtered voltage-source inverter with inverter-current feedback control', *IEEE Transactions on Power Electronics*, 33(7), pp. 6248–6261. Available at: https://doi.org/10.1109/TPEL.2017.2740946.
48. Yang, J. et al. (2015) 'The hybrid-cascaded DC-DC converters suitable for HVdc applications', *IEEE Transactions on Power Electronics*, 30(10), pp. 5358–5363. Available at: https://doi.org/10.1109/TPEL.2015.2420666.
49. Yaniv, O. and Nagurka, M. (2003) 'Robust PI controller design satisfying sensitivity and uncertainty specifications', *IEEE Transactions on Automatic Control*, 48(11), pp. 2069–2072. Available at: https://doi.org/10.1109/TAC.2003.819646.
50. Zhang, Y., Liu, J., and Zhang, C. (2013) 'Improved pulse-width modulation strategies for diode-assisted Buck–boost voltage source inverter', *IEEE Transactions on Power Electronics*, 28(8), pp. 3675–3688.
51. Zhou, H.Q. et al. (2008) 'Digital design of combined PI and state feedback controller for non-linear stochastic systems', *IMA Journal of Mathematical Control and Information*, 25(3), pp. 305–322. Available at: https://doi.org/10.1093/imamci/dnm030.

3 Advanced Power Converters for Renewable Energy Systems

Jayachandran Divya Navamani and Anbazhagan Lavanya

3.1 INTRODUCTION

In this chapter, an extensive survey of studies related to high step-up topologies is done. In a variety of applications, high step-up topology is crucial for the power conditioning system. In the literature, various topologies are derived to overcome the limitations encountered in the already existing topologies. This chapter includes a detailed literature review on the state of art technologies in deriving the topologies of high step-up converters. The various uses for high gain dc–dc converters are also covered. A comprehensive review of various applications with the different operating voltages of the converter is presented, and a comparison of a few topologies with their applications is also given in this chapter to have an overview of the various techniques employed to derive the new topologies. Furthermore, a review of various articles published in IEEE and IET journals is also summarized. This study will help the readers to identify the merits and demerits of the already reported topologies in the literature.

The number of articles published on high step-up topologies is increasing year by year. Figure 3.1 shows the articles published on high step-up topologies in IEEE and IET journals for the last two decades. It is noted that the number of journal publications on the high step-up topologies has increased drastically in the last decade. Figure 3.1 displays the percentage of high step-up converter-derived topologies that are most common in the literature. From the study, it is observed that at present, the interest in developing different topologies is on the rise. Also, it is noted from Figure 3.1 that most of the researchers are focusing their attention on deriving high gain converters by integrating a coupled inductor. This literature survey outlines the scope of the research prevailing in this area.

High gain topologies play a significant role in the green energy system. In a microgrid system, it is typically utilized to raise the voltage levels of low-voltage sources, such as solar PV, batteries, supercapacitors, and fuel cells, to 400 V. PV arrays connected in parallel need a high step-up converter to raise the low voltage to

DOI: 10.1201/9781003496946-3

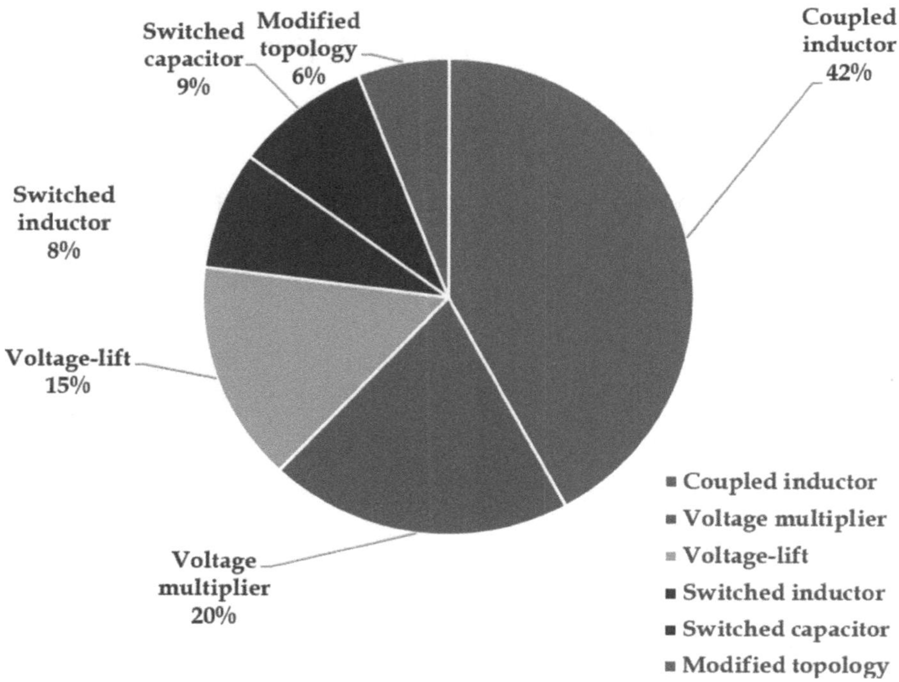

FIGURE 3.1 Share of research articles on high gain dc–dc converters with different high gain techniques.

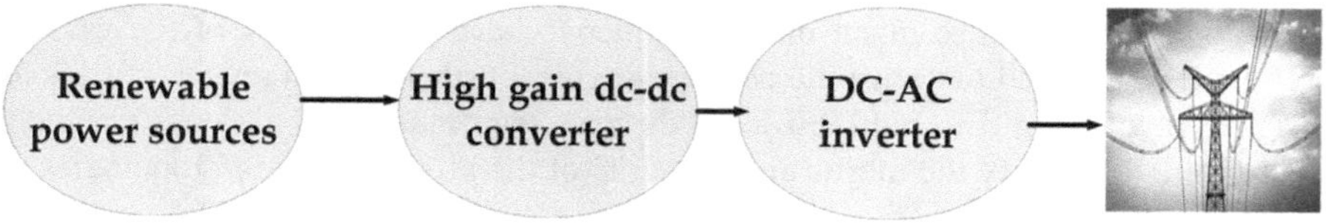

FIGURE 3.2 Flow diagram of the grid-connected distributed energy system.

400 V, which is the high voltage needed for the PV inverter. The block diagram of the system for producing renewable power is shown in Figure 3.2. A DC microgrid is utilized in a distributed generation system to supply users with high-quality power. However, the major limitation in the distributed generation system is the use of a low-voltage energy source. By placing a high gain dc–dc converter between the dc bus and the low-voltage source, this problem can be solved.

High step-up converter topologies are derived based on cascading, coupled inductor, voltage multiplier (VM), switched inductor, switched capacitor, voltage lift, Z source, Y source, quasi-Z source, quasi-Y source, floating, and three state switching cell. High gain topologies are preferred for battery powered and solar/fuel cell applications which require ripple-free input current converters to increase their lifetime. High step-up converters with an input inductor are recommended for those uses.

For DC distributed systems, many key features like soft switching operation, optimized maximum power point tracking technique, and equal power distribution are achieved in the PV converter module [1]. Full-bridge modular multilevel converters are used in offshore wind power integration and transmission in [2] to balance the reactive power.

Table 3.1 presents a comparison of a few high gain topologies reported in the literature. This comparison makes it abundantly evident that power converters take up a significant amount of space in a renewable energy system. It is more advantageous to derive single switch converters than multiple switch converters by avoiding the complexity in switching techniques and control strategies. Optimizing the size of the passive components and utilizing fewer inductors will result in a smaller and lighter converter. It is clear from the discussion in this section that the majority of

TABLE 3.1
Comparison of Few Topologies in the Literature

Ref.	Voltage Gain	No. of Switches/Diode	Application	Merits	Demerits
[3]	$\frac{N+1}{1-D}$	2-Switches 3-Diodes	Microgrid	1. Multiport converter 2. Draws continuous current	Controller design is quite complicated due to two switches
[4]	$\frac{2+N}{1-D}$	3-Switches 2-Diodes	Clean-energy powered system	1. Voltage clamping 2. Synchronous rectification 3. Soft switching	Two switches are clamped to high-side voltage
[5]	$\frac{1+D}{1-D}$	2-Switches 2-Diodes	PV and fuel cell	1. Ripple-free input current 2. Less ripples in output voltage	Number of components is high
[6]	$\frac{N+1}{1-D}$	2-Switches 4-Diodes	Stand-alone solar PV system	Partial shading effects on the solar energy source are reduced	This system is not suitable for high power levels
[7]	$\frac{N+1}{1-D}$	2-Switches 2-Diodes	PV system architecture	1. MVAC grid connection is achieved with one step-up transformer 2. Soft switching	Converter gain is just 9.3 for a duty cycle of 0.71
[8]	$\frac{2N+1}{1-2D}$	1-Switch 4-Diodes	PV module	1. Leakage inductance energy is recycled 2. Switch voltage stress is less	Input current is discontinuous
[9]	$\frac{3M+4D+2}{1-D}$	1-Switch 5-Diodes	Renewable energy system	Voltage stress across the semiconductor devices is reduced severely	An increase in VMCs deteriorates the conversion efficiency
[10]	$\frac{2}{1-D}$	4-Switches	Hybrid power system	The control strategy is implemented easily	The voltage gain is not too high

applications call for a high step-up dc–dc converter that satisfies the aforementioned specifications. Design requirements for high step-up dc–dc converters for various applications are high efficiency, low voltage stress across the semiconductor devices, low voltage ripple, low cost, simplicity, low electromagnetic interference, low ripple input current, high voltage gain, control of the dc–dc converter with respect to input and load variations, flexible arrangement of equipment, lightweight, small volume, and so on.

In [11], it is noted that still, conventional quadratic boost converter topology is used to derive high step-up converters with various high gain cells. In the following section, we will explore the topologies derived from a quadratic boost converter.

3.2 TRADITIONAL QUADRATIC BOOST CONVERTER

Figure 3.3 presents the circuit configuration of a typical quadratic boost converter. The total component count is eight including the output capacitor, C_o. The gain of this converter is

$$G_{\text{QB(CCM)}} = \frac{V_0}{V_{\text{in}}} = \frac{1}{(1-D)^2} \tag{3.1}$$

With a 50% duty cycle in continuous conduction mode, a typical boost converter can deliver a voltage conversion ratio of 4. This will not be enough to raise the low-voltage sources' voltage to the necessary level. The conventional quadratic boost converter is not appropriate for high step-up applications requiring many microgrids because of

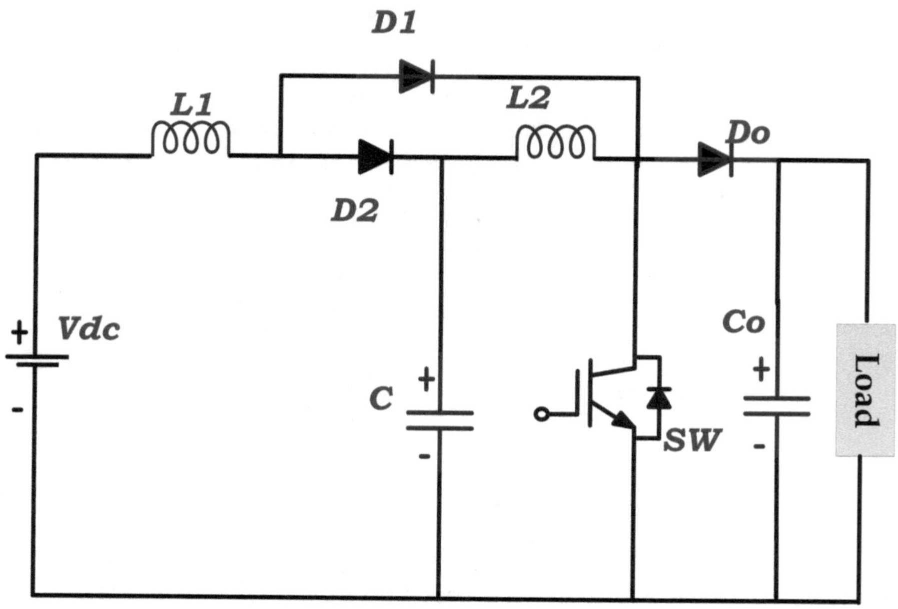

FIGURE 3.3 Conventional quadratic boost converter.

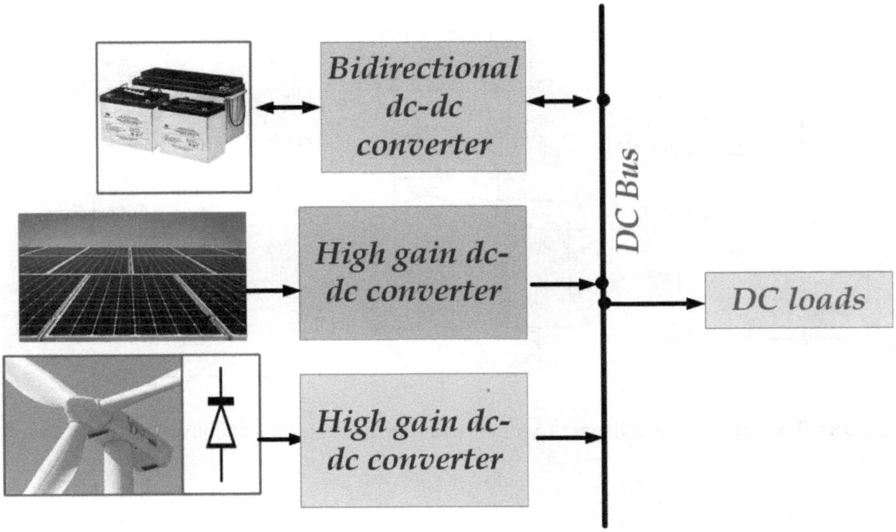

FIGURE 3.4 Block diagram of DC microgrid architecture.

its limited voltage gain. In the proposed converter, the voltage gain is increased over that of a quadratic converter by using a coupled inductor [12].

Combining an extended duty ratio boost converter with the conventional quadratic boost converter yields an ultra-high gain (UHG) [13]. By combining a coupled inductor with quasi-resonant operation [15] and a quadratic boost converter with a switched capacitor cell [14], ultra step-up operation can be accomplished. Due to the input capacitor's existence, the quadratic boost converter preserves its inherent qualities of single power switch operation, small ripple continuous input current, and quadratic voltage gain, which puts less strain on the output capacitor [16].

Figure 3.4 presents the dc microgrid architecture, which depicts the significance of high step-up converters. The required voltage level in the dc bus is 380 V–750 V to convert the dc voltage to a single or three phase ac voltage. All the renewable sources are intermittent in nature and produce voltage of low value. From this, it is clearly understood that even for a hybrid power system, the presence of a high gain dc–dc converter is mandatory.

It can be deduced from the literature review that researchers are concentrating on the derivation of high step-up converters by employing the conventional quadratic boost converter as a foundation. We'll address the several ways to extend a quadratic boost converter to a high step-up converter in the following section.

3.3 DERIVATION OF A HIGH STEP-UP CONVERTER FROM A CONVENTIONAL QUADRATIC BOOST CONVERTER

This section examines the steady-state analysis of the high gain dc–dc converter using five high gain cells. Figure 3.5 explains how the high gain cells are integrated or incorporated to derive the high step-up topology.

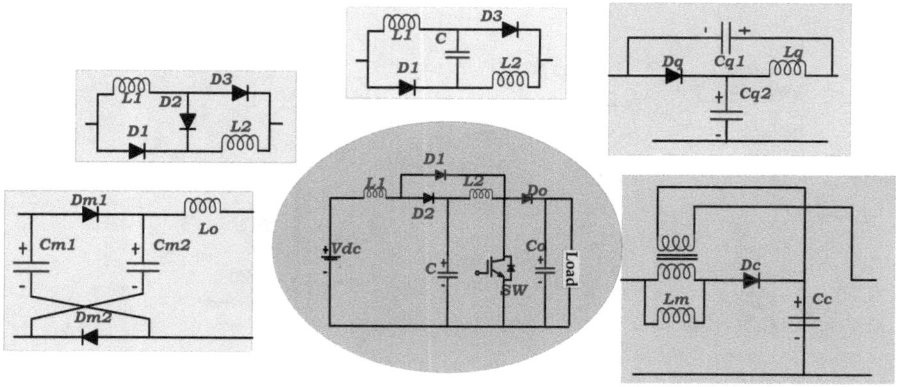

FIGURE 3.5 High gain cells incorporated with the quadratic boost converter.

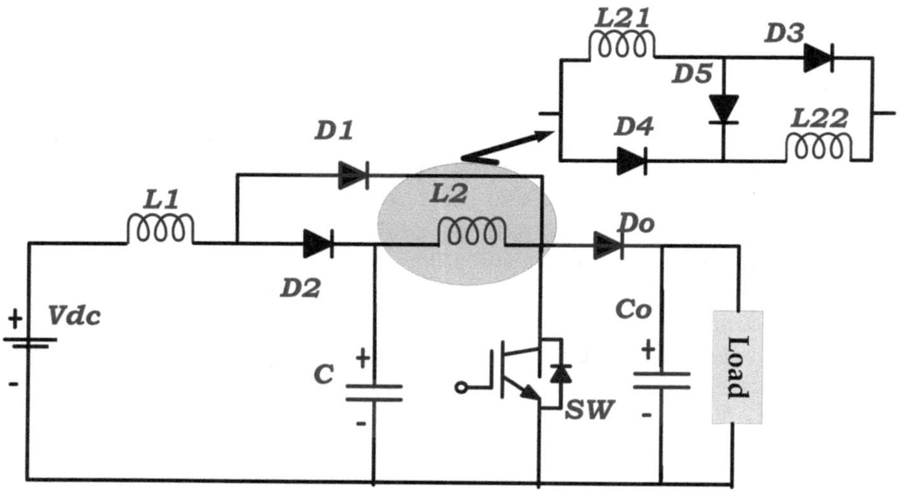

FIGURE 3.6 High gain quadratic boost converter with the switched-inductor cell.

3.3.1 High Gain Quadratic Boost Converter with a Switched-Inductor Cell

To create a high step-up converter that is compatible with renewable energy systems, various high gain cells are used. The circuit layout of the switched-inductor-based quadratic boost converter is shown in Figure 3.6. This design uses a switched-inductor cell, which consists of two passive parts (inductors) and three diodes, to replace the second inductor of a traditional quadratic boost converter.

The switched-inductor cell's inductors in mode I will be charged to the capacitor voltage V_c when switch SW is turned ON. The inductors will be connected in series to discharge the voltage to the load during the off state (mode II). As a result, the output voltage will increase. The voltage conversion ratio of the converter is obtained by steady-state analysis.

The voltage conversion ratio of the high gain switched-inductor cell-based quadratic boost converter is

$$G_{SLQB(CCM)} = \frac{V_0}{V_{in}} = \frac{1+D}{(1-D)^2} \tag{3.2}$$

3.3.2 High Gain Quadratic Boost Converter with a Switched-Inductor-Capacitor Cell

In this setup, the quadratic boost converter's second inductor is swapped out for a switched-inductor-capacitor cell. A capacitor takes the place of one diode in the prior high gain cell. This cell comprises two inductors, two diodes, and a capacitor. Figure 3.7 depicts the circuit configuration of the switched-inductor-capacitor-based high gain quadratic boost converter

In the operating mode, all the passive components (inductors and capacitor) will be charged together in parallel in on mode and the passive components are discharged in series fashion in off mode. By applying volt-second balance law, the voltage conversion ratio of the converter is derived as

$$G_{SLCQB(CCM)} = \frac{V_0}{V_{in}} = \frac{1+2D}{(1-D)^2} \tag{3.3}$$

3.3.3 High Gain Quadratic Boost Converter with a Diode-Capacitor Cell

The diode-capacitor cell arrangement of a high gain quadratic boost converter is shown in Figure 3.8. By substituting the output diode D_o, a diode-capacitor cell is added in this architecture. There are two diodes and two passive parts (capacitors) in this VM.

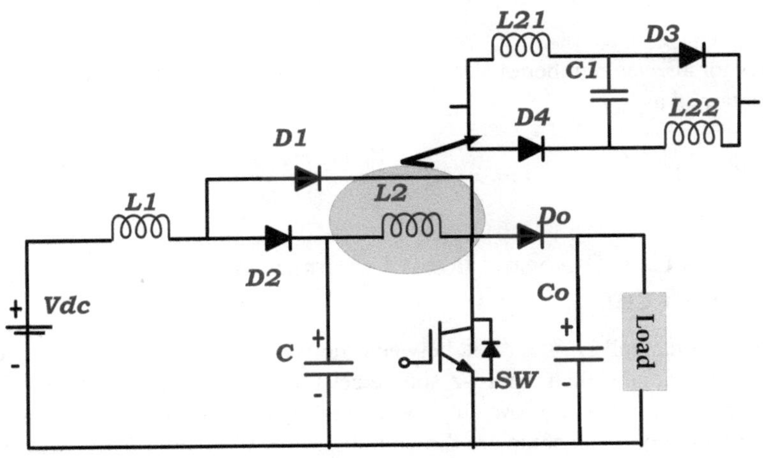

FIGURE 3.7 High gain quadratic boost converter with the switched-inductor-capacitor cell.

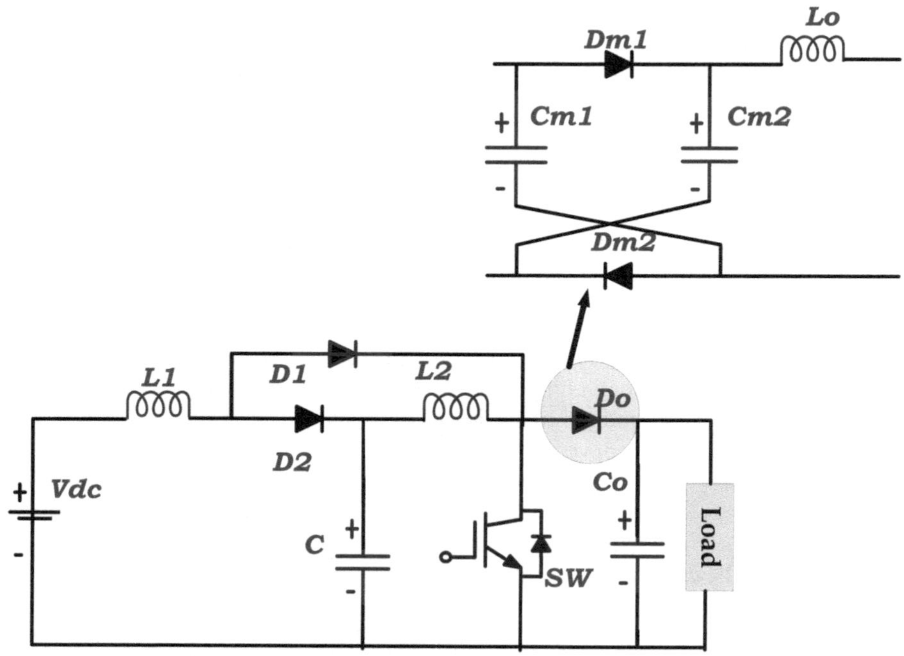

FIGURE 3.8 High gain quadratic boost converter with the diode-capacitor cell.

In this configuration, the expression for the voltage conversion ratio varies for odd and even numbers of VM cells. Consider n as the number of VM cells. For an odd number of VM cells, $n = 1, 3, 5\ldots$, the voltage conversion ratio of the converter is

$$G_{\text{VMQB(CCM)}} = \frac{V_0}{V_{\text{in}}} = \frac{(n+1)-D}{(1-D)^2} \tag{3.4}$$

Similarly for an even number of VM cells, $n = 2, 4, 6\ldots$, the voltage gain of the topology is obtained as

$$G_{\text{VMQB(CCM)}} = \frac{V_0}{V_{\text{in}}} = \frac{n+D}{(1-D)^2} \tag{3.5}$$

3.3.4 High Gain Quadratic Boost Converter with a Quasi-Z Source Cell

A quasi-Z source cell is integrated between the typical quadratic boost converter's second inductor and switch. Quasi-Z source cells have the primary benefit of having a high conversion ratio at a low duty cycle. The quasi-Z source cell and high gain quadratic boost converter setup are shown in Figure 3.9. The converter's voltage gain is displayed as

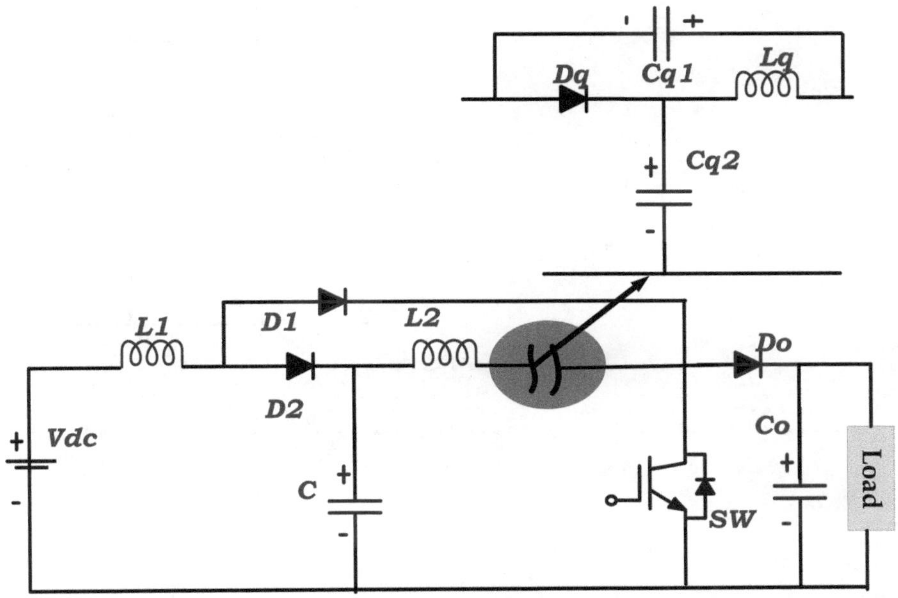

FIGURE 3.9 High gain quadratic boost converter with the quasi-Z source cell.

$$G_{\text{QZQB(CCM)}} = \frac{V_0}{V_{\text{in}}} = \frac{1}{(1-D)(1-2D)} \tag{3.6}$$

The operating duty cycle of this configuration is less than 0.5 duty cycle. With the operating duty cycle, it is viable to achieve a very high voltage gain.

3.3.5 High Gain Quadratic Boost Converter with a Coupled Inductor Cell

By turning the second inductor to a coupled inductor, this converter is derived from a traditional quadratic boost converter. Leakage inductance from the addition of a connected inductor will cause voltage spikes across the switch, SW. The addition of a passive or active clamp circuit can solve this problem.

The high step-up quadratic boost converter architecture with a coupled inductor cell is shown in Figure 3.10. The addition of a passive clamp circuit, which consists of a capacitor (C_c) and a diode (D_c), addresses the leakage inductance issue. This converter's voltage gain is obtained as

$$G_{\text{CIQB(CCM)}} = \frac{V_0}{V_{\text{in}}} = \frac{1+n(1-D)}{(1-D)^2} \tag{3.7}$$

In the above table, the expressions are obtained in the ON and OFF states of the converter. After the application of volt-second balance law, the derived voltage conversion ratio is also presented in Table 3.2.

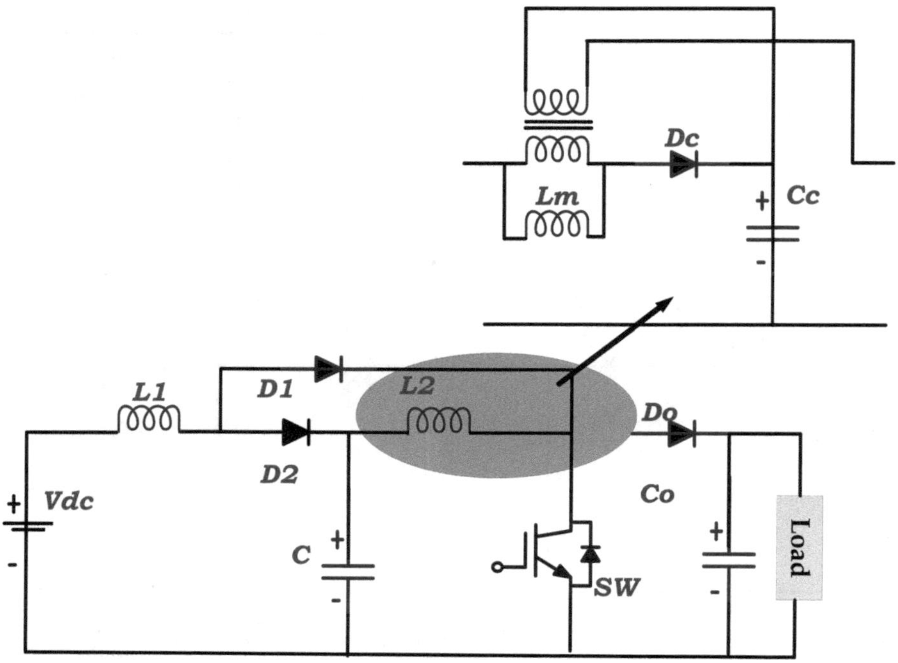

FIGURE 3.10 High gain quadratic boost converter with the coupled inductor cell.

TABLE 3.2
Steady-State Analysis of the Chosen Topology

Topology	ON State	OFF State	Voltage Gain
SLQB	$V_{L1} = V_{dc}$ $V_{L21} = V_{L22} = -V_C$	$V_{L1} = V_{dc} - V_C$ $V_{L21} = V_{L22} = \frac{V_C - V_O}{2}$	$\frac{V_O}{V_{dc}} = \frac{1+D}{(1-D)^2}$
SLCQB	$V_{L1} = V_{dc}$ $V_{L21} = V_{L22} = -V_C$	$V_{L1} = V_{dc} - V_C$ $V_{L21} = V_{L22} = \frac{V_C - V_O}{3}$	$\frac{V_O}{V_{dc}} = \frac{1+2D}{(1-D)^2}$
VMQB	$V_{L1} = V_{dc}$ $V_{L2} = -V_C$ $V_{LO} = 2V_{CM1} - V_O$	$V_{L1} = V_{dc} - V_C$ $V_{L2} = V_C - V_{CM1}$ $V_{LO} = V_{CM1} - V_O$	$\frac{V_O}{V_{dc}} = \frac{1+D}{(1-D)^2}$
QZQB	$V_{L1} = V_{dc}$ $V_{L2} = V_C + V_{Cqz}$ $V_{Lqz} = V_{cqz}$	$V_{L1} = V_{dc} - V_C$ $V_{L2} = V_C - V_{Cqz}$ $V_{Lqz} = V_{cqz} - V_O$	$\frac{V_O}{V_{dc}} = \frac{1}{(1-D)(1-2D)}$
CIQB	$V_{L1} = V_{dc}$ $V_{Lm} = -V_C$	$V_{L1} = V_{dc} - V_C$ $V_{Lm} = V_C - V_{CC}$ $V_0 = V_{CC} + V_{N2}$ $= V_{CC} + nV_C$	$\frac{V_O}{V_{dc}} = \frac{1+n(1-D)}{(1-D)^2}$

3.4 COMPARISON OF TOPOLOGIES DERIVED FROM A QUADRATIC BOOST CONVERTER

The comparison of the high gain dc–dc converter and the traditional boost converter is the focus of this section (Table 3.3).

The SLQB [17] and VMQB [18] converters can provide a voltage gain of six at a 50% duty cycle. The gain of the converter in [19] can be raised to 8 by simply substituting a capacitor for the diode in the switched-inductor cell.

With a QZQB converter [20], the gain of 8 is achievable with a 40% of duty cycle. The CIQB converter in [21] can provide a voltage gain of 8 with a turns ratio of 2. Total components of the CIQB converter are 8, which is less compared to other topologies.

This configuration creates an UHG dc–dc converter by combining switched-inductor and diode-capacitor VM cells. The UHG quadratic boost converter schematic circuit with two high gain cells is displayed in Figure 3.11. The sole method available for deriving the voltage gain of the UHG topology is steady-state analysis. It is crucial to examine the acquired converter's operating principle before starting the steady-state analysis. The modes of operation and the steady-state analysis of

TABLE 3.3
Comparison of High Gain Topologies Chosen for the Study

Topology/Parameters	SLQB	SLCQB	VMQB	QZQB	CIQB
Gain for a 0.5 duty cycle	6	8	6	8 ($D=0.4$)	8 ($n=1{:}2$)
Total number of components	12	12	12	12	10

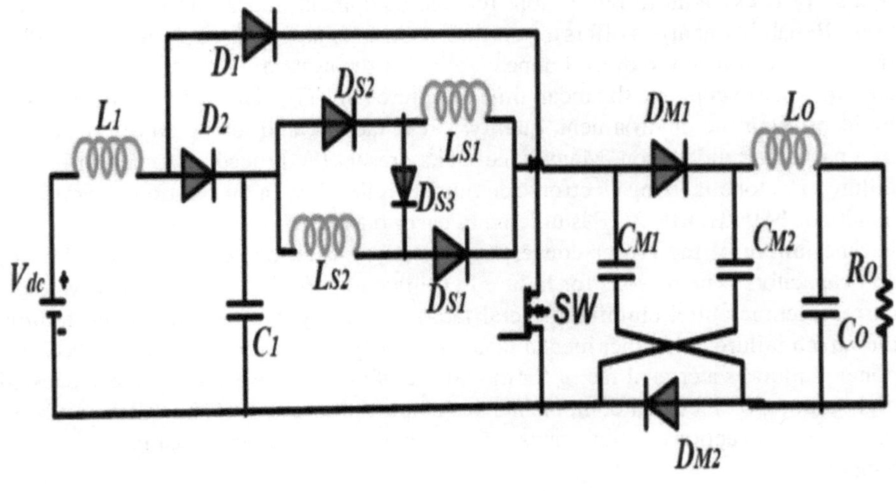

FIGURE 3.11 UHG quadratic boost converter with two high gain cells.

UHG topology are covered in the next section. During a single switching period, the UHG converter's operational modes are separated into modes I and II.

The generalized voltage conversion gains for even and odd numbers of high gain cells are

$$G_{\mathrm{UHG}(n=2,4,6\ldots)} = \frac{[(n+1)-D][1+D]}{[1-D]^2} \tag{3.8}$$

$$G_{\mathrm{UHG}(n=1,3,5\ldots)} = \frac{[n+D][1+D]}{[1-D]^2} \tag{3.9}$$

3.5 RELIABILITY ANALYSIS

In the highly competitive market, a promise of equipment quality and reliability is inevitable. Most of the dc–dc converters in various applications are used continuously for a longer period because dc–dc converters play a significant role in their power supply unit. This frequent usage of the equipment makes the reliability factor a major concern in the design of the equipment. In those applications, reliability is particularly essential since failures in switching converters can lead to a breakdown of the entire system. Over the years, consumers have been demanding more reliable power electronic products. Simultaneously, these products grow more and more complex. Considering these factors, it is essential to design a converter that ensures fault-free operation over a longer period. Failure analysis provides the reason for failures, which leads to development in the design of fault-free equipment and thus helps to enhance the reliability of electronic equipment.

Reliability of equipment can be determined by finding the failure rate of the devices. In most practical environments, dc–dc converters frequently endure numerous stresses, which are liable to damage them. Detailed analysis of failure rate and reliability is essential to build more reliable equipment under different load conditions. Reliability analysis offers a quantitative base for assessing equipment reliability. In order to estimate the overall dependability of the system, field data must be gathered in order to compute the mean time to failure (MTTF). The MTTF is determined by identifying the environment, quality, stress factors, and temperature boundaries of a particular application. Many handbooks present the procedure to determine the failure rates for analyzing electronic equipment reliability. In this section, the military handbook MIL-HDBK-217F is utilized to carry out the reliability prediction.

The failure of the power converter is mostly caused by semiconductor devices and capacitors. The reasons for failure in semiconductor devices are excess voltage, current, temperature, aluminum-metal faults, oxide-layer faults, wire-bond failure, die-attach failure, and other mechanical stresses. Hence, it is observed that the component failure is a crucial factor for the failure of dc–dc converters. In the suggested high gain topologies, ten components are employed to estimate the failure rates. In the following sections, failure rates of the elements are evaluated using a reliability handbook.

This section examines the reliability assessment of the CIQB converter through an analysis of the inductor, capacitor, switch, and diode failure rates. Figure 3.12a shows why the CIQB converter was selected from among the five topologies. With fewer components, the converter's voltage conversion ratio is seen to be high.

The power loss expression of the switch in the CIQB converter is

$$\frac{V_O^2}{R_O^2}\left\{\left(\frac{2-D+n(1-D)}{[1-D]^2}\right)^2 DR_{\text{ds(on)}} + f_S C_O \frac{R_O^2}{[1+n(1-D)]^2}\right\} \tag{3.10}$$

Using the expression (3.10), the power loss of the switch is calculated. Similarly, the power loss of the diode is obtained by the equation below:

$$P_D = V_f I_{D(\text{avg})} + I_{D(\text{RMS})}{}^2 R_f \tag{3.11}$$

The majority of the component failure rate [22] can be expressed generally as

$$\lambda_P = \lambda_b.\pi_{\text{T}}.\pi_Q.\pi_E \quad \text{Failures}/10^6\ \text{h} \tag{3.12}$$

Finally, the failure rate of all the components is calculated as

$$\lambda_T = \lambda_L + \lambda_C + \lambda_{SW} + \lambda_D \quad \text{Failures}/10^6\ \text{h} \tag{3.13}$$

The total failure rate of the CIQB converter is evaluated, and it is presented in Figure 3.12b. From the above evaluation of reliability analysis, it is observed that the MTTF of CIQB topology for PV application is higher compared to other topologies. Therefore, attributing higher appropriateness and reliability, the derived CIQB converter promises appropriate performance in perceptive industrial applications in which high MTTF is the major concern.

3.6 COST ANALYSIS

The most important problem encountered in designing converters for many applications is the volume of the power supply unit because its volume adds heat and increases the cost of the system. In this section, the work is focused on the increase in the power density of the dc–dc converter, which is connected between the low power source and the inverter in most of the applications. Several topologies reported in the literature have different advantages for various applications. However, a size comparison of the topologies is not yet fully explored. Hence, the major design criterion is to bring down the size of the converter. The cost and weight analyses are carried out among the derived topologies, and the optimal sizing of the most reliable converter is also discussed in this section.

In most of the applications, apart from high efficiency, minimization of noise, low cost, and high reliability, power electronics designers insist on strictly adhering to the minimum height and space for the converters.

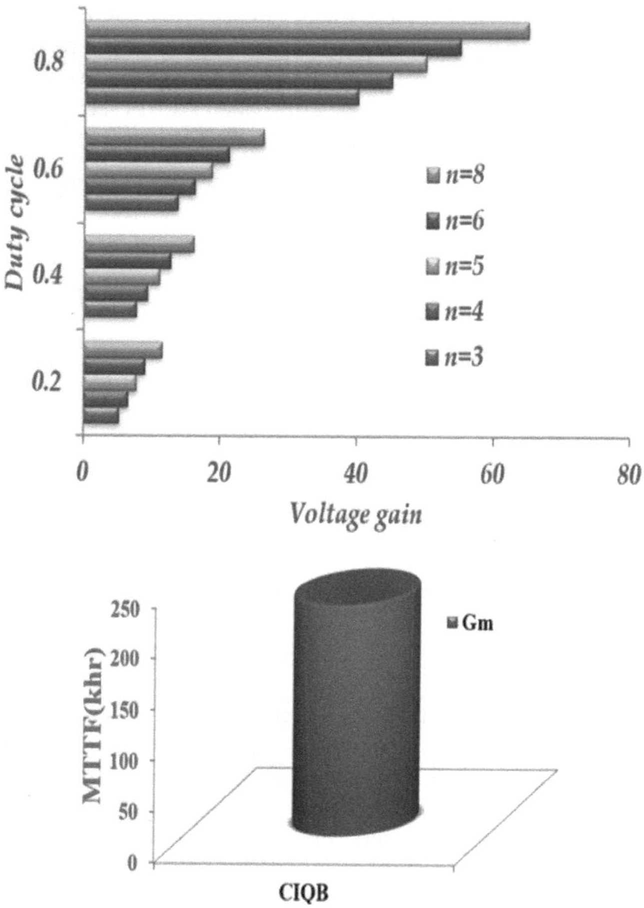

FIGURE 3.12 (a) Voltage gain of the CIQB converter for various values of turns ratio. (b) MTTF of the CIQB converter.

The most important factors considered for the choice of the inductor are the type of mounting style, DC Resistance value, and type of core. Considering these factors, few inductors commercially available in the market are chosen and listed in Table 3.4. In the selection of inductors, surface-mount technology is preferred over the through-hole technology to reduce the height and space of the component. The factors considered for the selection of semiconductor devices are current rating, forward voltage drop, on-state resistance, and packaging. For dc–dc converters, the Schottky diode is the popular choice due to its low forward voltage drop and fast switching. Significant factors considered for the selection of components for dc–dc converters in automotive applications are low cost and less weight.

The two major sources for selection of components are Mouser Electronics and Digikey, where searches are carried out for passive components and semiconductor

TABLE 3.4
Inductors Compared for Weight, Cost, and DC Resistance

Inductor	Model Number	DC Resistance (mΩ)	Weight (g)	Mounting Style	Price (₹)	Inductor Type
15/17 uH	5500-183J (18 uH, 6.7 A)	22	13	SMD	302.94	Powered iron
	1120–220K-RC (22 uH, 7.2 A)	19	23.5	Through hole	275.7	Ferrite core
	ED0006- AL (20 uH, 7.2 A)	42	31	Through hole	374.30	Ferrite core
	B82559A0203A019 (20 uH, 8A)	14.7	11	SMD	374.95	Ferrite core
	PB2020.223NLT (19uH, 8A)	17	9.8	SMD	226	Ferrite core

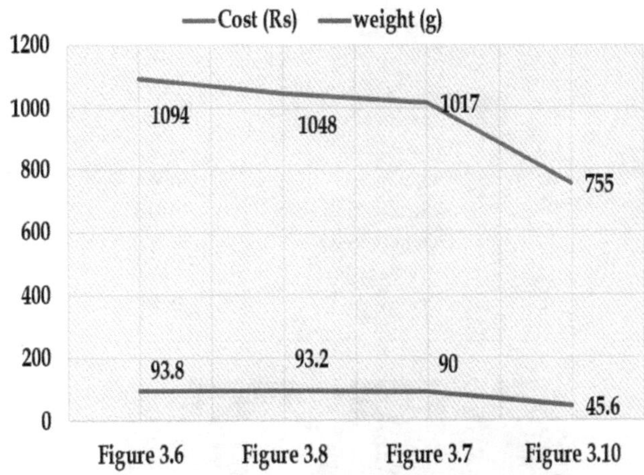

FIGURE 3.13 Cost and weight analysis of the chosen topologies.

devices of several ratings. In most of the coupled inductor-based topologies in the literature, ferrite cores are used to build the coupled inductor with primary and secondary turns. In [22], they used CF139 material, Cosmo ferrites for the coupled inductor. From the material datasheet of Cosmo ferrites, it is observed that the relative core loss is about 0.06-0.37 W/m^3. Considering several ferrite cores available in the market, ETD2910 is taken for analysis from Cosmo ferrites ltd, India, and its weight and cost are included in the comparison discussed below.

The graphical comparison is given in Figure 3.13. From the comparison, it is observed that the weight and cost of CIQB topology are the least compared to other derived topologies. It is also observed that inductors are generally the largest, most costly, and volumetrically inefficient component in power converters. Hence, improvement in inductor design has a greater impact on the cost and size of the converter.

3.7 HARDWARE RESULTS OF THE CIQB CONVERTER

The CIQB topology is validated by the construction and testing of a 40 W laboratory prototype. Table 3.5 provides the design specification for the experiment. An independent microcontroller, the PIC16F877A, uses an optically isolated TLP250 gate driver circuit to drive the metal-oxide-semiconductor field-effect transistor.

Using the aforementioned components, CIQB topology is implemented for a 40 W, xenon lamp application. The experimental waveforms of the CIQB converter for this power rating with a 12 V input voltage are given in Figure 3.14a–d. It can be seen that the maximum switch voltage is 44 V, which is much lesser than the output voltage, 96 V. This feature allows a low rated power switch with low ON state resistance to be utilized, which significantly decreases the conduction loss. Figure 3.14d illustrates

TABLE 3.5
Component Rating of CIQB Topology

Sl. No.	Components	Model/Rating
1	Switch	IRF840
2	Diode	IN4003
5	PIC microcontroller	PIC16F877A
6	Gate driver circuit	TLP250
7	Input voltage	12 V
8	Output power	40 W
9	Switching frequency	40 kHz
10	Output voltage	96 V
11	*n* (Turns ratio)	2
12	Duty cycle	0.5

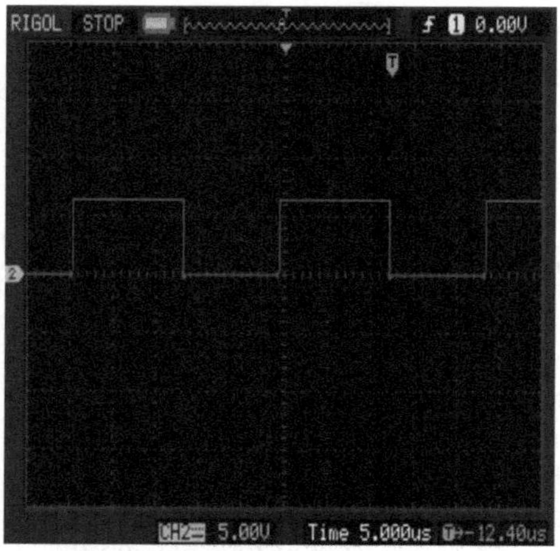

FIGURE 3.14 Hardware results of the CIQB converter. (a) Gate pulse. (b) Load voltage. (c) Maximum voltage across the switch. (d) Voltage across clamp diode Vcc.

(Continued)

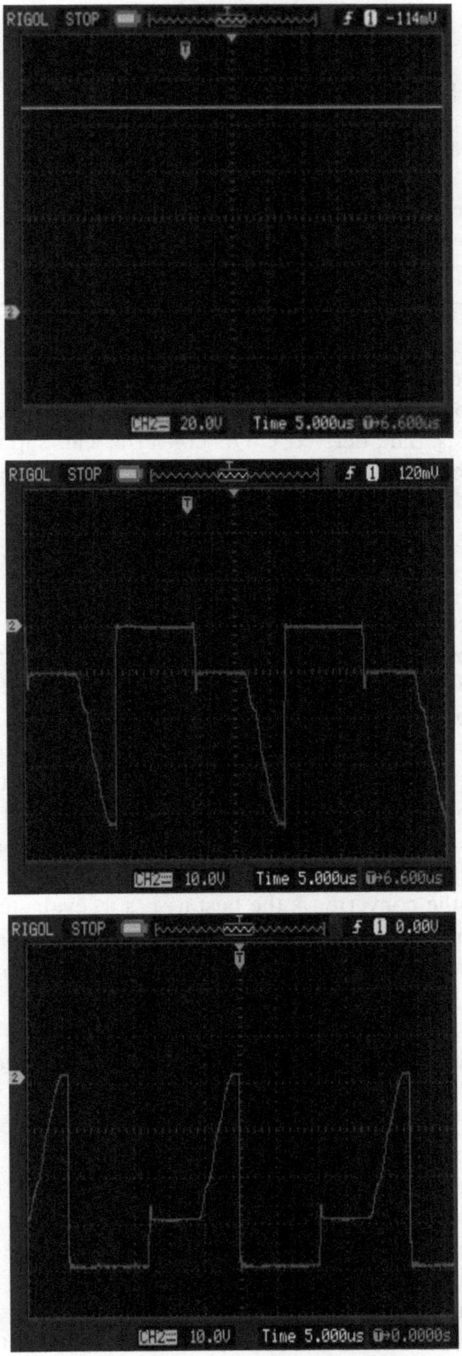

FIGURE 3.14 (*Continued*) Hardware results of the CIQB converter. (a) Gate pulse. (b) Load voltage. (c) Maximum voltage across the switch. (d) Voltage across clamp diode Vcc.

TABLE 3.6
Results of the CIQB Converter

CIQB Topology	Theoretical Voltage (V)	Simulated Voltage (V)	Experimental Voltage (V)
Input voltage	12	12	12
Output voltage	96	94	93
Switch voltage stress	48	47	44
Clamp diode voltage stress	48	47	44

the stress voltage across the clamp diode Dc. It is evident that the voltage stress across the clamp diode is equal to the switch voltage stress.

The presented prototype waveforms of the converter validate the CIQB topology operation and analysis. The derived CIQB converter can supply a voltage gain of eight to step-up the output voltage of battery to steady-state operation of the xenon lamp. The validation of CIQB topology's results is presented in Table 3.6.

3.7.1 Summary

The whole analysis in this chapter is carried out to derive the best and suitable topology for renewable power application. The voltage conversion ratio of the CIQB converter is high with less component count than the other derived topologies. Stresses on the semiconductor devices are also comparatively less. In addition to that, the total energy volume of the inductors of the CIQB converter is very low. Furthermore, the input current ripple is less due to the presence of the inductor in series with the input supply. Accordingly, it can be integrated with a low power supply like batteries and fuel cells, which require an input filter. This converter increases the reliability of these sources by reducing ripples in the input current without any additional filters.

The reliability of the converter is the best assess to evaluate its predictable lifetime. It is noticed from the results that the failure rate of the component depends on power losses and temperature. Observations from the analysis made are summarized as follows.

- The failure rate of the semiconductor devices is higher compared to other components in the converters.
- The failure rate of the semiconductor components is greatly affected by their parasitic elements and the temperature.
- The failure rate of the power switch depends on Rds(on) of the switch, which is closely related to the stresses on the component.
- The failure rate of the inductors is not only affected by the parasitic element but also depends on the radiating surface area, which is closely related to the heat dissipation.
- Factors that affect the failure rate of the capacitors are the voltage stress ratio and temperature.

Other inferences from the study include:

- The high gain dc–dc converter overcomes various problems encountered in the conventional boost converters. Due to the broad application of the high gain dc–dc converter in many power architectures, it attracts many researchers to design and derive efficient topologies for several applications.
- Validating the theoretical findings about the voltage gain and voltage stress across the semiconductor devices is the main goal of the investigative examination. The findings obtained from the hardware and the simulation waveform accord very well, confirming the recommended converters' preferred performance.
- Certain investigation in the literature suggests that the use of silicon carbide semiconductors improves the efficiency of high gain converters. For any application, efficiency and reliability are the important key performance metrics in performance analysis.
- In most of the applications, reliability is particularly significant because failures in the power converters result in a breakdown of the entire system. In order to prevent further damage as well as to enhance the system reliability, a fault-tolerant design of the converter is a superior option.
- Fault diagnosis is a significant design feature for high gain dc–dc converters. It can prevent dc–dc converters from further damage and gives useful information to prepare for remedial actions.
- Recently, in the design of power electronic converters, an optimal design has drawn a lot of attention. In most of the product development, the presizing approach is the most essential step to optimize the size and performance of the product.
- In future work, these ultra gain topologies can be analyzed thoroughly for any specific application, and their compatibility with the application can be studied in detail.

Future scope of this study can be explored as

- In this study, soft switching and controller design are not studied and discussed. The core objective of this research work is to analyze the performance indices of the high gain dc–dc converters.
- Moreover, the performance measures assessed in this chapter are helpful for examining the exactness of the converter developed. However, the entire analysis in this study is made with the help of steady-state analysis. Hence, the dynamic analysis can be performed to design the controller for the proposed topologies. There is a lot of scope in extending the dynamic analysis of the proposed topologies.
- All the derived topologies can be extended by adding a fault-tolerant feature to the structure. Fault diagnosis, identification, and rectification will be carried out by reconfiguring the topology.

- Further, the topologies can also be modified as multi-input dc–dc converters and energy management techniques can be studied by adding battery and other renewable energy sources.
- Finally, this chapter will be a guide to the researchers who are working in high gain dc–dc converters.

REFERENCES

1. K. Kanathipan and J. Lam, "High-Frequency Interlinking High-Voltage Gain PV Converter Modules with Embedded Power Balancing Technique for DC-Distributed System," *IEEE Journal of Emerging and Selected Topics in Power Electronics,* vol. 12, no. 2, pp. 1769–1781, April 2024, doi: 10.1109/JESTPE.2024.3369534.
2. C. Wang, A. Ali and F. Blaabjerg, "Composition and Control of a New Type of Hybrid Voltage-Source Converter Based on DRUs and FB-MMC for Large-Scale Offshore Wind Power Integration and Transmission," *IEEE Transactions on Power Electronics,* vol. 39, no. 5, pp. 5721–5732, May 2024, doi: 10.1109/TPEL.2024.3363617.
3. V.A.K. Prabhala, P. Fajri, V.S.P. Gouribhatla, et al., "A DC–DC Converter with High Voltage Gain and Two Input Boost Stages," *IEEE Transactions on Power Electronics,* vol. 31, no. 6, pp. 4206–4215, 2016.
4. R.-J. Wai, R.-Y. Duan, and K.-H. Jheng, "High-Efficiency Bidirectional DC–DC Converter with High-Voltage Gain," *IET Power Electronics*, vol. 5, no. 2, pp. 173–184, 2012.
5. S. Choi, V.G. Agelidis, J. Yang, et al., "Analysis, Design and Experimental Results of a Floating-Output Interleaved-Input Boost-Derived DC–DC High-Gain Transformer-Less Converter," *IET Power Electronics,* vol. 4, no. 1, pp. 168–180, 2011.
6. M. Das, and V. Agarwal, "Novel High-Performance Stand-Alone Solar PV System with High-Gain High-Efficiency DC–DC Converter Power Stages," *IEEE Transactions on Industry Applications*, 51(6), pp. 4718–4728, 2015.
7. H. Choi, M. Ciobotaru, and M. Jang, "Performance of Medium-Voltage DC-Bus PV System Architecture Utilizing High-Gain DC–DC Converter," *IEEE Transactions on Sustainable Energy*, vol. 6, no. 2, pp. 464–473, Apr 2015.
8. B. Poorali, A. Torkan, and E. Adib, "High Step-Up Z-Source DC–DC Converter with Coupled Inductors and Switched Capacitor Cell," *IET Power Electronics*, vol. 8, no. 8, pp. 1394–1402, 2015.
9. T. Nouri, S. H. Hosseini, E. Babaei, et al., "Generalised Transformerless Ultra Step-Up DC–DC Converter with Reduced Voltage Stress on Semiconductors," *IET Power Electronics*, vol. 7, no. 11, pp. 2791–2805, 2014.
10. C.-C. Lin, L.-S. Yang, and G.W. Wu, "Study of a Non-Isolated Bidirectional DC–DC Converter," *IET Power Electronics*, vol. 6, no. 1, pp. 30–37, 2013.
11. F. Li, J. He, D. Huang, P. Luo, and H. Jiang, "Synchronous Dual-Switch Ultrahigh Step-Up DC–DC Converter Based on Coupled Inductor and Voltage Multiplier for Photovoltaic Systems," *IEEE Transactions on Industrial Electronics,* vol. 71, no. 5, pp. 4807–4817, May 2024, doi: 10.1109/TIE.2023.3283699.
12. S.-W. Lee and H.-L. Do, "Quadratic Boost DC–DC Converter with High Voltage Gain and Reduced Voltage Stresses," *IEEE Transactions on Power Electronics,* vol. 34, no. 3, pp. 2397–2404, March 2019, doi: 10.1109/TPEL.2018.2842051.
13. A. Gupta, N. Korada, and R. Ayyanar, "Quadratic-Extended-Duty-Ratio Boost Converters for Ultra High Gain Application with Low Input Current Ripple and Low Device Stress," *IEEE Transactions on Industry Applications,* vol. 59, no. 1, pp. 938–948, January–February 2023, doi: 10.1109/TIA.2022.3207132.

14. N. Subhani, Z. May, M.K. Alam, I. Khan, M.A. Hossain, and S. Mamun, "An Improved Non-Isolated Quadratic DC-DC Boost Converter with Ultra High Gain Ability," *IEEE Access*, vol. 11, pp. 11350–11363, 2023, doi: 10.1109/ACCESS.2023.3241863.
15. M. Nikbakht, K. Abbaszadeh, S. Abbasian, H. Allahyari, and S. A. Gorji, "An Ultra-Step-Up Quadratic Boost DC–DC Converter Based on Coupled Inductors and Quasi-Resonance Operation," *IEEE Journal of Emerging and Selected Topics in Industrial Electronics*, vol. 4, no. 4, pp. 1096–1109, October 2023, doi: 10.1109/JESTIE.2023.3302706.
16. M. Ashok Bhupathi Kumar and V. Krishnasamy, "Quadratic Boost Converter with Less Input Current Ripple and Rear-End Capacitor Voltage Stress for Renewable Energy Applications," *IEEE Journal of Emerging and Selected Topics in Power Electronics*, vol. 10, no. 2, pp. 2265–2275, April 2022, doi: 10.1109/JESTPE.2021.3122354.
17. D.N. Jayachandran, V. Krishnasamy, and J. Ramiah, "Stability Analysis of a Novel Switched Inductor Based Quadratic Boost DC-DC Converter," *Advances in Electrical and Electronic Engineering,* vol. 15, no. 5, 2017, doi: 10.15598/aeee.v15i5.2436.
18. J. Divya Navamani, K. Vijayakumar, and R. Jegatheesan, "Study on High Step-up DC-DC Converter with High Gain Cell for PV Applications," *Procedia Computer Science*, vol. 115, pp. 731–739, 2017, doi: 10.1016/j.procs.2017.09.109.
19. J.D. Navamani, K. Boopathi, M.J. Sathik, A. Lavanya, Kitmo, and P. Vishnuram, "Analysis of Higher Dimensional Converter Using Graphical Approach," *IEEE Access*, vol. 11, pp. 75076–75092, 2023, doi: 10.1109/ACCESS.2023.3295996.
20. J. Divyanavamani, A. Lavanya, M.J. Sathik, et al., "Expandable Transformer-Less High-Gain DC–DC Converter Based on Quasi-Z Source and Multiplier Cells," *Sādhanā*, vol. 46, no. 22, 2021, doi: 10.1007/s12046-020-01551-7.
21. J. Divya Navamani, K. Vijayakumar, and R. Jegatheesan, "Reliability and Performance Analysis of a High Step-Up DC–DC Converter with a Coupled Inductor for Standalone PV Application," *International Journal of Ambient Energy*, vol. 41, no. 12, pp. 1327–1335, 2020, doi: 10.1080/01430750.2018.1517662.
22. M. Das and V. Agarwal, "Generalized Small Signal Modeling of Coupled-Inductor-Based High-Gain High-Efficiency DC–DC Converters," *IEEE Transactions on Industry Applications*, vol. 53, no. 3, pp. 2257–2270, 2017.

4 High Gain Boost Converters for Solar PV Applications

Jayanthi Kathiresan, Senthil Kumar Natarajan, and Gnanavadivel Jothimani

4.1 INTRODUCTION

In the current scenario, the primary need of a stakeholder from the energy market is to deliver reliable, secure, and quality power [1]. The contemporary electrical infrastructure features available in the transmission and distribution section face a lot of challenges. The cost involved in the expansion of the existing structure to meet the growing energy demand is huge. To overcome the said drawbacks, the power is generated locally and delivered to the consumer at the distribution side, which leads to the concept of the microgrid. So, the cost involved in the expansion of transmission infrastructure is reduced. The need for a microgrid is to generate electricity locally and reliably. Microgrids are defined as "electricity distribution networks including loads with distributed energy sources (such as distributed generators, storage devices, or controllable loads) that can be functioned in a regulated, organized way either while connected to the main power network or while islanded" by the International Council on Large Electric Systems (CIGRR, Fr.), C6.22 Working Group [2]. In India, nearly 20% of power demand is satisfied by solar PV. The government of India has taken steps to achieve 450 GW of solar power by 2030. To develop a DC microgrid set up using a solar PV array, it is mandatory to model a solar PV array, a controller to track the maximum power point, a grid-tied inverter, and their controlling strategy for DC voltage regulation and power flow in the microgrid.

DC–DC converters are typically divided into two groups: isolated and non-isolated. In isolated converters, transformers can be customized to meet specific voltage conversion ratio requirements, but these add weight to the system. These converters also experience significant voltage spikes, and the transformer's leakage inductance leads to greater power loss during switching. On the other hand, non-isolated DC–DC converters can achieve substantial gains using only passive components. They are more efficient and smaller, particularly non-isolated high gain converters. For applications connected to PV systems, non-isolated DC–DC converters are preferred because the converter stage significantly impacts the efficiency of the solar PV system.

The generation of low voltage power from renewable energy sources necessitates the use of high gain DC–DC converters to elevate the voltage to the required load level. Figure 4.1 outlines the diverse power converter topologies employed across

 DOI: 10.1201/9781003496946-4

various applications [3]. Meanwhile, Figures 4.2–4.4 present a series of non-isolated converters developed using the boost converter configuration. Thus far, the adjustment of the output voltage has primarily been achieved utilizing PWM boost converters. Nonetheless, it is important to acknowledge that boost converters are associated with certain drawbacks, including the necessity for large duty ratio selection, switch voltage stress, and the occurrence of high voltage spikes. The quadratic boost converter

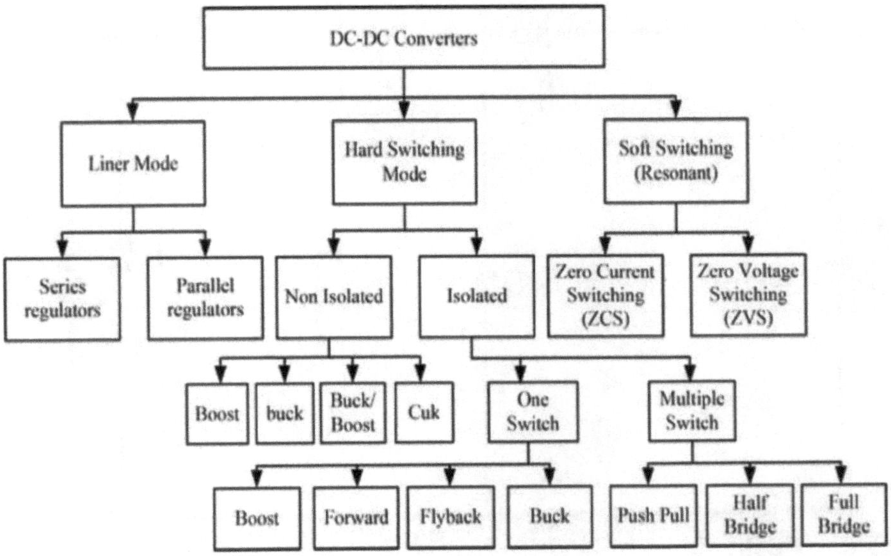

FIGURE 4.1 Power converter topologies [3].

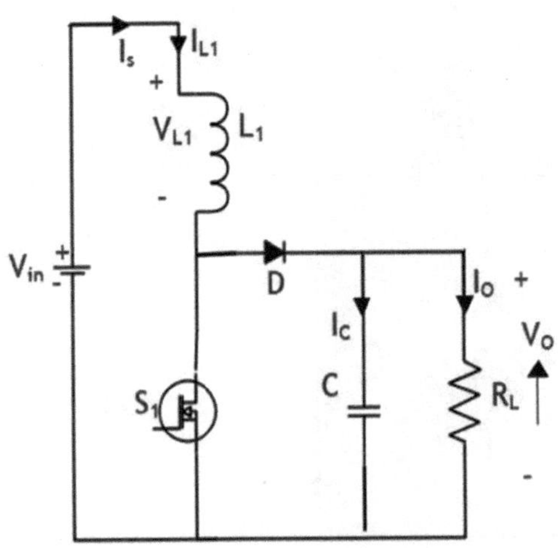

FIGURE 4.2 Conventional boost converter.

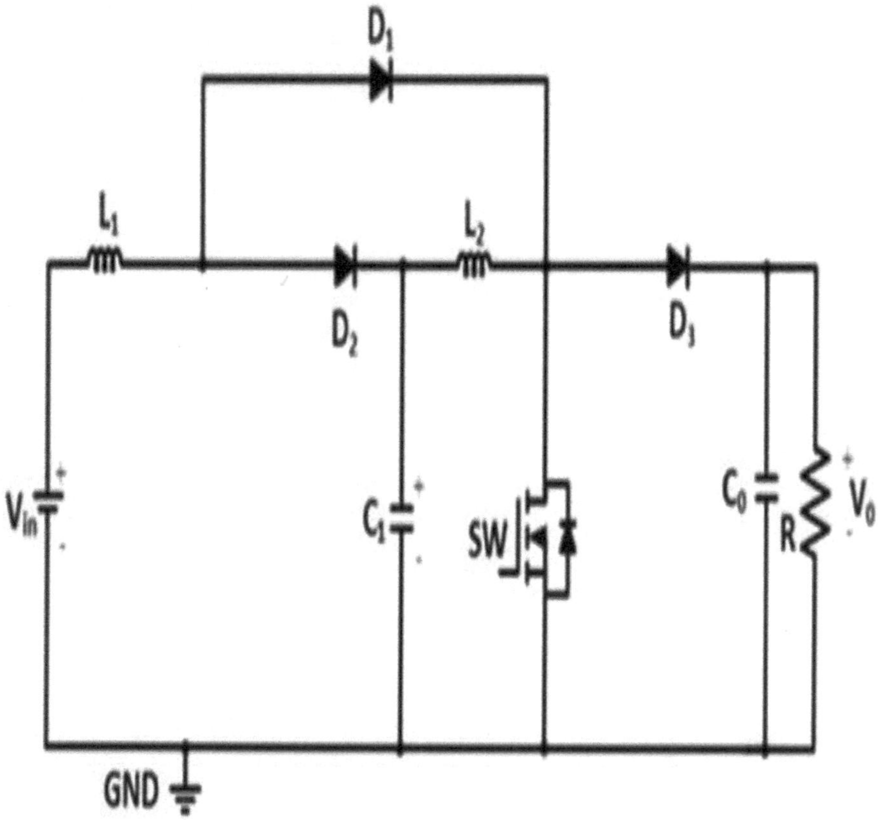

FIGURE 4.3 Quadratic boost converter [4].

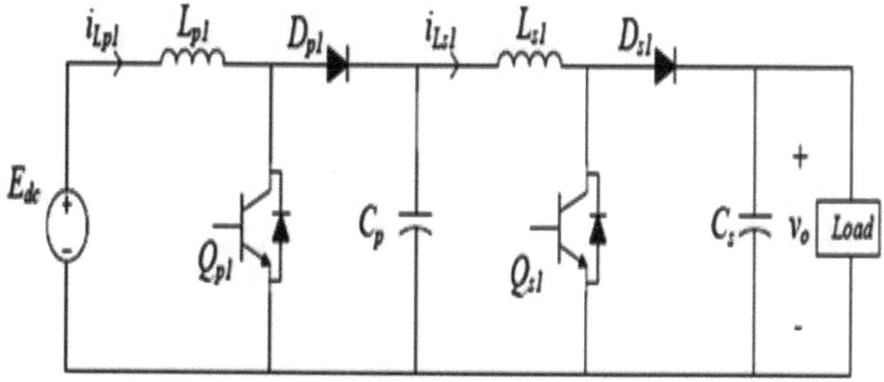

FIGURE 4.4 Cascaded boost converter [5].

is a specialized type of boost converter that utilizes a series connection of the source inductor, diode, inductor, and capacitors to achieve a higher voltage output. The output voltage of the converter is directly proportional to the source voltage and its quadratic gain. Cascaded boost converters, featuring multiple converters connected in series, have the capability to double the output voltage. In a bid to enhance efficiency and mitigate losses, a more streamlined cascade boost converter design with a reduced number of switches was introduced in [5]. This particular type of converter is employed in applications necessitating low voltage ripple and high voltage gain. Further details regarding the properties of these converters are available in Table 4.1.

Also, the converter voltage at the output side is improved by employing voltage boosting techniques such as a switched capacitor, switched inductor, coupled inductor, voltage lift, voltage multiplier, interleaved, cascaded, multilevel, etc. The voltage boosting techniques are shown in Figure 4.5.

TABLE 4.1
Features of Various Boost Converter Topologies

Converter Topology	Voltage Gain	Component Count	Characteristics
Boost converter	<5	4	Diode reverse recovery problem arises when the gain exceeds 5.
Cascaded boost converter	5	8	To achieve the desired outcome, two numbers of power devices, magnetic cores, and control circuits are necessary, resulting in a complex circuit. This setup may also lead to decreased stability.
Quadratic boost converter	4	8	Achieving a voltage gain of 10 or more is unattainable without encountering challenges related to extreme duty ratios and diode reverse recovery issues.
Integrated cascade boost converter	10	9	A greater gain can be attained solely through the use of a substantial turns ratio of 12.

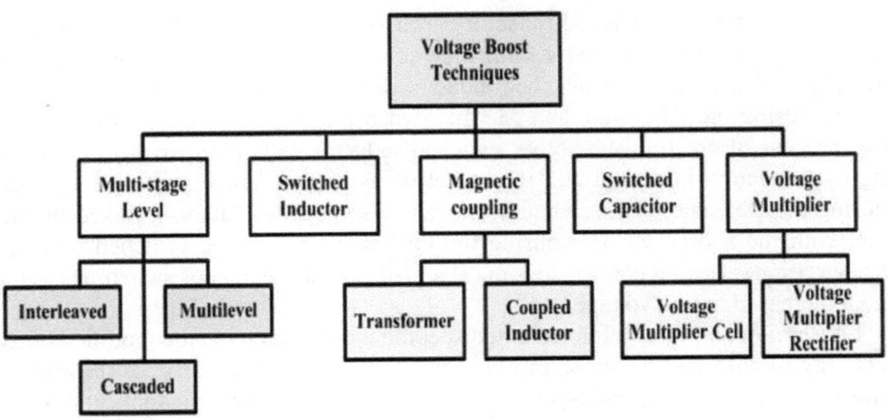

FIGURE 4.5 Voltage boosting techniques [6].

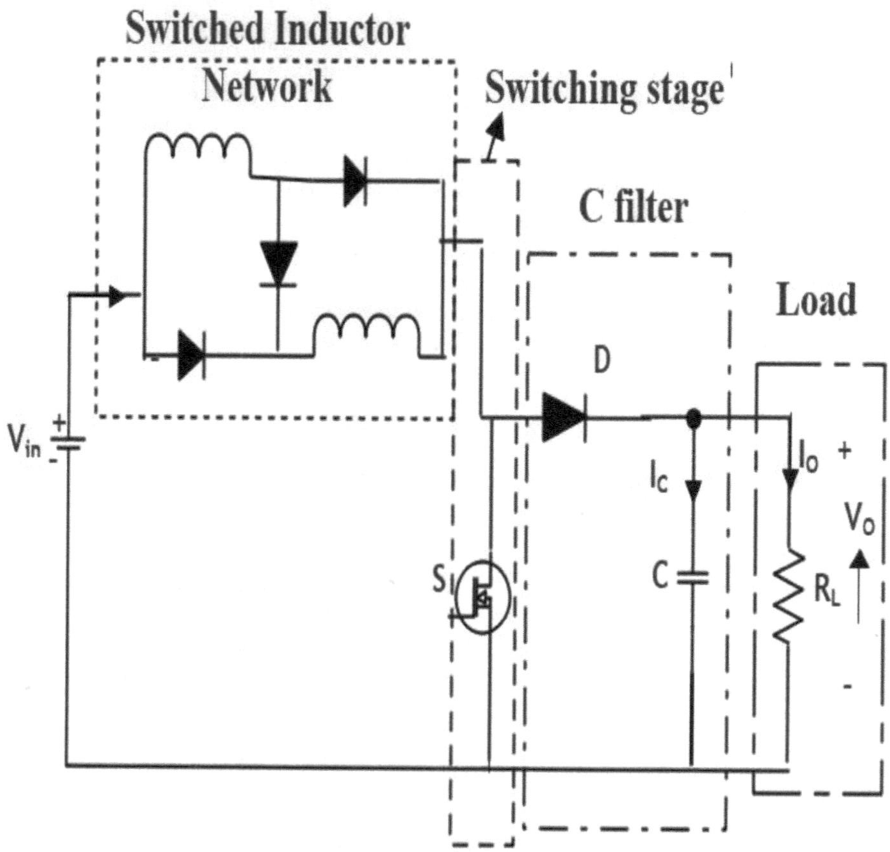

FIGURE 4.6 Switched-inductor-based boost converter [7].

In Figure 4.6, the switched-inductor-based boost converter [4,7] is designed by substituting the source inductor with a switched inductor cell. The output generates high voltage when the switched inductor cell's inductors discharge their stored energy during the OFF state and energize when the switch is in the ON position. The process of creating the active switched inductor cells [8,9] involves connecting the inductors in series with the switches, as depicted in Figures 4.7 and 4.8. During the ON state of the switches, the active switched inductor is linked in parallel with the source voltage; during the OFF state, the active switched inductor discharges in series. According to Lung et al. [10], the simultaneous switching results in an increased output voltage.

Lakshmi et al. [9] have implemented specific modifications to the modified boost topology to enhance the static gain and minimize the voltage drop of the switch. This particular topology is designed for utilization in DC microgrid applications and amplifies the static voltage gain through the functions of three power switches at two distinct duty ratios. Under full load conditions, a duty ratio of d_1=0.5 and d_2=0.35

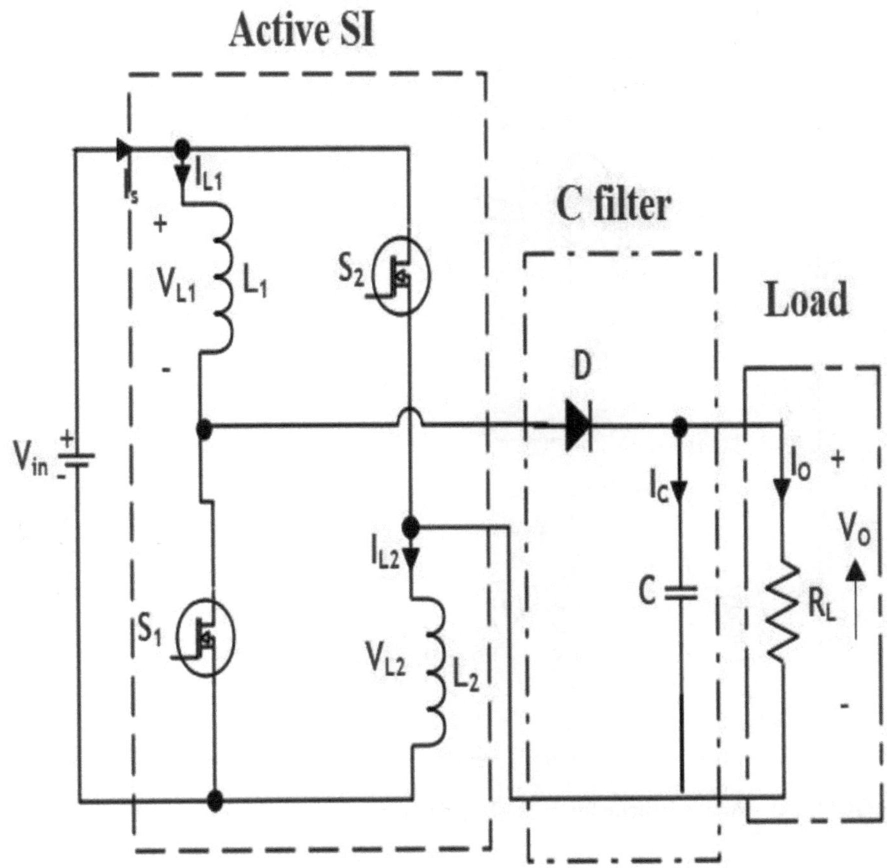

FIGURE 4.7 Active SI-based converter [8].

yields a voltage gain of 10, with an efficiency of 93.6%. Additionally, Kandala Saichaitanya et al. have examined the utilization of nominal or decreased duty ratios in the high gain converter topology to achieve an elevated voltage gain. The revised boost converter forms the cornerstone of this converter topology as well. The converter's three power switches are operated with duty ratios of d_1=0.5 and d_2=0.35 to achieve a voltage gain of 13. The switched capacitor-based boost converters are depicted in Figure 4.9a and b. These configurations play a pivotal role in enhancing the converter's output voltage.

Padmanaban Sanjeevikumar et al. [15] presented various high gain converter structures by combining different converters, such as SEPIC – Switched Inductor SEPIC, CUK-Switched Inductor CUK, Boost – Switched Inductor Boost for renewable energy applications. With the use of single power switch, six different output voltages with different conversion ratios are obtained with high efficiency. Ahmed Alzahrani et al. [16] presented an augmented boost converter by combining the

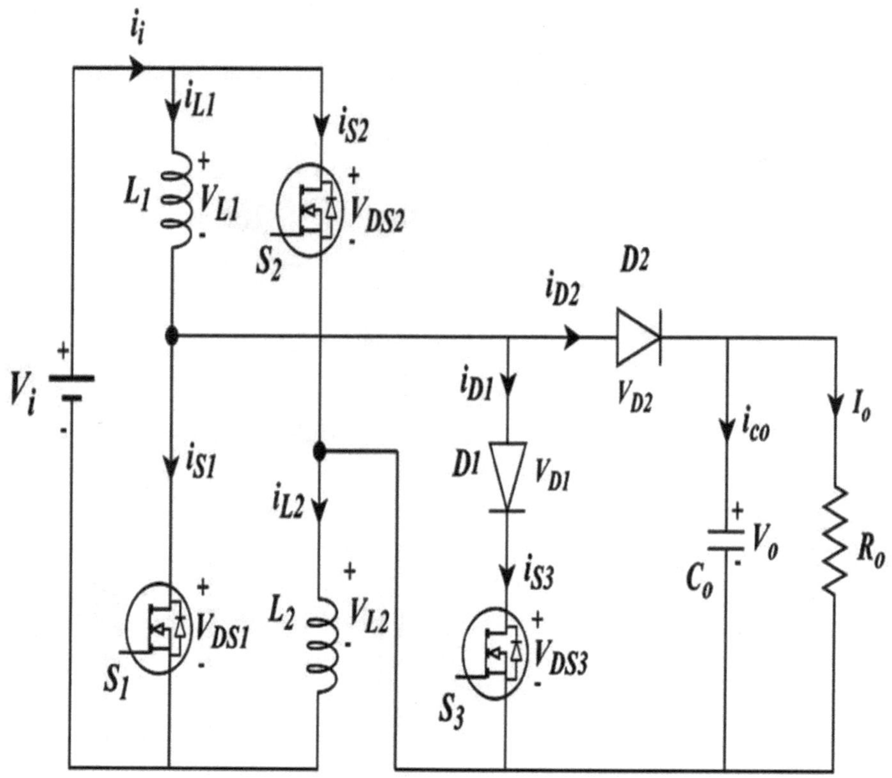

FIGURE 4.8 Active switched-inductor-based boost converter [9].

diode-capacitor-based voltage multiplier cell and Dickson and Cockroft-Walton to enhance the voltage. The designed topology offers a lot of advantages, such as reduced ripple current and reduced voltage across the switch, thereby achieving high efficiency. Lenon Schmitz et al. [17] discussed about a generalized step-up converter-based voltage gain cell topologies to improve the output voltage. The output voltage boosting gain cells are developed and added to the boost converter. Likewise, the pros and cons of converters designed using voltage boosting techniques have been elaborated by Mojtaba Forouzesh et al. [6]. The authors also discussed the applications of these converters in different domains.

Divya Navamani et al., Santosh Kumar Reddy et al., Milad Rezaie et al. [18–20]. discussed the power converter using a quadratic boost cell and diode-capacitor voltage multiplier cell. The addition of a voltage multiplier cell at the load side boosts the output voltage. The source inductors provide ripple-free source current suitable for fuel cell and solar PV-based applications. This converter conquers high voltage gain with low voltage stress on the switches, thereby achieving maximum efficiency. Marcos Antonio Salvador et al. [21] discussed about the development of a step-up converter structure suitable for non-fossil-source-based grid-connected systems.

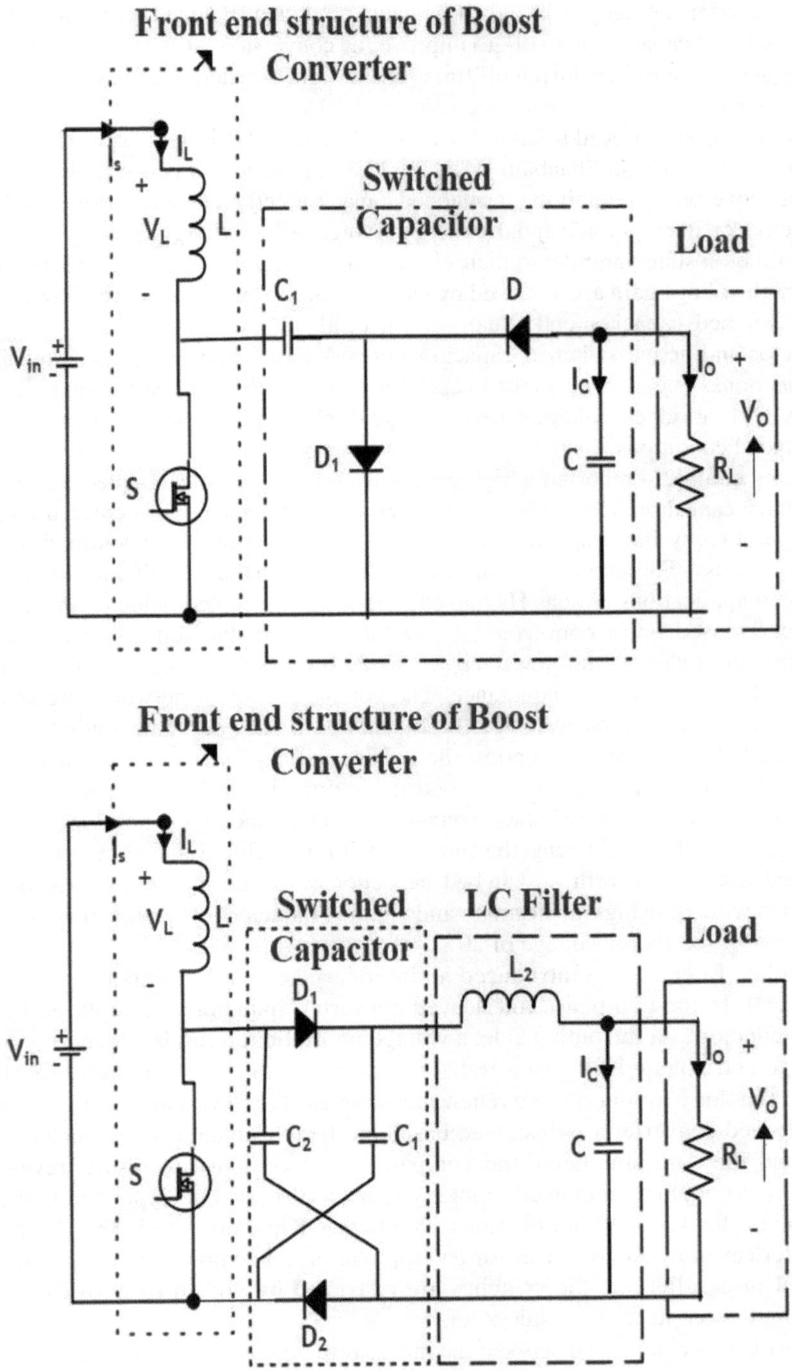

FIGURE 4.9 (a and b) Switched capacitor-based boost converters [11–14].

The converter topology is designed by using an active switched inductor and a passive switched capacitor network to improve the conversion ratio and to trim down the stress across the switch during off time. The designed structure achieves a maximum efficiency of 95.45% for the source voltage of 40 V.

Tohid Jalizadeh et al., Kalarathi et al., Shima Sadaf et al., Alencar Franco de Souza et al., Maryam Shaabani [11,12,14,22,23]. presented a two-switch augmented boost converter by combining a switched capacitor cell and energy storage cell. The inductors on the input side and the energy storage cell get charged, while the switches are in the on state and release their charge during the off state. High output voltage and high voltage gain are achieved by using simultaneous on/off of the power switch and switched capacitor cell. Yuanwei Gu et al. [24] reported an active switched inductor and active switched capacitor network-based high gain converter for PV applications. The active switched capacitor network lifts the conversion ratio effectively with abridged voltage stress. The peak efficiency of 95% is achieved under rated load conditions.

Gang et al. [25] reported a high gain boost topology with a coupled inductor and auxiliary capacitor. The capacitor in series with source voltage increased the input voltage, thereby boosting the static voltage gain. The output capacitors reduce the voltage stress. The continuous current in the input makes it suitable for solar PV, fuel cell applications. Waqas Hassan et al. reported a coupled inductor and switched capacitor-based boost converter for interfacing renewable sources. The clamping circuits are added to the power circuit to curb the voltage spike and recover the energy from the leakage inductance. The voltage multiplier networks are arranged in a stack configuration to realize high static gain. The secondary winding of the coupled inductor is meant to boost the voltage. Jessika Melo de Andrade et al. [13] discussed the modified active switched inductor and switched capacitor cell-based converter to attain elevated static voltage gain. This topology is developed from the boost converter by replacing the inductor with a modified active switched inductor and a capacitor with a switched capacitor cell. The maximum gain of 20 is attained without using transformers and coupled inductors. The efficiency of 95% is attained for the source voltage of 20 V.

Wuhna Li et al. [26] introduced an interleaved converter with a voltage multiplier cell. In the two-phase interleaved converter, inductors are replaced by coupled inductors. At the output side, a voltage multiplier circuit is added to boost the voltage and voltage gain. Anu Rahavi et al. presented an interleaved boost converter topology for interfacing renewable sources. Three variants of the two-phase interleaved converter are discussed based on the placement of the coupled inductor, and they are simulated and compared. The comparison results reveal that the direct-coupled interleaved topology reduced the current ripple and is the best candidate for fuel cell applications. Weerachat Khadmun et al. [27] introduced an interleaved boost converter for PV applications. The boost converters are connected in parallel and the switches are restricted by 90° shifting to each other, making the output voltage smoother.

Kuo-Ching et al. [28] discussed the interleaved high gain boost converter employing a voltage multiplier module. Boost structure with a coupled inductor constitutes a voltage multiplier module to improve the gain. Since the converter is interleaved, it offers minimal current swell and makes it apt for PV applications. The coupled inductor turns ratio decides the static gain of the converter. Ahmad Alzahrani et al. [16] constructed a step-up interleaved boost converter with a bi-fold Dickson voltage multiplier to interface renewable energy sources. The inductors in the interleaved stage store the energy and released it to the voltage multiplier cell to boost the output voltage. The modularity feature and continuous current in the input side of the converter find an application in solar PV. Sri Revathi et al. [29] presented a high gain boost converter topology by combining an interleaved converter with coupled inductors and a diode-capacitor multiplier to boost the voltage gain with minimal current ripple. But the design of a high-frequency coupled inductor with a control circuit makes the converter design complex.

An interleaved multilevel boost converter with voltage multiplier topology is presented by Mahajan Sagar Bhaskar et al. [30] to boost the voltage gain. Two different power sources connected at the input side make them suitable for renewable energy applications. The voltage gain can be increased further by adding a voltage multiplier at the output side. Vinod Kumar Bussa et al. reported an interleaved boost converter with a reduced number of components. This converter operating in three different duty regions gives rise to different voltage gains. The voltage stress and current stress across the elements are different and make the converter cost-effective. The pros and cons of various boosting techniques are tabulated in Table 4.2.

A lot of literature studies are available to explore the design of non-isolated DC–DC converters. The major important parameters considered for selecting the converter for solar PV application are continuous current at the input side, less current ripple, and high efficiency. Also, the smaller number of components makes the converter control scheme simpler and cost-efficient. From the literature survey, it has been found that there is a need for improving the existing DC–DC converter for use in DC microgrids in terms of voltage gain, reduced switch voltage, efficiency, etc.

TABLE 4.2
Pros and Cons of Voltage Boosting Techniques

Voltage Boosting Techniques	Merits	Demerits
Switched capacitor [11–14]	Small size, lightweight circuits, high power density, easy to be integrate, fast dynamic response	Inrush current during start-up, sensitive to the ESR of the capacitor
Voltage multiplier [16,20,28,31]	Very high voltage ability with simple topology, cell-based structure	High voltage stress on components, need several cells with high ratings for very high voltage applications
Magnetic coupling [24,25,32–38]	flexible in boost capacity due to the tuneable turns ratio of magnetic coupling,	Hefty voltage spikes due to leakage inductance, need exact coupled magnetic design, bulky
Multistage/level [30,39]	Modular structure, reliable and efficient, high current/voltage level	Outsized components, fairly large, massive and costly, efficiency degrades with the number of stages/levels

This chapter proposes a M-HGBC with the following objectives:

- To model and analyze the M-HGBC for the DC Microgrid setup.
- To design the DC Microgrid system by integrating solar PV with an alternating current (AC) power grid.
- To find the suitability of the proposed converter in the DC microgrid setup, and to perform a performance study in the DC microgrid setup for various operating conditions.

4.2 CIRCUIT DIAGRAM AND MODES OF OPERATION

The proposed M-HGBC is developed from the boost converter [9]. The circuit diagram of the proposed converter is shown in Figure 4.10. The following assumptions are made in the circuit while designing. All circuit components are ideal. The equivalent series resistance (ESR) of the inductor and capacitor and voltage across the diode are neglected. The number of turns in the inductors L_1 and L_2 is made the same so that the voltages across the inductors are the same.

Three power switches S_1, S_2, and S_3, two identical inductors L_1 and L_2, capacitors C_1, filter capacitor C_0, and a diode D_1 are used to construct the converter structure. This type of converter is designed in such a way that charges the inductors when the power switches S_1 and S_2 are ON with the gating signal of V_{gs1}. During the ON state, the inductors are connected in parallel with the source voltage. The voltages across the inductors are the same as the input voltage. During the OFF state, the power switch S_3 is turned ON by using the gating signal of V_{gs2}, and the inductors are connected in series with the input source and thus boost the output voltage.

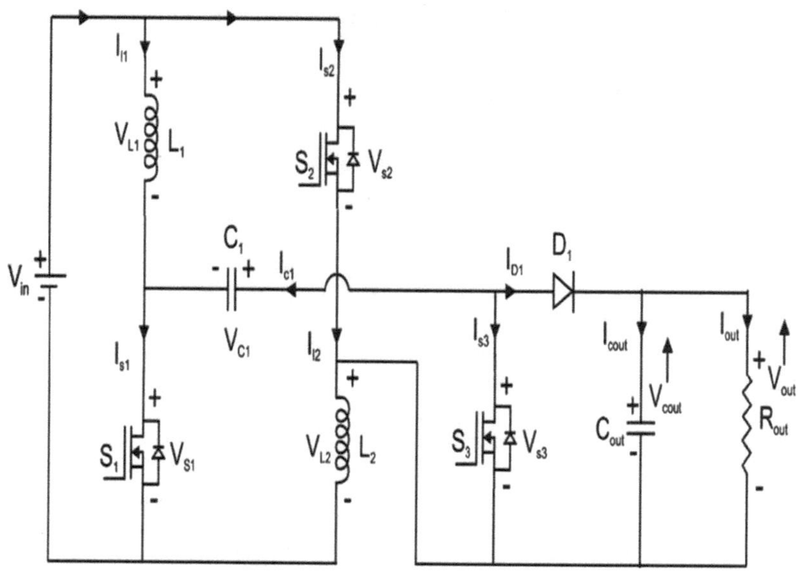

FIGURE 4.10 Circuit diagram of the proposed modified high gain DC–DC boost converter.

4.3 OPERATING MODES AND STEADY-STATE ANALYSIS

The operating mode of the proposed converter (M-HGBC) is identified from the voltage and current waveforms of the converter and is shown in Figure 4.11. There are three operating modes. In mode 1, the power switches S_1 and S_2 are ON/OFF with the gate pulse V_{gs1}. The gate pulse V_{gs2} is used to ON/OFF the power switch S_3. The following assumptions are made in the converter to ensure continuous condition mode. The number of turns in Inductor L_1 and L_2 is assumed to be the same.

$$L_1 = L_2 = L \tag{4.1}$$

Mode 1 ($t_0 < t < t_1$): During mode 1 operation, the power switches S_1 and S_2 are turned ON, and switch S_3 is turned OFF. The inductors L_1 and L_2 are in parallel with input

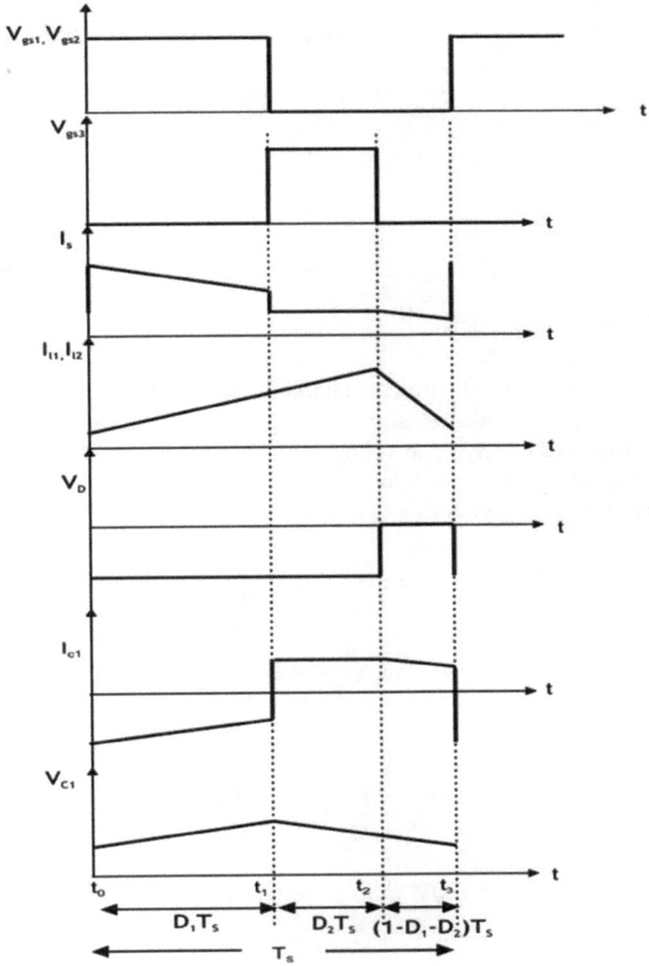

FIGURE 4.11 Voltage and current waveform in continuous conduction mode.

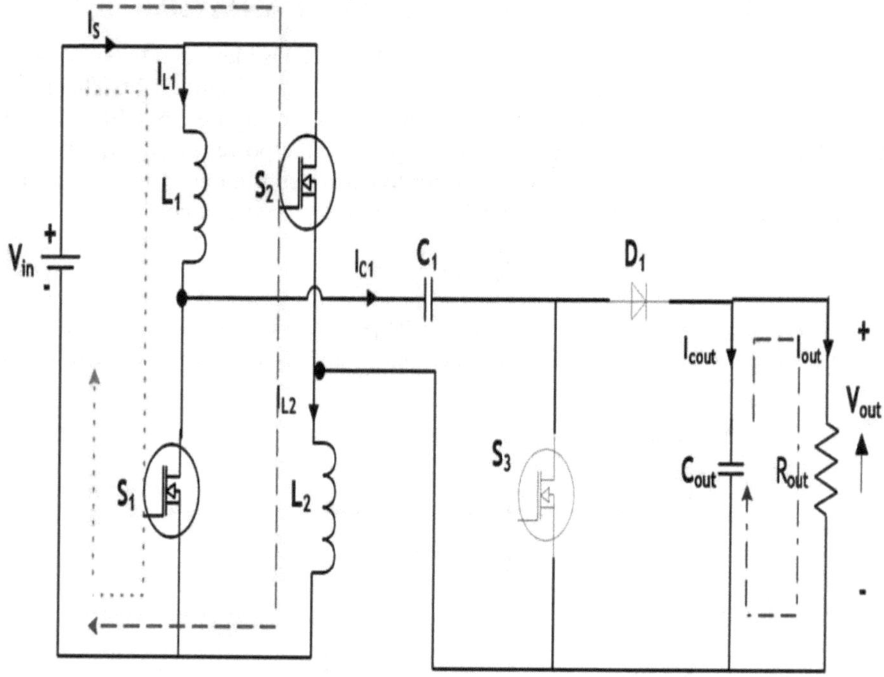

FIGURE 4.12 Mode 1 operation.

source voltage V_{in} and the inductors get charged with the voltage of V_{in}. The current flowing in the circuit in mode 1 is shown in Figure 4.12. The capacitor C_{out} delivers energy to the load resistor R_L. The voltage and current equations of the circuit elements are obtained by applying Kirchhoff's Voltage Law (KVL) and Kirchhoff's Current Law (KCL) and are expressed in Equations (4.1–4.8):

$$V_{L1} = L_1 \frac{di_{L1}}{dt} = L \frac{di_{L1}}{dt} \tag{4.2}$$

$$V_{L2} = L_2 \frac{di_{L2}}{dt} = L \frac{di_{L2}}{dt} \tag{4.3}$$

$$V_{\text{in}} = V_{L1} = V_{L2} = V_{C1} \tag{4.4}$$

$$\frac{dV_{c1}}{dt} = \frac{I_{L1}}{C_1} + \frac{I_{L2}}{C_1} - \frac{I_s}{C_1} \tag{4.5}$$

$$\frac{dV_{\text{out}}}{dt} = -\left(\frac{V_{\text{Cout}}}{C_{\text{out}} R_{\text{out}}} \right) \tag{4.6}$$

$$V_{\text{out}} = V_{\text{Cout}} \tag{4.7}$$

$$I_{\text{cout}} + I_{\text{out}} = 0 \tag{4.8}$$

Mode 2 ($t_1 < t < t_2$): In mode 2, the power switches S_1 and S_2 are turned off, and switch S_3 is turned ON. The inductors L_1 & L_2 are now in series with the source voltage. The inductor L_1 and L_2 now gets charged and current flows through V_{in}, L_1, L_2. By applying KVL and KCL to Figure 4.13, the voltage and current expressions in mode 2 are given in Equations (21.9–21.14):

$$I_{c1} = I_{L1} \tag{4.9}$$

$$V_{\text{in}} + V_{c1} = V_{L1} + V_{L2} \tag{4.10}$$

$$I_{L1} = I_{L2} = I_{c1} \tag{4.11}$$

$$\frac{dI_{L1}}{dt} = \frac{V_{\text{in}} + V_{c1}}{L_1 + L_2} \tag{4.12}$$

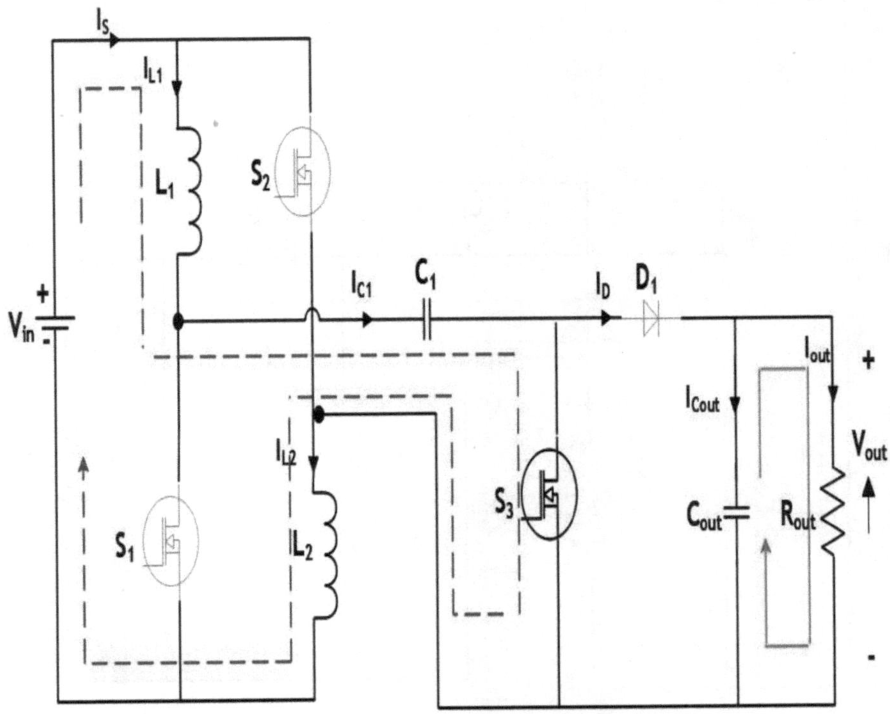

FIGURE 4.13 Mode 2 operation.

$$\frac{dI_{L2}}{dt} = \frac{V_{in} + V_{c1}}{L_1 + L_2} \tag{4.13}$$

$$\frac{dV_{c1}}{dt} = -\left(\frac{I_{L1}}{C_1}\right) \tag{4.14}$$

Mode 3 ($t_{3<}t<t_4$): In mode 3, the power switches S_1, S_2, and S_3 are in the OFF state, and the diode D_1 becomes forward biased. The inductor current flows through capacitor C_1, diode D_1, filter capacitor $C_{out,}$ and inductor $L_{2,}$ as shown in Figure 4.14.

The current through the circuit elements is obtained by applying KCL to Figure 4.14 and is given in Equation (4.15):

$$I_{L1} = I_{c1} = I_{L2} \tag{4.15}$$

By applying KVL, the voltage across the circuit parameters is given in Equations (4.16–4.19):

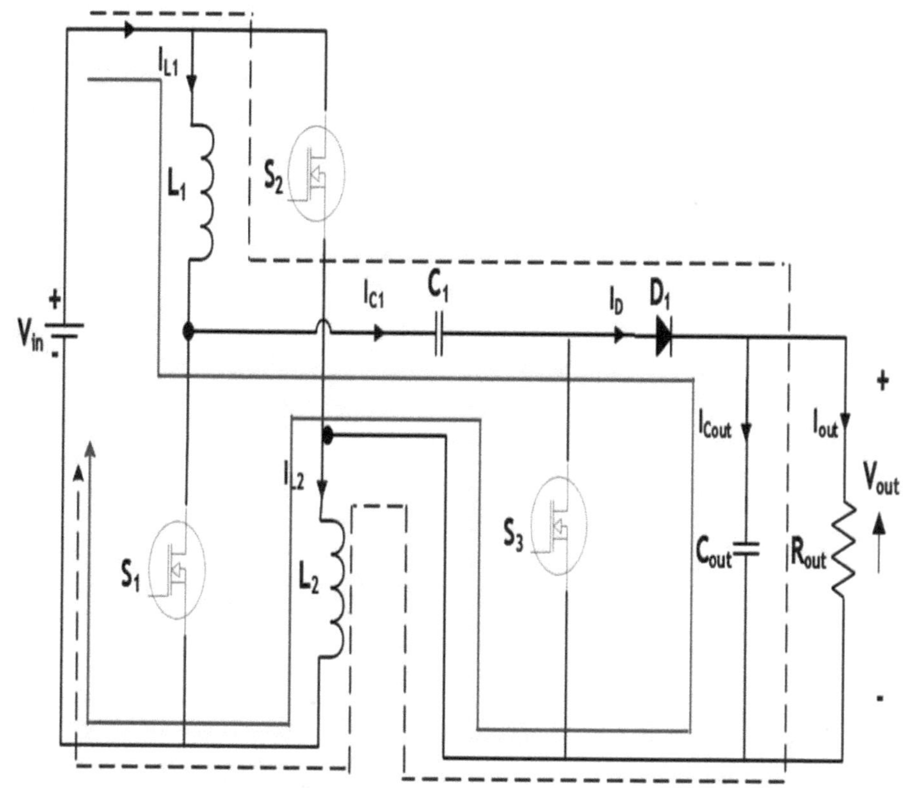

FIGURE 4.14 Mode 3 operation.

$$V_{in} - V_{L1} + V_{C1} - V_{out} - V_{L2} = 0 \tag{4.16}$$

$$\frac{di_{L1}}{dt} = \frac{V_{in} + V_{C1} - V_{out}}{2L_1} \tag{4.17}$$

$$\frac{di_{L2}}{dt} = \frac{V_{in} + V_{C1} - V_{out}}{2L_2} \tag{4.18}$$

$$\frac{dV_{C1}}{dt} = \frac{I_{L2}}{C_1} \tag{4.19}$$

By using the state-space averaging method,

$$\int_0^{d_1T_s} \left(\frac{dI_l}{dt}\right)^{I} dt + \int_0^{d_2T_s} \left(\frac{dI_l}{dt}\right)^{II} dt + \int_0^{(1-d_1-d_2)T_s} \left(\frac{dI_l}{dt}\right)^{III} dt = 0 \tag{4.20}$$

where *I, II,* and *III* are operating modes of the converter.

On substituting the values and simplification, the voltage gain is obtained and given in Equations (4.21 and 4.22):

$$\int_0^{d_1T_s} \left(\frac{V_{in}}{L}\right) dt + \int_0^{d_2T_s} \left(\frac{V_{in} + V_{c1}}{2L}\right) dt + \int_0^{(1-d_1-d_2)T_s} \left(\frac{V_{in} + V_{c1} - V_{out}}{2L}\right) dt = 0 \tag{4.21}$$

$$\frac{V_o}{V_{in}} = \frac{2}{1 - d_1 - d_2} \tag{4.22}$$

4.4 DESIGN AND SELECTION OF INDUCTOR AND CAPACITOR

In the proposed M-HGBC, the inductor design is based on the duty ratio (d_1, d_2), switching frequency (f_s), and load resistor. To operate the converter in continuous conduction mode, the value of the inductor is designed using Equation (4.23).

$$L_1 = L_2 = \frac{(1 - d_1 - d_2)^2 (d_1 + d_2) R_L}{2f_s} \tag{4.23}$$

The selection of capacitor C_1 and C_0 values depends on duty ratios (d_1, d_2), switching frequency (f_s), ripple voltage of the capacitors (ΔV_{C1}, ΔV_{C0}), and load resistor. The value of capacitor C_1 is selected by using Equation (4.24):

$$C_1 = \frac{I_s(1-d_1)}{\Delta V_{C1} f_s} \tag{4.24}$$

Similarly, the value of C_{out} is selected using Equation (4.25):

$$C_{out} = \frac{I_{out}(1-d_1-d_2)}{\Delta V_{CO} f_s} \tag{4.25}$$

4.5 STATE-SPACE ANALYSIS OF THE CONVERTER

The state-space modeling of the converter is done to identify the stability of the proposed converter. The generalized state-space average modeling equations of the DC–DC converter over a while can be expressed and given in Equations (4.26 and 4.27):

$$\frac{dx(t)}{dt} = Ax(t) + Bu(t) \tag{4.26}$$

$$y(t) = Cx(t) + Du(t) \tag{4.27}$$

where $A = A_1d_1 + A_2d_2 + A_3(1-d_1-d_2)$, $B = B_1d_1 + B_2d_2 + B_3(1-d_1-d_2)$, $C = C_1d_1 + C_2d_2 + C_3(1-d_1-d_2)$, $D = D_1d_1 + D_2d_2 + D_3(1-d_1-d_2)$, and d is duty ratio.

The state-space equation of the proposed converter-HGBC in mode I operation is obtained from Equations (4.2–4.6) and Equations (4.28 and 4.29).

$$\begin{pmatrix} \frac{di_{L1}}{dt} \\ \frac{di_{L2}}{dt} \\ \frac{dV_{C1}}{dt} \\ \frac{dV_{Cout}}{dt} \end{pmatrix} = \begin{pmatrix} 0 & 0 & 0 & 0 \\ 0 & 0 & 0 & 0 \\ \frac{1}{C_1} & \frac{1}{C_1} & 0 & 0 \\ 0 & 0 & 0 & \frac{-1}{R_{out}C_{out}} \end{pmatrix} \begin{pmatrix} i_{L1} \\ i_{L2} \\ v_{C1} \\ V_{Cout} \end{pmatrix} + \begin{pmatrix} \frac{1}{L_1} & 0 \\ \frac{1}{L_2} & 0 \\ 0 & \frac{-1}{C_1} \\ 0 & 0 \end{pmatrix} \begin{pmatrix} V_{in} \\ I_i \end{pmatrix} \tag{4.28}$$

$$y(t) = \begin{pmatrix} 0 & 0 & 0 & 1 \end{pmatrix} \begin{pmatrix} i_{L1} \\ i_{L2} \\ v_{C1} \\ V_{Cout} \end{pmatrix} \tag{4.29}$$

Similarly, the state-space equation in mode II of M-HGBC is obtained from Equations (4.12–4.14) and is given in Equations (4.30 and 4.31):

$$\begin{pmatrix} \frac{di_{L1}}{dt} \\ \frac{di_{L2}}{dt} \\ \frac{dV_{C1}}{dt} \\ \frac{dV_{Cout}}{dt} \end{pmatrix} = \begin{pmatrix} 0 & 0 & \frac{1}{2L_1} & 0 \\ 0 & 0 & \frac{1}{2L_2} & 0 \\ \frac{1}{C_1} & 0 & 0 & 0 \\ 0 & 0 & 0 & \frac{-1}{R_{out}C_{out}} \end{pmatrix} \begin{pmatrix} i_{L1} \\ i_{L2} \\ v_{C1} \\ V_{Cout} \end{pmatrix} + \begin{pmatrix} \frac{1}{2L_1} & 0 \\ \frac{1}{2L_2} & 0 \\ 0 & 0 \\ 0 & 0 \end{pmatrix} \begin{pmatrix} V_{in} \\ I_i \end{pmatrix} \tag{4.30}$$

$$y(t) = \begin{pmatrix} 0 & 0 & 0 & 1 \end{pmatrix} \begin{pmatrix} i_{L1} \\ i_{L2} \\ v_{C1} \\ V_{Cout} \end{pmatrix} \tag{4.31}$$

The state-space equation in mode III of M-HGBC is obtained from Equations (4.17 to 4.19) and is given in Equations (4.32 and 4.33):

$$\begin{pmatrix} \frac{di_{L1}}{dt} \\ \frac{di_{L2}}{dt} \\ \frac{dV_{C1}}{dt} \\ \frac{dV_{Cout}}{dt} \end{pmatrix} = \begin{pmatrix} 0 & 0 & \frac{1}{2L_1} & \frac{1}{2L_1} \\ 0 & 0 & \frac{1}{2L_2} & \frac{1}{2L_1} \\ \frac{1}{C_1} & 0 & 0 & 0 \\ \frac{1}{C_{out}} & 0 & 0 & \frac{-1}{R_{out}C_{out}} \end{pmatrix} \begin{pmatrix} i_{L1} \\ i_{L2} \\ v_{C1} \\ V_{Cout} \end{pmatrix} + \begin{pmatrix} \frac{1}{2L_1} & 0 \\ \frac{1}{2L_2} & 0 \\ 0 & 0 \\ 0 & 0 \end{pmatrix} \begin{pmatrix} V_{in} \\ I_i \end{pmatrix} \tag{4.32}$$

$$y(t) = \begin{pmatrix} 0 & 0 & 0 & 1 \end{pmatrix} \begin{pmatrix} i_{L1} \\ i_{L2} \\ v_{C1} \\ V_{Cout} \end{pmatrix} \tag{4.33}$$

On solving, the averaged state-space matrix for the proposed M-HGBC is expressed in Equations (4.34 and 4.35):

$$\begin{pmatrix} \frac{di_{L1}}{dt} \\ \frac{di_{L2}}{dt} \\ \frac{dV_{C1}}{dt} \\ \frac{dV_{Cout}}{dt} \end{pmatrix} = \begin{pmatrix} 0 & 0 & \frac{1-d_1-d_2}{2L_1} & \frac{1-d_1-d_2}{2L_1} \\ 0 & 0 & \frac{1-d_1-d_2}{2L_2} & \frac{1-d_1-d_2}{2L_2} \\ \frac{1-2d_1}{C_1} & \frac{1-2d_2}{C_1} & 0 & 0 \\ \frac{-1+d_1+d_2}{C_{out}} & 0 & 0 & \frac{1-2d_1-2d_2}{R_{out}C_{out}} \end{pmatrix} \begin{pmatrix} i_{L1} \\ i_{L2} \\ v_{C1} \\ V_{CO} \end{pmatrix}$$

$$+ \begin{pmatrix} \frac{-1+2d_2+2d_1}{2L_1} & 0 \\ \frac{-1+2d_1+2d_2}{2L_2} & 0 \\ 0 & \frac{-d_1}{C_1} \\ 0 & 0 \end{pmatrix} \begin{pmatrix} V_{in} \\ I_i \end{pmatrix} \tag{4.34}$$

$$y(t) = \begin{pmatrix} 0 & 0 & 0 & 1 \end{pmatrix} \begin{pmatrix} i_{L1} \\ i_{L2} \\ v_{C1} \\ V_{Cout} \end{pmatrix} \tag{4.35}$$

The stability of the converter is analyzed from the transfer function obtained by considering the specification and design values given in Table 4.3. From the function, it is inferred that the converter is stable.

TABLE 4.3
Designed Parameters of the Proposed High Gain Boost Converter

Parameters	Simulated Values
Input voltage (V_{in})	36 V–48 V
Load voltage (V_0)	380 V
Switching frequency (f_s)	21 kHz
Inductor L_1 and L_2	360 μH
Capacitor C_1	50 μF
Output filter capacitor C_0	3,600 μF
Load power	1 kW

4.6 SIMULATION RESULTS AND ANALYSIS

The converter-HGBC is specifically designed to uphold a load voltage of 380 V. The circuit parameters of the converter are computed using Equations (4.23–4.25) and meticulously detailed in Table 4.3. This design is tailored to accommodate a load power of 1 kW. The Simulink diagram of the proposed converter in a closed-loop configuration is visually represented in Figure 4.15. To regulate and maintain the load voltage, a proportional integral (PI) controller is employed, with the kp and ki values computed using the Z-N method. With the controller in place, the converter effectively sustains a load power of 1 kW and a load voltage of 380 V for an input of 48 V. The simulated waveforms, inclusive of input voltage, current, power, load voltage, current, and power, are elaborately illustrated in Figure 4.16. The controller demonstrates its efficiency in regulating the output voltage, achieving the load voltage within less than 0.05 second.

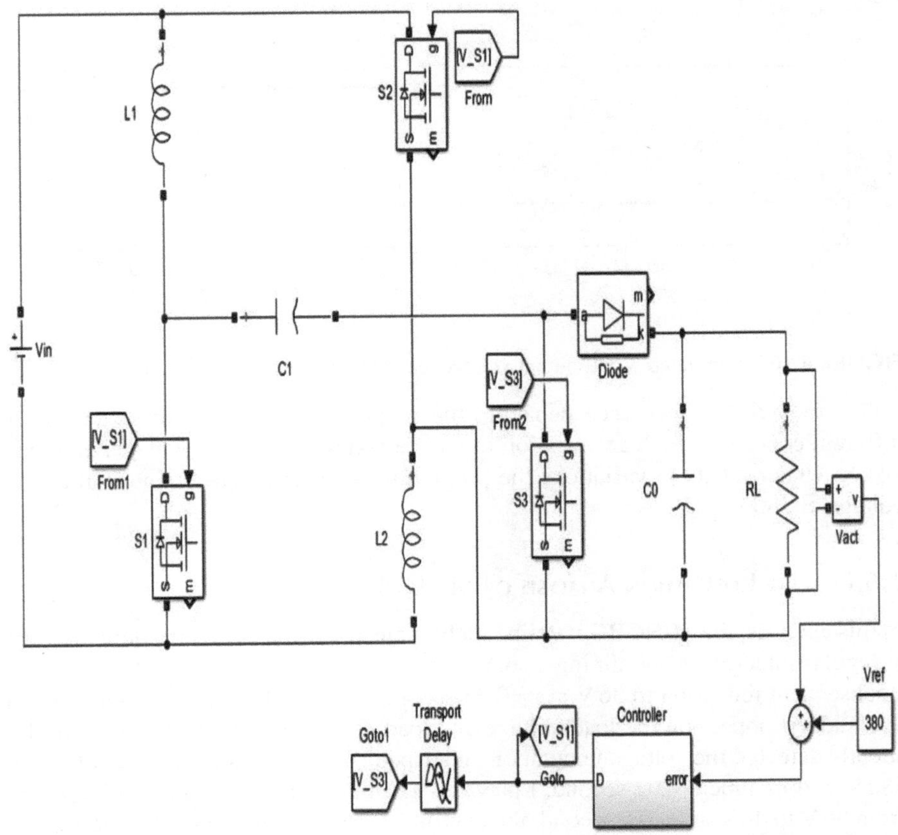

FIGURE 4.15 Closed-loop Simulink model of the proposed M-HGBC.

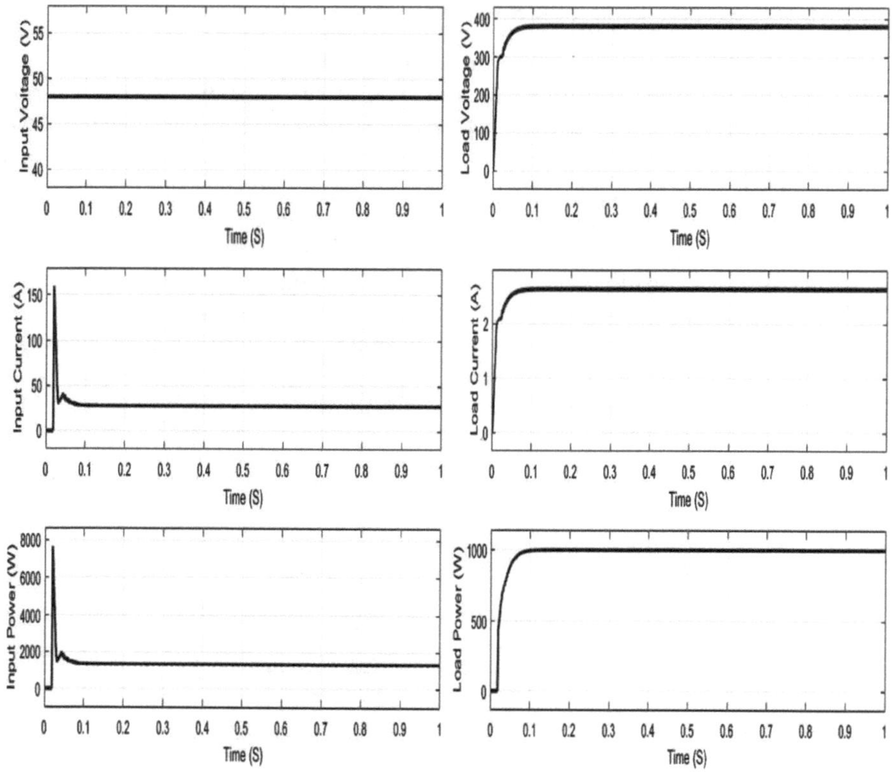

FIGURE 4.16 Simulated input, output voltage, current, and power waveforms.

To study the performance behavior, the proposed converter is analyzed under different conditions such as variation in source voltage, reference voltage, and load power. Under all these variations the proposed converter has to maintain the load voltage of 380 V.

4.6.1 Line Regulation Analysis of M-HGBC

In this analysis, the M-HGBC is deliberately subjected to variations in source voltage at regular intervals. Maintaining a source voltage of 48 V from $t=0$ to 0.25 second, a subsequent reduction to 36 V at $t=0.25$ second, sustained until $t=0.5$ second, led to a sudden impact on the load voltage and load power. Nonetheless, the controller adeptly detected the voltage fluctuation and expediently regulated the load voltage to 380 V within a mere 0.02 second. Likewise, when the source voltage was increased from 36 V to 48 V at $t=0.5$ second, the controller promptly identified the change and issued a control signal to uphold the load voltage at 380 V. The corresponding variations in source voltage and the consequential fluctuations in load voltage and load power are meticulously detailed in Figure 4.17 and accurately tabulated in Table 4.4.

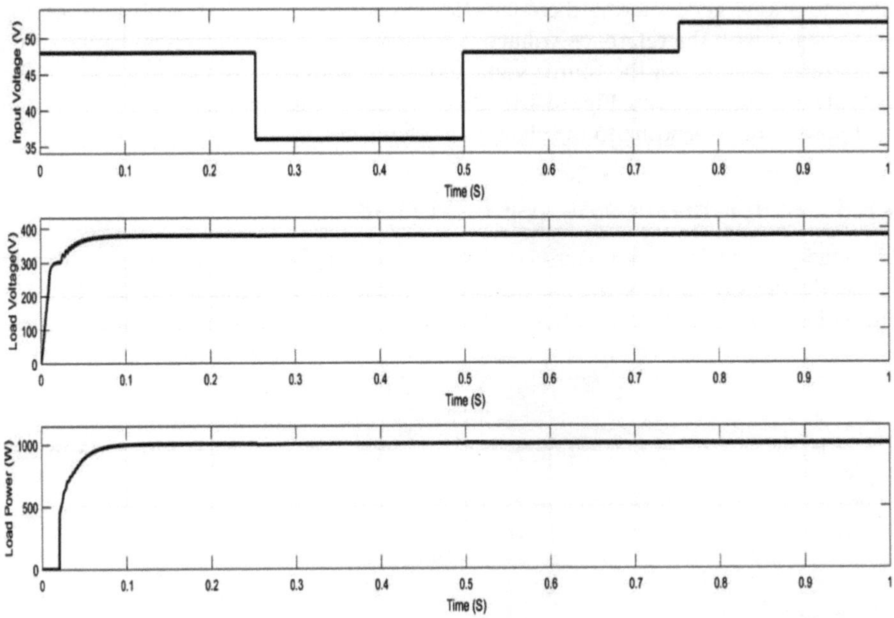

FIGURE 4.17 Performance analysis of the converter under source voltage variation: (i) input voltage, (ii) load voltage, and (iii) load power.

TABLE 4.4
Line Regulation

Time (s)	Input Voltage (V)	Input Current (A)	Input Power (W)	Output Voltage (V)	Output Current (A)	Output Power (W)	Efficiency (%)
0–0.25	48	22.47	1,079	380	2.63	1,000	92.67
0.25–0.5	36	33.27	1,198	380	2.63	1,000	83.47
0.5–0.75	48	22.47	1,079	380	2.63	1,000	92.67
0.75–1	52	4.80	1,134	380	2.63	1,000	88.18

4.6.2 Reference Voltage Variation Analysis of M-HGBC

The converter, in conjunction with the controller, has demonstrated its capability to maintain the load voltage at 380 V despite disturbances affecting the converter reference voltage. During the analysis, a deliberate adjustment of the converter reference voltage from 380 V to 420 V at time $t=0.25$ second was observed, resulting in fluctuations in both load voltage and load power. The controller promptly detected the voltage variance and issued a control signal to the converter to align with the reference voltage, successfully tracking the variation in less than 0.02 second. Subsequently, at time $t=0.5$ second, the reference voltage was reduced from

420 V to 300 V, necessitating the controller to take nearly 0.05 seconds to align the load voltage with the reference voltage. Figure 4.18 visually illustrates the variation in reference voltage for the source voltage of 48 V, with corresponding data detailed in Table 4.5. Additionally, Figure 4.18 also depicts the fluctuation in load voltage and load power corresponding to the change in reference voltage.

4.6.3 Load Regulation Analysis of M-HGBC

The load power of the converter is deliberately adjusted at specific time intervals to assess the performance of the converter in conjunction with the controller. The primary objective is for the controller to effectively regulate and uphold the load voltage

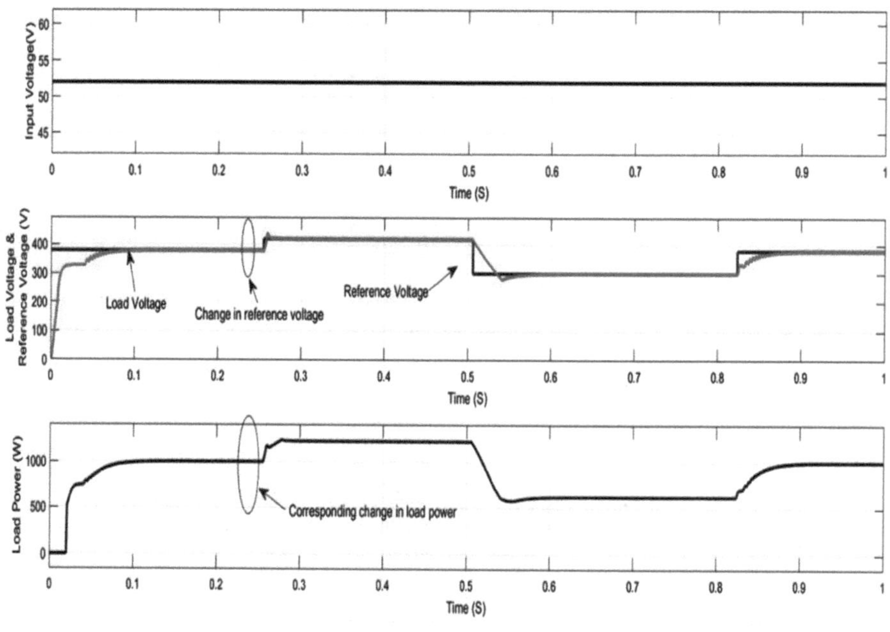

FIGURE 4.18 Performance analysis of the converter under reference voltage variation: (i) input voltage, (ii) load voltage and reference voltage, and (iii) load power.

TABLE 4.5
Reference Voltage Variation

Time (s)	Input Voltage (V)	Reference Voltage (V)	Load Voltage (V)	Load Current (A)	Load Power (W)
0–0.25	48	380	380	2.63	1,000
0.25–0.5		420	420	2.85	1,200
0.5–0.81		300	300	1.73	520
0.81–1		380	380	2.63	1,000

at 380 V, particularly in response to any disturbances. Figure 4.19 provides a visualization of the simulated waveform of the converter during load power variation, with corresponding data tabulated in Table 4.6. Initially, at time $t=0$ second, a consistent load power of 0.25 kW is maintained. Subsequently, at time $t=0.5$ second, the load power is incrementally increased from 0.25 to 0.5 kW, resulting in fluctuations in the load voltage. Upon detection of the voltage variance, the controller promptly issues a control signal to restore and stabilize the load voltage within the specified parameters in less than 0.01 second. Furthermore, at time $t=1$ second, the load power is further elevated from 0.5 to 0.75 kW. This sudden increase in load power prompts further fluctuations in load voltage; however, the controller adeptly responds by generating the necessary control signal to effectively maintain the load voltage within the desired range.

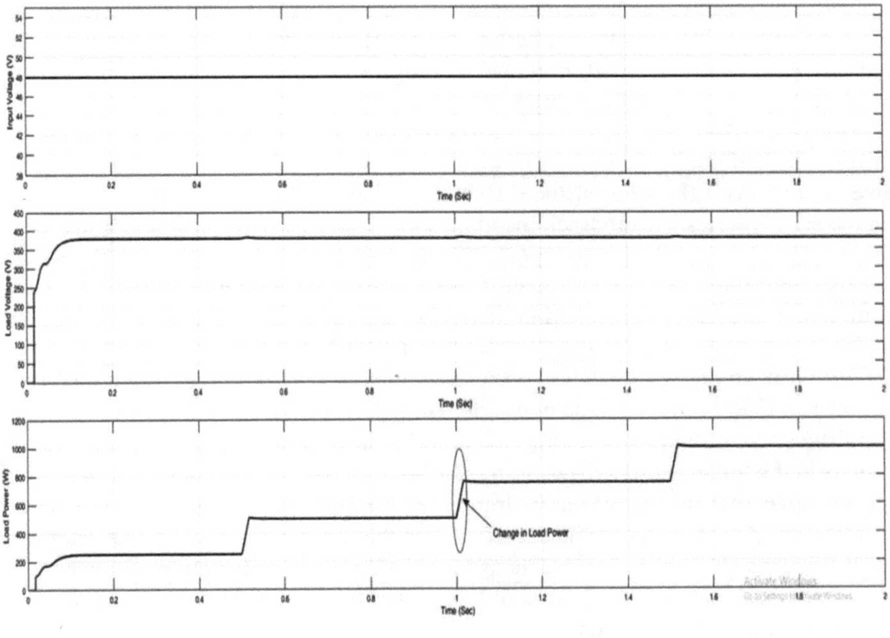

FIGURE 4.19 Performance analysis of the converter under load power variation: (i) input voltage, (ii) load voltage, and (iii) load power.

TABLE 4.6
Load Regulation

Time (s)	Input Voltage (V)	Load Voltage (V)	Load Current (A)	Load Power (W)
0–0.25	48	380	0.657	250
0.25–0.5		380	1.315	500
0.5–0.75		380	1.974	750
0.75–1		380	2.632	1,000

4.7 Performance Analysis of M-HGBC in the Solar PV Fed DC Microgrid

Based on the analysis presented in the preceding sections, it has been observed that the controller effectively regulates the load voltage within 0.02 second. This analysis supports the conclusion that the converter, when equipped with a controller, is well-suited for DC microgrid applications. To assess its suitability for such applications, the converter is linked with a renewable source, specifically the solar PV array. Considering the intermittent nature of the source, the variability in its output voltage and power deems it unsuitable for load connections. Therefore, to optimize the utilization of power generated by the solar PV array, the tracking of the maximum power point is essential. For this purpose, an incremental conductance MPPT algorithm is employed, successfully achieving maximum power tracking from the source. Subsequently, the converter is integrated with the AC power grid through a three-phase inverter to analyze the power balance within the system and to regulate the DC grid link voltage. The converter's performance within the DC microgrid system is evaluated under various conditions, including the connection of both DC loads and AC loads.

The solar PV-based DC microgrid system has been designed and is operated in grid-connected mode. The power electronic circuit is crafted to interconnect renewable sources with the grid on the distribution side. The power electronic converter on the source side serves to track the maximum power from the renewable source. A three-phase voltage source inverter (VSI) functions as a grid-tied converter, connecting the renewable source with the grid. Within the system, household DC loads and commercial DC loads are linked in the DC grid link, while AC loads are connected in the distribution network at the point of common coupling (PCC). In this study, a DC grid link voltage of 700 V has been selected to obviate the need for a transformer on the grid side. In line with the power demanded by the load and the power extracted from the source, the power management and controller unit (PMCU) issues gating signals to the three-phase VSI. Within the PMCU, a voltage controller, utilized in the outer loop, is responsible for maintaining the DC grid link voltage at 700 V. The real and reactive power controller, deployed within the inner loop, supports the establishment of power flow in the microgrid system, facilitating the injection of power from the source to the loads or the extraction of power from the grid to meet the required power demand from the loads.

4.7.1 DC Microgrid Structure

The block diagram illustrating the solar PV fed DC microgrid system is presented in Figure 4.20. The specifications for the DC microgrid system being studied have been compiled in Table 4.7. The solar PV array serves as the input source, with two 5 kW PV arrays linked in parallel using the proposed high gain converter topologies. The fluctuating air conditions can cause variations in the power output of solar PV cells. Using a MPPT method is essential for effectively extracting the peak power produced by solar PV. To track the maximum power from the solar PV array, an incremental conductance MPPT control method has been used in this study. Table 4.8 contains more information about the solar PV module's specifications.

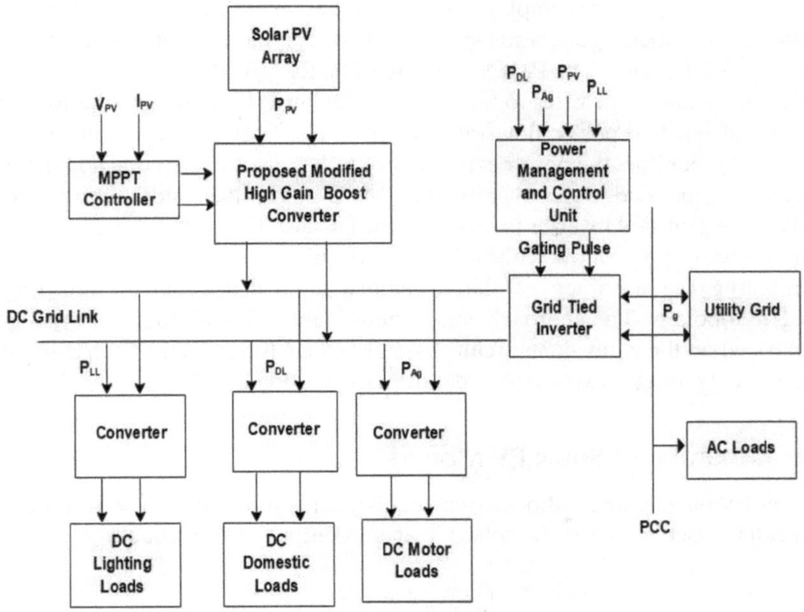

FIGURE 4.20 Block diagram of the solar PV fed DC microgrid system.

TABLE 4.7
Parameters of the DC Microgrid System under Study

Parameters	No. of Units Connected	Values
Solar PV array	2	5 kW
DC motor load	10	110 W, 24 V
Constant power load	5	48 V, 250 W
Lighting load	5	500 W
AC load	3	2 kW, 3 kW, 1 kW
DC grid link voltage	–	700 V

TABLE 4.8
KYOCERA Solar KC200GT – Solar PV Module Parameters

Parameters	Values
Maximum power	200.143 W
Open circuit voltage (V_{oc})	32.9 V
Short-circuit current (I_{sc})	8.21 A
Voltage at the maximum power point (V_{mp})	26.3 V
Current at the maximum power point (I_{mp})	7.61 A
Cells per module	54

The grid-tied inverter is employed to establish an interface between the AC power grid and the distributed generation system. The three-phase AC power grid is considered to be a stiff source. The PMCU is responsible for providing the necessary gating pulse to the grid-tied inverter to maintain the DC grid link voltage and to regulate the real and reactive power flow into the grid. The PMCU utilized in the system continuously monitors the power extracted from the renewable source and the power demanded by the load. Subsequently, the PMCU determines and issues the control signals to the grid-tied inverter to maintain the DC grid link voltage and to ensure the real and reactive power flow within the grid system.

Validating the accuracy of the developed controller requires modeling the microgrid topology. The approach entails building the DC microgrid energy source model based on the equivalent circuits using MATLAB Simulink. Each subsystem's mathematical model is explained in detail in the section that follows.

4.7.2 Modeling of Solar PV Module

The solar PV panel's single-diode circuit equivalent is shown in Figure 4.21. Equation (4.36) can be used to define the solar PV array's output current equation:

$$I_{\mathrm{pv}} = I_{\mathrm{ph}} N_p - I_0 N_p \left(e^{\left(\frac{V + IR_s \left(\frac{N_s}{N_p} \right) q}{n^* K^* T^* N_s} \right)} - 1 \right) - \frac{V + IR_s \left(\frac{N_s}{N_p} \right)}{R_p \left(\frac{N_s}{N_p} \right)} \tag{4.36}$$

where I_{pv} is the current from solar PV, I_{ph} is the photon current, N_s is the series connected module count, N_p is the parallel connected module count, k is the Boltzmann constant (1.38×10^{-23} J/K), q is the electron charge (1.6×10^{-19}C), n is the ideality factor (1 or 2), T is the surface temperature of the solar panels (K), and I_0 is the reverse saturation current. Equation (4.37) expresses the reverse saturation current I_0's temperature-dependent characteristic, which is found in Equation (4.36), as

$$I_0 = I_{\mathrm{rs}} \left(\frac{T}{T_r} \right)^3 \exp\left(\frac{qE_g}{kn} \left(\frac{1}{T_r} - \frac{1}{T} \right) \right) \tag{4.37}$$

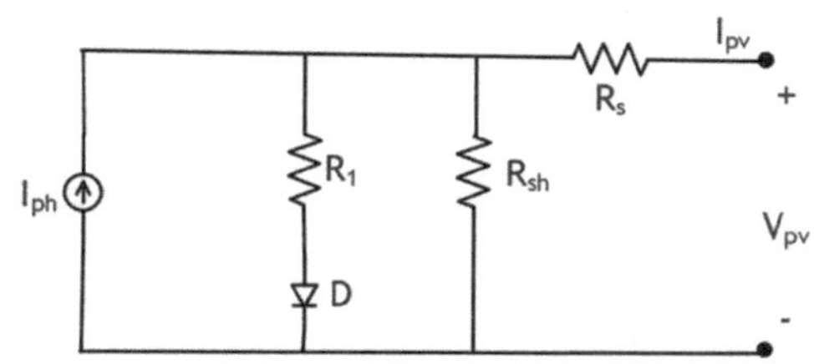

FIGURE 4.21 Equivalent circuit of the solar PV system.

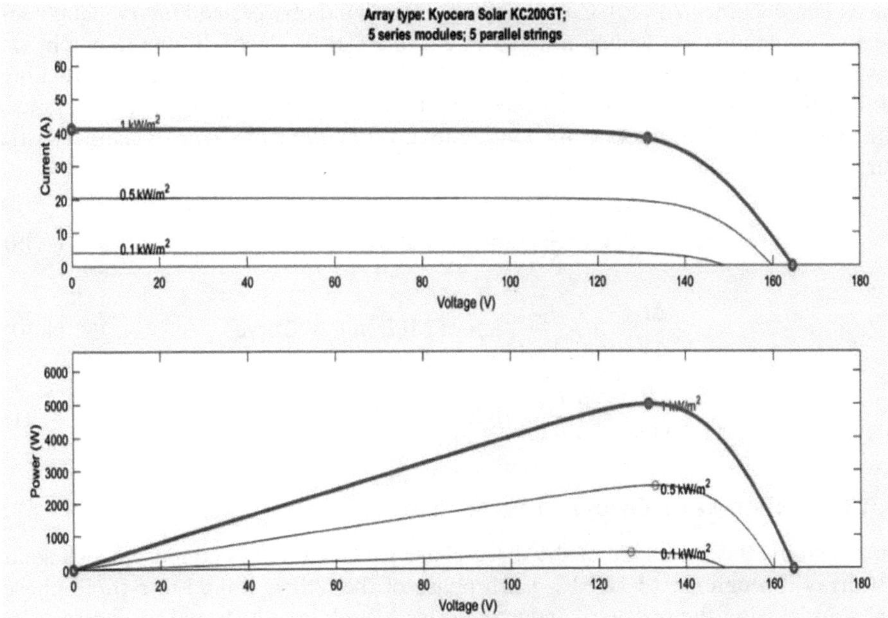

FIGURE 4.22 *P-V* and *I-V* characteristics of the solar PV array.

where T_r is the reference temperature (K), I_{rs} is the reverse saturation current at temperature T_r (K), and E_g is the energy bandgap given in Equation (4.38):

$$I_{ph} = \left(I_{sc} + \propto (T - T_r) * \frac{S}{100} \right) \tag{4.38}$$

where I_{sc} is the short-circuit current at reference temperature T_r and irradiance intensity 1 kW/m^2, α is the short-circuit current temperature coefficient, and S is the irradiance level (kW/m^2). Figure 4.22 shows the *P-V* and *I-V* characteristics curve for the solar PV array under various irradiance conditions. From the figure, it is inferred that, for 1 kW/m^2 irradiance, the utmost power extracted by the PV array is 5 kW, the maximum voltage (V_{mp}) is about 132 V, and the maximum current (I_{mp}) is 38 A.

4.7.3 MPPT Algorithm

The fundamental concept of MPPT is to optimize the operational point of the converter to attain the maximum power output from the solar module or array. To enhance system efficiency, it is crucial to implement effective power conditioning strategies. A key approach to maximizing system output involves improving the MPPT of the converter by monitoring the voltage and current of the PV array and adjusting the operative point to the maximum power point. Various MPPT techniques of different complexities, including the open circuit technique, short-circuit current technique, ripple correlation technique, perturb and observe technique, and incremental conductance technique, have been documented [40]. Currently, the perturb and observe and incremental conductance methods are widely utilized in commercial products. In our research, we have adopted the incremental conductance MPPT method due to its accurate tracking of maximum power, particularly under conditions of rapid change in solar irradiance. This method involves tracking the maximum power point from the *P-V* characteristic curve using specific governing equations:

$$\frac{\Delta I_{PV}}{\Delta V_{PV}} = \frac{I_{PV}}{V_{PV}}, \text{ at MPP} \tag{4.39}$$

$$\frac{\Delta I_{PV}}{\Delta V_{PV}} > -\frac{I_{PV}}{V_{PV}}, \text{ left of MPP in P-V Curve} \tag{4.40}$$

$$\frac{\Delta I_{PV}}{\Delta V_{PV}} < -\frac{I_{PV}}{V_{PV}}, \text{ right of MPP in P-V Curve} \tag{4.41}$$

4.7.4 Structure of Grid-Tied Inverter System

In this study, the input side of the three-phase grid-tied VSI is connected to a solar PV array through the M-HGBC. Each phase of the VSI is linked to a three-phase AC source using the resistor – inductor series branch, as illustrated in Figure 4.23. For analytical purposes, the AC source is assumed to be a stiff source, with balanced sinusoidal voltage sources V_{sa}, V_{sb}, and V_{sc} free from harmonics. Sine pulse width modulation is employed to generate the gating signals (m_a, m_b, and m_c) for VSI. The output voltage of the inverter is denoted as V_{ta}, V_{tb}, V_{tc}, with i_a, i_b, and i_c representing the currents flowing through the three-phase system. When interconnected with the AC grid, the grid-tied inverter is modeled as a current source. The grid-tied inverter can inject or draw real and reactive power while simultaneously preserving the DC grid connection voltage, depending on the control algorithm that is used.

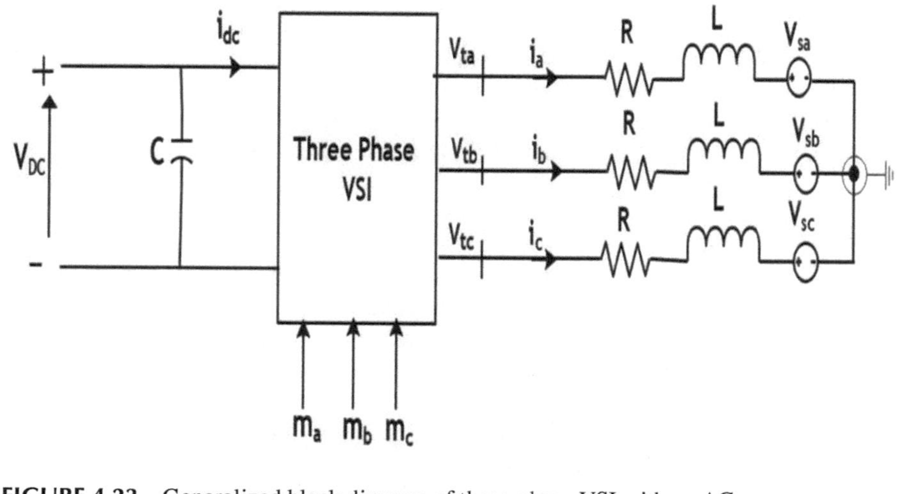

FIGURE 4.23 Generalized block diagram of three-phase VSI with an AC power source.

4.7.5 Control Strategies in DC Microgrid Setup

The DC side of the grid-tied inverter is interconnected with the solar PV system, facilitating the exchange of instantaneous real power and reactive power of the AC grid as per the load power requirements. The VSI regulates actual and reactive power using two different techniques: voltage mode control and current mode control. Real and reactive powers are controlled in voltage mode control based on the phase angle and magnitude of the VSI voltage at the terminal concerning the PCC voltage. On the other hand, when V_{tabc}'s magnitude and phase angle roughly match those of V_{sabc}, autonomous regulation of real and reactive powers is possible. Notably, this method does not provide direct control over the VSI line currents. Conversely, the current control method utilizes a specific scheme to regulate the VSI line current through the VSI terminal voltage. This strategy enables the management of real and reactive power flow based on the phase angle and magnitude of the VSI line current in relation to the PCC voltage. As seen in Figure 4.24, the power control unit is responsible for managing the DC voltage regulation as well as the actual and reactive power flow within the system.

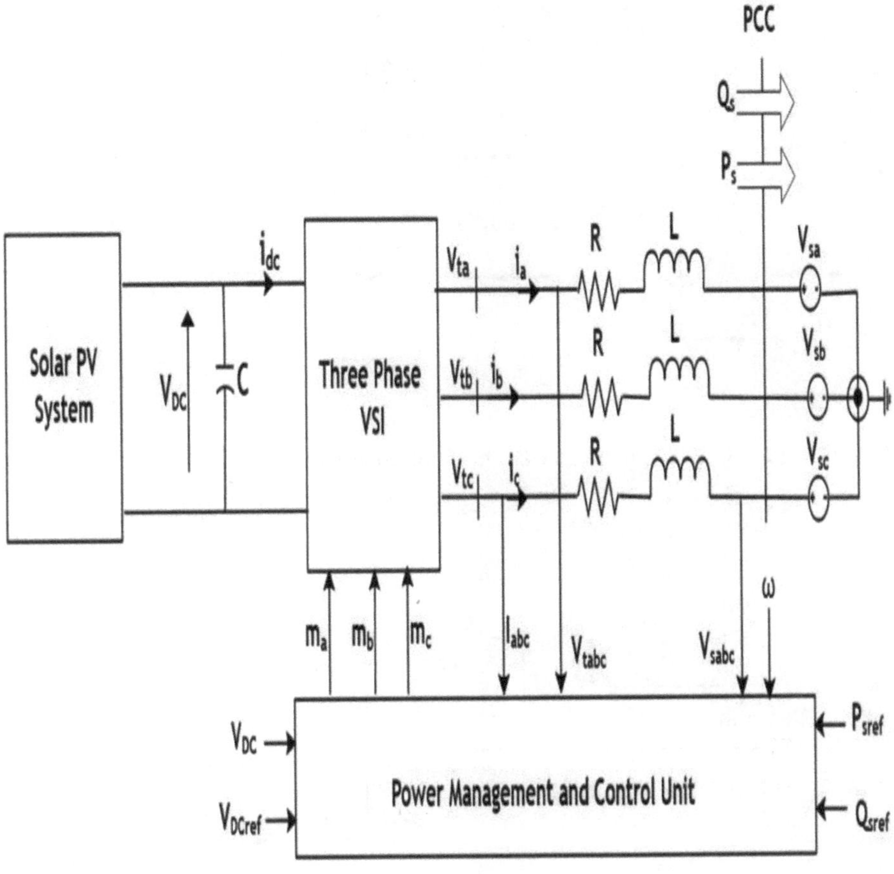

FIGURE 4.24 Schematic diagram of the power controller unit.

The power management and control unit comprises two essential controllers: the DC voltage controller and the real and reactive power controller. The primary function of the DC voltage controller is to regulate the grid link voltage (V_{DC}) by utilizing a feedback mechanism. The process outlined in Figure 4.25 shows how the DC voltage controller generates the control signal needed to acquire the benchmark active power (P_{sref}) signal, which is then fed into the real and reactive power controller.

Upon receiving the P_{sref} signal from the DC voltage controller, the i_{dref} and i_{qref} currents are computed using Equations (4.46 and 4.47) as illustrated in Figure 4.26. The current components i_d and i_q regulate the actual power $P_s(t)$ and reactive power $Q_s(t)$. Before being sent to the VSI, the feedback and feed-forward signals are first converted into the dq frame and then into the abc frame.

Following these control actions, the real and reactive powers delivered to the AC grid at the PCC are then determined:

$$P_s(t) = \frac{3}{2}\left[V_{sd}(t)i_d(t) + V_{sq}(t)i_q(t)\right] \tag{4.42}$$

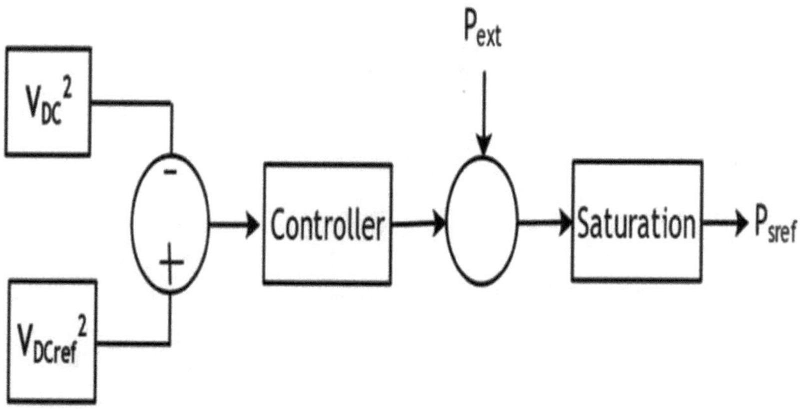

FIGURE 4.25 Block diagram of the DC grid link controller.

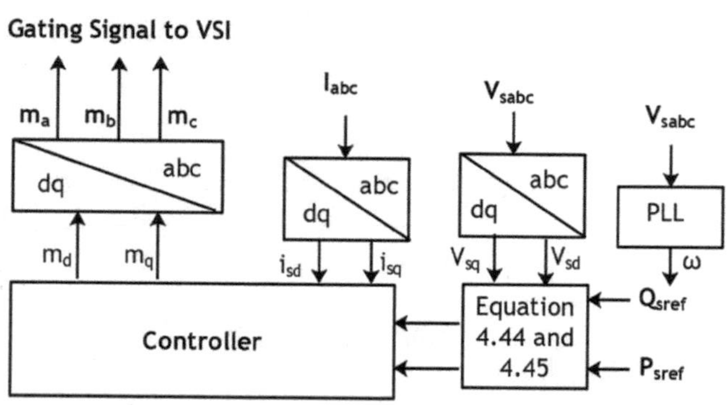

FIGURE 4.26 Block diagram of the real and reactive flow controller.

$$Q_s(t) = \frac{3}{2}\left[-V_{sd}(t)i_q(t) + V_{sq}(t)i_d(t)\right] \tag{4.43}$$

where V_{sd} and V_{sq} are the AC system dq frame voltage components and cannot be controlled by the VSI. In a steady-state condition, $V_{sq} = 0$ and thus power equations can be rewritten as

$$P_s(t) = \frac{3}{2}\left[V_{sd}(t)i_d(t)\right] \tag{4.44}$$

$$Q_s(t) = \frac{3}{2}\left[-V_{sd}(t)i_q(t)\right] \tag{4.45}$$

Therefore, based on Equations (4.44 and 4.45), the real power $P_s(t)$ and $Q_s(t)$ can be controlled by i_d and i_q, respectively. The reference current in *dq* frame is written as

$$i_{dref}(t) = \frac{2}{3V_{sd}} P_{sref}(t) \tag{4.46}$$

$$i_{qref}(t) = \frac{2}{3V_{sq}} Q_{sref}(t) \tag{4.47}$$

The modulation index in the *dq* frame is determined as

$$m_d = \frac{2}{V_{DC}}\left(V_d - L\omega_0 i_q + V_{sd}\right) \tag{4.48}$$

$$m_q = \frac{2}{V_{DC}}\left(V_q + L\omega_0 i_d + V_{sq}\right) \tag{4.49}$$

The gating signals for the VSI are given after transforming the modulation index from a *dq* frame to abc frame. Thus, the power flow and voltage regulation are performed in the proposed DC microgrid system.

4.7.6 Analysis of Power Flow in the DC Microgrid System When the Solar PV System Is Interconnected with the Grid

In this case, a three-phase VSI is used for coupling the PV array with the electrical grid. To maximize the power output, a grid-tied inverter connects a 10 kW solar PV array to the suggested high gain converter. The system draws its maximum power from the source, which is fed into the grid because there is no load attached to it. The converter tracks a maximum power of approximately 9.7 kW at an irradiance of 1,000 W/m^2. The control algorithm within the three-phase inverter calculates the generated power and the power required for the load. The difference between these powers is utilized to create the reference current signal, which in turn governs the gating signal after requisite transformations. Subsequently, the real power of 9.5 kW is injected into the grid based on the gating pulse, with the circuit maintaining a DC

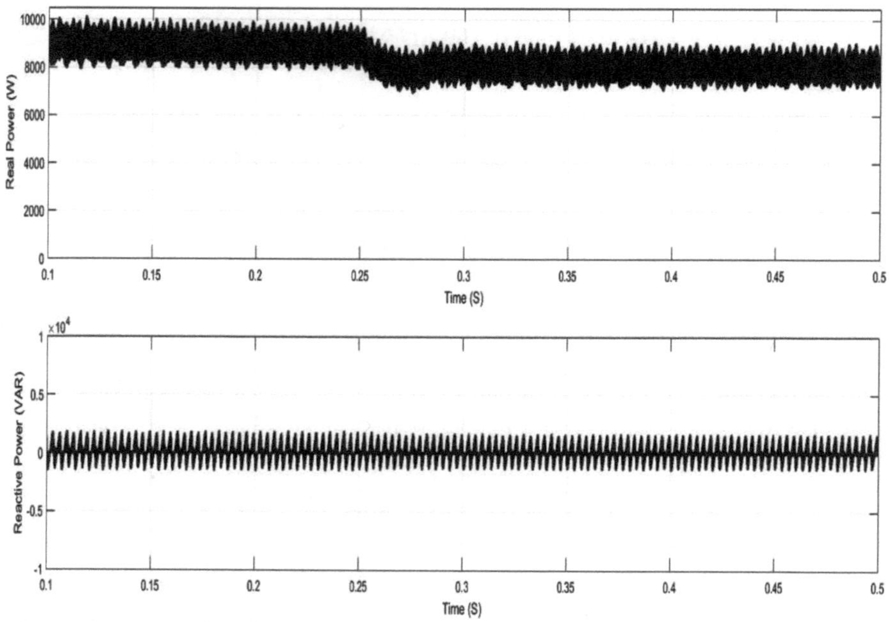

FIGURE 4.27 Simulated waveforms of real power and reactive power injected into the grid.

TABLE 4.9
Power Flow in the System When Solar PV Is Interconnected with the Grid

Time (s)	Irradiance (W/m²)	Power Extracted from the Source (kW)	Net Power Injected into the Grid (kW)
$0<t<0.25$	1,000	9.7	+9.7
$0.25<t<0.5$	800	8	+8

grid link voltage of 700 V. The power flow in the system is illustrated in Figure 4.27, with the corresponding values tabulated in Table 4.9. At time $t=0.25$ second, the irradiance is adjusted from 1,000 to 800 W/m², prompting the MPPT controller to optimize the power output to 8 kW from the solar PV array, which is then injected into the grid.

4.7.7 Analysis of Power Flow in the DC Microgrid System When DC Loads Alone Are Connected to the System

The DC loads of 110 W/24 V BLDC motor load of 10 numbers, 250 W/48 V of domestic lighting loads of 5 numbers, and 700 V/500 W of lighting loads of 5 numbers are connected in the DC grid link. The total DC load power requirement is about 4.85 kW. The solar PV system's irradiation is kept constant at 1,000 W/m^2. The maximum power of 9.7 kW is extracted from the source. The PMCU computes the power requirement of the system. The computational power is lower when compared to its peak power. Thus, 4.85 kW of power is delivered to the load by the solar PV array, while 4.85 kW of net power is added to the grid. At time $t=0.25$ second, the solar PV insolation is varied from 1,000 to 800 W/m^2. The variance causes the power drawn from the source to decrease, resulting in a net power injection into the grid of 3.8 kW once the load demand is satisfied. Figure 4.28 and Table 4.10 demonstrate the results of this calculation.

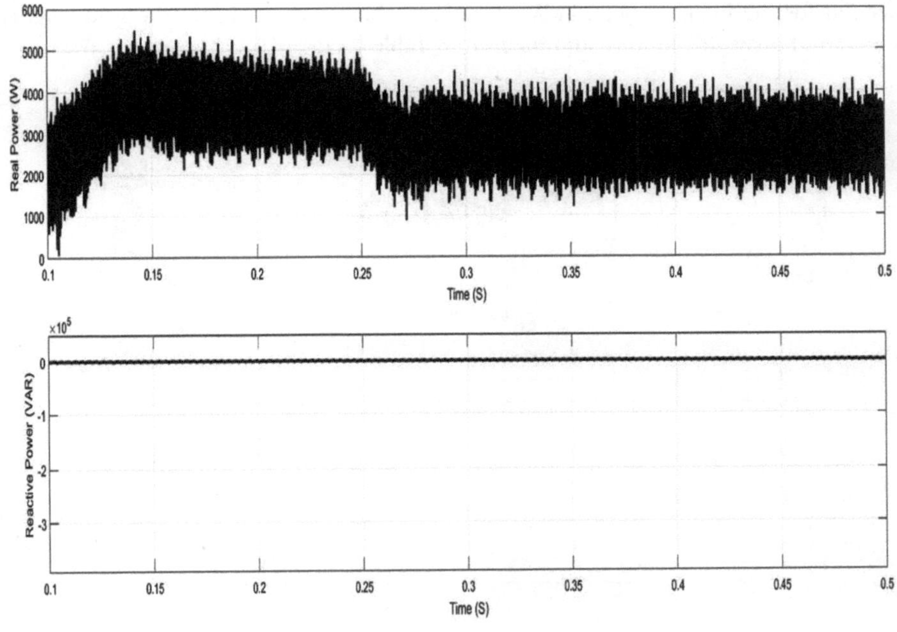

FIGURE 4.28 Simulated waveforms of real power and reactive power injected into the grid when DC loads connected in the system.

TABLE 4.10
Power Flow in the System When Solar PV and DC Loads Are Interconnected with the Grid

Time (s)	Irradiance (W/m^2)	Power Demanded by the Load (kW)	Power Extracted from the Source (kW)	Net Power Injected into the Grid (kW)
$0<t<0.25$	1,000	4.85	9.7	+4.85
$0.25<t<0.5$	800		8	+3.15

4.7.8 Analysis of Power Flow in the DC Microgrid System When AC Loads Alone Are Connected to the System

In this section, AC loads of 2, 3, and 1 kW are coupled to the grid at PCC. The total energy demanded by the load is 6 kW. The AC loads of various power ratings are connected at different time intervals. Initially, an AC load of 1 kW is connected at PCC. For the solar insolation of 1,000 W/m², the maximum power extracted is about 9.7 kW. The net power of 8.7 kW is injected into the grid. At time $t=0.2$ second, 3 kW load is connected at PCC. So a net power of 6.7 kW is injected into the grid after satisfying the load requirement. At time $t=0.35$ second, 2 kW load is connected in the system. Currently, the power generated by the source is greater than the power needed for the load. About 7.7 kW of power is being injected into the grid; this is shown in Figure 4.29 and is summarized in Table 4.11.

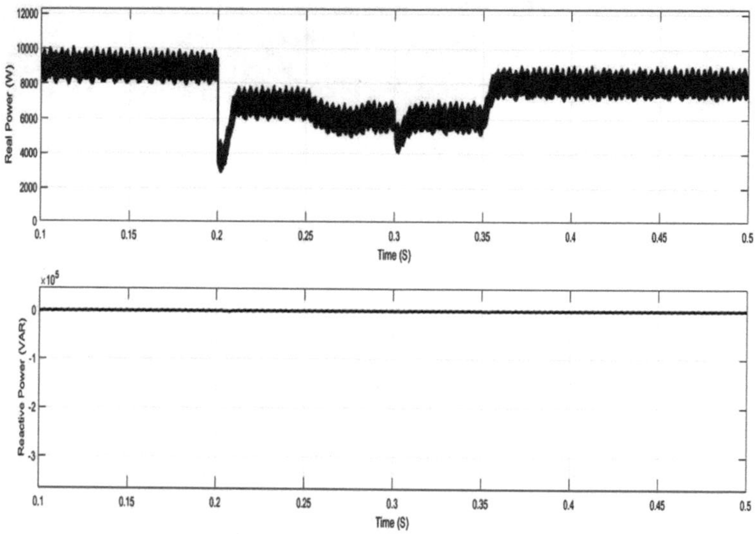

FIGURE 4.29 Simulated real power and reactive power injected into the grid when AC loads are connected in the system.

TABLE 4.11
Power Flow in the System When Solar PV and AC Load Are Interconnected with the Grid

Time (s)	Irradiance (W/m²)	Power Demanded by the Load (kW)	Power Extracted from the Source (kW)	Net Power Injected into the Grid (kW)
$0<t<0.2$	1,000	1	9.7	+8.7
$0.2<t<0.3$	1,000	3	9.7	+6.7
$0.3<t<0.35$	800		8.9	+5.9
$0.35<t<0.5$	1,000	2	9.7	+7.7

4.7.9 Analysis of Power Flow in the DC Microgrid System When DC and AC Loads Are Connected to the System

The DC loads of 4.85 kW and AC loads of 6 kW are connected to the system at different time intervals. The solar insolation is maintained at 1,000 W/m^2; the maximum power of 9.7 kW is extracted from the source. At time $t=0$ second, DC loads of 4.85 kW are connected to the system. The control algorithm computes whether the load demand is met by the renewable source or not. The power required by the load is less than the amount of energy retrieved; 9.7 kW of power from the solar PV array meets the 4.85 kW power demand, and an additional 4.85 kW of power is fed into the grid as excess. At time $t=0.2$ second, AC loads are also added to the system at PCC. Now, the load power requirement is about 10.85 kW. Here, extracted power is 9.7 kW but the demanded power is 10.85 kW. Also, at time $t=0.25$ second, the solar insolation is changed from 1,000 to 800 W/m^2. As a result, the solar PV array's power output drops from 9.7 to 8.86 kW. Because of this, the solar PV array's output is unable to meet the load power need, and 1.15 kW of extra power must be pulled from the grid to keep the system's power balance. The 2 kW AC load is cut off from the system at $t=0.4$ second. At this point, the load power need is about 8kW. This load power requirement is satisfied by the power generated by the solar PV array alone. Hence, neither the power drawn nor the power injection to the grid takes place, as demonstrated in Figure 4.30, and the values are tabulated in Table 4.12. The simulated DC grid link is depicted in Figure 4.31.

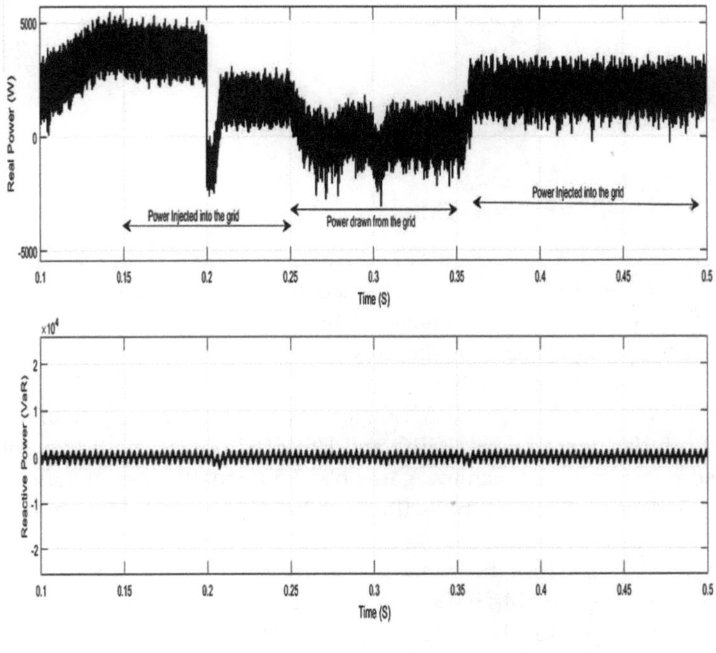

FIGURE 4.30 Simulated waveforms of real power and reactive power injected into the grid when both DC & AC loads are connected in the system.

TABLE 4.12
Power Flow in the System When Solar PV, DC, and AC Loads Are Interconnected with the Grid

Time (s)	Type of Load	Load Status	Load Power (kW)	Power Flow Status
$0<t<0.2$	DC	Connected	4.85	Injected into the grid (+ 4.85 kW)
$0.2<t<0.25$	DC+AC	Connected	10.85	Drawn from the grid (−1.15 kW)
$0.25<t<0.35$	DC+AC	Connected	10.85	Drawn from the grid (−1.99 kW)
$0.35<t<0.5$	DC+AC	Connected	7.85	Inject power to the grid (+1.01 kW)

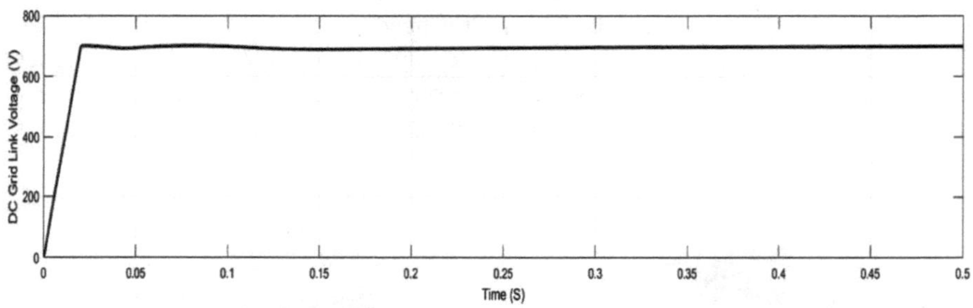

FIGURE 4.31 Simulated waveform of DC grid link voltage.

4.8 COMPARATIVE ANALYSIS

The proposed M-HGBC structure is compared with existing boost converter and modified boost converter topologies. The comparison factors of those converters are shown in Table 4.13. When comparing the voltage gain, the proposed M-HGBC accomplishes a gain of 13.33 for the duty ratio of $d_1=0.5$ and $d_2=0.35$, while the conventional boost converter achieves 6.66 for the duty ratio $d=0.85$, and the high gain modified boost converter (HGMBC) attains 10 for the duty ratio of $d_1=0.5$ and $d_2=0.35$. Both the proposed M-HGBC and HGMBC operate with two duty ratios, $d_1=0.5$ and $d_2=0.35$, which achieve gains of 13.33 and 10, respectively. The static voltage ratio against the duty ratio is illustrated in Figure 4.32. It is implied that the suggested M-HGBC has increased gain.

The voltage stress across the switches is decreased by operating the switches *S*1, *S*2, and *S*3 of the proposed M-HGBC with a lower duty ratio than the boost converter. Comparing the suggested M-HGBC to the traditional boost converter, the voltage stress is lowered by 56.31%. In a similar vein, Figure 4.33 shows that the voltage stress across the switch of the M-HGBC is 22.42% less than that of the HGMBC.

TABLE 4.13
Comparative Analysis of the M-HGBC with Existing Converters

Parameters	Conventional Boost Converter	Modified Boost Converter	HGMBC	Proposed M-HGBC
Voltage gain	$\frac{1}{1-d}$	$\frac{1+d}{1-d}$	$\frac{1+d}{1-d_1-d_2}$	$\frac{2}{1-d_1-d_2}$
Voltage across the switch	V_o	$\frac{V_0+V_{in}}{2}$	$\frac{V_0+V_{in}}{2}, V_o$	$\frac{V_0-V_{in}}{2}, V_o$
Voltage across the diode	V_o	$V_{in}+V_o$	$V_{in}, V_{in}+V_o$	$\frac{V_o}{2}, V_0$
No. of power switches	1	2	3	3
No. of inductors	1	2	2	2
No. of capacitors	1	1	1	2
No. of diodes	1	1	2	1

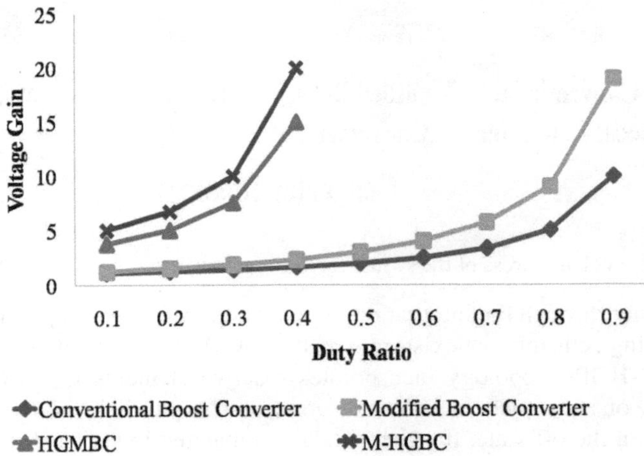

FIGURE 4.32 Duty ratio versus voltage gain.

When compared to converters with the same topology, the research shows that the suggested M-HGBC has a higher voltage gain and lower voltage stresses across the switches. Therefore, because of its increased efficiency and less voltage stress across the switches, the suggested M-HGBC is appropriate for solar PV-based DC microgrid systems.

4.9 SUMMARY

The boost converter topology is commonly utilized to elevate the essential voltage of the load to be driven. However, voltage gain is hampered by the diode reverse recovery problem. As a remedy, numerous modifications are necessary in the

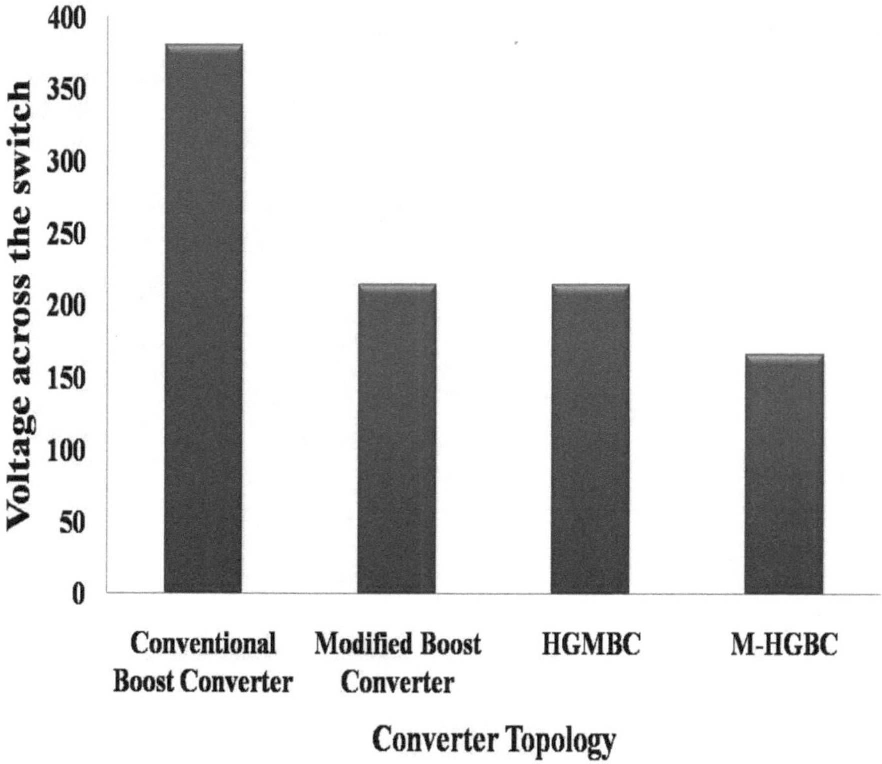

FIGURE 4.33 Voltage stress of the switch for various converters.

topology, either through the integration of voltage boosting techniques or the addition of boosting cells into the existing structure. A M-HGBC topology is discussed here. The M-HGBC topology incorporates reactive elements in series with the switch. In the on state, these elements are connected in parallel with the source and get charged. In the off state, the elements are connected in series with the source and discharge their stored energy. In contrast to a traditional boost converter, this enhancement amplifies the static gain ratio while concurrently alleviating voltage drop on both the switch and the diode. The applicability of M-HGBC in a microgrid system is explored through the use of a controller. The proposed M-HGBC topology, derived from the HGMBC, realizes a voltage conversion ratio of 40 for the duty ratios of d_1=0.5 and d_2=0.45. State-space analysis of the M-HGBC is conducted to assess the stability of M-HGBC. In addition, an analysis is conducted under different scenarios to determine how well the converters perform with the controller. If there are voltage fluctuations in the converter, a controller named PI is utilized to control and preserve the voltage at the load. The suggested converter reaches an optimal efficiency of 93% under rated load conditions. To assess the suitability of the converter in a DC microgrid environment, the M-HGBC is fed from the solar PV array. The M-HGBC is used to track the solar PV array's maximum power because the source is reliant on nature. The M-HGBC is used in conjunction with an incremental conductance MPPT method to maximize energy extraction from the source. Additionally, a three-phase grid-tied inverter is employed to interface the solar PV with the AC grid, allowing for the connection of DC loads and AC loads to study real and reactive power flow and the regulation of the DC grid voltage. Real and reactive power flow analysis is performed by connecting and disconnecting loads at different intervals. Based on the control algorithm, the grid-tied inverter efficiently preserves power flow in the DC microgrid system, proving its capacity to do so.

REFERENCES

1. Nikos Hatziargyriou, 2014, 'Guest Editorial Special Section on Microgrids for Sustainable Energy Systems', *IEEE Transactions on Sustainable Energy*, vol. 5, no. 4, pp. 1309–1309.
2. Stephon A. Roosa, 2021, *Fundamentals of Microgrids - Development and Implemen tation*, CRC Press, Taylor & Francis Group.
3. Pandav Kiran Maroti et al., 2022, 'The State-of-the-Art of Power Electronics Converters Configurations in Electric Vehicle Technologies', *Power Electronic Devices and Components*. DOI: 10.1016/j.pedc.2021.100001
4. Javed Ahmad, Mohammad Zaid, Adil Sarwar, Mohd Tariq, and Zeeshan Sarwer,2020, 'A New Transformerless Quadratic Boost Converter with High Voltage Gain', *Smart Science*, vol. 8, no. 3, pp. 163–183.
5. Muhammad Sulthon, 2017, 'A New Simplified Cascaded Boost Converter for High Ratio and Low Ripple Application', in *4th International Conference on Electric Vehicular Technology (ICEVT)* October 2–5, 2017, Bali, Indonesia.
6. Mojtaba Forouzesh, Yam P. Siwakoti, Saman A. Gorji, Frede Blaabjeg, Brad Lehman, 2017, 'Step-Up DC-DC Converters: A Comprehensive Review of Voltage Boosting Techniques, Topologies and Applications', *IEEE Transactions on Power Electronics*, vol. 32, no. 12, pp. 9143–9178.
7. Jothimani Gnanavadivel, Kathiresan Jayanthi, Sridhar Vasundhara, Kannan Velramakrishnan Swetha, Kannan Jeya Keerthana, 2023, 'Analysis and Design of High Gain DC-DC Converter for Renewable Energy Applications', *Automatika*, vol. 64, no. 3, pp. 408–44.
8. Yugal Kishor, Ramnarayan Patel, Lalit Kumar Sahu, 2023, 'An Improved Z-Source-Derived Converter for Efficient PV Integration', *International Journal of Electronics*, vol. 111, no. 12, pp. 2041–2059.
9. Manirao Lakshmi, Sathish Hemamalini, 2018, 'Non Isolated High Gain DC-DC Converter for DC Microgrids', *IEEE Transactions on Industrial Electronics*, vol. 65, no. 2, pp. 1205–1212.
10. Lung-Sheng Yang, Tsorng-Juu Liang, Jiann-Fuh Chen, 2009, 'Transformerless DC-DC Converter with High Step-Up Voltage Gain', *IEEE Transaction on Industrial Electronics*, vol. 56, no. 8, pp. 3144–3152.
11. Mahalingam Kalarathi, Jothimani Gnanavadivel, Kathiresan Jayanthi, 2024, 'High Boost DC-DC Converter Based on Switched Inductor, Switched Capacitor and Voltage Multiplier Cell', *Iranian Journal Science Technology Transaction Electrical Engineering*. https://doi.org/10.1007/s40998-024-00702-4.
12. Alencar Franco de Souza, Fernando Lessa Tofoli, Enio Roberto Ribeiro, 2020, 'Switched Capacitor DC-DC Converters: A Survey on the Main Topologies', *Design Characteristics, and Applications. Energies*, vol. 14, no. 8, pp. 2231.

13. Jessika Melo de Andrade, Roberto Francisco Coelho, Telles Bruneli Lazzarin, 2020, 'High Step-Up DC-DC Converter Based on Modified Active Switched Inductor and Switched Capacitor Cells', *IET Power Electronics*, vol. 13, no. 14, pp. 3127–3137.
14. Maryam Shaabani, Amin Mirzaei, Mahdi Rezvanyvardom, Farshad Khosravi, Saman A. Gorji, 2023, 'A Hybrid Switched-Inductor/Switched-Capacitor DC-DC Converter with High Voltage Gain Using a Single Switch for Photovoltaic Application', *Energies*, vol. 16, pp. 5524–5544.
15. Padmanaban Sanjeevikumar, Sagar Bhaskar Mahajan, Patrick Wheeler, Frede Blaabjerg, Marco Rivera, Rishi Kulkarni, 2016, 'X-Y Converter Family: A New Breed of Buck-Boost Converter for High Step-Up Renewable Energy Applications', in *IEEE International Conference of Automatica*, pp. 1–8, Curico, Chile.
16. Ahmad Alzahrani, Mehdi Ferdowsi, Pourya Shamsi, 2019, 'High-Voltage-Gain DC–DC Step-Up Converter with Bifold Dickson Voltage Multiplier Cells', *IEEE Transactions on Power Electronics*, vol. 34, no. 10, pp. 9732–9742.
17. Lenon Schmitz, Denizar C. Martins, Roberto F. Coelho, 2017, 'Conception of High Step-Up DC-DC Boost Based Converters', in *Brazilian Power Electronics Conference*, pp. 1–6, Juiz de Fora, Brazil.
18. Jayachandran Divya Navamani, Krishnasamy Vijayakumar, Ra Jegatheesan, 2018, 'Non-Isolated High Gain DC-DC Converter by Quadratic Boost Converter and Voltage Multiplier Cell', *Ain Shams Engineering Journal*, vol. 9, no. 4, pp. 1397–1406.
19. Padala Lakshmi Santosh Kumar Reddy, Yeddula Pedda Obulesu, 2023, 'A Non-Cascading DC/DC Quadratic Boost Converter with High Voltage Gain for PV Applications', *International Journal of Electronics*, vol. 110, no. 10, pp. 1756–1777.
20. Milad Rezaiel, Vahid Abbasi, 2020, 'Effective Combination of Quadratic Boost Converter with Voltage Multiplier Cell to Increase Voltage Gain', *IET Power Electronics*, Vol. 13, no. 11, pp. 2322–2333.
21. Marcos Antonio Salvador, Telles Bruneli Lazzarin, Roberto Francisco Coeiho, 2018, 'High Step-Up DC-DC Converter with Active Switched Inductor and Passive Switched Capacitor Networks', *IEEE Transactions on Industrial Elecronics*, vol. 65, no. 7, pp. 5644–5654.
22. Tohid Jalilzadeh, Naghi Rostami, Ebrahim Babaei, Mohammad Maalandish, 2018, 'Non-Isolated Topology for High Step-Up DC-DC Converter', *IEEE Journal of Emerging and Selected Topics in Power Electronics (Early Access)*, vol. 11, no. 1, pp. 1154–1168.
23. Shima Sadaf, Mahajan Sagar Bhaskar, Mohammad Meraj, Atif Iqbal, Nasser Al-Emadi, 2021, 'A Novel Modified Switched Inductor Boost Converter with Reduced Switch Voltage Stress', *IEEE Transaction on Industrial Electronics*, vol. 68, no. 2, 1275–1289.
24. Yuanmao Ye, Shikai Chen, Yong Yi, 2021, 'Switched -Capacitor and Coupled Inductor based high Step-Up Converter with Improved Voltage Gain', *IEEE Journal of Emerging and Selected Topics in Power Electronics*, vol. 9, no. 1, pp. 754–764.
25. Gang Wu, Xinbo Ruan, Zhihong Ye, 2018, 'Non-Isolated High Step-up DC-DC Converter Adopting Auxiliary Capacitor and Coupled Inductor', *Journal of Modern Power Systems and Clean Energy*, vol. 6, no. 2, pp. 384–398.
26. Wuhua Li, Yi Zhao, Xiangning He, 2010, 'Interleaved Converter with Voltage Multiplier Cell for High Step-Up and High Efficiency Conversion', *IEEE Transactions on Power Electronics*, vol. 25, no. 9, pp. 2397–2408.
27. Weerachat Khadmun, Wanchai Subsingha, 2013, ' High Voltage Gain Interleaved DC Boost Converter Application for Photovoltaic Generation System', *Energy Procedia*, vol. 34, pp. 390–398.
28. Kuo-Ching Tseng, Chi-Chih Huang, We-Yuan Shih, 2013, 'A High Step-Up Converter with a Voltage Multiplier Module for a Photovoltaic System', *IEEE Transactions on Power Electronics*, vol. 28, no. 6, pp. 3047–3057.

29. B. Sri Revathi, Prabhakar Mahalingam, 2019, 'Interleaved High Gain DC-DC Converter for Integrating Solar PV Source to DC Bus', *Solar Energy*, vol. 188, pp.924–934.
30. Mahajan Sagar Bhaskar, Dhafer J. Almakhles, Sanjeevikumar Padmanaban, Frede Blaabjerg, Umashankar Subramanina, Dan M. Ionel, 2020, 'Analysis and Investigation of Hybrid DC-DC Non-Isolated and Non-Interleaved Multilevel Boost Converter for High Voltage Step-Up Applications', *IEEE Access*, vol. 9, pp. 87309–87328.
31. Jothimani Gnanavadivel, Kathiresan Jayanthi, Monikanthan Nivethihi, Saravana Perumal Jeyachthra, Ravichandran Illakiya, 2023, 'Implementation of Non-Isolated High-Gain DC-DC Converter for Electric Vehicle Application', *International Journal of Electronics*, pp. 1–22.
32. Saeed Pourjafar, Hossien Shayeghi, Seyed Majid Hashemzadeh, Farzad Sedaghati, Mohammad Maalandish, 2020, 'A Non-Isolated High Step-Up DC–DC Converter Using Magnetic Coupling and Voltage Multiplier Circuit', *IET Power Electronics*, vol. 14, no. 9, pp. 1541–1723.
33. Vijay Joseph Samuel, Gna Keerthi, Prabhakar Mahalingam, 2019, 'Coupled Inductor-Based DC–DC Converter with High Voltage Conversion Ratio and Smooth Input Current', *IET Power Electronics*, vol. 13, no. 4, pp. 733–743.
34. Tao Jin, Jiaqi Lin, Haibin Li, Xu Yan, Yusen Weng, Xingkui Mao, 2023, 'A Novel Three-Winding Coupled Inductor-Based High Step-Up DC-DC Converter for Renewable Energy Application', *IEEE Journal of Emerging and Selected Topics in Power Electronics*, vol. 11, no. 4, pp. 4477–4490.
35. Arash Imanlou, Reza Behka, Morteza Nazari-Heris, And Gevork B. Gharehpetian, 2023, 'A New High Voltage Gain Active Switched-Inductor Based High Step-Up DC–DC Converter with Coupled-Inductor', *IEEE Access*, vol. 11, pp. 56749–56765.
36. Mingfa Yang, Yusen Weng, Haibin Li, Jiaqi Lin, Xu Yan, Tao Jin, 2023, 'A Novel Three-Winding Coupled Inductor-Based High-Gain DC–DC Converter with Low Switch Stress and Continuous Input Current', *IEEE Transactions on Power Electronics*, vol. 38, no. 12, pp. 15781–15791.
37. Tsorng-Juu Liang, Peng Luo, Kai-Hui Chen, 2022, 'A High Step-Up DC–DC Converter with Three-Winding Coupled Inductor for Sustainable Energy Systems', *IEEE Transactions on Industrial Electronics*, vol. 69, no. 10, pp.10249–10258.
38. Pouria Sarvghadi, Ali Yazdian Varjani, Mahdi Shahparasti, 2022, 'A High Step-Up Transformerless DC–DC Converter with New Voltage Multiplier Cell Topology and Coupled Inductor', *IEEE Transactions on Industrial Electronics*, vol. 69, no. 10, pp. 10162–10171.
39. Manish Kumar Dwivedi, Raasakannu Jayapragash, 2023, 'Multi-Stage Switched-Inductor Switched-Capacitor High Gain DC-DC Converter', *International Journal of Electronics Letters*, vol. 11, no. 4, pp. 381–398.
40. Musong L. Katche,Augustine B. Makokha, Siagi O. Zachary and Muyiwa S. Adaramola, 2023, 'A Comprehensive Review of Maximum Power Point Tracking (MPPT) Techniques Used in Solar PV Systems, *Energies*, vol.16,no. 5, pp.2206–2229.

5 Improving Solar Power Efficiency Using Zeta Converter and MPPT Control

Sivagama Sundari and Raghul Kumar

5.1 INTRODUCTION

Power converters are electronic devices capable of converting electricity between different forms and are essential to modern power systems. These are important because they modify voltage and frequency and regulate the amount of electricity that flows. Power converters are found in everything from phone chargers to electric cars. Without developments in semiconductors, their development would not have been feasible. These developments produced strong electronic switches with fast operation and the ability to withstand high voltages. MOSFETs, IGBTs, and thyristors are examples of the specific switches that work the magic in power converters. These are available in numerous designs, or topologies, to suit different purposes, including converting AC to DC, DC to DC, and vice versa. In the end, power converters are revolutionary for contemporary power systems since they transform and regulate electricity for a variety of applications in an effective and dependable manner. The bidirectional DC-DC converter, which enables energy to move between DC sources with differing voltages in both directions, is a prime example. This technology finds application in a wide range of fields, including energy storage, renewable energy systems, and electric cars.

Bidirectional DC-DC converters are becoming increasingly popular due to the increase in renewable energy sources like solar and wind power. This is due to the fact that these converters effectively and dependably convert the DC output from wind turbines and solar panels into AC electricity that the grid can use. They accomplish this by regulating the duty cycle, or on/off time, of unique switches inside the converter. Bidirectional DC-DC converters provide more, even if their primary job is to transfer power in both directions. They can control voltage, eliminate dirty power (conditioning), and even guard against electrical problems. As an example, consider the Zeta converter. With the use of a unique design known as "Zeta architecture", this kind of bidirectional converter effectively steps up or down voltage with just one inductor by combining features from other converter kinds.

Known for its "buck and boost" modes, the Zeta converter is a power electronics workhorse that handles bidirectional power transfer in electric vehicles, renewable

DOI: 10.1201/9781003496946-5

energy, and battery storage. With just two switches, a special Zeta inductor, and a control circuit, this efficient converter can handle both voltage increase and decrease. It replaces older, bulkier options with a simple yet effective solution for modern power systems. DC-DC bidirectional Zeta converters solve various shortcomings of previous versions. They don't experience excessive switch stress at high settings, in contrast to certain DC-DC converters. Furthermore, compared to conventional solutions, Zeta converters obtain better voltage gain. Although power flow can be reversed using buck-boost converters, their control circuitry can be intricate. An easier and more reliable solution is provided by the Zeta converter.

According to the International Energy Agency, more people are using renewable energy sources like solar energy. In 2020, renewable energy was used in 90% of newly built power plants, producing 28% of global electricity! The capacity of solar power has grown eighteen times since 2010. Solar panels are fantastic since they are cheap to maintain, widely available, and good for the environment. They can be put to use even in the absence of the electrical grid. The truth is that solar panels function best in direct sunlight. Shade can significantly reduce their efficacy. To optimize power even when there is shade, researchers are approaching this challenge in novel ways. The optimization technique for a PV system's efficient operation and enhancing solar power efficiency is the main topic of this study.

5.2 LITERATURE REVIEW

Solar photovoltaic (PV) has become the most widely used method of power generation, when compared to other renewable technologies. Government support, lower-cost materials, and improved manufacturing processes are making solar panels more accessible and appealing [1]. But there's a catch: because it depends on sunlight and temperature, the quantity of electricity they generate varies [2]. To overcome this barrier and enhance the electricity produced by solar panels, sophisticated equipment is needed that tracks the ideal output level [3,4].

The purpose of this study was to compare the performance of DC-DC converters in systems that are connected to the power grid with the MPPT controllers that employ the Perturb and Observe (P&O) algorithm [5]. Zeta converters provide the best output to maximize PV module output power and are compatible with commercial grids, according to their research using Matlab simulation. Zeta converters have been studied by researchers for use in small-scale MPPT PV systems. Using models and experiments, they evaluated the converter's performance in both constant and fluctuating conditions [6]. Furthermore, a technique to increase the power output of solar panels has been proposed that combines the Incremental Conductance MPPT algorithm with a Zeta converter [7]. A mathematical model of the corresponding Zeta converter has been developed using the state space averaging technique. These converters are selected to ensure that throughout the input voltage range, the output voltage remains consistent [8].

Additionally, a brand-new Zeta converter based on the MPPT algorithm was introduced in [9] for battery charging by the extraction of the maximum power from PV panels. According to the research, a solar energy system should have a single-phase asynchronous motor, inverters, and solar panels in addition to a Zeta converter. The voltage from the solar panels would be increased and regulated with the aid of this Zeta converter [10,11]. In order to determine the optimal power converter topology in terms of efficiency and energy output, an MPPT system has been presented [12]. A hybrid energy storage system that combines lead-acid batteries with a superconducting magnetic energy storage system is proposed in this research as a solution for an unstable grid-connected microgrid. The goal of this mixture is to keep the microgrid's power supply steady. The suggested system's efficiency is enhanced by the use of a dual input single output Zeta (di-Zeta) converter to connect the two hybrid power sources [13,14]. A novel kind of Zeta converter that provides low duty cycle high voltage boost by using independent inductors instead of coupled ones is presented. This design has multiple advantages over a standard Zeta converter: it requires fewer parts, has a high efficiency, and maintains a constant input current, which makes it perfect for use in LED lights and electric vehicles [15]. SAZZ and Zeta, two DC-DC converters, are compared in this study. Using both the Incremental Conductance MPPT and P&O algorithms, the simulations demonstrate that the SAZZ converter system operates more steadily and smoothly than the Zeta converter system.

In a prior study, the use of a particular converter (HOA with a boost converter) in a battery-powered solar system was investigated. With a constant voltage from the battery, this arrangement functioned nicely. However, the converter's ability to maintain a constant voltage output was compromised when it was directly linked to the power grid (grid-tied), which reduced the solar power's overall performance. Investigating a different kind of converter (Zeta converter), this research aims to solve this problem in grid-tied systems without batteries. One benefit of this converter is that it can continue to provide a steady voltage output even when the grid is unstable. Because of an innovative harvesting method, it also functions well when the solar panels are partially shaded.

5.2.1 Existing Studies and Their Limitations

5.2.1.1 Zeta Converter Topology – Strengths

a. **Versatile Operation**: The literature has commended the Zeta converter for its versatility in a range of applications due to its capacity to step-up or step-down voltages. Research indicates that it offers consistent voltage control even in the face of input voltage variations, which is especially advantageous for solar power systems since sunshine conditions vary often.
b. **Decreased Input Current Ripple**: Research to date shows that the Zeta converter contributes to this reduction in input current ripple, which is essential to preserving the life of solar panels by shielding them from high-frequency oscillation damage.

5.2.1.1.1 Limitations

a. **Ignorance of Real-World Issues**: A lot of research on the Zeta converter concentrates on how well it performs in ideal scenarios, such as steady-state operations, and ignores issues like partial shading, temperature fluctuations, and component aging. For instance, under non-ideal conditions, a converter's efficiency may drop dramatically, although simulations frequently ignore this.

b. **Component Losses and Temperature Control**: There is insufficient consideration given to the real losses brought on by heat problems, parasitic effects, and switching components. Studies conducted in the real world frequently omit thermal management techniques, which are crucial for avoiding overheating-related efficiency losses, especially in high-power applications.

c. **Cost and Implementation Complexity**: Despite the Zeta converter's potential efficiency, its design complexity and cost are rarely thoroughly discussed. Performance and cost must frequently be carefully balanced in real-world applications, and this is especially true when scaling up from small-scale systems to bigger ones, like solar farms.

5.2.1.2 MPPT Algorithms – Strengths

a. **Effectiveness in Energy Harvesting**: The capacity of MPPT algorithms, in particular P&O and Incremental Conductance, to optimize energy harvest from solar panels has been extensively researched. These algorithms enhance solar system efficiency significantly by adjusting the panels' operating point to match the MPP under different temperature and irradiance circumstances.

b. **Implementation Ease**: P&O and other simple MPPT algorithms are popular because they are useful for a wide range of solar power applications and are reasonably simple to implement in both hardware and software.

5.2.1.2.1 Limitations

a. **Suboptimal Performance in Dynamic Environments**: The algorithms work well under stable situations, but they break down in dynamic environments, including those with quickly changing weather. Research has demonstrated that oscillations around the MPP caused by algorithms like as P&O can result in energy losses. The algorithm finds it difficult to monitor the correct power point in situations that are prone to partial shading, which makes this problem especially severe.

b. **Sluggish Convergence Rates**: When there are large fluctuations in the external environment, the current algorithms typically have sluggish convergence rates. Because the system takes some time to adapt to the new MPP – especially in the morning and evening when light levels are changing quickly – this leads to inefficiencies in energy acquisition.

c. **Computational Complexity**: To increase accuracy, more sophisticated MPPT algorithms have been investigated. These include ones based on fuzzy logic, neural networks, or genetic algorithms. These approaches, however, are typically too complex to be implemented in low-cost or small-scale solar systems due to their high processing power requirements. Previous research frequently emphasizes these advantages without fully addressing the difficulties in implementing such algorithms in practical, budget-constrained settings.

5.2.1.3 Combined Zeta Converter and MPPT Algorithms – Strengths

a. **Complementary Functionality**: Research utilizing MPPT algorithms in conjunction with Zeta converters exhibits encouraging outcomes. Voltage levels are stabilized by the Zeta converter, and system efficiency is maximized using the MPPT algorithm. When combined, they can enhance system performance as a whole, particularly in low-voltage solar applications.
b. **Enhanced Energy Harvesting**: It has been demonstrated that combining these two technologies will increase the efficiency of energy harvesting, especially in small-scale solar systems like home solar arrays. These experiments demonstrate how voltage regulation and energy tracking work together to enhance system performance.

5.2.1.3.1 Limitations

a. **Simulation versus Real-World Performance**: Studies that combine MPPT algorithms with Zeta converters frequently have this drawback because they mostly rely on simulation-based analysis. Such real-world inefficiencies (e.g., component aging, electromagnetic interference (EMI), power losses) are generally ignored in favor of ideal components in these research. It is challenging to establish a solid prediction about these systems' practical performance because of the discrepancy between simulation findings and real-world performance.
b. **Long-Term Dependability and Degradation**: The long-term dependability of these systems has not been extensively studied. Although solar power systems should last for many years, research frequently ignores the consequences of component deterioration over time. There is insufficient coverage of elements like the deterioration of semiconductors, the aging of capacitors, and the long-term durability of the MPPT algorithms under different weather scenarios.
c. **Scalability Issues**: MPPT algorithms and Zeta converters perform admirably in smaller, isolated systems, but scaling is still a major problem. The performance of these technologies in bigger systems, such as solar farms or grid-tied installations, has not received much attention. Growing larger can present additional difficulties such harmonic distortion, grid synchronization problems, and higher needs for thermal control.

5.2.1.4 General Research Gaps

a. **Environmental Impact and System Sustainability**: The majority of this field's study mostly concentrates on short-term efficiency benefits, ignoring the system components' long-term sustainability or their effects on the environment. Research should take into account the effects of materials used in MPPT controllers and Zeta converters on the overall sustainability of solar systems, particularly with regard to production and disposal at the end of life.
b. **Cost-Benefit Analysis**: The studies that are currently in publication do not provide thorough cost-benefit assessments. It can be costly to implement sophisticated converters and control algorithms, thus, study should determine whether the higher efficiency is worth the extra expense, especially in emerging or low-income areas where affordability is crucial.

The paper is structured as follows: Section 5.1 presents the introduction, followed by the literature review and existing studies and their limitations in Section 5.2. Section 5.3 outlines the Zeta converter topology, Section 5.4 details the Zeta converter-based MPPT algorithm for PV, Section 5.5 outlines the MPPT, Section 5.6 details the HOA, and Section 5.7 details efficiency and losses. Finally, Section 5.8 discusses the results, and Section 5.9 offers a conclusion.

5.3 ZETA CONVERTER TOPOLOGY

One kind of flexible DC-DC converter is the Zeta converter. They are helpful in a variety of applications because, in contrast to some converters, they can both step-up and step-down voltages. These innovative designs are inspired by buck and boost converters that already exist. Their ability to store and then transfer energy in two different phases using inductors and capacitors is their secret. An inductor receives power from the input in the first phase, while a capacitor transfers the stored energy to the output in the second phase. The ability of Zeta converters to maintain steady current on both the input and output sides is one of their best features. Numerous benefits result from this smooth flow, including reduced electrical noise (EMI), increased dependability, and even lower current ripple. Zeta converters also have the option of isolating the input and output for even more versatility in various applications.

The core principle of the Zeta converter involves transforming DC voltage to another DC voltage by interchanging energy between the inductors and capacitors is facilitated by a coupling capacitor and switching inductors. A switch regulates the rate at which this energy exchange occurs. This Zeta design must be used in continuous conduction mode to maximize energy exchange efficiency and overall converter efficiency. Choosing the correct component values is crucial when designing the Zeta converter to ensure optimal performance according to the provided specifications. The essential components of the Zeta converter include the switching transistor, inductors, diode, and input/output filter capacitors. Selecting the right capacitance values for the input and output filter capacitors is essential to lowering input and output voltage ripple. C_o has to have a suitably big capacitance to adequately attenuate the excessive output voltage ripple caused by the output-connected diode's pulsing behavior.

A different version of the Cuk converter is depicted in Figure 5.1, where the positions of the transistor and inductor *L*1 are reversed. The term "inverse SEPIC" refers to this design's resemblance to the SEPIC converter.

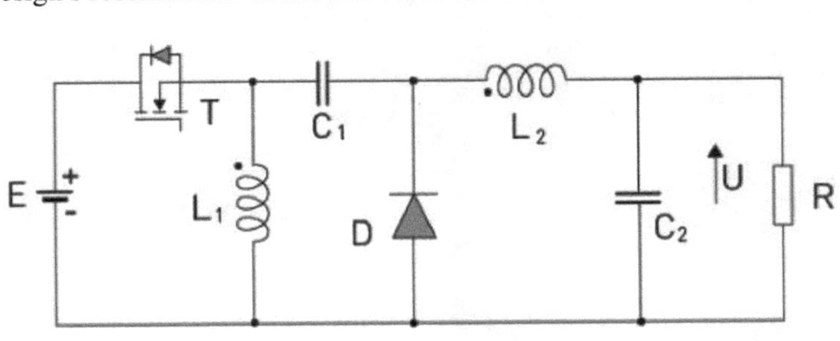

FIGURE 5.1 Zeta converter.

The average voltage across the inductors cancels out to zero when the circuit reaches a steady state. This indicates that the DC power source's voltage and the average voltage across the capacitor are equal. Any slight variations in the voltage of the capacitor (an AC component) can be disregarded due to its size. The equivalent circuit of the SEPIC converter during the "tonne" period is shown in Figure 5.2. The voltages across inductors $L1$ and $L2$ become throughout this period of time (the remainder of the statement stays the same).

$$U_{L1} = E$$

$$U_{L2} = U_{L1} + U_{C1} - U = E$$

So that their currents linearly increase.

Figure 5.3 shows the equivalent circuit of the SEPIC converter during the period OFF. In this period, the voltages across inductors $L1$ and $L2$ are:

$$U_{L1} = U_{C1} = -U$$

$$U_{L2} = -U$$

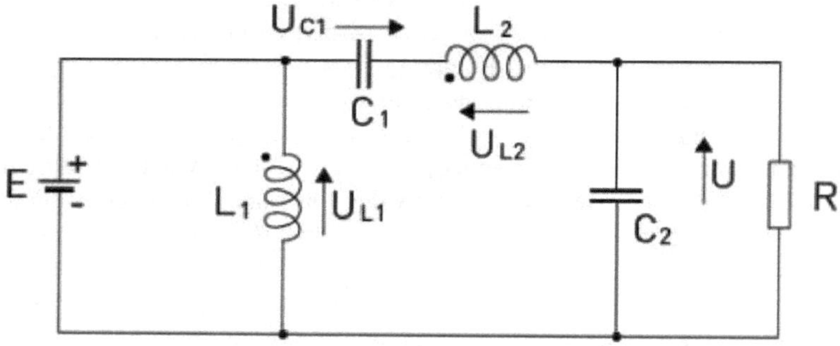

FIGURE 5.2 Zeta converter – interval t_{ON}.

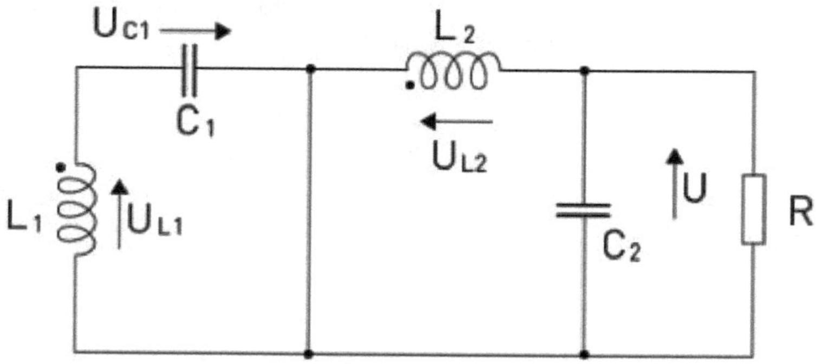

FIGURE 5.3 Zeta converter – interval T_{off}.

In contrast to the SEPIC converter that operates with a continuous input current, the Zeta converter's input current decreases steadily, that is, linearly, during a particular operational phase. But the output current doesn't change; it's still constant. The Zeta's inductors can be magnetically connected for smoother operation, just as the SEPIC converter. By utilizing inductors with comparable inductance values, this coupling is accomplished. This connection reduces the alternating component, or variations, in the current passing through each individual inductor.

In the steady state, the average voltage value on the inductor is equal to zero:

$$E.T.d = U.T.(1-d)$$

$$\Leftrightarrow U = E.\frac{d}{1-d}$$

which means that this converter operates as a buck-boost converter.

For modifying voltage levels in electrical circuits, the Zeta converter (DC-DC converter) is an effective instrument. In contrast to more straightforward converters, it employs a fourth-order, more intricate architecture that integrates components from both boost and buck converters. Several essential elements are needed for this special design to work properly: like tiny electromagnets, two inductors ($L1$ and $L2$) store and release energy. To even out voltage variations, two capacitors ($C1$ and $C2$) function as miniature rechargeable batteries. A diode (D) regulates the direction of electrical current flow by functioning as a one-way valve. Like a quick-acting gatekeeper, a switch (S), usually an IGBT or MOSFET, controls the flow of power.

The input voltage source is coupled to one end of the first inductor ($L1$) in a standard Zeta converter configuration. The switch (S), which regulates the current flowing through the inductor, is linked to the other end of $L1$. With its cathode attached to the output side and its anode linked to the common node between $L1$ and S, the diode (D) is wired in parallel with the switch. The output voltage node and the diode's cathode are linked to the second inductor ($L2$). With their common node connected back to the input voltage source, the capacitors ($C1$ and $C2$) are connected in series between the output voltage node and ground. A Zeta converter's switch (S) cycles between ON and OFF states to regulate the amount of energy that passes through the circuit. Inductor $L1$ receives a charge from the input voltage source when the switch is turned ON, and its magnetic field stores energy. In addition, the diode (D) undergoes reverse biasing, and capacitor $C2$ facilitates the flow of energy from inductor $L2$ to the output load. The current in $L1$ continues to flow via the diode (D) after the switch is turned off, charging the inductor $L2$ and discharging the capacitor $C1$. Next, through capacitor $C2$, the energy stored in $L2$ is transmitted to the output load.

5.3.1 Continuous and Discontinuous Conduction Modes

Like other DC-DC converters, Zeta converters operate primarily in two modes: continuous mode (CCM) and discontinuous mode (DCM). Depending on the purpose of the converter, the amount of electricity it must manage, and the desired level of efficiency, select the appropriate mode.

Continuous Mode (CCM): This mode maintains a steady current flow via the inductors of the converter. Because it generates a smoother output voltage and is more efficient, it is frequently chosen for high-power applications. It is also easier to regulate and analyze CCM. But in this mode, the inductors may need to be larger and lose more energy.

Discontinuous Mode (DCM): The current flow in the inductors (*L*1 and *L*2) isn't always constant in the DCM. It can be more efficient at low loads and allows for smaller inductors, making it a desirable option for low-power applications or scenarios with varying power needs. The drawbacks are a more intricate control mechanism and a less stable output voltage.

Selecting the Correct Mode: Depending on the intended usage of the converter, the best mode (CCM or DCM) should be selected. In order to attain great efficiency at various load levels, some even blend the two modes or alternate between them depending on the power requirements.

5.3.2 Design Selections and Considerations

To guarantee appropriate component selection and system performance, a number of factors and computations are taken into account when building a dependable and efficient Zeta converter. The following considerations should be made when building a Zeta converter: establish the input and output voltage levels required for your application to make sure you get the right components and figure out the duty cycle (*D*) of the switch.

a. **Details of the Input and Output Voltages**: Required load and output power: the size of the required inductors and capacitors will depend on the output power and load requirements of your application, which will also affect the converter's conduction mode (DCM or CCM).
b. Calculate and select the appropriate inductor values (*L*1 and *L*2) depending on the required load, the intended conduction mode, and input/output voltage specifications. This will affect the efficiency, ripple in the output voltage, and transient response of the converter.
c. Calculate the duty cycle (*D*) of the switch, which is the ratio of the ON time to the overall switching period. This will provide you with the input-output voltage relationship that your converter needs to operate properly.
d. **Selection of Capacitors**: To maintain a steady output voltage and reduce voltage fluctuations, choose the input (C_{in}) and output (C_{out}) capacitors depending on the required output voltage ripple, transient response, and converter stability.
e. **Switching Frequency**: Select a suitable frequency by taking into account elements like EMI, component size, and efficiency. While using smaller inductors and capacitors may be possible at higher switching frequencies, doing so may also increase switching losses and EMI.
f. Select a control strategy (current mode, voltage mode, or digital control) to adjust the output voltage or current of the converter. The intricacy of the

control circuit and how well it performs under various operating situations will depend on your control strategy.

g. **Thermal Management**: To maintain converter efficiency and guarantee long-term dependability, ensure appropriate thermal management by calculating power losses and ensuring sufficient heat dissipation.

5.4 ZETA CONVERTER-BASED MPPT ALGORITHM FOR PV

Figure 5.4 illustrates the block diagram of a Zeta converter-based MPPT algorithm designed for PV applications. To get the optimal solution, this algorithm employs a novel method known as HOA. HOA takes its cues from the natural behavior of horses, particularly from their abrupt accelerations when pursuing an object. This method aids the algorithm in maximizing power output, particularly in the event of variable or unequal sunshine. This proposed method demonstrates superior performance compared to conventional techniques across various weather scenarios. It boasts reduced calculation time, rapid convergence, and minimal oscillations once the MPP is reached. Through a comparison with other methods like the P&O technique, the effectiveness of the HOA becomes evident. The solar power generated is stored in a battery for future usage. The Zeta converter, functioning as a buck-boost converter, ensures a stable voltage output, thus maintaining the stored power at a consistent level. Because of its ability to step up and down, the Zeta converter is incredibly efficient. Similarly, the MPPT algorithm, although excellent for optimizing energy output, may necessitate sophisticated control systems when combined with the Zeta converter, adding complexity to the entire configuration. In order to maximize system efficiency, synchronizing the outputs of wind or hybrid systems and optimizing power flow are necessary when integrating the Zeta converter.

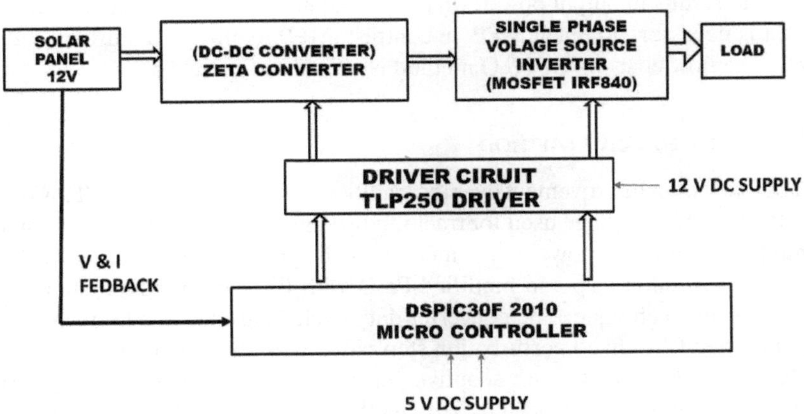

FIGURE 5.4 Block diagram of the Zeta converter-based MPPT algorithm for PV.

5.5 MAXIMUM POWER POINT TRACKING

A controller using MPPT can be used to get the most power output possible out of any PV panel. Generally speaking, the greatest power can be extracted using one of two methods. There are two approaches: indirect technique and direct technique. Under the indirect approach, fixed voltage, short circuit current, and open circuit voltage are included. This approach is made simple by the incorporation of extremely basic assumptions and periodic estimation. Only the operating voltage can be changed under various irradiation levels in the fixed voltage mode of approach to obtain bigger MPP voltages. However, because of seasonal variations in temperature and irradiance, such as winter and summer, the approach's end output is not producing a profitable outcome. Conversely, the open circuit voltage style of approach is the indirect strategy. This is one of the simplest and most straightforward methods for obtaining the maximum output voltage when compared to other methods. Multiplying the open circuit voltage by the constant *K* will yield the maximum output voltage. However, because the open circuit voltage value must be determined monotonically with respect to changes in the irradiation level, the value of *K* is determined under assumptions, which results in a drop in overall efficiency.

5.5.1 P&O Algorithm

With the use of very few parameters and a straightforward feedback design, this most well-known algorithm was developed. This algorithm perturbs the cell voltage cyclically and then compares the received output power to the initial power. The process is repeated if the received power is increased, else it is reversed. The output power that the cell provides determines how much the output voltage increases and decreases. The cell's operating point on the left side of the MPP displays a rise in voltage that results in output power; this circumstance indicates that a disturbance to the right is necessary to reach MPP. In contrast, MPP requires a disturbance toward the left. The flow chart of the P&O method is shown in Figure 5.5.

5.5.2 Improved P&O Method

This approach is an improvement over the traditional P&O methodology. This method scales the reference voltage used for tracking by 0.8 of the Voc of the PV modules in order to prevent wasting power while measuring the global MPP. Modified P&O with adaptive perturbation step and modified P&O with fixed perturbation step are two subtypes of this technique. The converter duty cycle is used as the perturbed signal in the modified P&O fixed perturbation step rather than the module voltage.

Since the updated P&O with adaptive perturbation step makes use of a variable duty ratio, it is more sophisticated. This method has certain drawbacks despite its great performance, such as a high computational burden against accuracy trade-off and a dependence on predetermined constants.

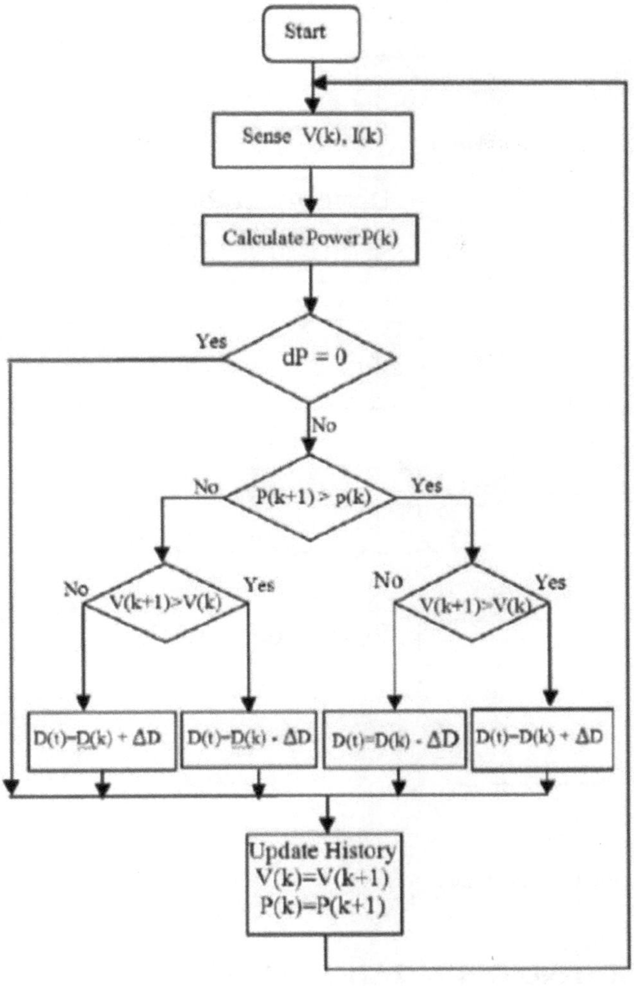

FIGURE 5.5 Flowcharts of the algorithm.

5.5.3 Constant Voltage

The MPP is considered to have a set voltage value in the constant voltage technique, which is the same as the value recorded under the standard test conditions specified by the manufacturer. Usually, this constant voltage falls between 72% and 80% of the V_{oc}. A feedback control loop is then utilized to modify the MPPT converter's duty cycle using this reference voltage. All that's needed to build a constant voltage MPPT control is measuring the array voltage. It can be put into practice with both digital and analog electronics. The flow chart is shown in Figure 5.6.

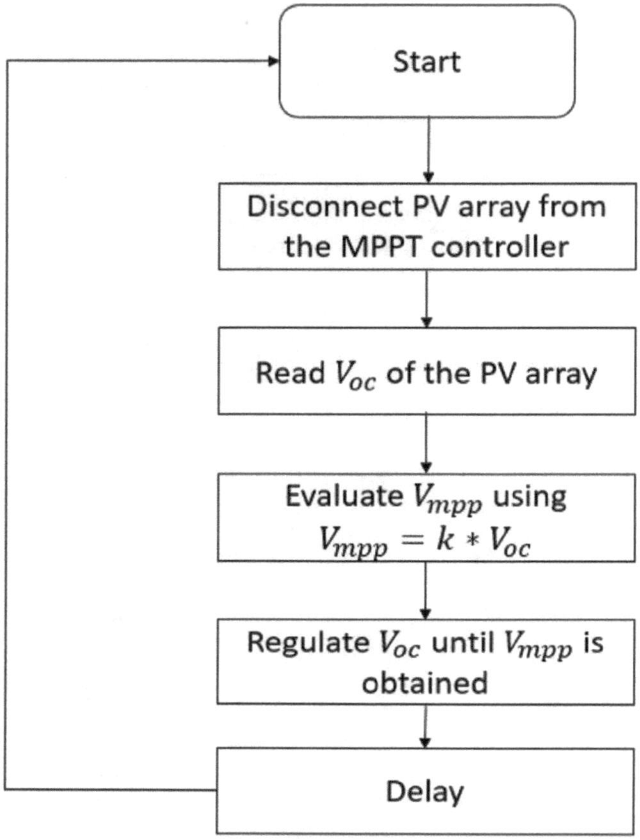

FIGURE 5.6 Flowcharts of the algorithm.

5.6 HORSE HERD OPTIMIZATION ALGORITHM (HOA)

This explanation shows how the ideal solution is tracked by the HOA. It accomplishes this by taking into account the duty cycle (or on/off ratio of the signal) and the transient value (or transitory voltage level). HOA comprises four horses identified as P1, P2, P3, and P4. Employing four horses ensures the optimal performance of the HOA, as altering the number of horses can impact the effectiveness of the PV system. The fitness of each horse is evaluated based on the power output. Movement of the horses is dictated by their equivalent power at specific points, with Point P2 being allocated the output due to its highest fitness value. A horse's mobility is determined by its matching power at a given position. Since Point P2 has the highest fitness value, P2 receives the output. The abrupt increase in power over the first 45–55 ms is a result of its extremely effective searching and exploration capabilities.

Due to its great efficiency, there is a significant surge in power in the first 45–55 ms. The exploration phase lasts for 20 ms, starting at zero. P2 exchanges take place every 50 ms. For a brief while, the remaining horses show decreased fitness, but the controlled output in terms of the selected converter's DC does not change. Conversely, other horses start searching for the ideal solution. The horse continues to search until and unless it finds a better option when it gets wedged in, which represents the minima of its own personal best. The flowchart for the HOA is shown in Figure 5.7.

5.7 EFFICIENCY AND LOSSES

Power converter design must prioritize efficiency because it has a direct impact on power dissipation, thermal management, and system performance as a whole. A Zeta converter's efficiency can be defined as the percentage difference between its input and output powers. Comprehending and reducing the several losses that transpire within the converter is crucial for maximal efficiency.

a. Conduction losses are experienced by power electronic components such as switches, diodes, and inductors when current flows through them. Due mostly to the resistive nature of the components, these losses are reliant on the square of the current flowing through them. Choosing components with

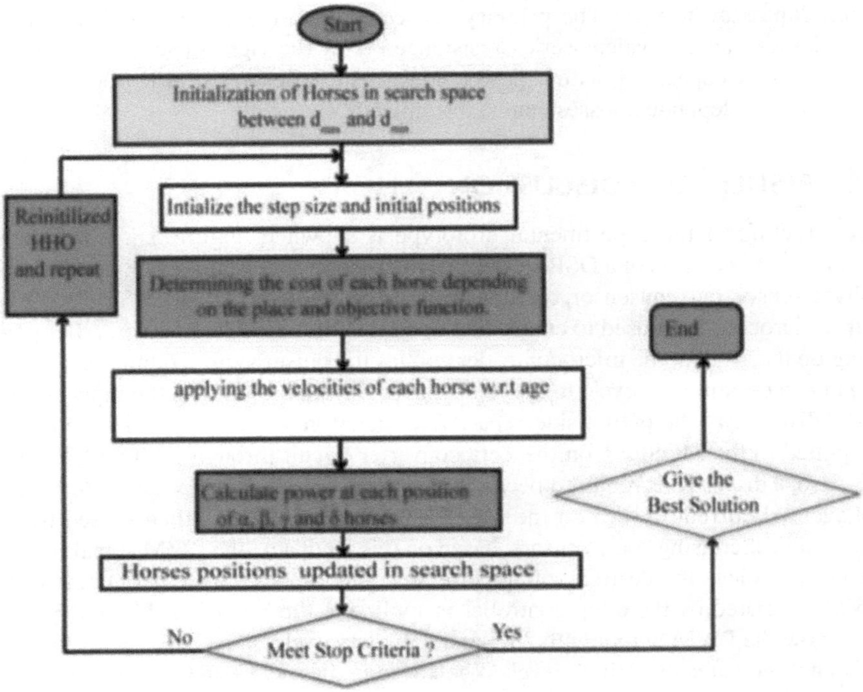

FIGURE 5.7 Flowcharts of the algorithm.

reduced resistance or enhancing the circuit's architecture to reduce parasitic resistances are two ways that designers might reduce conduction losses.

b. **Switching Losses**: When power switches switch from the ON to the OFF position, switching losses take place. Power dissipation occurs during these transitions because voltage and current are present at the same time. Switching losses are influenced by the switching frequency, the speed at which voltage and current change during transitions, and the characteristics of the switching devices. Switching losses can be decreased by designers by using switches with faster switching rates or by improving gate driving circuits to shorten transition times.
c. **Diode-Reverse Recovery Losses**: These losses are related to how a diode behaves as it goes through the reverse recovery procedure. Reverse current flow occurs momentarily as the diode passes from the conducting to the blocking state. This current flow may result in significant power dissipation at high frequencies. To minimize reverse recovery losses, designers can use Schottky diodes, also known as fast recovery diodes, which have extraordinarily short reverse recovery periods.
d. **Core Losses in Magnetic Components**: The alternating magnetic fields cause core losses in the magnetic cores of transformers and inductors. The converter's form, core material, and operating frequency all affect these losses. Designers can lower core losses by choosing the right core materials and improving the magnetic component designs.
e. **Capacitor Losses**: The primary causes of capacitor losses are dielectric losses and equivalent series resistance. With the right dielectric materials and capacitors with a lower equivalent series resistance, these frequency-dependent losses can be minimized.

5.8 RESULTS AND DISCUSSION

The structure of the experimental prototype is shown in Figure 5.8. The experimental model consists of a DSPIC30F2010 microcontroller, solar panel, optoisolator, voltage sensor, current sensor, driver circuit, and MOSFET IRF840. DSPIC30F2010 is the microcontroller used to control the operation of the Zeta converter. After powering up the system, the microcontroller applies the pulse-width modulation (PWM) signal at a certain duty cycle to the Zeta converter which boosts up the input voltage and delivers it at the output side. The PWM signal generated by a microcontroller is applied to the MOSFET on the Zeta converter circuit through a MOSFET driver IC. Also, a driver IC is required because the microcontroller cannot generate enough voltage and current to turn on the MOSFET. Output voltage is then sensed by the microcontroller using some sensors; based on this feedback, the PWM signal is modulated to achieve the desired voltage level and single-phase inverter input. Inverter PWM generated by the microcontroller is applied to the MOSFET. Figures 5.9 and 5.10 show the PWM output with 25% and 75% duty cycles. The duty cycle of a digital signal indicates its activity level. When a signal has a 75% duty cycle, it is on for three of the four time intervals and off for the fourth. On the other hand, a 25% duty cycle denotes that the signal is only active for one of every four intervals, with the

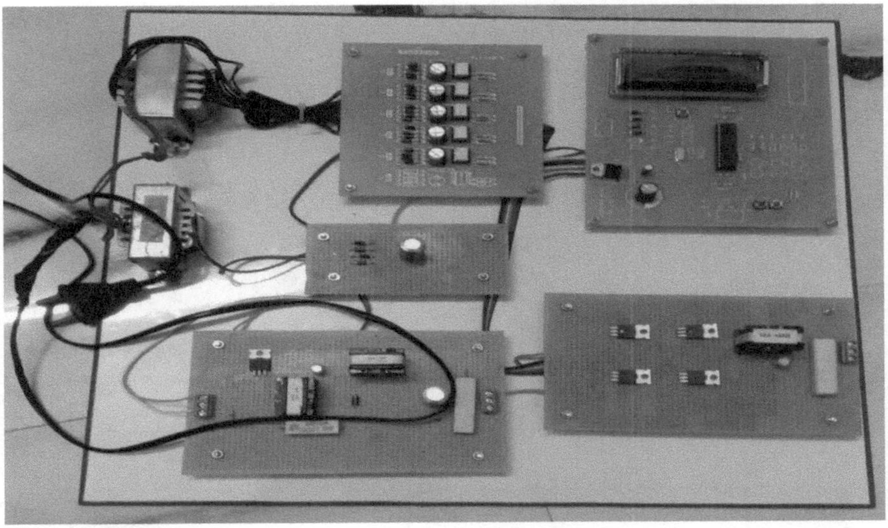

FIGURE 5.8 Structure of the experimental prototype.

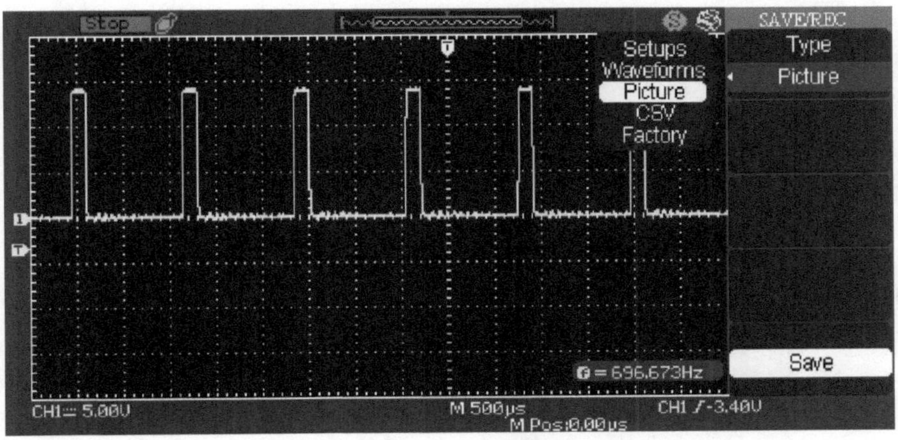

FIGURE 5.9 PWM output with a 25% duty cycle.

remaining time being off. Increasing the frequency of high pulses results in a higher average voltage and accelerates the operation of the converter. The proposed system also doesn't exhibit steady-state oscillation, which lowers power losses. The voltage output of the new system is displayed as a graph in Figure 5.11. The graph shows that the system raises the voltage from 20V to 100V with effectiveness. Table 5.1 shows the parameters of the solar panel. The output is tested with an input voltage source and enables accurate testing and verification of the MPPT algorithm's and the Zeta converter design's performance in a simulated environment.

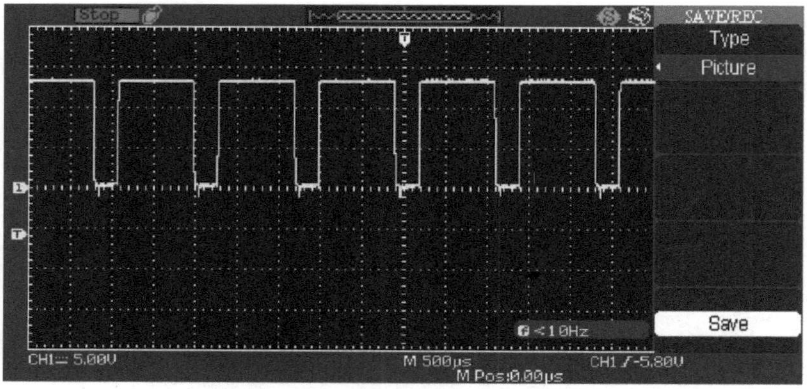

FIGURE 5.10 PWM output with a 75% duty cycle.

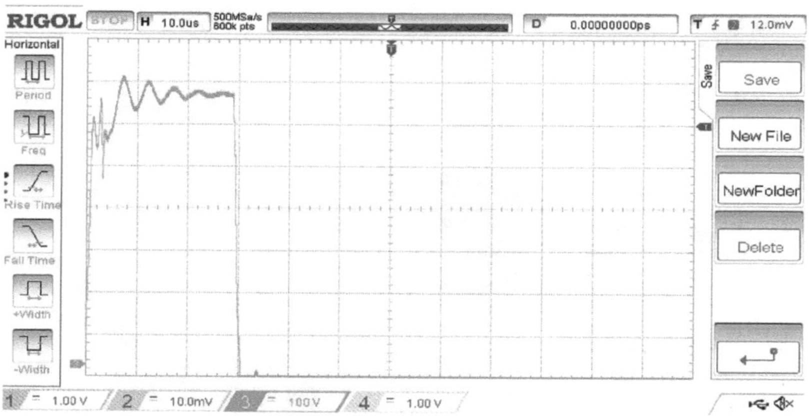

FIGURE 5.11 Hardware output of the proposed system.

TABLE 5.1
Parameters of the Solar Panels

Parameters		**Values**
DC side	MPP voltage	17.7V
	MPP current	6.5 A
	Open circuit voltage	22.2V
	Short circuit current	7.97 A
	Temperature	25°C
	Solar irradiance	1,000 W/m^2
AC side	Rated grid voltage	230V
	Rated grid current	0.5 A
	Switching frequency	20 KHz
	Inductor filter	0.9 μH

5.9 CONCLUSION

In order to create a maximum power extraction technique for PV applications, this paper introduces a novel approach HOA. In this context, the HOA MPPT is a new MPPT that is being evaluated for effectiveness. Additionally, compared to the standard algorithm, the proposed function-based MPPT algorithm reacts to variations in solar irradiation and load more quickly. The proposed algorithm also doesn't exhibit steady-state oscillation, which lowers power losses. The experimental results demonstrate that the proposed MPPT in the Zeta DC/DC converter network quickly converges to achieve maximum power.

5.9.1 Future Research Directions

1. **Advanced Control Algorithms for MPPT:** Even while MPPT methods like Incremental Conductance and P&O are frequently utilized, they have drawbacks, especially in situations when the environment is changing quickly, like in partial shading. Future work may concentrate on creating more intelligent artificial intelligence-driven algorithms. For instance, machine learning techniques might be used to create algorithms that more rapidly and precisely estimate the MPP and adjust to temperature and sunshine variations. These might be trained to react to various environmental factors, which would lead to improved performance in difficult scenarios where conventional MPPT algorithms break down.
2. **Integration with Energy Storage Systems**: An increasing number of solar power systems are being coupled with energy storage devices, such as supercapacitors or batteries, to store extra energy for use in the absence of sunlight. To enable effective cycles of charging and discharging, research might examine how Zeta converters can be optimized for smooth connection with storage systems. In addition to increasing system dependability, this would facilitate the installation of solar power in isolated or off-grid areas where grid electricity is not available.
3. **Optimization of Zeta Converter Design**: By concentrating on lowering component size, raising power density, and enhancing thermal management, Zeta converter design can be made better. Future studies may look at the application of novel materials, which give greater efficiency at higher frequencies and temperatures, such as silicon carbide or gallium nitride (GaN) semiconductors. Additionally, to enable more effective functioning in solar power systems seen in real environments, researchers could examine methods to reduce losses and optimize the switching frequency.
4. **Hybrid Converter Systems**: Increased flexibility and efficiency may be possible with hybrid power converters that combine the advantages of Zeta converters with other converter topologies, such as SEPIC or Cuk converters. These systems might be more suited for multi-source renewable energy systems if they were made to operate across a wider range of input and output voltages. For example, combining wind turbines and solar panels

into a single power system and increasing overall efficiency would be best accomplished with a hybrid converter.

5. **DC Microgrid Applications**: DC microgrids could be the way of the future for solar power, especially in places where integrating renewable energy sources is difficult. Zeta converters have the potential to be extremely important in DC microgrid voltage stabilization, which guarantees effective distribution of solar electricity to nearby loads. To make microgrids more robust and able to manage variations in power supply, future studies could investigate how to enhance the connection between Zeta converters, the DC grid, and other distributed energy resources.
6. **Enhanced Real-Time Performance Evaluation**: While simulation models are the main focus of many current studies, real-time performance testing of MPPT algorithms and Zeta converters under various environmental circumstances is also necessary. More advanced testing environments, such as hardware-in-the-loop simulation platforms, that replicate real-world circumstances and enable quick prototyping and assessment of novel converter designs and MPPT algorithms could be developed through research. This would contribute to bridging the knowledge gap between theoretical research and real-world application.

5.9.2 Practical Applications

1. **Residential Solar Power Systems**: In order to optimize the energy yield from rooftop solar panels, Zeta converters in conjunction with MPPT algorithms can be used in residential solar power systems. By increasing the efficiency of their solar systems and so reducing energy bills and fostering energy independence, this combination guarantees that homeowners may lessen their dependency on grid electricity.
2. **Off-Grid Solar Power Solutions**: Solar power systems using Zeta converters and MPPT algorithms can be a dependable source of energy in places with little to no access to grid electricity, such as rural or remote areas. These systems could be utilized to provide energy in previously unrequired regions, so enhancing quality of life and promoting sustainable development in homes, schools, or healthcare institutions.
3. **Electric Vehicle Charging Stations**: Solar-powered electric vehicle charging stations have an opportunity as electric vehicle usage rises. These stations can optimize solar energy utilization by integrating Zeta converters and MPPT algorithms, guaranteeing that the most power is extracted and supplied to refuel electric cars. This might help create a more sustainable transportation infrastructure and drastically lower the carbon impact of electric vehicle charging.
4. **Agricultural Applications**: In agricultural, solar electricity is being utilized more and more for operations including irrigation, machinery operation, and water pumping. Solar-powered agricultural systems can be made more efficient with Zeta converters and MPPT algorithms, enabling farmers

to use less fossil fuel and spend less on operating expenses. These devices can be especially helpful in places where grid electricity is scarce, allowing for more independent and sustainable farming methods.

5. **Portable Solar Generators**: During power outages, outdoor activities, and emergency response scenarios, portable solar generators are frequently utilized. Zeta converters' small size and high efficiency make them the perfect choice for lightweight, portable solar generators that must deliver consistent power under a variety of circumstances. In an emergency or off-grid scenario, these generators can power small gadgets, medical equipment, or communication gear.
6. **Solar-Powered Internet of Things Devices**: There is a growing need for compact, long-lasting solar-powered gadgets that can function independently as the Internet of Things expands. Zeta converters can power Internet of Things sensors and devices, ensuring they can operate independently without frequent battery replacements, when paired with effective MPPT algorithms. These systems could be used in smart cities, smart agriculture, or environmental monitoring, where a steady and dependable supply of electricity is essential.

REFERENCES

1. M. Obi and R. Bass, 'Trends and challenges of gridconnected photovoltaic systems', A review, *Renewable and Sustainable Energy Reviews*, Vol. 58, pp. 1082–1094, 2016.
2. M.S. Ibbini and A.H. Adawi, 'Analysis and design of a maximum power point tracker for a stand-alone photo voltaic system using simscape', *International Journal of Advanced Trends in Computer Science and Engineering*, Vol. 8(1), pp. 54–57, January–February 2019.
3. M.Z. Hossain, N.A. Rahim and J. Selvaraj, 'Recent progress anddevelopment on power DC-DC converter topology, control, design and applications', A review, *Renewable and Sustainable Energy Reviews*, Vol. 81, pp. 205–230, 2018.
4. S. Rajalakshmi and S. Titus, 'Optimal slewing mode converter-based energy management system for renewable energy sources', *International Journal of Advanced Trends in Computer Science and Engineering*, Vol. 9(3), pp. 2493–2502, May-June 2020.
5. A. Ambikapathy, G. Singh and P. Tiwari, 'Performance of different types of DC-DC converter with P&O algorithm of MPPT', *International Journal of Smart Home*, Vol. 10(6), pp. 277–286, 2016.
6. Raj, S.R. Arya and J. Gupta, 'Solar PV array-based DC-DC converter with MPPT for low power applications', *Renewable Energy Focus*, Vol. 34, pp. 1–11, July 2020.
7. M. Kaouane, A. Boukhelifa and A. Cheriti, 'Implementation of incremental-conductance MPPT algorithm in a photovoltaic conversion system based on DC-DC ZETA converter', in *8th International Conference on Modelling, Identification and Control*, 2016, pp. 612–617, Algiers, Algeria.
8. K. Sudarsana Reddy, B. Sai Teja Reddy, K. Deepa and K. Sireesha, 'An optimized controller for zeta converter-based solar hydraulic pump', *Lecture Notes in Electrical Engineering*, 778, pp. 401–418, 2021.
9. P. Suresh, A. Sureshkumar and P. Vishnuram, 'Photovoltaic battery charger based on novel zeta converter', *International Journal of Pure and Applied Mathematics*, Vol. 118(19), pp. 2683–2694, Feb 2018.

10. Malhotra, S. Vij, P. Gaur and C. Malhotra, 'Design, analysis and performance of zeta converter in renewable energy systems', in *IEEE International Conference on Computing for Sustainable Global Development,* 2016, pp. 2487–2491, New Delhi, India.
11. M.S. Sivagama Sundari and D. Periyasamy, 'PSO based selective harmonic elimination PWM for Cascaded H-bridge multilevel inverter', *IOP Conference Series: Materials Science and Engineering*, Vol. 1012(2021), 012054, doi: 10.1088/1757-899X/1012/1/012054.
12. M. Birane, A. Derrouazin, M. Aillerie, A. Cheknane and C. Larbes, 'Evaluation and performance of different topologies of converters with efficient MPPT in a photovoltaic system', *Journal of Electrical Systems*, Vol. 16(3), pp. 308–319, 2020.
13. K. Aseem and S. Selvakumar, 'High temperature superconducting material based energy storage for solar-wind hybrid generating systems for fluctuating power management', *Materials Today: Proceedings*, Vol. 42, pp. 1122–1129, 2021.
14. K. Aseem and S. Selvakumar, 'Di-zeta converter-based solar–wind generating system with a novel SMES battery hybrid energy management', in *International Conference on Communication, Circuits and Systems*, IC3S 2020, pp. 443–450, Bhubaneswar, India.
15. P.L. Santosh Kumar Reddy, Y.P. Obulesu, S. Singirikonda, M. Al Harthi, M.S. Alzaidi and S.S.M. Ghoneim, 'A non-isolated hybrid zeta converter with a high voltage gain and reduced size of components', *Electronics*, Vol. 11(3), p. 483, 2022.

6 Hybrid Sliding Mode Control for Solar PV Systems with Cuk *Converter*

Booma Jayapalan, Kanimozhi Kannabiran, and Angalaeswari Sendraya Perumal

6.1 INTRODUCTION AND LITERATURE SURVEY

The extensive use of fossil fuels has greatly diminished their availability, making it crucial to seek alternative energy sources. Renewable energy sources such as wind, solar, and tidal energy offer promising solutions, with solar energy being the most abundant and cost-effective. The non-linear and volatile nature of PV arrays hinders optimal power generation, requiring an appropriate DC-DC converter and MPPT controller to maximize energy output. Various MPPT control techniques have been explored, including look-up table methods, hill climbing, P&O methods, improved P&O, incremental conductance methods, fuzzy logic-based methods, constant voltage methods, β methods, system oscillation methods, and ripple correlation methods. SMC has also been adopted to track the maximum power point (MPP), though it may lack robustness under varying operational conditions. The efficiency of PV systems is influenced by the choice of DC-DC converter. Previous research has analyzed various converters, including buck, boost, buck-boost, SEPIC, Cuk, and Zeta converters. The basic buck converter, commonly used for battery charging, suffers from high switching losses, noise, and component stress at high switching frequencies, negatively impacting performance. Non-inverting synchronous buck-boost converters experience switching voltage stress, which depends on circuit topology and can lead to power reduction and deterioration. Research on modifying conventional Cuk circuits has focused on reducing switching or conduction losses, component sizes, voltage or current stress, and improving transient responses. Approaches such as synchronous switch schemes and new synthesis procedures for soft-switching inverting buck-boost (IBB) converters have been explored. The efficiency of PV systems in terms of faster convergence and stability depends significantly on the choice of DC-DC converter.

To address these issues, a lossless soft-switching technique is introduced. Soft-switching techniques are classified into passive and active methods. Passive methods, which do not require extra switches or control circuitry, are less expensive,

DOI: 10.1201/9781003496946-6

more reliable, and achieve higher performance. For Pulse Width Modulation (PWM) converters, passive soft-switching techniques achieve zero-current turn-on and zero voltage turn-off, reducing switching losses. An MPPT controller is formulated using a sliding surface to achieve a robust MPPT approach. The robustness of this approach is validated through performance investigations under varying load and irradiance conditions. The PV system model with the hybrid SMC MPPT controller is verified through simulation and experimental validation. The proposed system's main advantage is its increased efficiency in generating maximum power using the non-inverting Cuk converter. The system is simulated under different climatic conditions. Results indicate that the proposed controller achieves faster convergence and improved system efficiency in reaching the MPP. The paper is organized as follows: Section 6.2 discusses the materials and methods for designing the passive soft-switched non-inverting DC-DC Cuk converter and PV system. Section 6.3 describes the proposed hybrid SMC approach for MPPT applied to the IBB. Section 6.4 presents the MATLAB®/Simulink® validation and experimental results. Finally, the conclusion is drafted in Section 6.5.

This paper presents [1] an integrated PV MPPT converter. The key advantage is the efficient tracking of the MPP, leading to optimized energy conversion from PV panels. However, the converter design could be complex and costly due to the integration requirements. A research gap exists in improving the cost-effectiveness of such integrated systems. Future work could explore the integration of more advanced MPPT algorithms and the use of more efficient and cost-effective materials. This foundational paper [2] on variable structure systems and SMC outlines the advantages of robustness and simplicity in control design. However, a key disadvantage is the phenomenon of chattering, which can lead to high-frequency oscillations in the system. The research gap lies in mitigating chattering while maintaining the robustness of the control system. Future research could focus on developing more sophisticated algorithms to reduce chattering without sacrificing performance. The paper [3] explores sliding motions in bilinear switched networks, emphasizing the control capabilities of such networks. The main advantage is the enhanced control precision; however, the complexity of the network could lead to stability issues. The research gap includes exploring stability under different network configurations. Future work could involve the development of simpler models or hybrid methods that combine the strengths of bilinear networks with other control strategies.

This work [4] discusses the properties and synthesis of passive lossless soft-switching PWM converters. The advantage is the reduction of switching losses, which improves efficiency. However, these converters may have limitations in high-power applications due to component stress. The research gap lies in scaling these c″onverters for higher power applications. Future research could investigate the integration of advanced materials or novel circuit designs to address these limitations. The paper [5] introduces a soft-switching boost converter with a flyback snubber designed for high-power applications. The key advantage is enhanced efficiency at high power levels. A disadvantage is the potential complexity of the circuit design, which could lead to difficulties in practical implementation. The research gap includes simplifying the design while maintaining performance. Future research could focus on developing more integrated solutions or exploring alternative soft-switching techniques. This study [6] examines a new passive lossless turn-off snubber, highlighting its potential to reduce switching losses. The advantage is the improved efficiency and reduced component stress. However, the design might not be easily scalable for various applications. The research gap is in enhancing the scalability and adaptability of the snubber design. Future research could explore modular designs or hybrid snubber techniques to increase versatility.

This paper [7] presents a systematic approach to developing single-stage soft-switching PWM converters, with the advantage of reduced complexity and improved efficiency. However, there may be limitations in dynamic performance under varying load conditions. The research gap involves improving load adaptability while maintaining efficiency. Future work could involve integrating adaptive control mechanisms to enhance performance across a wider range of operating conditions. The research [8] focuses on using the incremental conductance method for MPPT in PV systems. The advantage is accurate tracking under varying environmental conditions. A disadvantage is the potential for slower response times compared to other methods. The research gap includes improving response time without sacrificing accuracy. Future research could explore hybrid methods that combine incremental conductance with faster algorithms. This paper [9] introduces a phase-locked bidirectional converter with a pulse charge function for dual-voltage Power Nets, offering the advantage of efficient power management in automotive systems. However, the complexity of the design may limit its practical implementation. The research gap lies in simplifying the design for broader adoption. Future research could explore alternative architectures or integration strategies to reduce complexity. The study [10] provides an analysis of different DC-DC converter topologies for PV systems under varying climatic conditions. The advantage is the tailored optimization for specific environmental factors. However, a disadvantage is the potential difficulty in generalizing the findings to other regions. The research gap includes extending the applicability of the optimized designs to a wider range of conditions. Future work could involve developing adaptive systems that can automatically adjust to different climates.

This book [11] provides a comprehensive overview of SMC theory and its applications. The advantage of SMC is its robustness and simplicity in implementation. However, the main drawback is the chattering phenomenon, which can cause wear and tear in mechanical systems. The research gap is in reducing or eliminating chattering while retaining the robustness of SMC. Future research could explore advanced algorithms and practical implementations to mitigate this issue. This paper [12] discusses a hybrid digital-analog SMC for PV applications, which offers the advantage of improved efficiency and faster response time. However, the complexity of integrating digital and analog components may increase the cost and design difficulty. The research gap lies in simplifying the integration process while maintaining or improving performance. Future work could focus on fully digital implementations that emulate the benefits of analog control. This research [13] introduces an interleaving technique aimed at reducing voltage ripple in series-connected PV systems. The advantage is the reduction of voltage ripple, leading to more stable power output. However, the technique may be limited by its scalability to larger systems. The research gap is in extending the technique to accommodate more extensive and complex PV systems. Future research could explore hybrid interleaving techniques or alternative topologies to overcome scalability issues. The paper [14] focuses on enhancing the power conversion efficiency of a DC-DC boost converter using a passive snubber circuit. The primary advantage is the reduction in switching losses, leading to improved efficiency. However, the design may not be universally applicable to all converter topologies. The research gap includes exploring the applicability of the passive snubber circuit across different converter designs. Future research could involve developing adaptable snubber circuits that can be easily integrated into various systems. This work [15] presents a novel real-time simulation technique for PV systems using Real Time Digital Simulator (RTDS). The advantage is the ability to accurately model and simulate PV systems in real time, which is crucial for system design and testing. However, the high cost of RTDS systems may limit their accessibility. The research gap lies in making real-time simulation more cost-effective. Future work could explore alternative simulation platforms or methods that offer similar accuracy at a lower cost.

This study [16] reviews non-isolated DC-DC converters for PV applications, discussing both current technologies and future trends. The advantage is the comprehensive analysis of various converter topologies. However, the paper may not cover the latest developments due to its publication date. The research gap is in updating the analysis with the most recent advancements. Future research could focus on emerging non-isolated converter technologies and their potential applications in PV systems. The paper [17] introduces a modified adaptive hill-climbing MPPT method, which offers faster tracking of the MPP in PV systems. The advantage is the increased efficiency in varying environmental conditions. However, the method may suffer from convergence issues under certain conditions. The research gap is in improving the convergence reliability without compromising tracking speed. Future research could explore hybrid MPPT methods that combine hill climbing with other algorithms to enhance performance.

This paper [18] presents a near maximum power-point operation design for PV systems. The advantage is the reduced complexity and cost compared to full MPPT systems. However, the trade-off is a slight reduction in overall efficiency. The research gap is in balancing cost and efficiency more effectively. Future work could involve optimizing near maximum power-point operation designs to minimize efficiency losses while maintaining low complexity and cost. This research [19] explores neural network-based MPPT methods for PV systems, particularly in rapidly changing environments. The advantage is the method's ability to adapt to varying conditions, ensuring optimal power output. However, the complexity of training and implementing neural networks can be a disadvantage. The research gap is in simplifying the implementation of neural networks for real-time applications. Future research could focus on developing more straightforward neural network architectures or training methods tailored for MPPT. The paper [20] presents a simulation and hardware implementation of an incremental conductance MPPT with a direct control method using a Cuk converter. The advantages are the precise control and high efficiency achieved with the Cuk converter. However, the complexity of the hardware implementation can be a drawback. The research gap is in simplifying the hardware design without sacrificing performance. Future research could explore integrating advanced materials or techniques to reduce the design complexity.

This paper [21] discusses the design of a sliding mode-controlled SEPIC converter for PV MPPT applications. The advantage is the robustness and efficiency of the SEPIC converter under SMC. However, the SEPIC converter may introduce additional complexity compared to simpler topologies. The research gap lies in reducing the complexity while maintaining the benefits of SMC. Future research could focus on developing hybrid control strategies that simplify the design without compromising performance. This paper [22] presents an adaptive neuro-fuzzy model for a multilevel inverter in grid-connected PV systems. The advantage is the improved adaptability and performance of the inverter under varying conditions. However, the complexity of the neuro-fuzzy model may limit its practical implementation. The research gap includes simplifying the model while maintaining its adaptability. Future work could involve developing more straightforward fuzzy logic systems or integrating machine learning techniques to enhance performance. This research [23] focuses on a wide-range charge controller for solar sensors, emphasizing its adaptability to different conditions. The advantage is the broad operational range of the charge controller, which ensures efficient energy harvesting. However, the design may be too specialized, limiting its applicability to other PV systems. The research gap is in making the design more generalizable. Future research could explore integrating this technology into more versatile systems that can be applied to a wider range of PV applications. The paper [24] addresses the efficiency challenges in PV systems by combining a passive lossless Cuk converter with hybrid SMC. Traditional Cuk converters are known for their continuous input/output currents but suffer from switching losses and voltage stress. Soft-switching techniques, particularly passive methods, are explored to minimize these issues without complex control circuits. SMC is employed for its robustness in MPPT, ensuring efficient energy conversion under varying conditions. The paper contributes by validating the improved performance of this combined approach through both simulations and experiments, advancing the reliability and efficiency of PV systems.

The paper [25] discusses a DC/DC converter designed for ultra-low input voltages in energy harvesting applications. The advantage is the ability to operate efficiently with minimal input, making it suitable for low-energy environments. However, the converter may struggle with higher power applications due to its specialized design. The research gap includes extending the operating range of such converters to handle higher power levels. Future research could focus on hybrid designs that combine the benefits of ultra-low voltage operation with the capability to handle more significant power demands. This paper [26] presents a modified P&O MPPT method that reduces steady-state oscillations while improving tracking efficiency. The advantage is the enhanced stability and efficiency of the MPPT process. However, the method may still face challenges in rapidly changing conditions. The research gap is in further improving the adaptability of the P&O method to dynamic environments. Future research could involve integrating this method with other MPPT techniques to create a more versatile solution. The paper [27] introduces an advanced variable step size incremental conductance MPPT algorithm for PV systems, offering improved tracking speed and efficiency. The advantage is the adaptability to varying environmental conditions. However, the complexity of the algorithm may increase the computational burden. The research gap includes simplifying the algorithm for real-time applications without sacrificing performance. Future research could explore hybrid algorithms that balance complexity and efficiency. This research [28] applies genetic algorithms and robust control to MPPT for PV systems, providing the advantage of global optimization for tracking the MPP. However, the computational requirements of genetic algorithms may limit their real-time applicability. The research gap lies in reducing the computational load while maintaining the effectiveness of the algorithm. Future work could involve developing more efficient genetic algorithms or combining them with faster MPPT methods. The paper [29] focuses on analyzing the chattering phenomenon in SMC, offering insights into its causes and potential mitigation strategies. The advantage is a deeper understanding of chattering, which is crucial for improving SMC applications. However, the paper may not provide practical solutions to eliminate chattering. The research gap includes developing more practical and implementable solutions to reduce or eliminate chattering. Future research could focus on hybrid control methods that retain the robustness of SMC while minimizing chattering. This paper [30] presents an SMC for a PV system based on a Cuk converter. The advantage is the enhanced robustness and efficiency of the PV system. However, the SMC may introduce complexity in the control design and implementation. The research gap is in simplifying the SMC design while maintaining its benefits. Future research could explore integrating SMC with other control strategies to simplify the design and improve performance.

This paper [31] combines SMC with a P&O MPPT method for controlling a boost converter in PV systems. The advantage is the robust and efficient control of the boost converter, ensuring optimal power output. However, the combined method may introduce additional complexity in tuning and implementation. The research gap includes simplifying the integration of SMC and P&O for practical applications. Future research could focus on developing a more streamlined control algorithm that retains the benefits of both methods. The paper [32] presents an improved MPPT algorithm specifically designed for use with a Cuk converter in PV applications. The advantage is the enhanced tracking efficiency and stability, particularly under varying environmental conditions. However, the reliance on the Cuk converter may limit the algorithm's applicability to other converter topologies. The research gap lies in making the algorithm more versatile across different converter designs. Future research could explore adapting the algorithm for other types of converters or hybrid systems. This paper [33] discusses the application of SMC for PV systems using a Cuk converter, highlighting the advantages of robust control and improved efficiency. However, the complexity of the SMC design and the specific focus on Cuk converters may limit broader applicability. The research gap is in generalizing the SMC approach to other types of converters and simplifying the control design. Future research could focus on developing a more universal SMC framework that can be applied to various PV system configurations. The paper [34] explores enhancing wind energy conversion systems using a quasi-Z source inverter combined with a modified fuzzy logic-based hill-climbing search algorithm. Quasi-Z source inverter offers advantages like improved voltage boost and reliability in wind energy systems. Traditional MPPT methods, like hill-climbing search, often face challenges such as slow response to changing wind conditions. Integrating fuzzy logic with hill-climbing search aims to improve tracking efficiency, ensuring maximum power extraction from wind energy. The study demonstrates that this hybrid approach enhances the performance of wind energy systems, contributing to more efficient renewable energy technologies. This paper [35] addresses the optimization of wind energy efficiency within Internet of Things (IoT)-driven smart power systems by employing a modified fuzzy logic control approach. In the context of smart grids, where IoT technologies enable real-time monitoring and control, optimizing energy conversion and management is crucial. Traditional control methods may struggle with the dynamic and complex nature of smart grids. Fuzzy logic control, known for handling uncertainties and non-linear systems, offers an effective solution. By modifying the fuzzy logic approach, the study aims to enhance the responsiveness and efficiency of wind energy systems within IoT frameworks. The proposed method improves the overall energy management in smart grids, contributing to sustainable power system development. The paper [36] focuses on the development of a novel control scheme tailored for aquaponics systems within bioenvironments. Aquaponics, a sustainable method combining aquaculture and hydroponics requires precise control to maintain optimal conditions for both fish and plants. Traditional control systems often struggle with the complexity of managing the interdependent variables in such systems. The integration of advanced control schemes, particularly those utilizing AI, offers improved adaptability and efficiency. This study presents a novel control approach designed to enhance the stability and productivity of aquaponics systems, thereby contributing to the advancement of sustainable agricultural practices.

6.2 MATERIALS AND METHODS

To address the limitations inherent in SEPIC converters when utilized for MPPT in PV systems, a novel approach is proposed. SEPIC converters are known for their significant output ripple current, which negatively impacts power tracking precision and overall system efficiency. To mitigate these issues, a passive lossless soft-switched non-inverting Cuk converter is designed. This modified converter is specifically tailored to reduce output ripple current, minimize switch voltage stress, and enhance the power conversion efficiency of the PV system. The design incorporates a soft-switched Cuk converter equipped with a single capacitor turn-off snubber, aimed at optimizing component count and reducing overall design costs. This configuration ensures minimal voltage stress across the switch, thereby significantly reducing power losses. Additionally, the P&O MPPT algorithm is employed to ensure rapid and maximum power extraction from the PV module, even under fluctuating solar irradiance and temperature conditions. This implementation not only optimizes power output but also extends the operational lifespan of the PV system. The block diagram illustrating the proposed converter implementation is presented in Figure 6.1.

6.2.1 Revised Inverting Buck-Boost Converter Topology

Figure 6.2 provides the schematic of the conventional non-isolated Cuk converter, which serves as the foundation for the modified design. The converter includes two inductors (L1 and L2), two capacitors (C1 and C2), a MOSFET (active switch) labeled as *S*, and a diode *D*. This topology is selected for its ability to provide continuous input

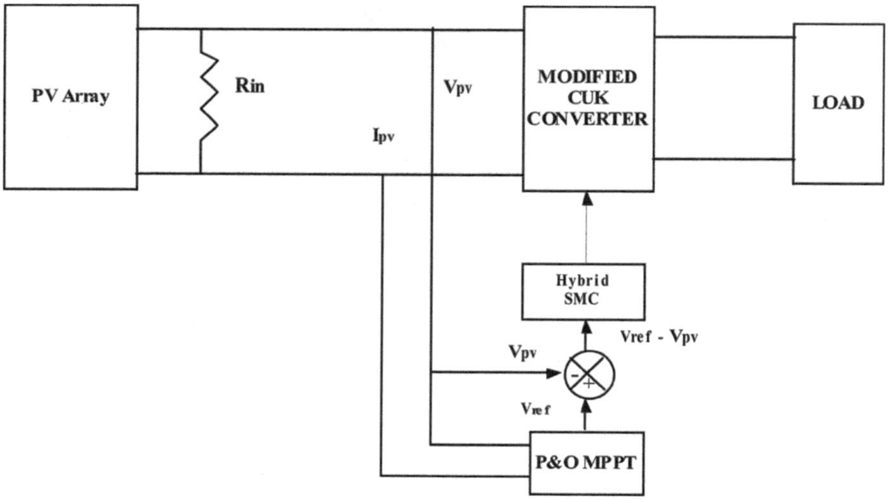

FIGURE 6.1 Passive lossless IBB converter.

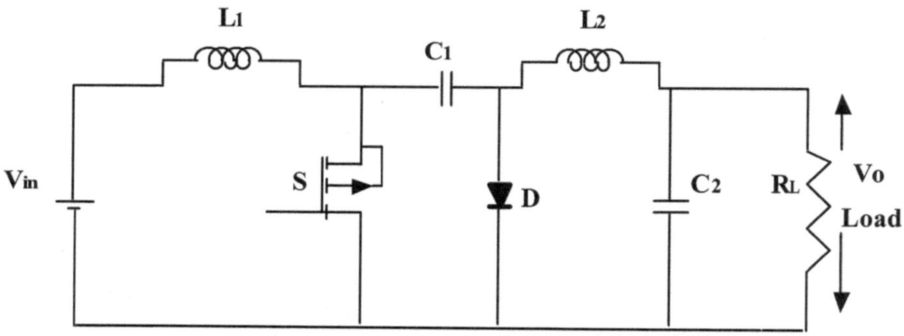

FIGURE 6.2 Circuit topology of the Cuk converter.

and output currents, which is critical for minimizing ripple and improving efficiency in PV applications. The modifications introduced in this design further enhance these characteristics, making the converter more suitable for high-performance PV systems.

The IBB converter offers significant advantages over traditional converters due to the presence of input and output inductors, which result in a filtered current on both sides. This filtering capability minimizes ripple current, thereby enhancing the overall efficiency of the system. However, when applied to PV arrays, the non-linear characteristics of the PV source present a challenge. Conventional hard-switched Cuk converters struggle to achieve maximum efficiency and effective power tracking due to the voltage stress and switching losses encountered by the switches during operation.

In a typical switching cycle, the switches endure stress as a result of the overlap between current and voltage, leading to energy losses. Passive soft-switching

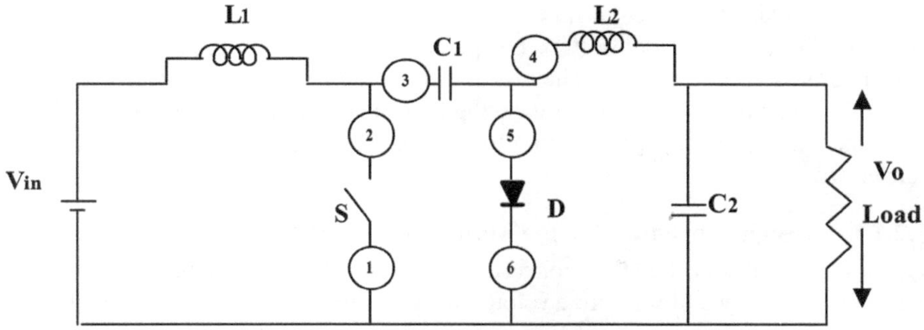

FIGURE 6.3 C_r and L_r positioning in the Cuk converter circuit.

techniques offer a superior alternative to active switching because they do not require additional switches or complex control circuits. This method enhances efficiency by minimizing switching losses and stress on the components. This approach provides a streamlined solution, and the same principles can be extended to circuits with multiple active switches. In a circuit with a single active switch, the inclusion of a zero-current loop (ZCL) and a Zero Voltage Crossing (ZVC) is sufficient to ensure "0"-current turn-on and "0" voltage turn-off of the switch. The optimal placements for the turn-on inductor (Lr) and the turn-off capacitor (Cr) are illustrated in Figure 6.3.

To further enhance the system, additional lossless passive components can be integrated to recover energy from the ZCL and ZVC. This modular approach allows for the incorporation of any number of passive components, providing flexibility in design and scalability for various applications. The interconnected components work synergistically to reduce switching losses and improve the overall efficiency of the converter. Figure 6.3 illustrates the potential locations for a set of ZCLs denoted as Lr. The primary function of Lr is to facilitate the "0" current turn-on of the active MOSFET switch (S). The connections between terminals labeled (1, 2, 3, 4, 5, 6) represent the potential positions where the ZVC can be placed. These specific points are critical in determining the optimal placement for the ZVC within the circuit, enabling efficient energy recovery and voltage stabilization. These passive elements are incorporated into a traditionally hard-switched converter to enable soft-switching, where ZCL and ZVC act as the turn-on and turn-off devices, respectively.

In addition to these passive components, a set of snubber diodes (Ds) is included to form a snubber cell, which plays a crucial role in limiting both switch current and switch voltage. The passive turn-on and turn-off snubber cells are designed to slow down the *di/dt* (rate of change of current) and *dv/dt* (rate of change of voltage) transitions of the switch (S). This controlled deceleration effectively recovers the energy stored in the ZCL and ZVC without loss, thereby maintaining manageable voltage stress across the switches and diodes. Consequently, this design reduces switching losses significantly. These functions are performed during the switching interval, which is influenced by both the switching frequency of the converter and the dimensions of the soft-switching components. Among the six potential locations shown in Figure 6.3, the optimal position for Lr is determined to be at link 4. The number of Lr inductors corresponds to the number of active switches (S) in the circuit. The ZVC is optimally placed between terminals 2 and 3. The ZVC subcircuit includes diodes Ds1, Ds2, Ds3, and a capacitor (Cs). The passive elements are meticulously designed to form a passive snubber cell. The layout of the modified IBB, featuring this snubber cell, is shown in Figure 6.4. This configuration ensures that voltage stress across the switch is minimized, contributing to the overall efficiency and reliability of the converter.

6.2.1.1 Design Procedure for the Modified Converter Topology

The design of the modified Cuk converter topology follows a systematic approach to optimize efficiency and minimize voltage stress (Figure 6.4). The steps are outlined as follows:

Location of ZCL:
The first step involves determining the optimal placement of the ZCL, denoted as Lr. Inserting resonant inductance Lr into the loop results in a reduction of switch voltage. There are six potential locations for Lr to achieve zero-current turn-on of the MOSFET switch. After thorough analysis, the diode branch at location number 5, as shown in Figure 6.3, has been identified as the optimal point with the least voltage stress.

Inductor Charging and Voltage Spike Mitigation:
When the switch (S) opens, the inductor Lr in series with the diode begins to charge. To ensure the circuit operates without energy loss, an alternative pathway is necessary to minimize voltage spikes and effectively handle the charging and discharging of the ZCL energy. Table 6.1 outlines six possible locations for capacitors, and the optimal placement for Cr is indicated in Figure 6.4.

Formation of the Snubber Cell:
The snubber cell is framed by Cr, Cs, and Ds, as depicted in Figure 6.4. Utilizing graph theory, the ZVC subcircuit is derived from a hard-switched topology. The ZVC must reset when the active switch turns on. The final design of the passive lossless Cuk converter topology, including Lr and Cr as the ZCL and ZVC, respectively, is illustrated in Figure 6.4. In this configuration, Ds_3 and Cr constitute the ZVC

TABLE 6.1
Potential Capacitor Placement with the Connection Diagram

Link Number	Connection
L1–2	Linking terminal 1 with terminal 2
L2–3	Linking terminal 2 with terminal 3
L3-3	Linking terminal 3 with terminal 3
L4–6	Linking terminal 4 with terminal 6
L5–6	Linking terminal 5 with terminal 6
L6-6	Linking terminal 6 with terminal 6

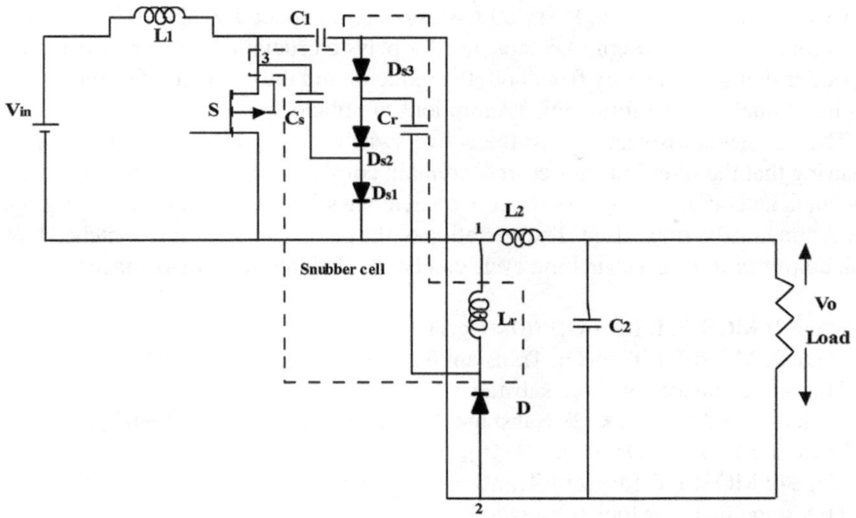

FIGURE 6.4 Passive lossless circuit topology for the non-inverting Cuk converter.

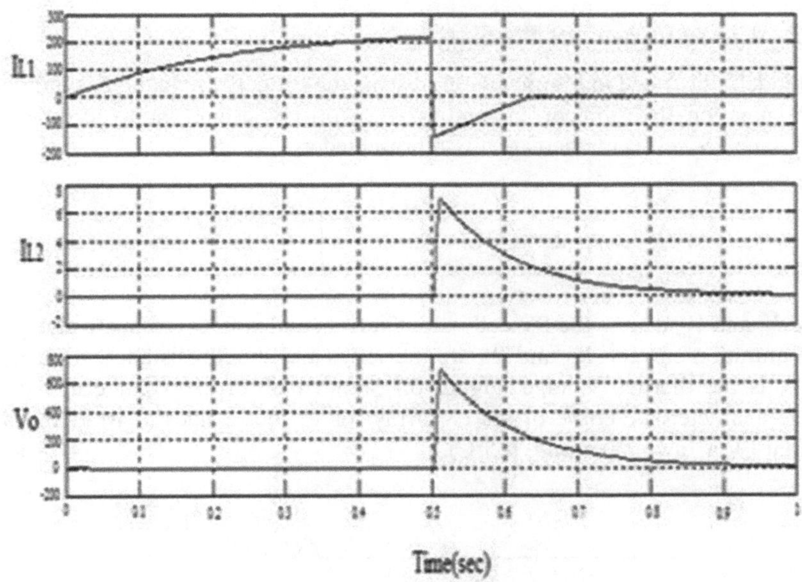

FIGURE 6.5 Inductor current dynamics in the modified Cuk converter.

subcircuit. Lr is responsible for transferring energy stored in Cr to Cs, which in turn recovers the energy from Lr and Cr. This design achieves efficient energy recovery and minimizes losses, contributing to the overall enhanced performance of the converter as shown in Figure 6.5.

Phases of Operation for the Passive Lossless Soft-Switched PWM IBB:

In this converter design, the capacitor Cs plays a crucial role as a relatively large capacitor that stores energy from both the inductor and the capacitor. For the purpose of circuit analysis, the following assumptions are made:

The transients associated with the active switch turning on and off are minimal, ensuring that the filter inductor currents remain constant when these transitions occur. The input and output voltages of the converter, along with other filter capacitor voltages, are assumed to be ripple-free. The operation of the passive lossless soft-switched PWM Cuk converter during a switching cycle can be divided into four major stages:

Stage 1: MOSFET Turn-Off ($t0 < t < t1$)
Stage 2: MOSFET Turn-On Transient ($t1 < t < t4$)
This stage consists of three substages:
Substage 2.1 ($t1 < t < t2$), Substage 2.2 ($t2 < t < t3$), Substage 2.3 ($t3 \leq t \leq t4$)
Stage 3: MOSFET On-Time ($t4 \leq t \leq t5$)
Stage 4: MOSFET Turn-Off Transient ($t5 \leq t \leq t9$)
This stage includes four substages:
Substage 4.1 ($t5 \leq t \leq t6$), Substage 4.2 ($t6 \leq t \leq t7$), Substage 4.3 ($t7 \leq t \leq t8$), Substage 4.4 ($t8 \leq t \leq t9$):

6.2.2 Configuration of PV System

The model, illustrated in Figure 6.6, represents the PV cell with a diode symbolizing the PN junction diffusion current, and two resistors (series and shunt) accounting for losses. The PV cell is mathematically represented as:

$$I_{pv} = I_{ph} - I_o\left(e^{\frac{(V_{pv}+R_{se}I_{pv})}{V_T}} - 1\right) - V_{pv} + \frac{R_{se}I_{pv}}{R_{sh}} \tag{6.1}$$

where I_{pv} and V_{pv} denote the PV cell output current and voltage, respectively. I_o is the dark saturation current; R_{se} and R_{sh} are the series and shunt resistance, respectively; and V_T is the thermal voltage. Figure 6.7 shows the current–voltage and power–voltage characteristic curve of the PV array for various ranges of irradiances at a constant PV array temperature (25°C).

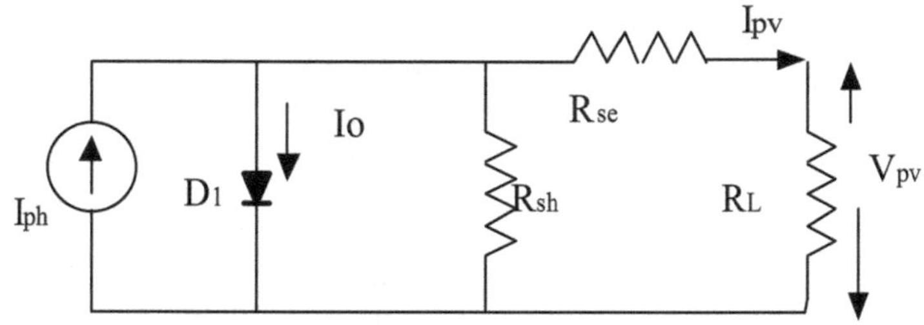

FIGURE 6.6 Formulation of the single-diode PV cell model.

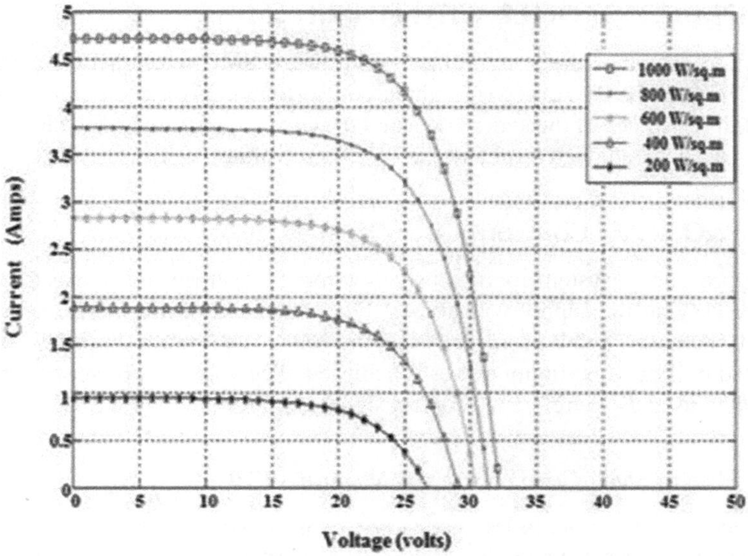

FIGURE 6.7 VI behavior of PV array across multiple irradiation scenarios.

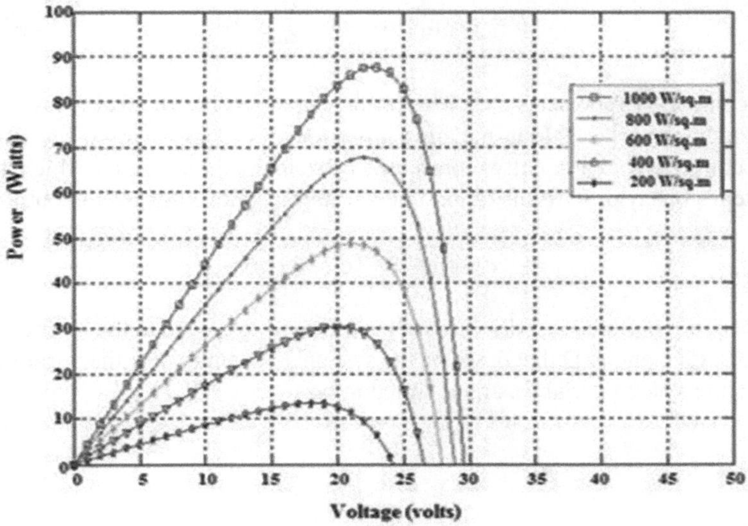

FIGURE 6.8 *P–V* characteristics of PV array across multiple irradiation scenarios.

A solar array is designed to achieve a maximum power output of 85.2 W, with a corresponding voltage at the MPP of 28.4 V and a current at the MPP of 9.1 A. Under a predefined irradiance condition of 1,000 W/m², the PV cell consistently reaches its MPP. The characteristics of the PV cell, demonstrating this MPP performance, are illustrated in Figures 6.7 and 6.8.

6.3 MPPT CONTROLLER WITH HYBRID SMC

This section discusses the design of a hybrid digital SMC-based MPPT controller. The modeling of the PV source is grounded in Norton's equivalent circuit, where the differential resistance of the source (R_{in}) and the current generated by the PV (I_{pv}) are functions of the irradiance level and the PV output voltage.

6.3.1 P&O MPPT Controller

In this approach, the system continuously monitors the voltage and adjusts the output power by varying the supply voltage. As the voltage and power increase, the duty cycle also increases, and it decreases when the voltage and power drop. This process is repeated until the maximum power is achieved. The voltage corresponding to the MPP is then used as the reference control signal (V_{ref}) for the hybrid SMC.

6.3.2 Hybrid SMC Control of the Modified IBB

The output of the converter is assumed to be a constant voltage source, similar to that used in battery chargers. The SMC for the IBB converter is modeled using state-space representation. The sliding surface for higher-order converters can be defined as:

$$S = f(x_1, x_2 \ldots x_N) = \sum_{i=1}^{N} k_i x_i \tag{6.2}$$

In this context, N represents the order of the converter. For the Cuk converter used in this study, the order is four, meaning it is a fourth-order converter. The dynamics of such converters are typically represented by state-space equations, which describe the system's behavior in terms of its state variables, control inputs, and outputs.

$$\dot{x} = Ax + \dot{B} + (Cx + D)u \tag{6.3}$$

The procedure outlined in prior research is followed to calculate the matrices.

A A, B, C C, and D D that describe the system's dynamics. For the purpose of this analysis, the system variables are assumed to be:

$$\begin{aligned} x_1 &= i_{L1} \\ x_2 &= v_{pv} - v_{ref} \\ x_3 &= v_{C2} \\ x_4 &= i_{L2} \end{aligned} \tag{6.4}$$

The differential equations for the IBB converter when $u = 1$ are

$$\dot{x}_1 = \frac{x_1}{L_1} + \frac{1}{R_{diff} L_1} x_2 + \frac{v_{ref}}{L_1} - i_{pv}$$

$$\dot{x}_2 = -\frac{x_1}{L_1} - v_{ref}$$
$$\dot{x}_3 = \frac{x_4}{C_2} \tag{6.5}$$
$$\dot{x}_4 = \frac{x_3}{L_2}$$

The differential equations for the IBB converter when $u=0$ are

$$\dot{x}_1 = \frac{x_1}{L_1} + \frac{1}{R_{in}L_1}x_2 - \frac{x_4}{C_2} - i_{pv} + \frac{v_{ref}}{L_1} - \frac{v_o}{C_2}$$
$$\dot{x}_2 = -\frac{x_1}{L_1} - v_{ref}$$
$$\dot{x}_3 = \frac{v_o}{C_2} \tag{6.6}$$
$$\dot{x}_4 = -\frac{x_1}{L_1} - \frac{1}{R_{in}L_1}x_2 - \frac{v_{ref}}{C_2R_{in}} + \frac{i_{pv}}{C_2}$$

The sliding surface S for the proposed system is represented by

$$S = i_{L1} - i_{ref} = i_{L1} - k_p \cdot (v_{ref} - v_{pv}) + k_i \cdot \int (v_{ref} - v_{pv}(t))dt \tag{6.7}$$

The operating condition for the MOSFET S is defined as follows: when u equals 1, S is triggered into the ON state, and when u equals 0, S shifts to the OFF state. This defines the dynamic behavior of the system's variables associated with the converter.

$$x = \left[i_{L1}, v_{pv} - v_{ref}, i_{L2}, v_{C2}\right]^T \tag{6.8}$$

when $u=1$, S goes to the ON state and $u=0$, S goes to the OFF state. The system variables for the converter are:

The matrices A, B, C, and D are calculated from the differential equation (6.5).

$$A = \begin{bmatrix} \frac{1}{L_1} & \frac{1}{R_{in}L_1} & 0 & 0 \\ \frac{1}{L_1} & 0 & 0 & 0 \\ 0 & 0 & 0 & \frac{1}{C_2} \\ 0 & 0 & -\frac{1}{L_2} & 0 \end{bmatrix} \quad B = \begin{bmatrix} \frac{v_{ref}}{L_1} \\ 0 \\ 0 \\ 0 \end{bmatrix} \tag{6.9}$$

$$C = \begin{bmatrix} 0 & 0 & 0 & -\frac{1}{C_2} \\ 0 & 0 & 0 & 0 \\ 0 & 0 & 0 & \frac{1}{C_2} \\ \frac{1}{L_1} & -\frac{1}{L_1 R_{in}} & 0 & 0 \end{bmatrix} \quad D = \begin{bmatrix} -\frac{v_0}{C_2} \\ 0 \\ \frac{v_0}{C_2} \\ -\frac{v_{ref}}{C_2 R_{in}} + \frac{i_{pv}}{C_2} \end{bmatrix}$$

From the matrix representation presented, it can be deduced that the derivative of I_{L1} influences the control signal u that is directed to the MOSFET switch S. For a digital representation of the signal, the hybrid Sliding Mode (SM), as detailed in Equation (6.7), can be depicted as follows:

$$\begin{aligned} i_{L1} &> k_p \cdot (v_{ref} - v_{pv}) + k_i \cdot \int (v_{ref} - v_{pv}(t))dt + \frac{H}{2}, \quad u = 0 \\ i_{L1} &< k_p \cdot (v_{ref} - v_{pv}) + k_i \cdot \int (v_{ref} - v_{pv}(t))dt - \frac{H}{2}, \quad u = 1 \end{aligned} \tag{6.10}$$

In this configuration, V_{ref}, which is derived from the MPPT controller, is expressed in a digital format, and Tc denotes the sampling interval. This approach simplifies the digital implementation of Equation (6.7). Utilizing Euler's backward transformation, Equation (6.10) is adapted into a digital format, now designated as Equation (6.11).

$$\begin{aligned} e(k) &= v_{(k)ref} - v_{(k)pv} \\ I_{(k)ref} &= I_{(k-1)ref} + (k_p + k_i \cdot T_c) \cdot e_{(k)} + k_p \cdot e_{(k-1)} \end{aligned} \tag{6.11}$$

The transformations outlined in Equation (6.11) facilitate the conversion of the analog form of SM, as expressed in Equation (6.7), into a hybrid representation, designated as Equation (6.12).

$$\begin{aligned} i_{L1} &> I_{(k)ref} + \frac{H}{2} \rightarrow u = 0 \\ i_{L1} &< I_{(k)ref} + \frac{H}{2} \rightarrow u = 1 \end{aligned} \tag{6.12}$$

The hybrid sliding mode system is implemented using the PIC 16F877A microcontroller, which features two high-speed analog comparators. This microcontroller is selected due to its sufficient memory capacity, robust computational power, and superior analog-to-digital converter functionality.

6.4 EVALUATION OF HYBRID SMC IN PV SYSTEMS

This simulation, detailed as per the specifications in Table 6.2, revealed that the PV voltage response is predominantly determined by the SM controller gains rather than by the differential input resistance of the PV array or the characteristics of

TABLE 6.2
IBB Specifications

Parameter	Nominal Value
L_1	15 μH
C_1	22 μF
L_2	15 μH
C_2	44 μF
L_r	5 μH
C_r	100 nF
C_s	3.3 nF

the DC-DC converter. The simulations were aimed at exploring the system's performance under varied reference voltages and solar irradiance levels.

6.4.1 Simulation Insights

The hybrid SMC-based PV system incorporated a Cuk converter and was simulated according to the parameters in Table 6.2. The system's performance was evaluated under four different configurations:

a. Standard Cuk Converter with PWM Control
b. Standard Cuk Converter with Hybrid SM Control
c. Enhanced Cuk Converter with PWM Control
d. Enhanced Cuk Converter with Hybrid SM Control

These simulations were carried out at various levels of solar irradiation to determine the system's effectiveness and stability under different environmental conditions. The results for the standard IBB converter with hybrid SM control, shown in Figure 6.9, demonstrate that the system reached a maximum power output of 150 W within 100 ms, stabilizing at 110 W by 350 ms. As depicted in Figure 6.10, the voltage stress on the switch reduced from 200 V at 50 ms to 175 V at 200 ms, indicating a notable decrease. This reduction in voltage stress, however, led to increased losses, especially when the irradiance level dropped from 1,000 to 500 W/m², causing the power output to fall from 150 to 120 W. The efficiency of a PV system using a standard Cuk converter tends to be relatively low. Figure 6.11 depicts the output voltage and power generated by the proposed Cuk converter with hybrid SMC control. The output voltage profile ranges from 150 to 170 V, with the system achieving rapid convergence at 160 ms. This represents a significant improvement of 50 V compared to the standard Cuk converter. The enhanced performance of the proposed system is evident in its faster settling time and quicker convergence, achieving stability in just 160 ms.

The output power profile of the modified Cuk converter varies, reaching a peak of 165 W at 100 ms and decreasing to 155 W at 200 ms. Figure 6.12 illustrates the voltage stress experienced by switch *S*. In this modified design, the voltage stress is

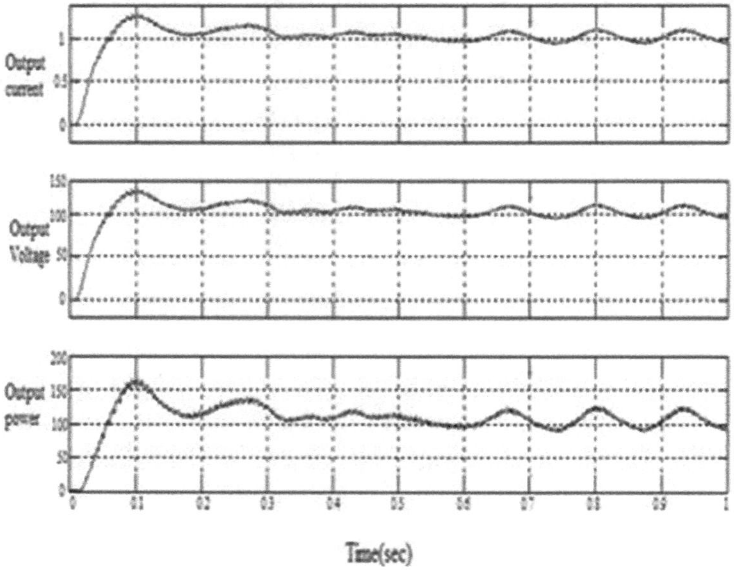

FIGURE 6.9 Output of the standard IBB converter with PWM control.

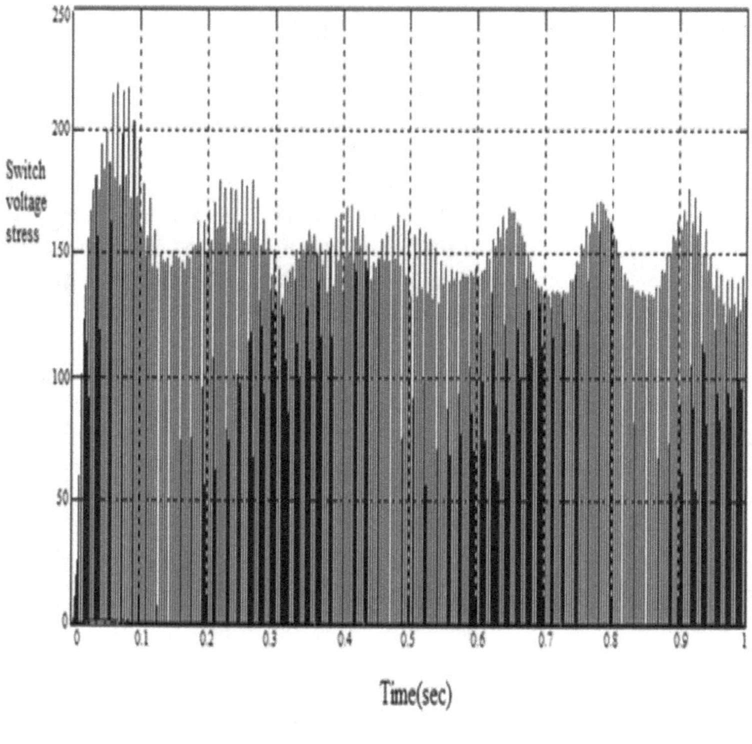

FIGURE 6.10 Switch voltage stress characterization in standard IBB converters.

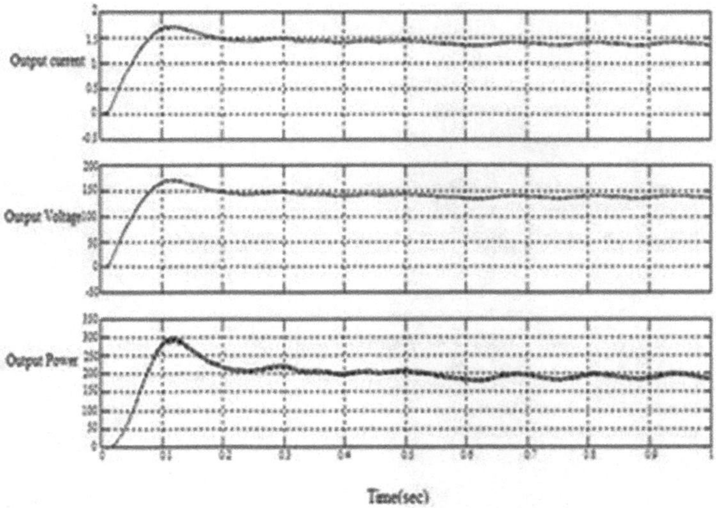

FIGURE 6.11 Output of the enhanced IBB converter.

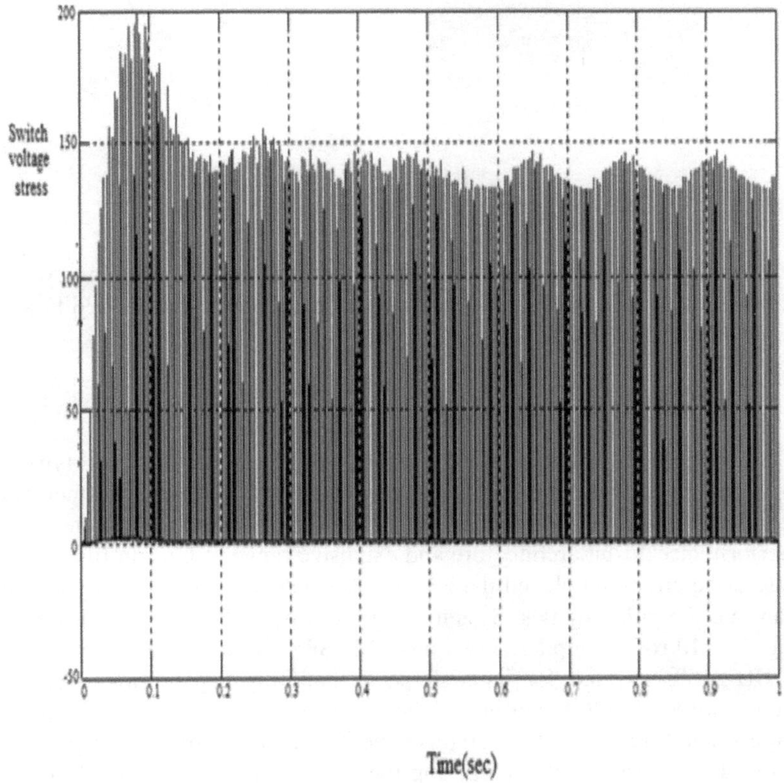

FIGURE 6.12 Switch voltage stress profile in the modified IBB converter.

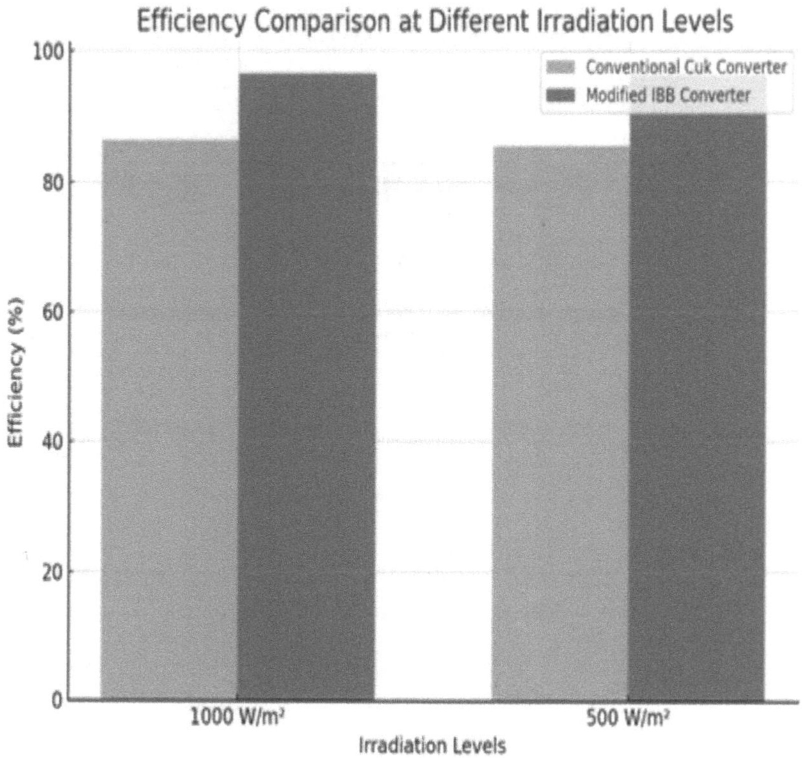

FIGURE 6.13 Efficiency comparison at different irradiation levels.

reduced to 150 V and remains constant throughout the entire switching cycle. This substantial reduction in stress leads to improved peak power, minimized losses, and an overall efficiency increase to 96.5% (Figure 6.13).

6.5 EXPERIMENTAL VALIDATION

The experimental validation was carried out using a laboratory prototype of the converter integrated with a hybrid SM controller. The hardware implementation of the hybrid SM controller utilized the PIC 16F877A microcontroller, known for its high-performance 16-bit architecture and extensive range of built-in functionalities. The gate drive circuit employed the HCPL 2232 optocoupler to effectively amplify and translate the pulse signals. System programming was conducted in Embedded C using the MikroC Compiler. The Power MOSFET used in the proposed system is the IRFP460, selected for its high power-handling capabilities. The snubber cell diodes are MUR3060, chosen for their fast recovery time and high efficiency. The experimental results, which confirm the simulation findings, are presented in Figures 6.14–6.16. In a study comparing the performance of a hybrid SMC MPPT

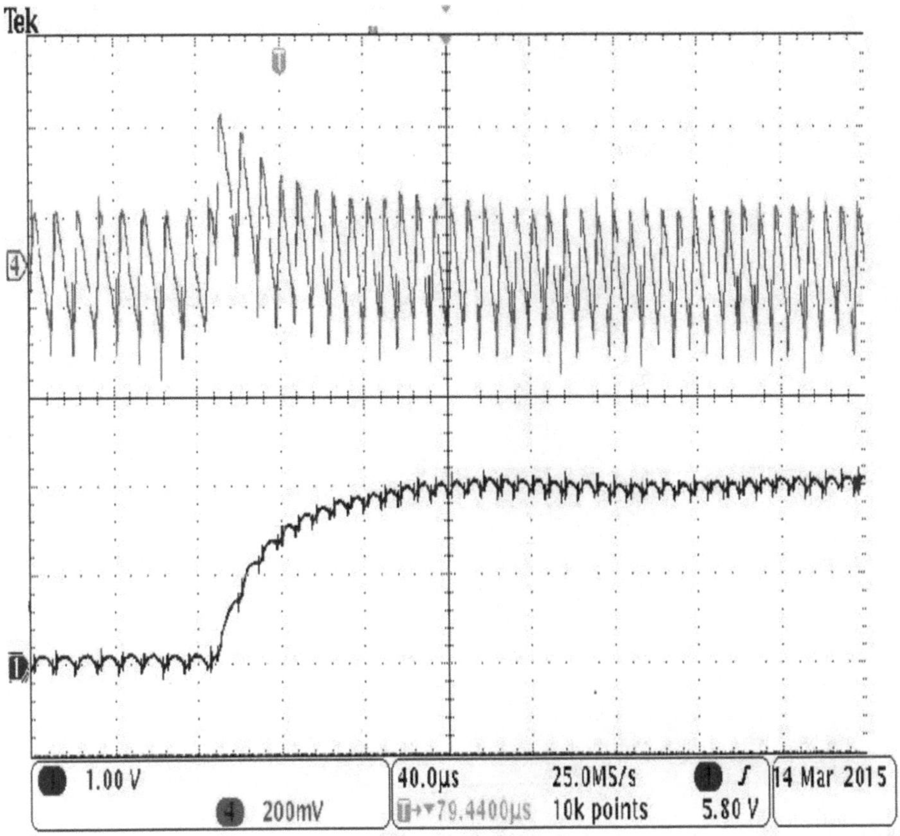

FIGURE 6.14 Input voltage step response analysis.

with an optimal switching surface against a standard PWM-based MPPT, the response to a sudden step change in input irradiation was analyzed. This change was simulated by introducing a step change in current, with the corresponding step response illustrated in Figure 6.14. For the MPPT algorithm, the sliding mode MPPT utilized a step size of 0.2, while the PWM-based MPPT used a smaller step size of 0.02. Both controllers operated with a step time of 30 ms per step. The results revealed a significant difference in convergence times between the two approaches. The sliding mode MPPT quickly converged within 10 ms, while the PWM-based MPPT took considerably longer, stabilizing only after 180 ms. The initial state was set with a high reference voltage (V_{ref}) of 17.4 V, higher than the voltage at the MPP of approximately 11.7 V. The initial low current of 0.85 A eventually increased and stabilized at the MPP current of 2.73 A. During the transient phase, the voltage remained nearly constant. Figure 6.15 illustrates the steady-state response of the PV voltage and current. At lower voltages, the PV panel operates as a current source, ultimately reaching the MPP at 28.4 V, as shown in Figure 6.16.

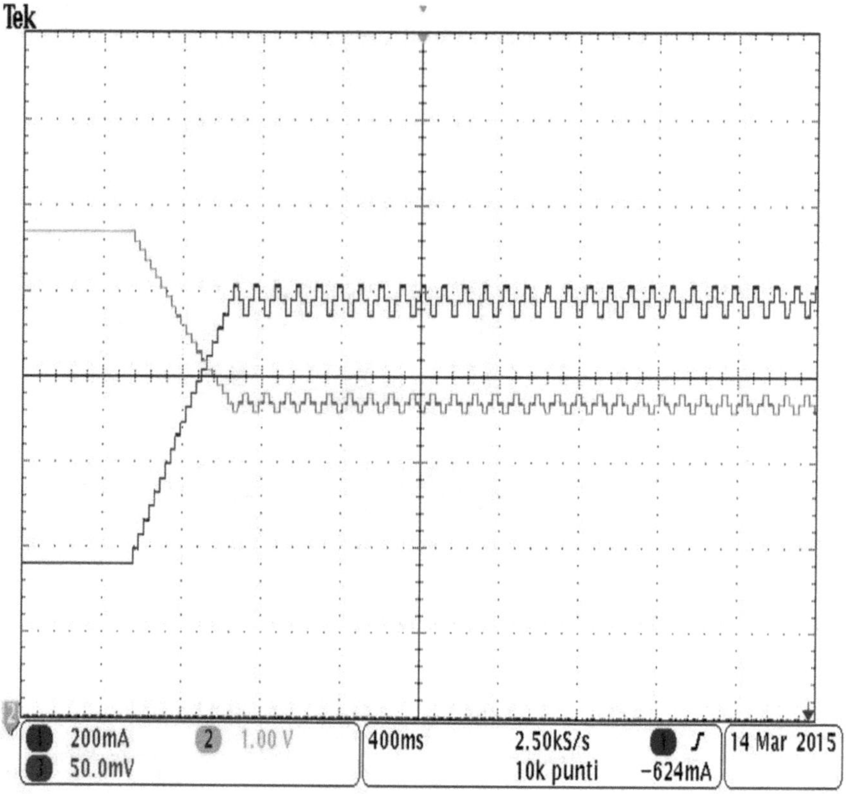

FIGURE 6.15 Steady-state performance of PV voltage and current.

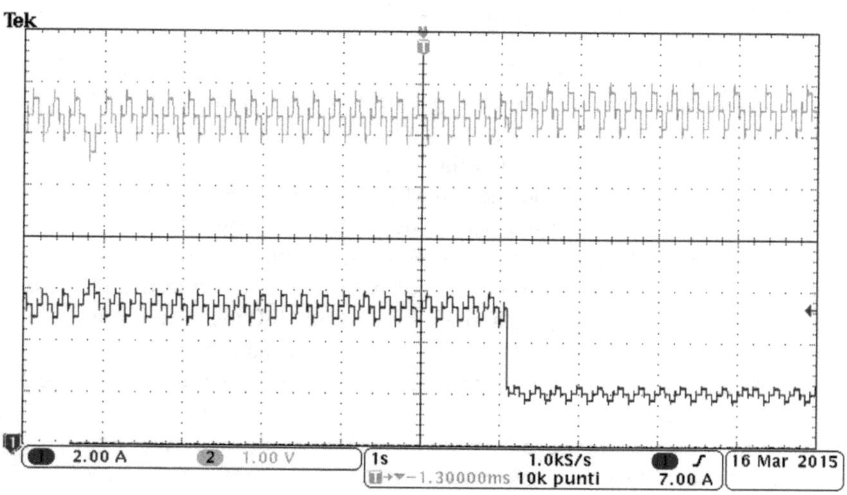

FIGURE 6.16 PV voltage and current for various irradiation changes.

6.6 SUMMARY AND IMPLICATIONS

This paper outlined the design process for a passive lossless non-inverting Cuk converter, developed for integration with a hybrid SMC MPPT controller. The proposed converter, which incorporates a snubber cell, successfully reduces voltage stress on the active switch during both turn-on and turn-off events, leading to a significant improvement in the conversion efficiency of the PV system. Simulation results clearly indicate that the modified non-inverting Cuk converter, when paired with the hybrid SM controller, achieves minimal ripple, lower switch voltage stress, and reduced conduction losses, all of which contribute to enhanced efficiency. The primary benefit of the modified non-inverting Cuk converter is that the inclusion of a snubber cell with passive components mitigates component stress and reduces the size and weight of the inductors. To validate the proposed converter's effectiveness for PV system applications, a laboratory prototype was constructed. The simulation and experimental results confirm the efficiency and reliability of the proposed Cuk converter in solar energy systems.

REFERENCES

1. J.H. Enslin, M.S. Wolf, D.B. Snyman, W. Swiegers, Integrated photovoltaic maximum power point tracking converter, *IEEE Transactions on Industrial Electronics*, 1997, 44(6), 769–773. DOI: 10.1109/41.644004
2. V.I. Utkin, Variable structure systems—present and future, *Automation and Remote Control*, 1984, 44(9), 1105–1120. DOI: 10.1134/S106422691011019X
3. H. Sira-Ramirez, Sliding motions in bilinear switched networks, *IEEE Transactions on Circuits and Systems*, 1987, 34(9), 919–933. DOI: 10.1109/31.24267
4. K.M. Smith, K.Y. Ma Smedley, Properties and synthesis of passive lossless soft-switching PWM converters, *IEEE Transactions on Power Electronics*, 1999, 14(5), 890–899. DOI: 10.1109/63.792686
5. T.-F. Wu, Y.-D. Chang, C.-H. Chang, J.-G. Yang, Soft-switching boost converter with a flyback snubber for high power applications, *IEEE Transactions on Power Electronics*, 2012, 27(3), 1108–1119. DOI: 10.1109/TPEL.2011.2179381
6. C. Munoz, Study of a new passive lossless turn-off snubber, *Proceedings of IEEE International Conference on Industrial Electronics, Control, and Instrumentation (CIEP)*, Morelia, Mexico, 1998, pp. 147–152. DOI: 10.1109/CIEP.1998.740317
7. T.F. Wu, S.A. Liang, A systematic approach to developing single-stage soft-switching PWM converters, *IEEE Transactions on Power Electronics*, 2001, 6(5), 581–593. DOI: 10.1109/63.1054158
8. G.J. Kish, J.J. Lee, P.W. Lehn, Modeling and control of photovoltaic panels utilizing the incremental conductance method for maximum power point tracking, IET Renewable Power Generation, 2012, 6(4), 259–266. DOI: 10.1049/iet-rpg.2011.0188
9. L.R. Chen, C.M. Young, N.Y. Chu, C.S. Liu, Phase locked bidirectional converter with pulse charge function for 42-V/14-V dual-voltage PowerNet, *IEEE Transactions on Industrial Electronics*, 2011, 58(4), 2045–2048. DOI: 10.1109/TIE.2010.2069075
10. M.A. Farahat, H.M.B. Metwally, A.A.E. Mohamed, Optimal choice and design of different topologies of DC-DC converter used in PV systems at different climatic conditions in Egypt, *Renewable Energy*, 2012, 43, 393–402. DOI: 10.1016/j.renene.2011.11.015
11. C. Edwards, S. Spurgeon, *Sliding Mode Control: Theory and Applications*, CRC Press, 1998., London, UK.

12. E. Mamarelis, G. Petrone, G. Spagnuolo, An hybrid digital-analog sliding mode controller for photovoltaic applications, *IEEE Transactions on Industrial Informatics*, 2013, 9(3), 1094–1103. DOI: 10.1109/TII.2012.2214774
13. B. Hu, S. Sathiakumar, A new interleaving technique for voltage ripple cancellation of series-connect photovoltaic systems, *Journal of Circuits, Systems, and Computers*, 2014, 23(1). DOI: 10.1142/S0218126614500455
14. J. Yun, H.-J. Choe, Y.-H. Hwang, Y.-K. Park, B. Kang, Improvement of power-conversion efficiency of a DC-DC boost converter using a passive snubber circuit, *IEEE Transactions on Industrial Electronics*, 2012, 59(4), 1808–1814. DOI: 10.1109/TIE.2011.2162181
15. M. Park, I.K. Yu, A novel real-time simulation technique of photovoltaic generation systems using RTDS, *IEEE Transactions on Energy Conversion*, 2004, 19(1), 164–169. DOI: 10.1109/TEC.2003.822306
16. M.H. Taghvaee, M.A.M. Radzi, S.M. Moosavain, H. Hizam, M.H. Marhaban, A current and future study on non-isolated DC-DC converters for photovoltaic applications, *Renewable Energy*, 2013, 17, 216–227. DOI: 10.1016/j.renene.2012.03.028
17. W. Xiao, W.G. Dunford, A modified adaptive hill climbing MPPT method for photovoltaic power systems, *Proceedings of 35th Annual IEEE Power Electronics Specialists Conference*, Aachen, Germany, 2004, pp. 1957–1963. DOI: 10.1109/PESC.2004.1355339
18. B.J. Huang, F.S. Sun, R.W. Ho, Near maximum power-point operation (nMPPO) design of photovoltaic power generation system, *Solar Energy*, 2006, 80(9), 1003–1020. DOI: 10.1016/j.solener.2005.10.011
19. Y.H. Liu, C.L. Liu, J.W. Huang, J.H. Chen, Neural network-based maximum power point tracking methods for photovoltaic systems operating under fast changing environments, Solar Energy, 2013, 89, 42–53. DOI: 10.1016/j.solener.2012.11.013
20. A. Safari, S. Mekhilef, Simulation and hardware implementation of incremental conductance MPPT with direct control method using cuk converter, *IEEE Transactions on Industrial Electronics*, 2011, 58(4), 1154–1161. DOI: 10.1109/TIE.2010.2048834
21. E. Mamarelis, G. Petrone, G. Spagnuolo, Design of sliding mode controlled SEPIC for PV MPPT applications, *IEEE Transactions on Industrial Electronics*, 2014, 61(9), 3387–3398. DOI: 10.1109/TIE.2013.2274817
22. P. Karuppusamy, A.M. Natarajan, K.N. Vijeyakumar, An adaptive neuro-fuzzy model to multilevel inverter for grid connected photovoltaic system, *Journal of Circuits, Systems, and Computers*, 2015, 24(2). DOI: 10.1142/S0218126615500668
23. C.-S.A. Gong, L.-X. Chang, A wide range charge controller for solar sensor, *Journal of Circuits, Systems, and Computers*, 2015, 24(5). DOI: 10.1142/S021812661550070X
24. K. Kanimozhi, A. Shunmugalatha, Photovoltaic systems with passive lossless cuk converter using hybrid sliding mode control, *Journal of Circuits, Systems and Computers*, 2016, 25(05), 1650036. DOI: 10.1142/S0218126616500365
25. Z. Yang, J. Wang, Y. Li, Y. Yang, A 20-mV input DC/DC converter for energy harvesting applications, *Journal of Circuits, Systems, and Computers*, 2015, 24(1). DOI: 10.1142/S021812661550070X
26. J. Ahmed, Z. Salam, A modified P&O maximum power point tracking method with reduced steady-state oscillation and improved tracking efficiency, *IEEE Transactions on Sustainable Energy*, 2016, 7(4), 1506–1515. DOI: 10.1109/TSTE.2016.2598720
27. A. Loukriz, M. Haddadi, S. Messalti, Simulation and experimental design of a new advanced variable step size incremental conductance MPPT algorithm for PV systems, *ISA Transactions*, 2015, 58(1), 1–9. DOI: 10.1016/j.isatra.2015.05.017
28. M. Dahmane, J. Bosche, A. El-Hajjaji, X. Pierre, MPPT for photovoltaic conversion systems using genetic algorithm and robust control, *Proceedings of the American Control Conference*, Washington, DC, 2013, pp. 6595–6600. DOI: 10.1109/ACC.2013.6580726

29. A. Levant, Chattering analysis, *IEEE Transactions on Automatic Control*, 2010, 55(6), 1380–1389. DOI: 10.1109/TAC.2010.2048556
30. S. Kumar, et al., Sliding-mode controller for a photovoltaic system based on a cuk converter, *IEEE Transactions on Industrial Electronics*, 2019, 66(4), 3221–3230. DOI: 10.1109/TIE.2018.2856129
31. J.M. Rodriguez, et al., Boost converter control of PV system using sliding mode control with perturb and observe MPPT, *IEEE Transactions on Power Electronics*, 2020, 35(4), 3687–3696. DOI: 10.1109/TPEL.2019.2947763
32. M.B. Gomez, et al., Improved maximum power point tracking algorithm with cuk converter for PV applications, *Solar Energy Materials and Solar Cells*, 2020, 210, 110524. DOI: 10.1016/j.solmat.2020.110524
33. A.N. Prasad, et al., Sliding mode control for photovoltaic systems using a cuk converter, *IEEE Journal of Emerging and Selected Topics in Power Electronics*, 2021, 9(1), 234–245. DOI: 10.1109/JESTPE.2020.3010979
34. S. Ramasamy, B. Jayapalan, B. Pappan, G. Singaram, QZSI based high efficient wind energy tracking by modified fuzzy logic based hill climbing search (HCS) algorithm, *AIP Conference Proceedings*, 2023, 2831(1), 040004.
35. B. Jayapalan, R. Sathishkumar, I.A. Prakash, M. Venkateswaran, Optimizing wind energy efficiency in IoT-driven smart power systems using modified fuzzy logic control, *AI Approaches to Smart and Sustainable Power Systems*, L. Ashok Kumar (Ed.), IGI Global, Hershey, PA, 2024, pp. 250–273.
36. K. Kannabiran, J. Booma, S.S. Kumar, Design of novel control scheme for an aquaponics system in bioenvironment, *AI Approaches to Smart and Sustainable Power Systems*, L. Ashok Kumar (Ed.), IGI Global, Hershey, PA, 2024, pp. 314–334.

7 Comparative Study of Hybrid MPPT Methods for Solar PV Systems Under Partial Shading

Manimegalai Damodaran, Karthikeyan Madhu, and Gunasekari Rethinam

7.1 INTRODUCTION

The diminishing availability of nonrenewable energy sources and their associated high levels of atmospheric pollution have prompted many industries to shift their focus toward renewable energy sources [1]. Among these, PV solar energy stands out due to its numerous advantages, including silent operation, low maintenance costs, high reliability, abundant availability, and minimal environmental impact. Operating a solar power plant requires minimal human intervention [2].

Silicon is the preferred material for PV cell design due to its strong photoconductivity, low toxicity, cost-effectiveness, and high energy efficiency [3]. Silicon solar cells are categorized into three types: amorphous silicon, polycrystalline silicon, and monocrystalline silicon [4].

1. **Monocrystalline Silicon**: Known for its high efficiency and uniform appearance, made from a single, pure crystal structure.
2. **Polycrystalline Silicon**: Made from multiple silicon crystals melted together, less efficient but more cost-effective than monocrystalline.
3. **Amorphous Silicon**: Lacks a crystalline structure and is the least efficient, used in low-cost and flexible applications.

Monocrystalline silicon is favored by most PV manufacturers for its superior efficiency and performance, despite its higher initial cost [5]. A single solar cell produces low voltage (about 0.65–0.7 V), necessitating series and parallel connections for higher output voltages and currents. Solar PV arrays use triple-diode PV technology for high precision, increased efficiency, and a good fill factor, although achieving maximum power from the PV array is challenging due to the nonlinear nature of solar cell operation [6].

Standalone solar PV power distribution systems focus on designing MPPT systems, PV cell selection, and choosing appropriate DC–DC converters to increase supply

DOI: 10.1201/9781003496946-7

voltage and system efficiency [7]. Various MPPT techniques, categorized into soft computing, artificial intelligence, conventional, or swarm intelligence, help mitigate high installation costs [8]. The perturb and observe (P&O) approach adjusts the PV system's operating point using the *V–I* slope characteristic, but it has limited efficacy in dynamic conditions [9,10]. The incremental conductance approach continuously monitors the conductance characteristic of the *I–V* curve for higher power extraction capabilities, although it incurs higher implementation costs [11,12]. The hill climbing approach provides minimal oscillations of the PV operating point but struggles with high dynamic fluctuations and abrupt changes in solar light temperature [13].

Incremental resistance methodology [14] and ripple correlation controllers [15] enhance MPP tracking accuracy and minimize power ripples, respectively. The fractional voltage MPPT approach [16,17] is simple and fast but suffers from high switching losses and efficiency issues under varying sunshine conditions [18]. Fractional current controllers [19] improve PV system efficiency by integrating an MPPT block based on a sliding mode controller, though conventional controllers struggle with efficiency in changing conditions [20].

New MPPT controllers based on artificial intelligence, swarm intelligence, chaos theory, soft computing, differential evolution, metaheuristics, and bioinspired algorithms are being researched to overcome the disadvantages of solar PV systems [21]. Artificial intelligence controllers, such as artificial neural network (ANN) controllers [22,23], optimize PV voltage fluctuations but face challenges like complex data structures and slower MPP tracking speeds [24]. Fuzzy logic controllers handle nonlinear problems without requiring mathematical equations but are complex to design and optimize [25,26]. Genetic algorithms enhance fuzzy controllers [27,28] but have drawbacks like high computational cost and premature convergence [29,30].

Hybrid techniques combining genetic algorithms with ANFIS controllers optimize PV systems, despite drawbacks like the curse of dimensionality and reduced interpretability [31–34]. Rectifier-based converters offer high-voltage conversion and good power quality [35,36], while push-pull converters in PV-fed systems reduce voltage fluctuations [37]. Distributed MPPT methodology and forward DC–DC converters address mismatch issues and maintain ripple-free load voltage in PV systems [38,39]. The improved forward multiple input converter is robust, scalable, and reasonably priced but complex and prone to electromagnetic interference [40,41]. Thus, PV voltage may benefit from a nonisolated DC–DC circuit, which is depicted in Figure 7.1.

7.2 RELATED WORKS ON MPPT CONTROLLERS

Nonlinear output performance is inherent in all solar systems. To manage this, various MPPT approaches are used to determine the best duty signal for DC–DC converter circuits. A study [42] analyzed 66 articles on MPPT algorithms under uniform atmospheric conditions, finding conventional algorithms less effective under varying solar irradiation levels. The study investigated online and hybrid methods considering five parameters: PV module dependency, total number of iterations to change inverter duty cycles, controller design cost, error in finding MPP, and overall system efficiency [43]. Swarm intelligence techniques demonstrated quick responses in any weather.

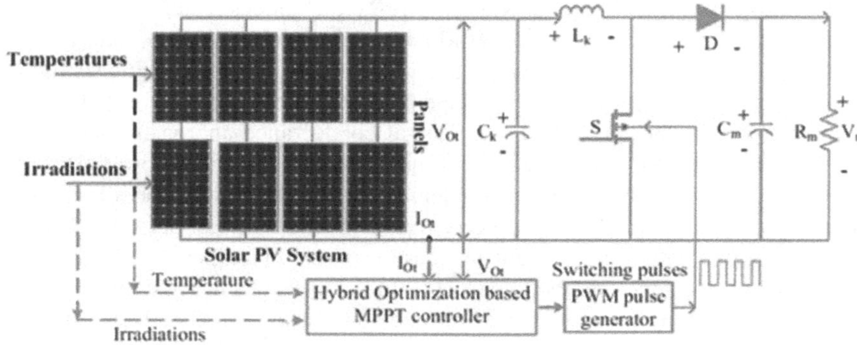

FIGURE 7.1 Hybrid optimization MPPT controller for PV systems with the proposed converter circuit.

Linear current control methods address the nonlinear behavior of PV systems by linearizing the *V–I* curves of interleaved DC–DC circuits, aligning the load line and PV's MPP at one point on the *P–V* slope [44]. This involves adjusting the differentiation of current with respect to power to zero to maximize peak power extraction [45]. In interleaved single-stage DC–DC circuits, the capacitor droop technique manages the intermittent power supply by finding the optimal duty signal using power equilibrium, resulting in stable output power and high-quality voltage [46].

An AC supply integrates with the PV system, where feedback from the DC-link inverter voltage and phase limits ensures steady grid power despite PV output fluctuations [47]. V_{link} acts as a reference voltage for the grid when supply current levels are high. The pulse generator block adjusts voltage and PV power curve slopes to find the MPP during abrupt changes in irradiation [48].

A solar PV system's operational temperature during inclement weather is controlled by adjusting the open-circuit voltage using the mean value theorem [49], offering simplicity, compact size, and quick reaction [50]. The gradient descent mathematical optimization approach locates the MPP of rooftop PV systems, reducing reliance on PV cell production and sensor needs while providing a quick static response [51,52]. Neural controllers integrated into P&O controllers in smart grid power networks improve convergence [53].

For hybrid PV/wind power systems, a neural network optimizes output parameters to find the global operating point, reducing the time to locate the solar MPP using the basic P&O method [54]. The weighted MPP finding method applies weighted points across the PV curves' search space to optimize convergence time and minimize MPP oscillations, benefiting from improved resilience and power conversion rates in cold weather [55].

Combining the P&O method with the flower pollination algorithm creates a hybrid controller for interleaved high-voltage conversion converters, enhancing convergence rates and predicting the global maximum power point (GMPP) during irregular weather [56]. However, the P&O method struggles with constant solar irradiation fluctuations, potentially causing computational difficulties [57]. Quadratic DC–DC converter circuits connected to the radial basis function power point identification controller improve solar PV system performance, meeting consumer load demands for high, stable voltages without distortions [58].

7.3 MATHEMATICAL MODELING OF THE PV SYSTEM UNDER PARTIAL SHADING CONDITION

A review of the available literature indicates that there are numerous PV cell technologies currently available on the market. Among these, monocrystalline PV modules are the most commonly used design technique, offering superior working efficiency compared to other technologies such as polycrystalline and thin film [59]. PV cells are connected in series to enhance system voltage, and in parallel to improve the current profile of the system. One-, dual-, and triple-diode configurations are among the circuit topologies that can be used in the production of PV cells [60].

This article uses the triple-diode approach to obtain accurate PV cell nonlinear curves. As seen in Figure 7.2, the triple-diode PV cell's operational structure is comparable to a diode's normal functioning. Through the usage of photon energy, electrons in the PV array transition from a higher energy band to a lower energy band when sunlight strikes it. The sun's warmth, the cross-sectional area of the cell, and the intensity of the sunlight all affect how much power PV cells can produce. Equation (7.1) can be used to compute the solar current based on the data shown in Figure 7.2. Benefits of the shunt element in the PV circuit include improving the performance of the entire PV network and absorbing leakage current [61].

$$I_0 = I_{\text{ph}} - I_{D_1} - I_{D_2} - I_{D_3} \tag{7.1}$$

$$I_0 = I_{\text{ph}} - I_1\left(e^{\frac{q(V_0 + I_0 R_{se})}{\eta_1 KT}}\right) - I_L \tag{7.2}$$

$$I_L = I_2\left(e^{\frac{q(V_0 + I_0 R_{se})}{\eta_2 KT}} - 1\right) - I_3\left(e^{\frac{q(V_0 + I_0 R_{se})}{\eta_3 KT}} - 1\right) \tag{7.3}$$

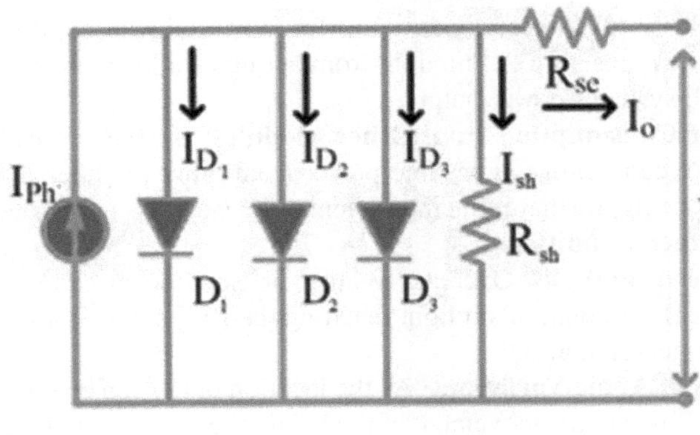

FIGURE 7.2 Solar cell with parallel resistances.

where V_0 and I_0 are the PV system supply voltage and current, respectively. In the context of the triple-diode PV cell methodology, the variables I_{sh} and I_{ph} represent the shunt current and photon current of the PV cell, respectively. The currents flowing through the three diodes in the circuit are denoted as I_{D_1}, I_{D_2}, and I_{D_3}. Equation (7.2) states that the photocurrent of the PV array steadily drops as the charge recombination current rises. Additionally, as the operating temperature (T) of the cell rises from one state to another, the power output from the PV array diminishes, which in turn decreases the overall lifespan of the PV array. In Equations (7.5) and (7.6), the parameters η_1, η_2, and η_3 represent the ideality factors for the three different diodes. A higher ideality factor indicates greater power loss in the PV array. If the circuit includes a shunt resistive element, the sunlight current can be derived as per the equations provided. This additional shunt element can absorb leakage currents and enhance the efficiency of the PV system [61].

$$I_0 = I_{\text{ph}} - I_{D_1} - I_{D_2} - I_{D_3} - I_{\text{sh}} \tag{7.4}$$

$$I_0 = I_{\text{ph}} - I_1\left(e^{\frac{q(V_0 + I_0 R_{\text{se}})}{\eta_1 KT}} - 1\right) - I_2\left(e^{\frac{q(V_0 + I_0 R_{\text{se}})}{\eta_2 KT}} - 1\right) - I_F \tag{7.5}$$

$$I_F = I_3\left(e^{\frac{q(V_0 + I_0 R_{\text{se}})}{\eta_3 KT}} - 1\right) - \frac{V_0 + I_0 R_{\text{se}}}{R_{\text{sh}}} \tag{7.6}$$

Partial shading and bad weather conditions can significantly impact the efficiency and performance of PV panels. Here are the primary issues caused by these conditions:

1. **Partial Shading**:
   - **Power Imbalance**: Shading on a portion of a PV module or a set of modules in a series string can create an imbalance in power flow, as some modules will generate less power due to reduced sunlight exposure.
   - **Bypass Diode Activation**: To counteract the negative effects of partial shading, bypass diodes (D_1, D_2, D_3) are integrated across each PV module or cell. These diodes allow current to flow around the shaded module, preventing the module from acting as a load and reducing the overall system's power output.
   - **Power Consumption from Other Modules**: In some cases, a shaded module can begin to consume power from other modules in the same series string, further reducing system efficiency.
2. **Bad Weather Conditions**:
   - **Reduced Sunlight**: Overcast weather or cloud cover can significantly reduce the amount of sunlight reaching the PV panels, leading to lower power generation.
   - **Sunlight Angle Variations**: As the location of the sun changes through the year due to the movement of the Earth, the angle of sunlight incident on the PV array varies, affecting efficiency.

- **Accumulation of Soil and Debris**: Dirt, dust, and other debris can collect on PV panel surfaces, decreasing their ability to absorb sunlight and thus decreasing overall output.

7.3.1 Shading Patterns

- **Pattern Analysis**: Different shading patterns (as illustrated in Figure 7.3a–c) can help analyze the impact of shading on the solar system.
- **Design Values**: Each solar module's design values, as explained in Table 7.1, provide critical information about how the system will respond to various shading patterns.

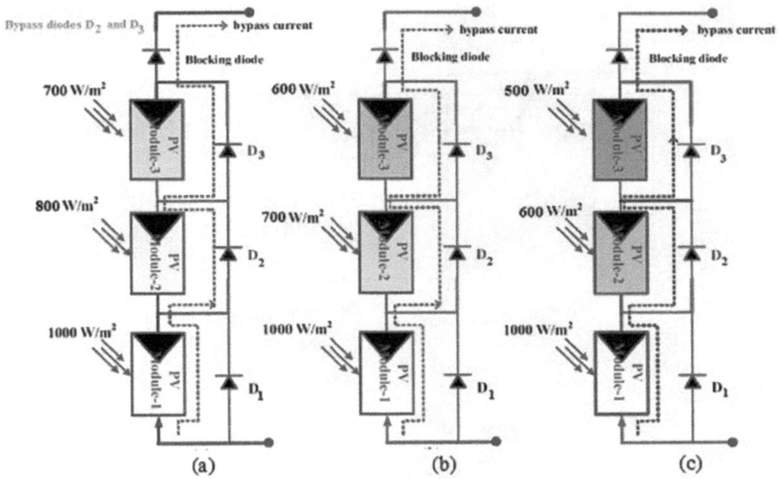

FIGURE 7.3 Various partial shading conditions: (a) Case (i), (b) Case (ii), and (c) Case (iii).

TABLE 7.1
Designed Parameters of the Shaded PV Module

Parameters	Representation	Values
Current at the maximum peak point	I_{MPP}	10.6 A
Voltage at the maximum peak point	V_{MPP}	42.5 V
Power at the maximum peak point	P_{MPP}	450 W
Short-circuit current	I_{SC}	12.2 A
Open-circuit voltage	V_{OC}	45 V
Diode operating conditions	η_1, η_2, η_3	0.8, 0.91, 0.98
Parallel resistance in the PV cell	R_{sh}	260.4 Ω
Series resistance in the PV cell	R_{se}	0.32 Ω
PV irradiance current	I_1, I_2, I_3	9.56 A
Diode saturation current	I_{ph}	1.5×10^{-10} A
No. of cells in the PV module	N_S	50

The presence of shading and bad weather conditions can create challenges for PV panels, affecting their performance and efficiency. Bypass diodes play a crucial role in identifying the impact of partial shading by allowing the flow of current around shaded modules. Let me know if you need any specific analysis or computations based on the provided data.

The performance of solar modules can vary significantly under different insolation (sunlight) conditions, as illustrated in Figures 7.4 and 7.5.

1. **Current (I_{MPP}), Maximum Power Point (P_{MPP}), and Voltage (V_{MPP}):**
   - Under constant irradiation, the maximum power point (P_{MPP}) is 920 W, with a current (I_{MPP}) of 10.6 A and voltage (V_{MPP}) of 120 V.

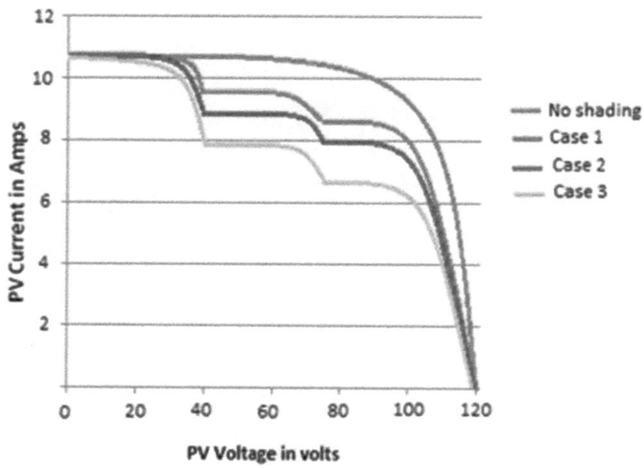

FIGURE 7.4 *I–V* characteristics.

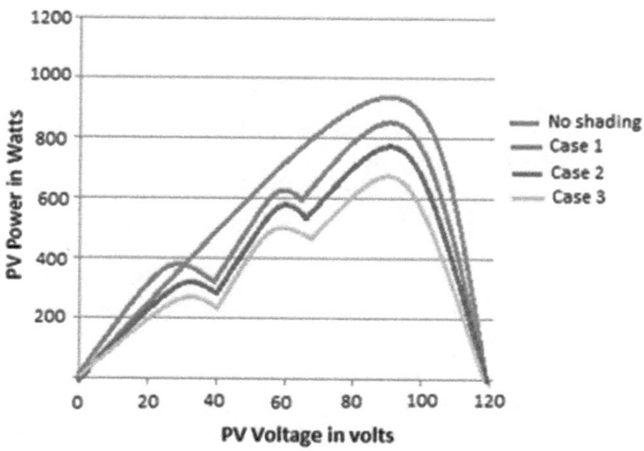

FIGURE 7.5 *P–V* characteristics.

2. **Insolation Conditions**:
   - Different insolation conditions for the PV modules:
     - **Case (i)**: Insolation of 1,000, 800, and 700 W/m^2. The corresponding PV parameters are power (868 W), voltage (120 V), and current (8.8 A).
     - **Case (ii)**: Insolation of 1,000, 700, and 600 W/m^2. The corresponding PV parameters are power (785 W), voltage (120 V), and current (8.1 A).
     - **Case (iii)**: Insolation of 1,000, 600, and 500 W/m^2. The corresponding PV parameters are power (672 W), voltage (120 V), and current (6.8 A).
3. **Operating Points**:
   - Cases (i) and (ii) are considered local points of the PV module, while case (iii) is the desired functioning point (global maximum) that can be reached using an MPPT controller.
4. **Impact on PV Performance**:
   - Different shading conditions lead to variations in the current, voltage, and power outputs of the PV system.
   - These variations can significantly reduce power generation during periods of lower insolation, such as night hours.
5. **Night Hours**:
   - During night hours, the generated power from the PV system is drastically reduced, affecting the overall energy supply.

7.4 COMPARATIVE ANALYSIS OF THE PROPOSED CONTROLLER WITH OTHER CONTROLLERS

PV cell efficiency is crucial for solar battery charging in cloudy conditions due to fluctuating light intensity. Consistent light intensity across PV cells enhances performance by balancing heating effects and increasing power output. Various methods for identifying optimal power points in PV systems exist, but conventional strategies have notable drawbacks, such as moderate tracking speed, applicability mainly in uniform sunlight conditions, high power consumption during perturbations, reliance on PV array manufacturing, and voltage ripples in converter-generated power. These methods also struggle with rapid changes in sunlight temperatures and varying sunlight angles. To address these challenges, hybrid optimized MPPT controllers have been studied for use in cloudy conditions. These controllers aim to maintain PV system efficiency despite variable sunlight angles and intensities, combining bioinspired MPPT controllers to keep cell output current within optimal limits.

7.4.1 P&O MPPT Controller

The P&O approach is a popular power point tracking controller for hybrid solar power networks in static insolation settings. It moves the PV operational point from a local to a global position on the *I–V* slope. Installation is simple, requiring only a few sensors to record PV voltage and current in various environmental conditions.

The P&O technique's advantage is its minimal dependence on PV array modeling. Adjusting the step constant in the MPPT controller can enhance PV system efficiency. When the PV's operating point is on the right side of the *P–V* curve, the variable constant is positive. Conversely, if the PV is closer to the true MPP on the left side of the *P–V* curve, the converter's duty cycle is increased.

7.4.2 ANFIS MPPT Controller

Neural networks face challenges such as high computational costs, extensive training requirements, and large datasets from solar systems. They also demand significant development time and skilled personnel. Their limited data-gathering capacity can hinder power optimization in PV systems. While recurrent neural networks excel in text creation and sequential tasks, they struggle with gradient issues like vanishing gradients. Fuzzy methods offer moderate accuracy with imprecise data but lack a specific approach for resolving membership function issues.

In real-time microgrid networks, systems show improved efficiency under abnormal PV conditions but may struggle with shaded modules. The ANFIS combines fuzzy logic with neural networks using the Takagi-Sugeno system, addressing the limitations of both approaches. ANFIS employs if-then rules to approximate PV functions and uses genetic algorithms to optimize network values. It is applied in energy management systems for smooth PV-fed battery operation and in PV/wind hybrid power networks to generate duty signals for power converters, mapping supply voltage, load voltage, and PV current.

7.4.3 PFO MPPT Controller

The pufferfish controller is an MPPT algorithm inspired by the defense mechanisms of pufferfish, which inflate into a spiny ball to deter predators. This natural behavior informs the PFO algorithm, which mimics predator–prey interactions to optimize solar energy extraction.

7.4.3.1 PFO Phases

1. **Exploration Phase (Predator Attack):**
   - **Predator Attack Simulation:** Models predators moving toward pufferfish.
   - **Position Update:** Updates candidates' positions in the search space.
   - **Objective Function:** Each candidate (predator) assesses others to find targets (pufferfish) with better objective function values using Equation (7.7).

$$P_{C_i} = \{X_r : F_r < r \neq p\}, \quad \text{where} \quad p = 1,2,3,\ldots,n \quad \& \quad r \in \{1,2,3,\ldots,n\} \quad (7.7)$$

   Here, P_{C_i}: This is the collection of potential pufferfish positions for the pth predator. It refers to the potential target locations for a specific predator during the optimization process. These targets are population candidates with better objective function values than the pth predator, X_r: This is the location of a population candidate that has a better objective function value than the pth predator. This candidate serves as a candidate pufferfish for the predator to target, and F_r: This is the objective function of the population candidate represented by X_r. It is used to compare different population candidates' performances and identify which ones have better (lower or higher, depending on the optimization goal) objective function values.

- **Random Target Selection:** Predators randomly select pufferfish targets.
- **New Position Calculation:** Uses Equation (7.8).

$$x_{p,q}^{L1} = x_{p,q} + \text{rand}_{p,q}.\left(P_{S_{p,q}} - R_{p,q}.x_{p,q}\right) \tag{7.8}$$

$$X_p = \begin{cases} X_p^{L1}, & F_p^{L1} \le F_p \\ X_p, & \text{else} \end{cases} \tag{7.9}$$

Here, P_{S_p}: chosen pufferfish for the randomly chosen ith predator from the P_{C_p} set (i.e., P_{S_p} is a parameter of the P_{C_p} set), $P_{S_{p,q}}$ is its qth element, x_p^{L1} is the new location estimated for the pth predator depending on the exploration stage of the PFO, $x_{p,q}^{L1}$ is its qth dimension, F_p^{L1} is its value of objective function, $\text{rand}_{p,q}$ are random values between 0 and 1, and $R_{p,q}$ are random numbers as 1 or 2.

- **Position Update:** Updates candidate positions if new values are better using Equation (4.3).

2. **Exploitation Phase (Pufferfish Defense):**
   - **Defense Simulation:** Models pufferfish inflating to deter predators.
   - **Predator Retreat:** Simulates predators moving away from pufferfish.
   - **Objective Function Improvement:** Compares new positions to previous ones using Equation (7.10).

$$x_{p,q}^{L2} = x_{p,q} + \left(1 - 2\ \text{rand}_{p,q}\right) \cdot \frac{UL_q - LL_q}{t} \tag{7.10}$$

$$X_p = \begin{cases} X_p^{L2}, & F_p^{L2} \le F_p \\ X_p, & \text{else} \end{cases} \tag{7.11}$$

Here, X_p^{L2} is the updated PFO site determined by using the second phase of the proposed PFO for the pth predator, $x_{p,q}^{L2}$ is its qth dimension, F_p^{L2} is its objective function value, $\text{rand}_{p,q}$ are random values between 0 and 1, and t is the epoch number.

- **Conditional Update:** Updates candidate positions if new values are better using Equation (7.11).

The algorithm iterates through these phases, updating candidate positions until it converges on an optimal solution. The best-performing candidate's position is reported as the solution.

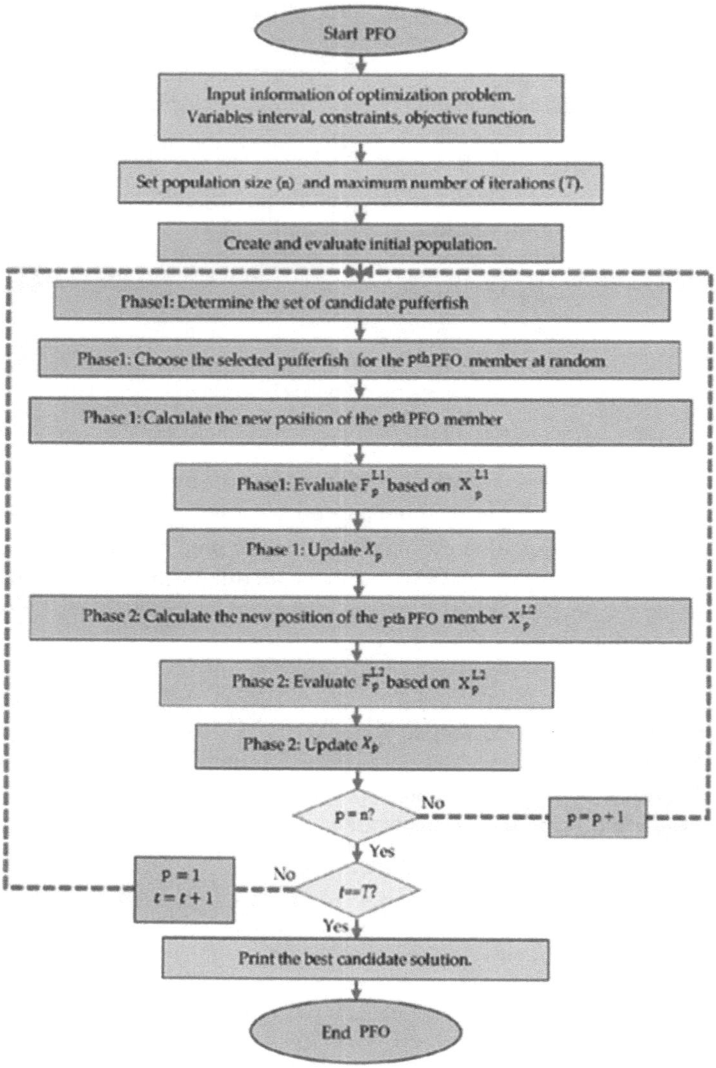

FIGURE 7.6 Flow chart of the PFO algorithm.

By alternating between broad exploration and focused exploitation, the PFO effectively navigates the solution space to optimize solar energy extraction. The flowchart in Figure 7.6 shows the procedures for putting the PFO algorithm into practice.

7.4.4 Proposed PFO-ANFIS MPPT Controller

Traditional methods for PV systems face limitations such as longer convergence times, slower MPP tracking, higher power supply distortions, expensive sensors, and increased size and cost. ANN has been applied to integrate renewable PV sources with the grid, offering benefits like simplified implementation, reduced installation area, and quicker response times. Data on open-circuit voltage, ideality factors, temperatures, and irradiations train the ANN using the Levenberg–Marquardt algorithm.

However, traditional neural networks struggle with locating the MPP precisely, increased complexity with more layers, and reduced effectiveness in handling nonlinear issues. Fuzzy logic is employed in hybrid grid-connected renewable energy systems for managing imprecise input variables and handling nonlinear challenges, but it may falter under partial shading and unreliable supply data.

This article examines integrating an enhanced pufferfish controller with ANFIS to identify the GMPP of PV under rapid irradiation changes. ANFIS, a hybrid of fuzzy logic and neural network methods, benefits from the strengths of both. The pufferfish controller selects membership functions at local PV peak points. The PFO controllers are supplied to the ANFIS system to reduce fluctuations around global maximum peak points once all local MPPs are discovered. Equations (7.12) and (7.13) describe a feedback loop between PFO and ANFIS controllers, improving overall efficiency and performance of the PV system in unfavorable climatic conditions.

$$\left|P_{\text{pr}} - P_{\text{pv}}\right| \leq \Delta_1 \tag{7.12}$$

$$\left|P_{\text{pr}} - P_{\text{pv}}\right| \geq \Delta_2 \tag{7.13}$$

where P_{pr} and P_{pv} are the previously calculated power at the PV array and the currently calculated power at the PV array. Δ_1 is the error value to the ANFIS controller. The selected value Δ_1 for this hybrid MPP finding value is 0.058. The output signal produced from the ANFIS is fed back to the PFO block depending on the chosen functional value. Similarly, Δ_2 is given to the pufferfish component for receiving the produced error value from ANFIS.

7.5 SIMULATION RESULTS AND DISCUSSION

The converter circuit powered by the solar system is designed and implemented using the Simulink MATLAB tool. The system's operation and installation are heavily influenced by the angle of sunlight. In order to enhance the solar system's power supply capacity, a converter is attached to the PV output since solar PV modules may only be able to deliver a limited amount of electricity during sudden changes in sun insolation.

The source-side capacitor (C_S) has a value of 220 μF and serves to filter out PV power ripples. It also provides protection for the switches during rapid supply voltage variations. The supply inductor (L_S) is chosen as 20 mH. The PV supply voltage is charged by this inductor when the MOSFET is operating in the forward bias mode. The inductor smoothens the supply voltage by providing voltage to the load when the switch is in the blocked position.

The benefits of a MOSFET include reduced gate-to-source voltage requirements, efficient operation at low voltages, quick response times, a high source resistance that limits gate current, and reduced damage risk. The load-side capacitor (C_l), chosen at 232 μF, maintains the load voltage and retains the PV panel's electrostatic energy for

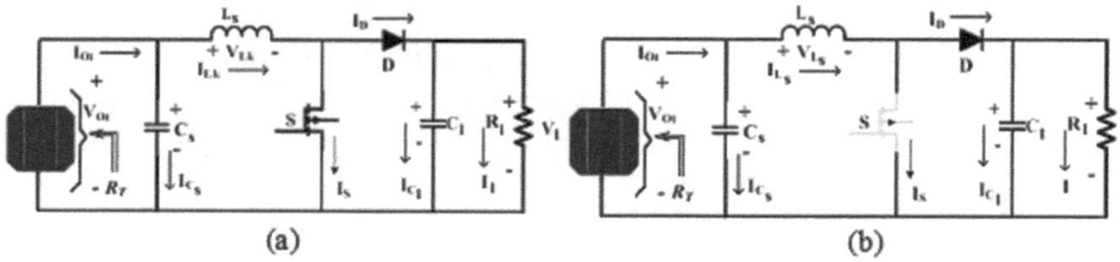

FIGURE 7.7 Proposed DC–DC converter with (a) ON and (b) OFF state.

quick temperature fluctuations in the sun. The load resistor (R_l) is connected in parallel to the load capacitor (C_l), as depicted in Figure 7.7.

7.5.1 Partial Shadings at Case (i)

Three shading cases are taken into consideration to evaluate the efficiency of the proposed regulating method. Below are the PV characteristic curves for each scenario under partially shadowed conditions.

In case (i) with partial shade, the PV system is kept at a constant temperature of 25°C while being exposed to three distinct irradiation levels and partial shading situations (1,000, 800, and 700 W/m^2). Figures 7.8–7.10 depict the simulation values of the PV system (current, voltage, duty cycle, and power of the DC–DC boost converter) using various MPPT algorithms under the first partial shadow pattern.

These findings show that, independent of shading patterns, the suggested MPPT technique depending on the hybrid heuristic algorithms PFO-ANFIS may effectively track the actual maximum power. The PFO-based MPPT approach outperforms the PFO method and provides a better response with less distortion of waveform harmonics. In comparison to the P&O, ANFIS, and PFO approaches, the PFO tracker efficiently extracts the highest global power in a short amount of time and with minimal oscillations. The GMPP may be tracked in 58 ms by PFO and 292 ms by ANFIS, respectively. In contrast, the P&O and ANFIS methods fail to track the GMPP and get stuck at the first peak of the *P–V* slope.

7.5.2 Partial Shading at Case (ii)

For case (ii), partial shading is examined under conditions of solar radiation at levels of 1,000, 700, and 600 W/m^2 with a constant temperature of 25°C. In this setup, the PV curve features one local peak at 0.49 kW and one global peak at 0.92 kW, situated at the second peak of the *P–V* curve.

The PFO and ANFIS controllers can effectively recognize and track the GMPP in 600 and 278 ms, respectively. However, the P&O and ANFIS controllers converge at power values of 0.75 and 0.72 kW, respectively.

Figures 7.11–7.13 provide details of the comparison of simulation results for different MPPT controllers under the second case of partial shading conditions. These figures illustrate how various controllers perform in terms of efficiency and effectiveness in tracking the GMPP under these shading conditions.

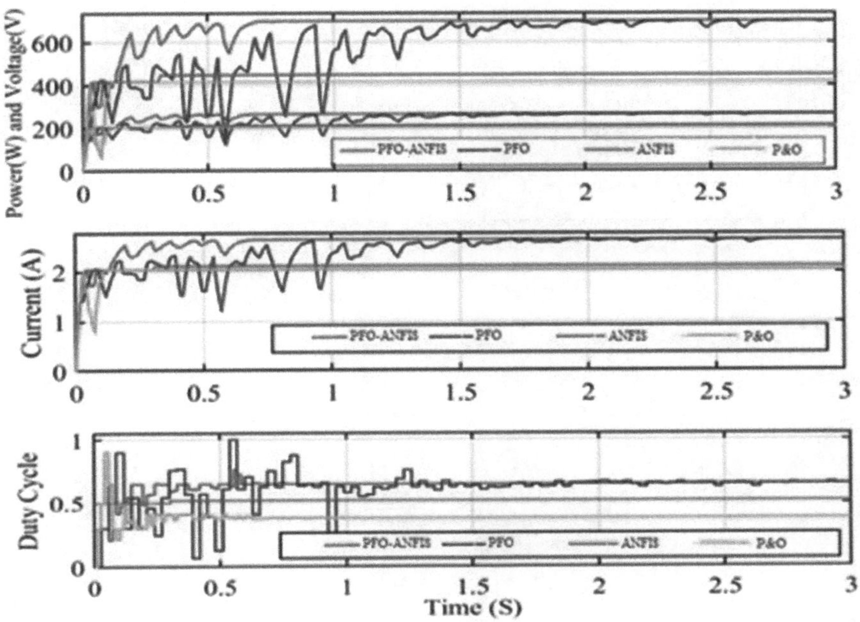

FIGURE 7.8 The output voltage, power, current, and duty cycle for PFO-ANFIS at case (i) of PSCs.

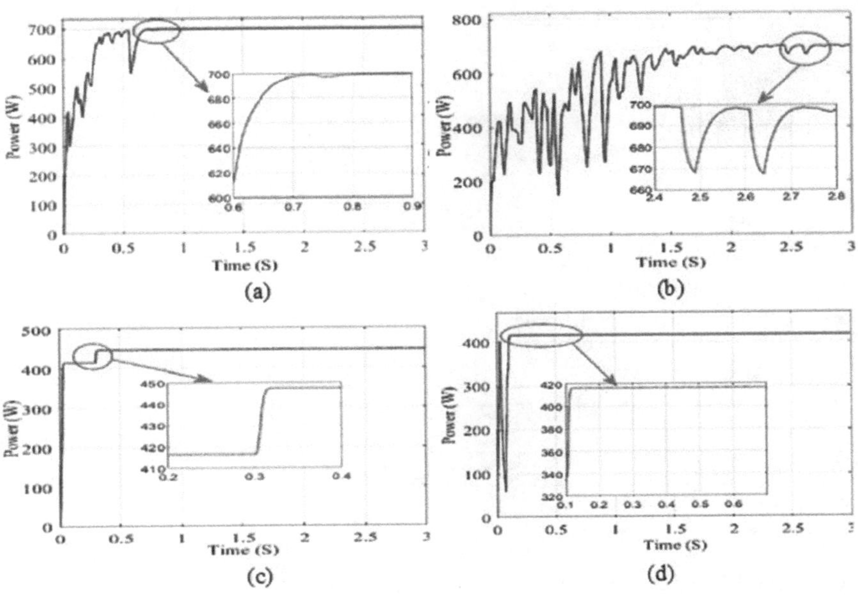

FIGURE 7.9 Output power of MPPT controllers (a) PFO-ANFIS, (b) PFO, (c) ANFIS, and (d) P&O for the PV system under PSC at case (i).

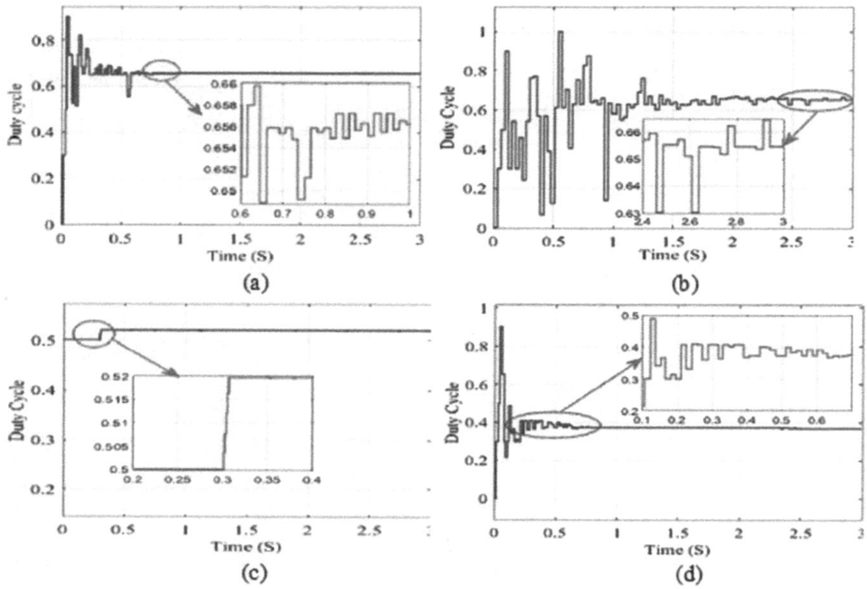

FIGURE 7.10 Duty cycle of MPPT controllers (a) PFO-ANFIS, (b) PFO, (c) ANFIS, and (d) P&O for the PV system under PSC at case (i).

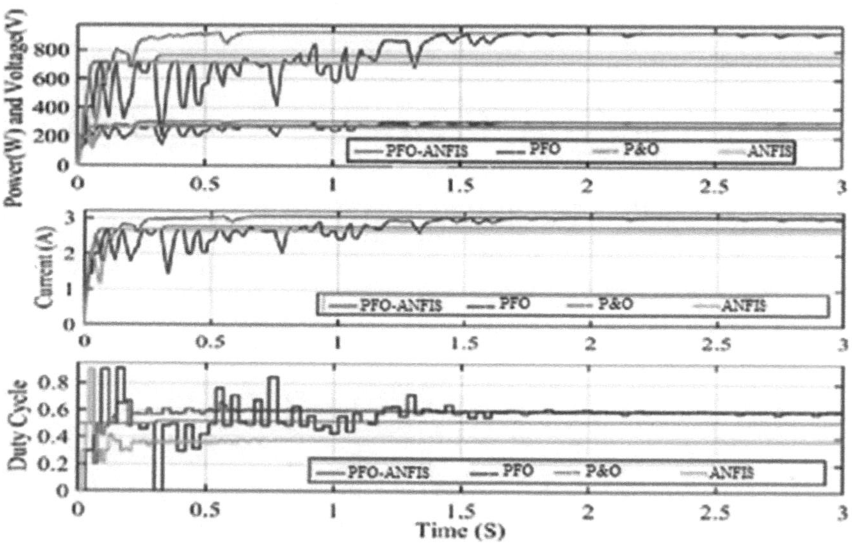

FIGURE 7.11 The output voltage, power, current, and duty cycle between the PFO-ANFIS at case (ii) of PSCs.

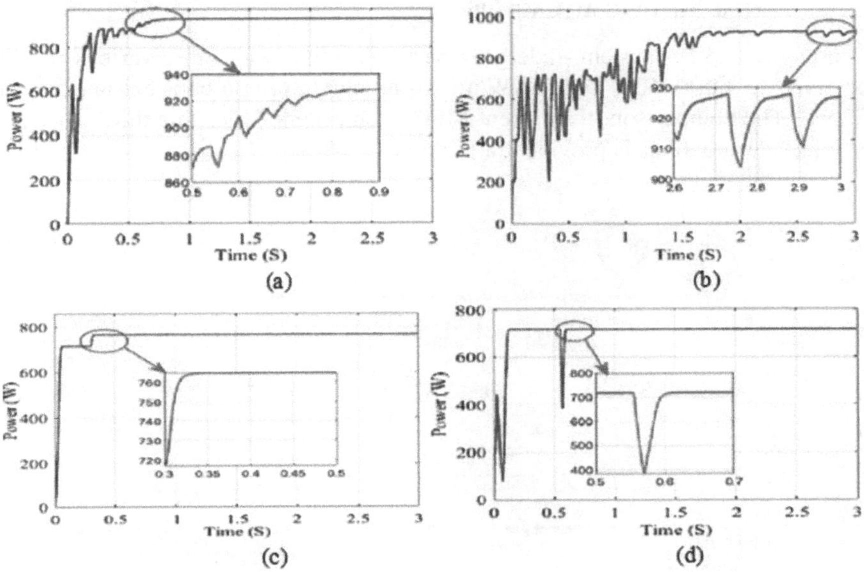

FIGURE 7.12 Output power of MPPT controllers (a) PFO-ANFIS, (b) PFO, (c) ANFIS, and (d) P&O for PV system under PSC at case (ii).

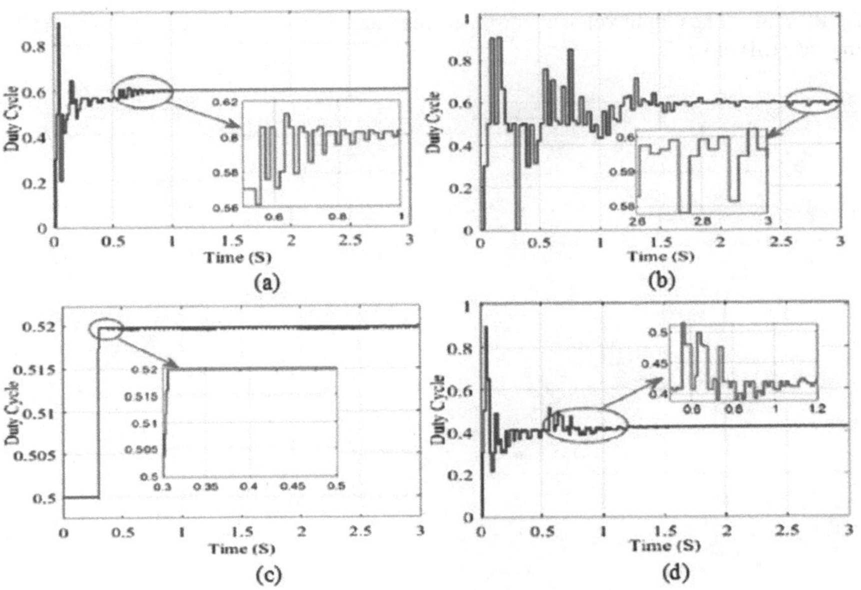

FIGURE 7.13 Duty cycle of MPPT controllers (a) PFO-ANFIS, (b) PFO, (c) ANFIS, and (d) P&O for the PV system under PSC at case (ii).

7.5.3 Partial Shading at Case (iii)

In case (iii), the PV system undergoes partial shading with three various irradiation levels at 1,000, 600, and 500 W/m^2 while maintaining a constant temperature of 25°C. The comparison of different MPPT controllers under this third pattern of partial shade situations is provided in Figures 7.14–7.16.

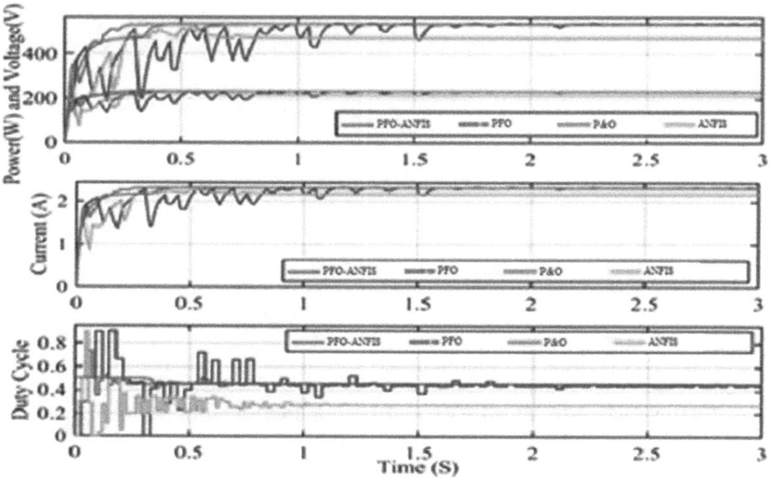

FIGURE 7.14 The output voltage, power, current, and duty cycle between the PFO-ANFIS at the case (iii) of PSCs.

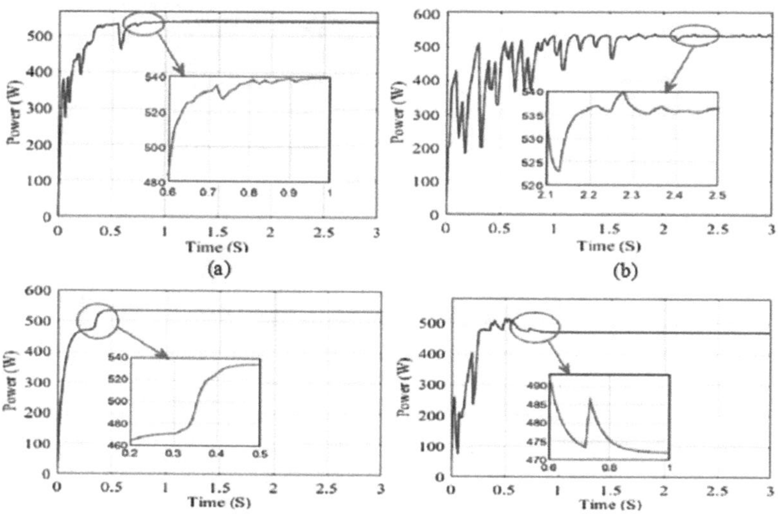

FIGURE 7.15 Output power of MPPT controllers (a) PFO-ANFIS, (b) PFO, (c) ANFIS, and (d) P&O for the PV system under PSC at case (iii).

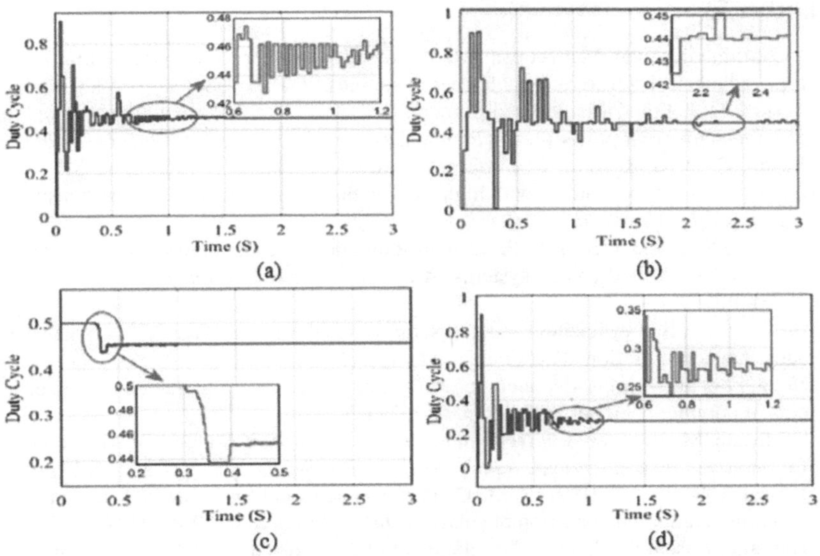

FIGURE 7.16 Duty cycle of MPPT controllers (a) PFO-ANFIS, (b) PFO, (c) ANFIS, and (d) P&O for the PV system under PSC at case (iii).

Both PFO-ANFIS and ANFIS algorithms, relying on search mechanisms, manage partial shading conditions effectively. However, PFO-ANFIS outperforms ANFIS in terms of speed; after partial shading occurs, PFO-ANFIS tracks the MPP in 311 ms with no fluctuations in either steady or transient states, while ANFIS takes 178 ms but exhibits more significant fluctuations.

Also, P&O and ANFIS methods tend to quickly become trapped at the local peak, limiting their ability to find and track the true GMPP.

7.6 CONCLUSION

The investigation of the PV system utilized the MATLAB Simulink tool to simulate the system under various sunlight shading conditions. The choice of a 3-diode model PV cell allowed for an accurate representation of the system's behavior under different partial shading conditions (PSCs). The current versus voltage curve for the triple-diode model PV exhibits nonlinearity, which was effectively, managed using the MPPT controller. The simulation results exposed that the proposed PFO controller-fed ANFIS approach provided superior peak output power from the PV system, with strong dynamic response and reduced steady-state distortions. This approach required fewer iterations and sensors for sensing PV parameters, leading to lower design costs and easier implementation. It also proved more efficient and particularly applicable for the rainy conditions experienced by PV systems. However, the high cost of PV power generation remains an undesirable aspect of current solar PV systems. To improve the supply's voltage profile, a nonisolated DC–DC converter was coupled to the PV system's output. This converter is advantageous due to its simplicity, high reliability, and cost-effectiveness.

REFERENCES

1. Mohammadi, A. & Mehrpooya, M. A comprehensive review on coupling different types of electrolyzer to renewable energy sources. *Energy* 158, 632–655 (2018).
2. Turney, D. & Fthenakis, V. Environmental impacts from the installation and operation of large-scale solar power plants. *Renew. Sustain. Energy Rev.* 15(6), 3261–3270 (2011).
3. Basha, C. H. H. & Rani, C. Different conventional and soft computing MPPT techniques for solar PV systems with high step-up boost converters: A comprehensive analysis. *Energies* 13(2), 371 (2020).
4. Sharma, V. & Chandel, S. S. Performance and degradation analysis for long term reliability of solar photovoltaic systems: A review. *Renew. Sustain. Energy Rev.* 27, 753–767 (2013).
5. Fu, F. et al. High-efficiency inverted semi-transparent planar perovskite solar cells in substrate configuration. *Nat. Energy* 2(1), 1–9 (2016).
6. Vella, E. et al. Ultrafast decoherence dynamics govern photocarrier generation efficiencies in polymer solar cells. *Sci. Rep.* 6(1), 29437 (2016).
7. Taghvaee, M. H. et al. A current and future study on non-isolated DC–DC converters for photovoltaic applications. *Renew. Sustain. Energy Rev.* 17, 216–227 (2013).
8. Hussaian Basha, C. H. & Rani, C. Performance analysis of MPPT techniques for dynamic irradiation condition of solar PV. *Int. J. Fuzzy Syst.* 22(8), 2577–2598 (2020).
9. Hussaian Basha, C. H., et al. Development of cuckoo search MPPT algorithm for partially shaded solar PV SEPIC converter. In: Das, K., Bansal, J., Deep, K., Nagar, A., Pathipooranam, P. and Naidu, R. (eds.), *Soft Computing for Problem Solving: SocProS 2018*, Volume 1. Springer, Singapore, 2020.
10. Femia, N. et al. Optimization of perturb and observe maximum power point tracking method. *IEEE Trans. Power Electron.* 20(4), 963–973 (2005).
11. Ambikapathy, A., Singh, G. & Tiwari, P. (eds). Smart switching algorithm between IC and PO algorithms for grid-connected PV system. In: *Advances in Smart Grid and Renewable Energy: Proceedings of ETAEERE-2016*. Springer Singapore, pp. 83–92 (2018).
12. Wang, F. et al. Analysis and optimization of flexible MCPT strategy in submodule PV application. *IEEE Trans. Sustain. Energy* 8(1), 249–257 (2016).
13. Jain, S. & Agarwal, V. A new algorithm for rapid tracking of approximate maximum power point in photovoltaic systems. *IEEE Power Electron. Lett.* 2(1), 16–19 (2004).
14. Mei, Q. et al. A novel improved variable step-size incremental-resistance MPPT method for PV systems. *IEEE Trans. Ind. Electron.* 58(6), 2427–2434 (2010).
15. Basha, C. H. & Rani, C. Design and analysis of transformerless, high step-up, boost DC-DC converter with an improved VSS-RBFA based MPPT controller. *Int. Trans. Electr. Energy Syst.* 30(12), e12633 (2020).
16. Kiran, S. R. et al. Reduced simulative performance analysis of variable step size ANN based MPPT techniques for partially shaded solar PV systems. *IEEE Access* 10, 48875–48889 (2022).
17. Baimel, D. et al. Improved fractional open circuit voltage MPPT methods for PV systems. *Electronics* 8(3), 321 (2019).
18. Hassan, A., Octavian, B. & Masoum, M. A. S. An improved genetic algorithm based fractional open circuit voltage MPPT for solar PV systems. *Energy Rep.* 9, 1535–1548 (2023).
19. Hosseini, E. & Shahgholian, G. Partial-or full-power production in WECS: A survey of control and structural strategies. *EPE J.* 27(3), 125–142 (2017).
20. Bendib, B., Belmili, H. & Krim, F. A survey of the most used MPPT methods: Conventional and advanced algorithms applied for photovoltaic systems. *Renew. Sustain. Energy Rev.* 45, 637–648 (2015).

21. Karami, N., Moubayed, N. & Outbib, R. General review and classification of different MPPT Techniques. *Renew. Sustain. Energy Rev.* 68, 1–18 (2017).
22. Liu, C. L. et al. An asymmetrical fuzzy-logic-control-based MPPT algorithm for photovoltaic systems. *Energies* 7(4), 2177–2193 (2014).
23. Monica, M., et al. PMSG based WECS: Control techniques, MPPT methods and control strategies for standalone battery integrated system. In *AIP Conference Proceedings*, Vol. 2405, No. 1. AIP Publishing (2022).
24. Awan, M. M. A. et al. Performance optimization of a ten check MPPT algorithm for an off-grid solar photovoltaic system. *Energies* 15(6), 2104 (2022).
25. Mehazzem, F., Andre, M. & Calif, R. Efficient output photovoltaic power prediction based on MPPT fuzzy logic technique and solar spatio-temporal forecasting approach in a tropical insular region. *Energies* 15(22), 8671 (2022).
26. Houssamo, I., Locment, F. & Sechilariu, M. Experimental analysis of impact of MPPT methods on energy efficiency for photovoltaic power systems. *Int. J. Electr. Power Energy Syst.* 46, 98–107 (2013).
27. Kofinas, P. et al. An Intelligent MPPT controller based on direct neural control for partially shaded PV system. *Energy Build.* 90, 51–64 (2015).
28. Adly, M. & Besheer, A. H. A meta-heuristics search algorithm as a solution for energy transfer maximization in stand-alone photovoltaic systems. *Int. J. Electr. Power Energy Syst.* 51, 243–254 (2013).
29. Sakeen, B., Bachache, N. K. & Wang, S. Frequency control of PV-diesel hybrid power system using optimal fuzzy logic controller. In: *2013 IEEE 11th International Conference on Dependable, Autonomic and Secure Computing*, Chengdu, China, 21–22 December 2013. IEEE (2013).
30. Perera, K. S., Zeyar, A. & Woon, W. L. Machine learning techniques for supporting renewable energy generation and integration: A survey. In: *Data Analytics for Renewable Energy Integration: Second ECML PKDD Workshop*, DARE 2014, Nancy, France. Springer International Publishing (2014).
31. Singh, R. & Rajpurohit, B. S. Performance Evaluation of Grid-Connected Solar Photovoltaic (SPV) System with Different MPPT Controllers. In: Hossain, J., Mahmud, A. (eds.), *Renewable Energy Integration. Green Energy and Technology*, Springer, Singapore, pp. 97–124 (2014).
32. Masoumi, A. P. et al. A new two-axis solar tracker based on the online optimization method: Experimental investigation. *Energy AI*, 14, 100284 (2023). Available at SSRN 4050437.
33. Haque, A. et al. Fault diagnosis of photovoltaic modules. *Energy Sci. Eng.* 7(3), 622–644 (2019).
34. Sridharan, N. V. & Sugumaran, V. Visual fault detection in photovoltaic modules using decision tree algorithms with deep learning features. *Energy Sour. Part A Recovery Util. Environ.* 1, 17 (2021).
35. Basha, C. H. H., Rani, C. & Odofin, S. A review on non-isolated inductor coupled DC-DC converter for photovoltaic grid connected applications. *Int. J. Renew. Energy Res.* 7(4), 1570–1585 (2017).
36. Bajaj, M. & Singh, A. K. Grid integrated renewable DG systems: A review of power quality challenges and state-of-the-art mitigation techniques. *Int. J. Energy Res.* 44(1), 26–69 (2020).
37. Lim, J.-W., Hassan, J. & Kim, M. Bidirectional soft switching push–pull resonant converter over wide range of battery voltages. *IEEE Trans. Power Electron.* 36(11), 12251–12267 (2021).
38. Kumar, N., Singh, B. & Panigrahi, B. K. PNKLMF-based neural network control and learning-based HC MPPT technique for multiobjective grid integrated solar PV based distributed generating system. *IEEE Trans. Ind. Inform.* 15(6), 3732–3742 (2019).

39. Merritt, N. R., Chakraborty, C. & Bajpai, P. An E-STATCOM based solution for smoothing photovoltaic and wind power fluctuations in a microgrid under unbalanced conditions. *IEEE Trans. Power Syst.* 37(2), 1482–1494 (2021).
40. Kishore, P. M. & Bhimasingu, R. Boost multi-port converter with simultaneous isolated DC, non-isolated DC and AC outputs. In: *IECON 2018–44th Annual Conference of the IEEE Industrial Electronics Society*, Washington, DC, 21–23 October 2018. IEEE (2018).
41. Basha, C. H., Rani, C. & Odofin, S. Analysis and comparison of SEPIC, Landsman and Zeta converters for PV fed induction motor drive applications. In: *2018 International Conference on Computation of Power, Energy, Information and Communication (ICCPEIC)*, Chennai, India, 28–29 March 2018. IEEE (2018).
42. Baba, A. O., Liu, G. & Chen, X. Classification and evaluation review of maximum power point tracking methods. *Sustain. Futures* 2, 100020 (2020).
43. El-Shahat, A. & Sumaiya, S. DC-microgrid system design, control, and analysis. *Electronics* 8(2), 124 (2019).
44. Irshad, S. M. & Ramesh, G. P. High gain power generation based on hybrid renewable energy for AC load application. *Indones. J. Electr. Eng. Comput. Sci.* 12(3), 1195–1202 (2018).
45. Babu, T. S. et al. Particle swarm optimization based solar PV array reconfiguration of the maximum power extraction under partial shading conditions. *IEEE Trans. Sustain. Energy* 9(1), 74–85 (2017).
46. Husain, M. A. et al. Comparative assessment of maximum power point tracking procedures for photovoltaic systems. *Green Energy Environ.* 2(1), 5–17 (2017).
47. Subramaniam, U. et al. A hybrid PV-battery system for ON-grid and OFF-grid applications—Controller-in-loop simulation validation. *Energies* 13(3), 755 (2020).
48. Haque, A. Maximum power point tracking (MPPT) scheme for solar photovoltaic system. *Energy Technol. Policy* 1(1), 115–122 (2014).
49. Goswami, A. & Sadhu, P. K. Degradation analysis and the impacts on feasibility study of floating solar photovoltaic systems. *Sustain. Energy Grids Netw.* 26, 100425 (2021).
50. Chiu, D. T. et al. Small but perfectly formed? Successes, challenges, and opportunities for microfluidics in the chemical and biological sciences. *Chem* 2(2), 201–223 (2017).
51. Kumar, M. et al. Comprehensive review of conventional and emerging maximum power point tracking algorithms for uniformly and partially shaded solar photovoltaic systems. *IEEE Access* 11, 31778–31812 (2023).
52. Das, H. S. et al. A comprehensive review on power conditioning units and control techniques in fuel cell hybrid systems. *Energy Rep.* 8, 14236–14258 (2022).
53. Kumar, A. & Kumar, P. Power quality improvement for grid-connected PV system based on distribution static compensator with fuzzy logic controller and UVT/ADALINE-based least mean square controller. *J. Mod. Power Syst. Clean Energy* 9(6), 1289–1299 (2021).
54. Nadimuthu, L. P. R. et al. Energy conservation approach for continuous power quality improvement: A case study. *IEEE Access* 9, 146959–146969 (2021).
55. Jiang, M. et al. A novel combinatorial hybrid SFL–PS algorithm based neural network with perturb and observe for the MPPT controller of a hybrid PV-storage system. *Control Eng. Pract.* 114, 104880 (2021).
56. Ram, J. P. et al. Performance enhancement of solar PV systems applying P&O assisted Flower Pollination Algorithm (FPA). *Solar Energy* 199, 214–229 (2020).
57. Basha, C. H. et al. Design of an LPF based slider controller for THD reduction in solar PV B-4 inverter. In: *2019 IEEE International Conference on Electrical, Computer and Communication Technologies (ICECCT)*, Coimbatore, India, 20–22 February 2019. IEEE (2019).

58. Udhay Sankar, V. et al. Application of wind-driven optimization for decision-making in economic dispatch problem. In: Das, K. N., Bansal, J. C., Deep, K., Nagar, A. K., Pathipooranam, P., Naidu, R. C. (eds.), *Soft Computing for Problem Solving: SocProS 2018*, Vol. 1. Springer Singapore (2020).
59. Hussaian Basha, C. H. et al. Mathematical design and analysis of photovoltaic cell using MATLAB/Simulink. In: *Soft Computing for Problem Solving: SocProS 2018*, Vol. 1. Springer, Singapore, pp. 711–726 (2020).
60. Murali, M., et al. Performance analysis of different types of solar photovoltaic cell techniques using MATLAB/simulink. In: *Proceedings of Fourth International Conference on Inventive Material Science Applications: ICIMA 2021*. Springer Singapore, pp. 203–215 (2022).
61. Osmani, K. et al. A novel MPPT-based lithium-ion battery solar charger for operation under fluctuating irradiance conditions. *Sustainability* 15(12), 9839 (2023).

8 Enhancing Grid-Connected Solar PV Systems with Multilevel Inverter Technology

Ezhilvannan Parimalasundar, Krishnan Suresh, Busireddy Hemanth Kumar, and Vanitha Selvaraj Chandrika

8.1 INTRODUCTION

The proliferation of renewable energy sources, particularly PV systems, has emerged as a pivotal solution in addressing the escalating energy demand and environmental concerns. Among the various configurations, three-phase grid-connected PV systems have garnered substantial attention owing to their efficacy in harnessing solar energy and integrating it seamlessly into the existing electrical grid infrastructure. These systems capitalize on the PV effect, wherein solar panels convert sunlight directly into electricity, thereby offering a sustainable and eco-friendly alternative to conventional fossil-fuel-based energy generation. The integration of PV systems with the electrical grid constitutes a paradigm shift toward a decentralized energy generation model, fostering energy independence and resilience while mitigating greenhouse gas emissions. In a three-phase grid-connected PV system, the generated solar power is synchronized with the grid's three-phase alternating current (AC) supply, facilitating efficient distribution and utilization across diverse industrial, commercial, and residential sectors. The inherent variability of solar irradiance necessitates sophisticated control algorithms and power electronics to ensure grid stability and optimal power flow. Furthermore, the deployment of maximum power point tracking (MPPT) algorithms enhances energy extraction by dynamically adjusting the PV system's operating point to match the varying solar conditions. The seamless integration of PV systems with the grid necessitates compliance with stringent regulatory standards and grid codes to ensure operational safety, reliability, and interoperability. Additionally, the advent of smart grid technologies and advanced communication protocols enables real-time monitoring, remote diagnostics, and predictive maintenance, thereby enhancing system performance and longevity. As the global transition toward renewable energy intensifies, three-phase grid-connected PV systems are poised to play a pivotal role in reshaping the energy landscape, fostering sustainability, and driving economic growth while mitigating the adverse impacts of climate change [1–3].

DOI: 10.1201/9781003496946-8

The integration of PV systems into three-phase grid-connected networks represents a pivotal advancement in renewable energy utilization, heralding a transition toward sustainable power generation. This integration, however, is not without its intricacies and challenges, ranging from technical complexities to regulatory frameworks. At its core, the fundamental premise of such systems lies in their ability to harness solar energy and convert it into electricity, which is then seamlessly fed into the existing electrical grid. Yet, the efficiency and reliability of this process are contingent upon a myriad of factors, including optimal system design, grid compatibility, and operational dynamics. From a technical standpoint, ensuring harmonious interaction between the PV system and the grid necessitates sophisticated control mechanisms to manage voltage fluctuations, frequency variations, and power quality issues. Moreover, the intermittency of solar irradiance poses a formidable challenge, demanding innovative solutions for grid stability and energy storage. Furthermore, the economic viability of three-phase grid-connected PV systems hinges not only on initial installation costs but also on long-term maintenance expenses and regulatory incentives. Regulatory frameworks play a pivotal role in facilitating or hindering the proliferation of such systems, with policies governing grid access, net metering, and feed-in tariffs significantly influencing investment decisions and market dynamics. Additionally, environmental considerations, such as land use implications and end-of-life disposal of PV modules, underscore the importance of adopting holistic approaches toward sustainable energy deployment. As the global community increasingly gravitates toward renewable energy sources to mitigate climate change and reduce dependency on fossil fuels, addressing the multifaceted challenges of three-phase grid-connected PV systems becomes imperative for realizing a greener and more resilient energy landscape [4–6].

The significance of grid-connected systems extends far beyond mere energy provision; they embody the transition toward a greener, more resilient future. However, amidst their promise lies a complex landscape of challenges and opportunities that demand meticulous exploration.

This essay embarks on a journey to dissect the intricacies of grid-connected systems, shedding light on their fundamental principles, operational dynamics, and the hurdles they encounter. By delving into the core components of these systems and elucidating the multifaceted issues they face, we aim to cultivate a comprehensive understanding of the present landscape and pave the way for informed discourse and innovative solutions. At its essence, a grid-connected system constitutes an interconnected network of power sources, transmission lines, substations, and distribution networks, engineered to deliver electricity to consumers seamlessly. Unlike standalone power systems, which operate in isolation, grid-connected systems are characterized by their integration into a larger network, facilitating the exchange of electricity between diverse sources and consumers. This interconnectedness forms the backbone of modern electricity distribution, enabling flexibility, scalability, and enhanced reliability. Central to the functionality of grid-connected systems are the diverse array of power sources they encompass. Traditionally, these systems relied heavily on centralized power plants fueled by nonrenewable resources such as coal, oil, and natural gas. However, the paradigm is rapidly shifting toward a more diversified and sustainable energy mix, with renewable sources like solar, wind, hydroelectric, and biomass gaining prominence. This transition reflects a global imperative to mitigate the adverse impacts of climate change, reduce carbon emissions, and foster energy independence. While the integration of renewable energy sources into grid-connected systems heralds a promising future, it also introduces a host of technical, economic, and regulatory challenges. Chief among these challenges is the intermittency and variability inherent in renewable energy generation. Unlike conventional power plants, which can maintain a consistent output, renewable sources are subject to fluctuations dictated by weather patterns, time of day, and other environmental factors. Effectively managing this variability poses a formidable task for grid operators, requiring sophisticated forecasting, balancing, and storage mechanisms to ensure grid stability and reliability. Moreover, the decentralized nature of renewable energy generation complicates traditional grid management strategies, necessitating a paradigm shift toward more flexible and adaptive approaches. The proliferation of distributed energy resources, such as rooftop solar panels, small-scale wind turbines, and battery storage systems, further exacerbates this challenge. While distributed energy resources offer significant benefits in terms of resilience, efficiency, and cost-effectiveness, their integration into grid-connected systems demands innovative solutions to address issues of voltage regulation, grid congestion, and interoperability. In addition to technical hurdles, grid-connected systems are also confronted with a myriad of economic and regulatory impediments. The transition toward renewable energy entails substantial upfront investments in infrastructure, technology, and grid modernization initiatives. While the long-term benefits of sustainability and resilience are undeniable, the short-term financial burdens can deter investment and slow the pace of adoption. Furthermore, regulatory frameworks often lag behind technological advancements, creating uncertainties and bottlenecks that hinder innovation and market competitiveness. Against this backdrop of challenges, however, grid-connected systems also offer a fertile ground for innovation and collaboration. The rise of smart grid technologies, advanced analytics, and digitalization tools holds the promise of revolutionizing the way we generate, transmit, and consume electricity. By harnessing the power of data, artificial intelligence, and Internet of Things, grid operators can optimize energy flows, enhance grid reliability, and empower consumers to actively participate in energy management decisions. Furthermore, novel financing mechanisms, such as green bonds, carbon pricing schemes, and public–private partnerships, are unlocking new avenues for investment in renewable energy infrastructure. These innovative approaches not only provide financial incentives for sustainable development but also foster a conducive environment for public–private collaboration and knowledge exchange [7–10].

However, the intermittent nature of solar irradiance and the fluctuating output of PV systems pose challenges for their effective integration into the power grid. To address these challenges and maximize the utilization of solar energy, advanced power electronic converters such as multilevel inverters have been widely investigated and adopted in PV system applications. Multilevel inverters offer several advantages over traditional two-level inverters, including improved voltage quality, reduced harmonic distortion, and increased power conversion efficiency. Multilevel inverters are power electronic devices used to convert DC (direct current) voltage into AC (alternating current) voltage with multiple levels of voltage output.

Unlike conventional two-level inverters, which produce output voltage levels that are either at the maximum or minimum of the DC bus voltage, multilevel inverters generate output voltage with several discrete levels, resulting in a staircase-like waveform. The key advantage of multilevel inverters lies in their ability to synthesize high-quality AC voltage with lower harmonic distortion, leading to improved performance and efficiency. There are several types of multilevel inverters, including diode-clamped (neutral-point-clamped), capacitor-clamped (flying capacitor), and cascaded H-bridge inverters. Each type has its unique topology and operating principles, offering different levels of voltage output and scalability. Diode-clamped and capacitor-clamped inverters are commonly used in medium-voltage applications, while cascaded H-bridge inverters are preferred for high-voltage and high-power applications. The integration of multilevel inverters with PV systems enables efficient power conversion and grid integration of solar energy. By utilizing multilevel inverters, the DC output voltage of PV arrays can be converted into high-quality AC voltage suitable for grid connection. Moreover, the multilevel topology allows for better utilization of PV system output voltage, reducing switching losses and improving overall system efficiency. Despite their numerous advantages, the widespread adoption of multilevel inverters in PV systems still faces several challenges, including cost, complexity, and reliability. Further research and development efforts are needed to address these challenges and optimize the performance of multilevel inverter-based PV systems. Future advancements in semiconductor technology, control algorithms, and system integration techniques are expected to further enhance the efficiency and reliability of multilevel inverter-based PV systems, paving the way for a sustainable and renewable energy future [11–14].

Sinusoidal pulse width modulation (SPWM) is one of the most widely used techniques for multilevel inverters. In SPWM, the reference sinusoidal waveform is compared with triangular carrier signals of varying frequencies. By adjusting the amplitude of the carrier signals, the modulation index can be controlled to generate the desired output voltage waveform. SPWM produces low harmonic distortion but suffers from voltage and current ripples at low modulation indices. This technique extends the concept of SPWM to multilevel inverters by using multiple carrier signals corresponding to each voltage level. Each carrier signal is compared with the reference waveform to generate individual switching signals for each level of the inverter. Multilevel carrier-based PWM offers improved harmonic performance and better utilization of the available voltage levels compared to SPWM. Selective harmonic elimination pulse width modulation (SHEPWM) aims to eliminate specific lower-order harmonics from the output voltage waveform by judiciously selecting the switching angles of the inverter. This technique requires solving a set of nonlinear equations to determine the optimal switching angles that satisfy the harmonic elimination criteria. SHEPWM offers precise harmonic control but may result in increased switching losses and complexity. SVPWM is a sophisticated PWM technique that utilizes the concept of space vectors to control the inverter output voltage. By dividing the reference voltage vector into six or seven active voltage vectors and zero vectors, SVPWM determines the appropriate switching states to approximate the desired voltage vector. This technique offers high modulation efficiency and low harmonic distortion but requires complex calculations and real-time implementation.

Hybrid PWM techniques combine the advantages of multiple PWM methods to achieve specific performance objectives. For example, combining SPWM with SHEPWM can result in hybrid PWM techniques that offer both harmonic elimination and low switching losses. These techniques are often tailored to meet the requirements of specific applications and may involve trade-offs between complexity, efficiency, and harmonic performance [15–17].

MPPT techniques in PV systems are crucial for optimizing energy extraction from solar panels. Among these techniques, PWM stands out as a widely used method. This essay delves into PWM techniques in MPPT, exploring their principles, applications, and effectiveness in enhancing PV system performance. PWM-based MPPT techniques operate by adjusting the duty cycle of a switching converter to maintain the PV array's operating point at the maximum power point (MPP). The primary goal is to match the PV array's impedance with the load impedance to ensure maximum power transfer. PWM controllers achieve this by periodically switching the voltage across the PV array to control the power flow. One of the prominent PWM techniques is the perturb and observe (P&O) algorithm. P&O adjusts the duty cycle based on the PV array's voltage and current characteristics. It perturbs the operating point by incrementing or decrementing the duty cycle and observes the resulting change in power. If the power increases, it continues in the same direction; otherwise, it reverses the direction. While simple to implement, P&O may suffer from oscillations around the MPP, especially under rapidly changing irradiance conditions. Another PWM technique is the incremental conductance method, which overcomes P&O's oscillation issue by considering the derivative of the power with respect to voltage. Incremental conductance compares the instantaneous conductance (ratio of voltage to current) with its incremental change to determine the direction of perturbation. This approach provides faster convergence to the MPP and exhibits improved performance under dynamic weather conditions. Fractional open-circuit voltage is a PWM technique based on the PV array's open-circuit voltage. Fractional open-circuit voltage tracks the MPP by adjusting the duty cycle to maintain the ratio of the output voltage to the open-circuit voltage at a predefined fraction. By directly utilizing the PV array's characteristics, fractional open-circuit voltage offers simplicity and robustness, making it suitable for low-cost MPPT implementations. Additionally, PWM techniques can be categorized into continuous and discontinuous modes based on the converter's operation. Continuous PWM operates with a fixed switching frequency, maintaining a continuous flow of current through the inductor. This mode ensures smoother operation but may lead to higher switching losses. Discontinuous PWM, on the other hand, allows the inductor current to fall to zero during each switching cycle, reducing switching losses but potentially introducing output voltage ripples [18–20].

Furthermore, PWM techniques can incorporate various control strategies to enhance MPPT performance. Proportional-integral-derivative control is commonly employed to regulate the duty cycle based on the error between the actual and desired operating points. Proportional-integral-derivative controllers adjust the duty cycle dynamically, considering both the present error and its integral and derivative over time. This enables faster response to changing environmental conditions and improved stability around the MPP. Another advanced control strategy is model predictive control, which utilizes a mathematical model of the PV array and converter dynamics to predict future system behavior. Model predictive control optimizes the duty cycle trajectory over a finite horizon while satisfying system constraints, such as input voltage and current limits. By explicitly considering the system's dynamics, model predictive control achieves superior MPPT accuracy and robustness compared to traditional control methods [21,22].

Multilevel inverters are gaining traction due to their ability to produce high-quality output waveforms with reduced harmonic distortion. Integrating advanced control techniques from machine learning, specifically artificial neural networks (ANNs), into multilevel inverter systems can enhance their performance in various applications, including three-phase grid integration. ANNs can be trained to predict future grid conditions such as voltage variations and load changes. This predictive capability can be utilized to adaptively adjust the switching patterns of the multilevel inverter in anticipation of these changes, ensuring optimal performance and stability. ANNs can be employed to optimize the operation of multilevel inverters by learning from historical data and real-time feedback. By continuously evaluating system parameters and grid conditions, ANNs can generate optimal control signals to minimize losses, improve efficiency, and enhance power quality. ANNs can be trained to detect and diagnose faults in the grid or the inverter system itself. By analyzing sensor data and waveform signatures, ANNs can accurately identify abnormalities such as voltage sags, swells, or short circuits, enabling rapid response and fault mitigation strategies. ANNs can facilitate adaptive control schemes that dynamically adjust the inverter's operating parameters based on changing grid conditions and load requirements. By continuously learning from the system's response and performance feedback, ANNs can optimize control strategies in real time to maintain stability and efficiency. ANNs can be used to mitigate harmonic distortion in the output waveform of multilevel inverters. By modeling the nonlinear characteristics of the system and the grid impedance, ANNs can generate compensating signals to cancel out harmonic components, thereby improving power quality and reducing electromagnetic interference. ANNs can enhance the robustness of multilevel inverter control by learning complex mappings between input variables and desired output responses. This enables the development of control strategies that are resilient to parameter variations, disturbances, and uncertainties in the grid and the system dynamics [23–25].

Multilevel inverters, such as nine-level configurations, offer significant advantages in power quality, efficiency, and reduced harmonic distortion compared to traditional two-level inverters. However, their implementation comes with inherent challenges related to cost and complexity. The primary cost drivers in multilevel inverters include semiconductor devices (such as insulated gate bipolar transistors [IGBTs] or metal oxide semiconductor field effect transistors [MOSFETs]), gate drive circuits, and passive components like capacitors and inductors. In a nine-level inverter, multiple levels require additional switching components arranged in series or cascaded H-bridge configurations. This increases the overall material cost compared to simpler inverters. Moreover, the need for precise control and cooling mechanisms adds to the bill of materials. Cost can also escalate due to the requirement for high-voltage handling capabilities of components, which are crucial in achieving the desired output voltage levels without compromising reliability. The design and control of multilevel inverters are more intricate than traditional inverters. They require sophisticated modulation techniques (such as PWM) to ensure accurate switching of multiple levels. The complexity intensifies with higher-level configurations, demanding precise synchronization and control of multiple switches in each phase leg. This complexity extends to the control algorithms, which must manage the switching patterns to achieve low harmonic distortion and high efficiency. Maintenance of multilevel inverters also presents challenges. The increased number of components and interconnections heightens the risk of failures, necessitating robust fault detection and protection mechanisms. Moreover, troubleshooting and repair can be more time-consuming and require specialized expertise. Despite these challenges, the benefits of multilevel inverters justify their use in various applications. They offer smoother output waveforms, reducing harmonic distortion and enhancing power quality. This feature is particularly advantageous in sensitive applications like renewable energy systems, motor drives, and grid-tied inverters where grid compliance and reliability are critical. The higher efficiency and reduced losses also contribute to energy savings over the long term, offsetting initial investment costs [26–30].

The following objectives are listed below for power conversion solutions for three-phase grid-connected PV systems with a nine-level inverter:

- Develop a novel control algorithm for grid-connected PV systems.
- Optimize power conversion efficiency under varying environmental conditions.
- Minimize harmonic distortion in the output voltage waveform.
- Enhance grid stability and reliability during transient events.
- Investigate the impact of grid voltage variations on inverter performance.
- Implement advanced modulation techniques to reduce switching losses.

8.2 PV INTERCONNECTED GRID SYSTEM

In a PV-connected grid system, sunlight is converted into electrical energy by PV panels. The output voltage from the PV panels is typically lower than the grid voltage, necessitating the use of a boost converter to raise the voltage to the required level for grid connection. A MPPT controller ensures that the PV system operates at its MPP, optimizing energy extraction from the panels. The DC power from the boost converter is then fed into an inverter, which converts it into AC power compatible with the grid. However, the AC output from the inverter often contains harmonics and other disturbances, which are mitigated by a filter before being fed into the grid. This filter helps ensure that the AC power injected into the grid meets the required quality standards, such as voltage magnitude, frequency, and waveform shape. Ultimately, the electricity generated by the PV system is seamlessly integrated into the grid, contributing to the overall energy supply while adhering to regulatory requirements and maintaining grid stability. Figure 8.1 illustrates the PV-connected grid system. The *I*–*V* (current–voltage) characteristics of a solar cell circuit describe its behavior under varying levels of illumination and temperature. At a given irradiance and temperature, the relationship between the current and voltage can be represented by a curve. Initially, the current increases linearly with voltage until it reaches

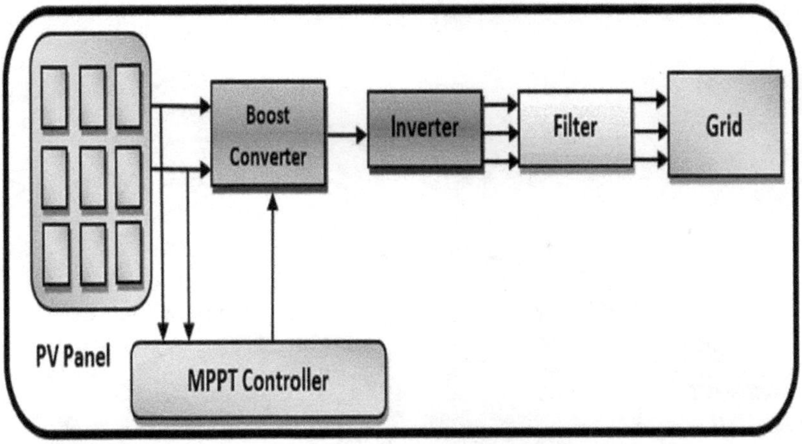

FIGURE 8.1 PV-connected grid system.

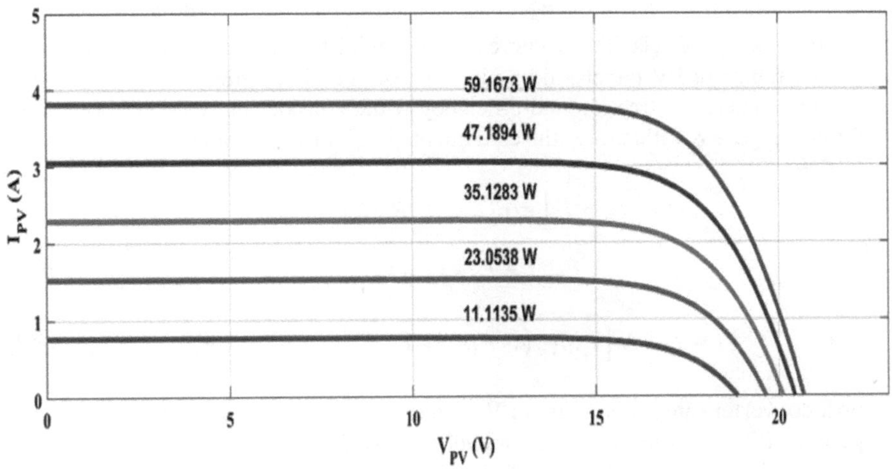

FIGURE 8.2 *I–V* characteristics of the PV panel.

a maximum value, known as the short-circuit current (I_{sc}). As voltage increases, the current decreases until it levels off at zero, corresponding to the open-circuit voltage (V_{oc}). The shape of the *I–V* curve reflects the solar cell's efficiency, saturation current, and series and shunt resistances, crucial for optimizing power output in PV systems. The following expressions (1.1–1.3) illustrate the diode current equation of the PV panel. Figure 8.4 illustrates the equivalent solar cell circuit.

Figure 8.2 illustrates the *I–V* (current–voltage) characteristics of a PV panel, depicting how its current output varies with different voltage levels under varying illumination and temperature conditions. This curve typically exhibits a knee point, representing the maximum power output of the panel. Figure 8.3 shows the corresponding *P–V* (power–voltage) characteristics, providing insight into the power output of the PV panel

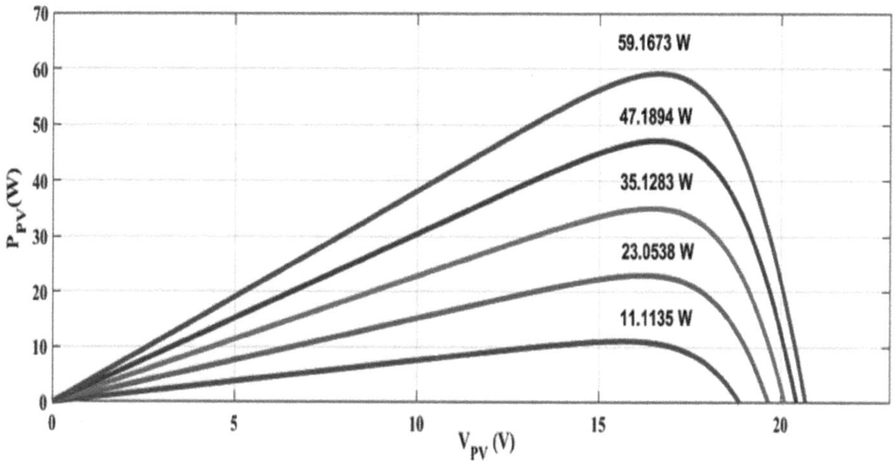

FIGURE 8.3 *P–V* characteristics of the PV panel.

at different voltage levels. These curves are essential for understanding the behavior and performance of PV panels, aiding in the design and optimization of PV systems to maximize energy extraction and efficiency while considering real-world operating conditions. Figure 8.4 illustrates the equivalent circuit of PV solar cell.

$$I_D = I_0\left[\exp\left(q(V + IR_s)KT\right) - 1\right] \tag{8.1}$$

$$I = I_L - I_D - I_{\text{sh}} \tag{8.2}$$

$$I = I_L - I_0\left[\exp\left(q(V + IR_s)KT - 1\right)\right] - (V + IR_s)/R_{\text{sh}} \tag{8.3}$$

A boost converter integrated with a PV system serves to elevate the low-voltage output generated by the solar panels to a level suitable for the subsequent inverter. This converter effectively increases the DC voltage, optimizing power extraction from the PV array under varying environmental conditions. Figure 8.5 illustrates a boost converter PV and inverter connected system, a fundamental setup in PV power generation. At its core is the boost converter, a DC–DC power converter tasked with increasing the low-voltage output from the PV panels to a level suitable for grid connection or storage. This boosted DC voltage is then fed into an inverter, which converts it into AC power, aligning it with the grid's requirements. The boost converter plays a crucial role in optimizing power extraction from the PV panels, especially under varying environmental conditions and irradiance levels. By dynamically adjusting the voltage to maximize power output, the system ensures efficient utilization of the available solar energy. The inverter, on the other hand, ensures seamless integration of the generated power into the grid, adhering to grid codes and standards for voltage, frequency, and waveform quality. Together, these components form a robust and efficient PV power generation system capable of contributing renewable

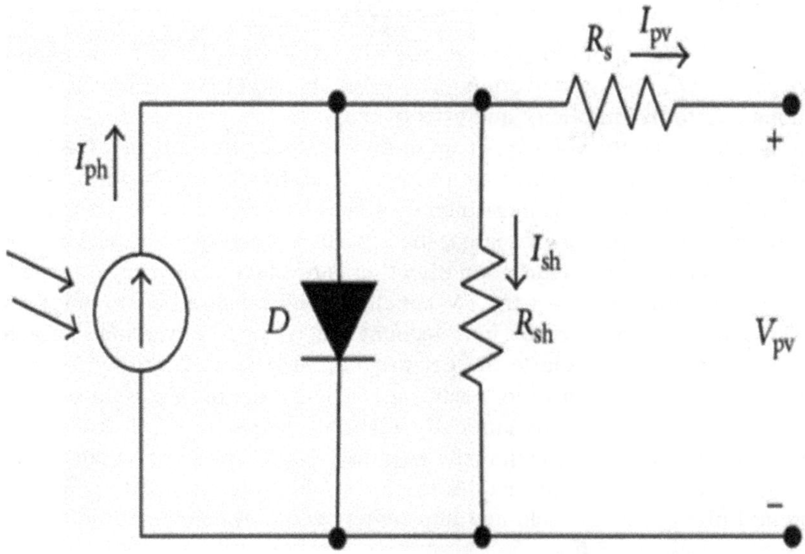

FIGURE 8.4 Equivalent solar cell circuit.

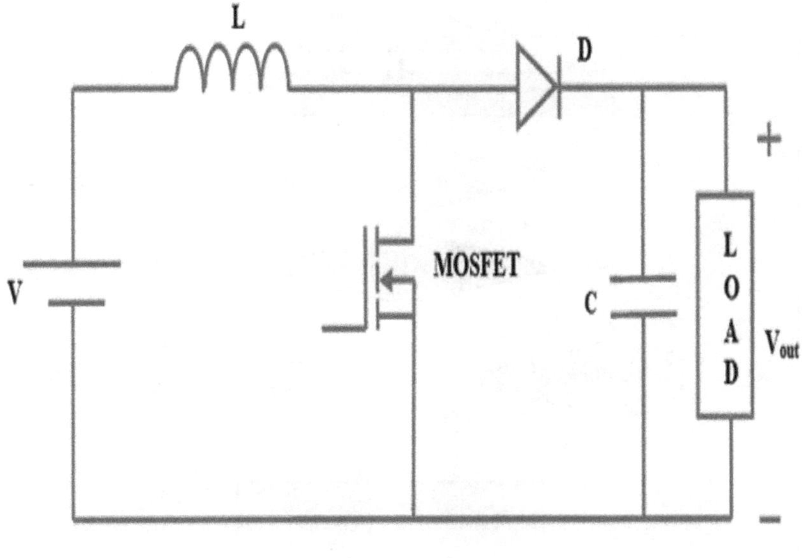

FIGURE 8.5 Boost converter.

energy to the grid. The boosted DC voltage is then supplied to the inverter, which converts it into AC power compatible with the grid or other AC loads. Together, this integrated setup facilitates efficient energy conversion and seamless integration of renewable solar power into the electrical grid.

8.3 MPPT TRACKING FOR THE PV SYSTEM

MPPT is crucial for optimizing the energy extraction from PV systems, especially when coupled with boost converters. The P&O algorithm is a widely used MPPT technique due to its simplicity and effectiveness. In the modified P&O algorithm for boost converters, the voltage perturbation is adjusted dynamically based on the operating point to enhance tracking accuracy. Initially, the algorithm perturbs the operating point by a small voltage increment and observes the resulting change in power output. If the power increases, the algorithm continues in the same direction; if it decreases, the perturbation direction is reversed. This process iteratively converges toward the MPP of the PV panel. In the case of a boost converter, the modified P&O algorithm takes into account the voltage conversion characteristics of the converter. It adjusts the perturbation step size according to the converter's voltage conversion ratio, ensuring efficient tracking across a wide range of operating conditions. By dynamically adapting the perturbation strategy based on the boost converter's behavior, the modified P&O algorithm enhances MPPT accuracy and efficiency. This enables the PV system to consistently operate at its MPP, maximizing energy yield and improving overall system performance which is illustrated in Figure 8.6.

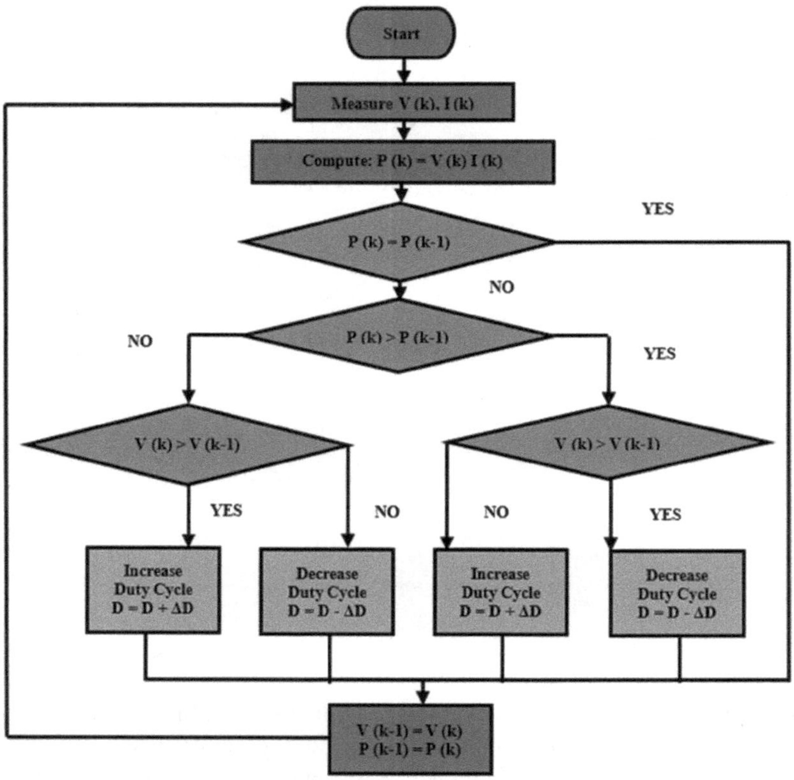

FIGURE 8.6 MPPT tracking of the PV system.

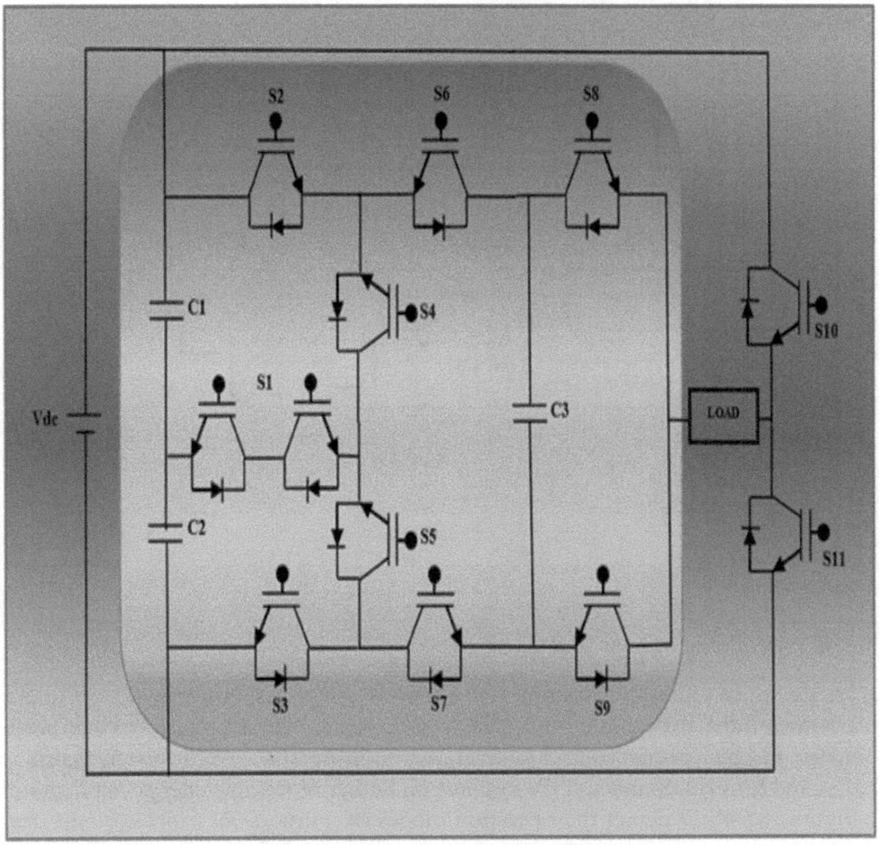

FIGURE 8.7 Nine-level inverter for the grid-connected system.

8.4 NINE-LEVEL MULTILEVEL INVERTER

Figure 8.7 illustrates a nine-level inverter designed for grid-connected systems, featuring three capacitors and 11 power switches. This configuration enables the generation of multiple voltage levels to produce a high-quality AC output waveform. The inverter utilizes three capacitors to create voltage levels at different intervals, while the 11 power switches control the flow of current to achieve the desired output voltage. By precisely switching the power switches in various combinations, the inverter can generate nine different voltage levels, enhancing the quality of the output waveform and reducing harmonic distortion. This topology offers advantages such as improved efficiency, reduced switching losses, and enhanced grid integration capabilities.

In Figure 8.8, the staircase nine-level output voltage ranges from $+2V_{dc}$ to $-2V_{dc}$, showcasing the ability of the inverter to produce multiple voltage levels. Each step in the staircase waveform corresponds to a specific voltage level generated by the inverter. By carefully controlling the switching sequence of the power switches, the

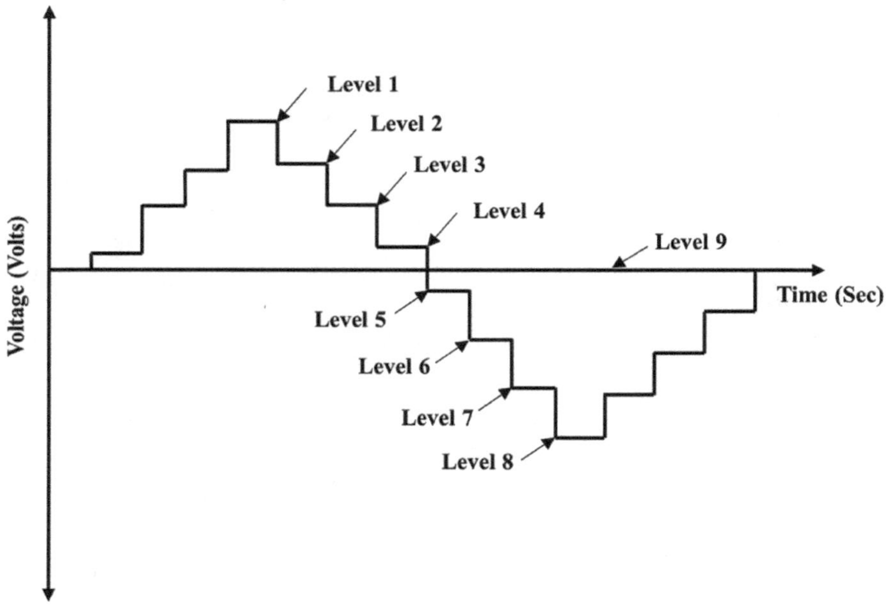

FIGURE 8.8 Staircase nine-level output voltage.

inverter achieves smooth transitions between voltage levels, ensuring a high-fidelity output waveform compatible with grid requirements. This staircase voltage profile demonstrates the precise control and versatility of the nine-level inverter, making it well-suited for grid-connected PV systems and other renewable energy applications.

Figures 8.9–8.17 depict the operation modes of a nine-level capacitor multilevel inverter, showcasing various output voltage levels ranging from $+2V_{dc}$ to $-2V_{dc}$. Each figure represents a specific operation mode, where the capacitor voltages are intelligently combined to achieve the desired output voltage. In Mode-I (Figure 8.9), the output voltage is at $+2V_{dc}$, while subsequent modes depict decreasing voltage levels down to $-2V_{dc}$ in Mode-IX (Figure 8.17). By selectively connecting the capacitors in different configurations using appropriate switching sequences, the inverter can produce precise voltage levels, facilitating high-quality AC output waveforms with reduced harmonic distortion.

8.5 RESULTS AND DISCUSSION

Figures 8.18–8.29 collectively illustrate the comprehensive operational aspects and performance characteristics of a nine-level inverter system integrated with a PV array for grid-connected applications. Figure 8.18 presents the irradiance levels incident on the PV system, crucial for determining solar power generation. Figures 8.19 and 8.20, respectively, depict the PV current and voltage outputs, which vary with changing irradiance and temperature conditions. Figure 8.21 showcases the maximum power received by the PV panel, indicating the optimal operating point for maximum energy extraction. Figure 8.22 illustrates the DC capacitor voltage, vital for the

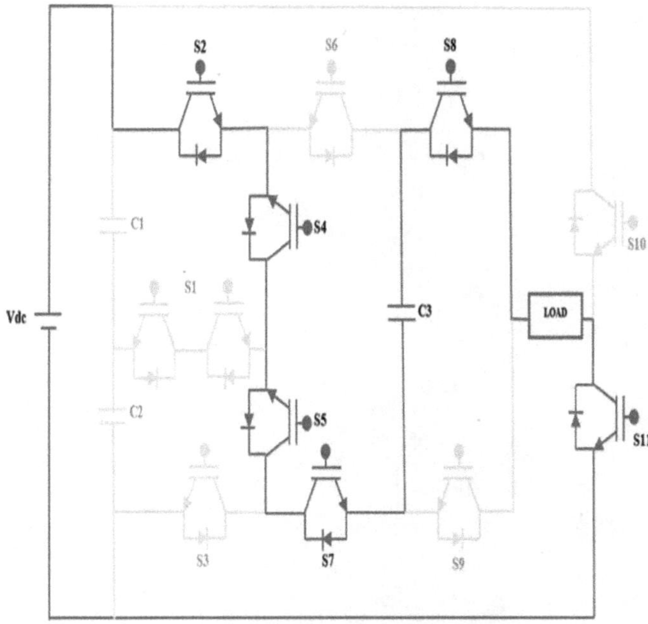

FIGURE 8.9 Mode-I operation of $+2V_{dc}$.

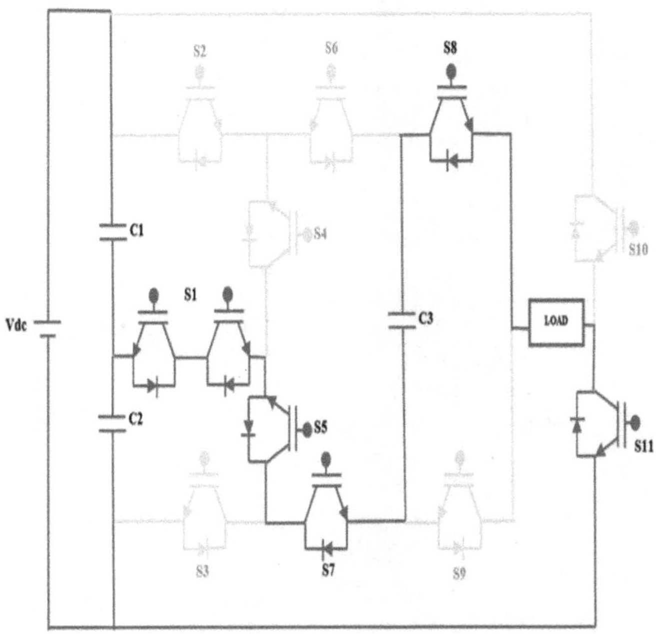

FIGURE 8.10 Mode-II operation of $+1.5V_{dc}$.

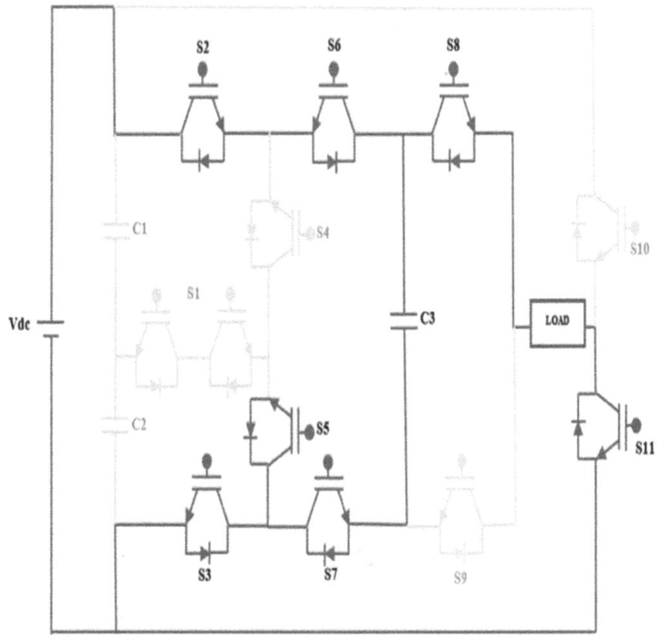

FIGURE 8.11 Mode-III operation of +1V_{dc}.

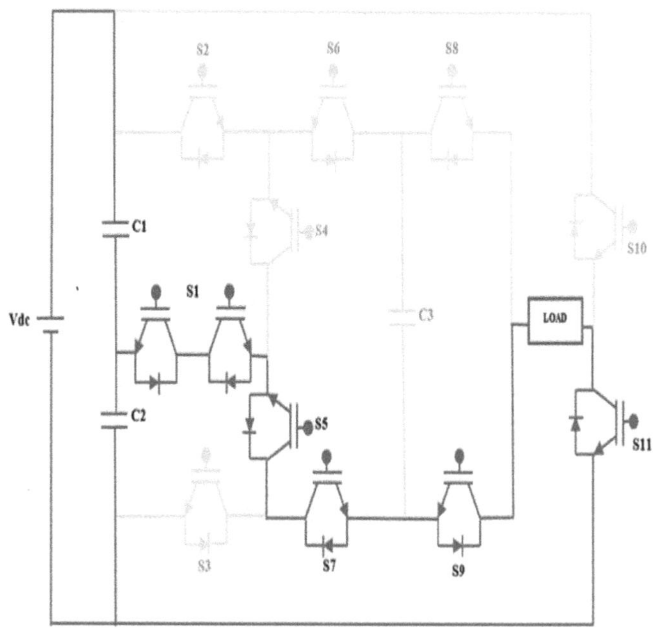

FIGURE 8.12 Mode-IV operation of +0.5V_{dc}.

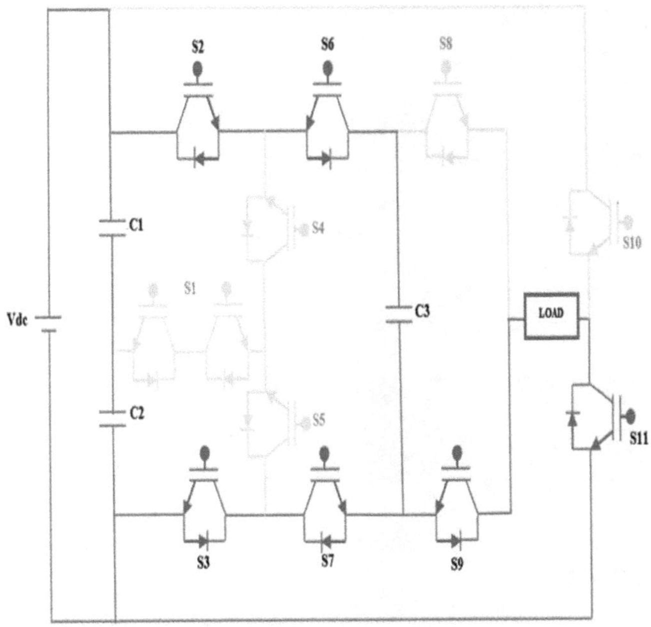

FIGURE 8.13 Mode-V operation of $+0V_{dc}$.

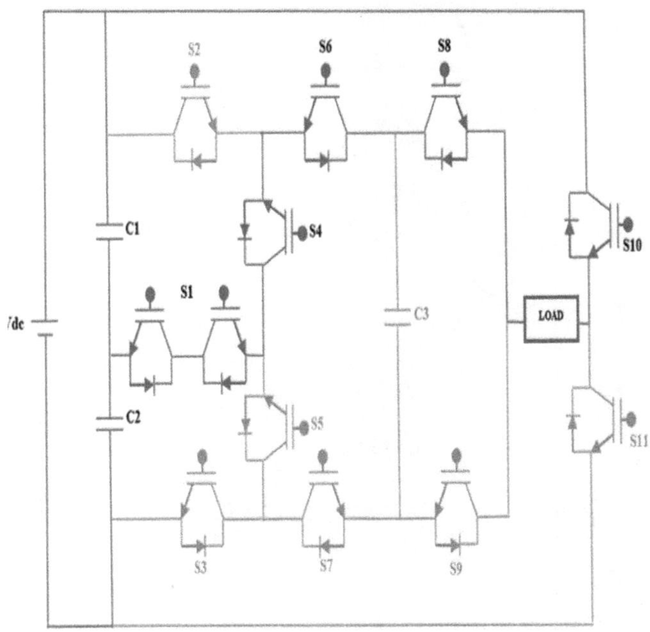

FIGURE 8.14 Mode-VI operation of $-0.5V_{dc}$.

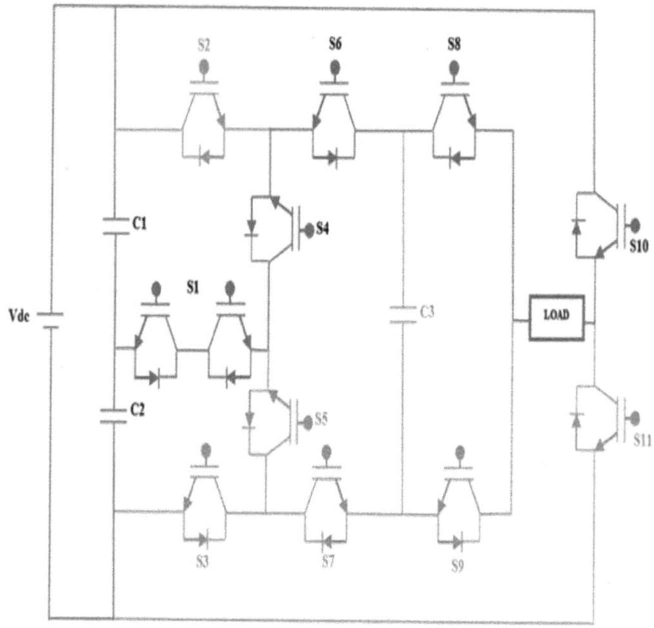

FIGURE 8.15 Mode-VII operation of $-V_{dc}$.

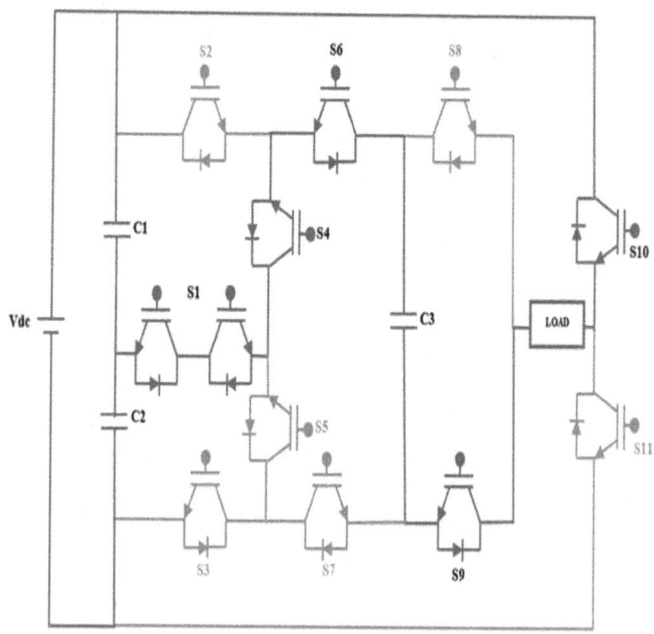

FIGURE 8.16 Mode-VIII operation of $-1.5V_{dc}$.

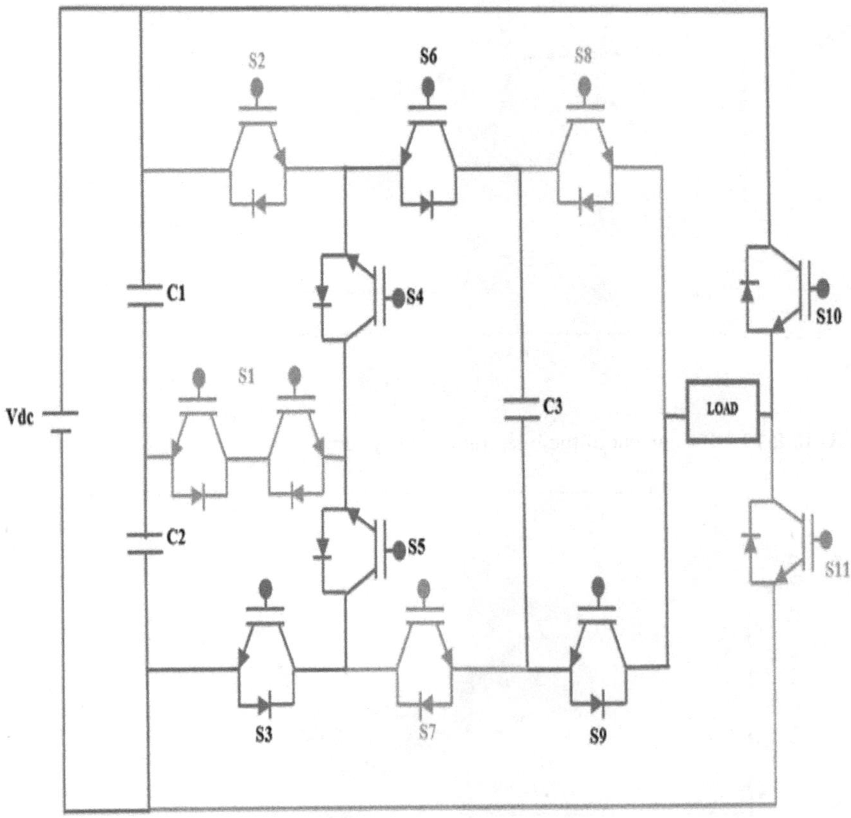

FIGURE 8.17 Mode-IX operation of $-2V_{dc}$.

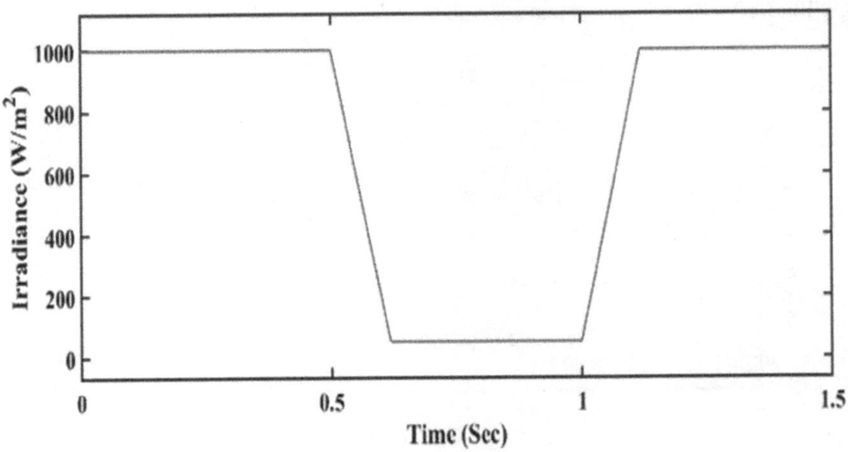

FIGURE 8.18 Irradiance of the PV system.

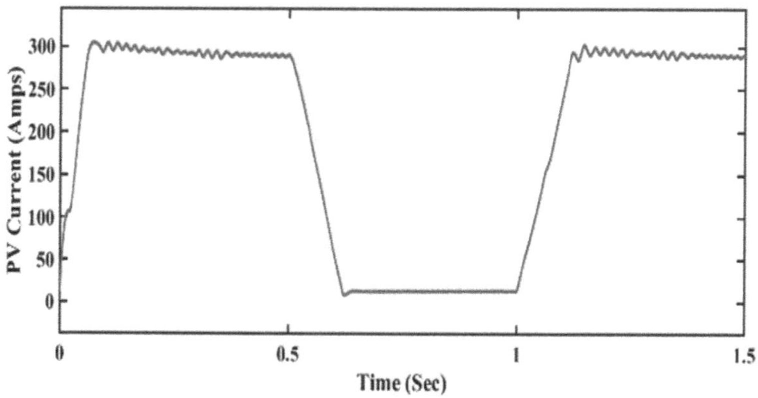

FIGURE 8.19 PV current of the interconnected system.

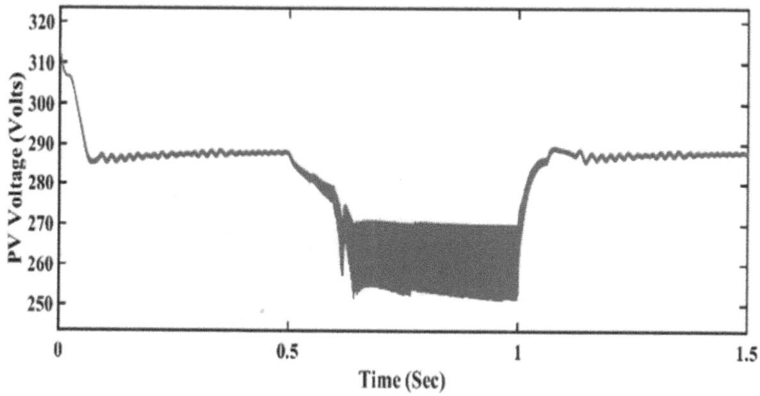

FIGURE 8.20 PV voltage system.

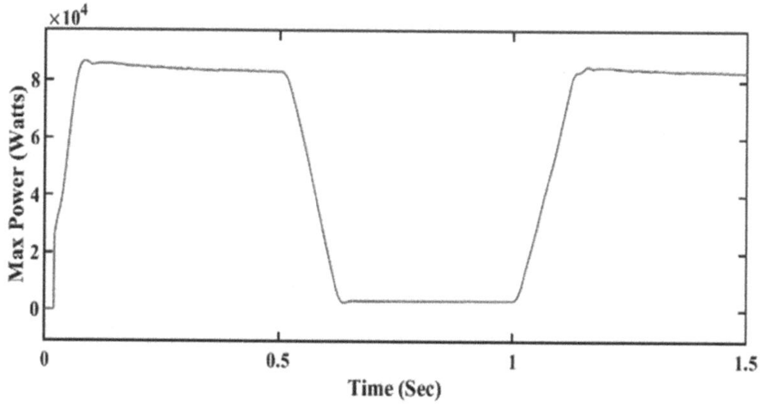

FIGURE 8.21 Maximum power received by the PV panel.

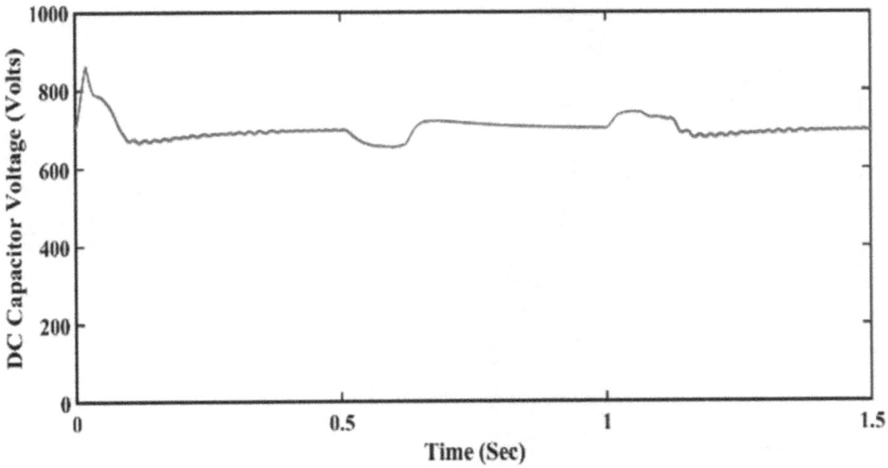

FIGURE 8.22 DC capacitor voltage to the inverter system.

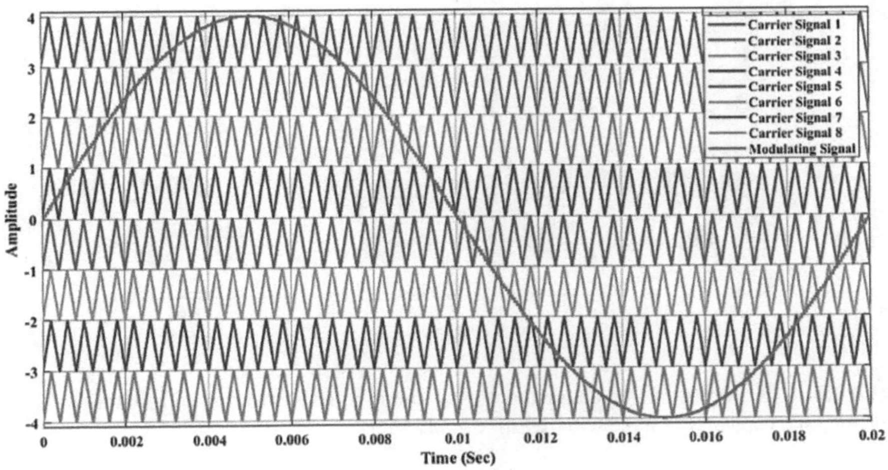

FIGURE 8.23 Nine-level inverter with the PWM technique.

boost converter's operation and ensuring stable input to the inverter. Figures 8.23 and 8.24 detail the performance and computational aspects of PWM techniques used in nine-level inverters. In these figures, triangular and carrier wave signals are employed as modulation strategies to control the switching of the inverter's semiconductor devices. Figure 8.23 typically illustrates the triangular PWM technique, where the width of the pulse is determined by comparing a triangular waveform with a reference signal. This method is straightforward in implementation and computationally lighter compared to more complex techniques, making it suitable for applications where moderate harmonic performance is acceptable. Conversely, Figure 8.24 depicts the carrier wave signal PWM technique, where a high-frequency carrier

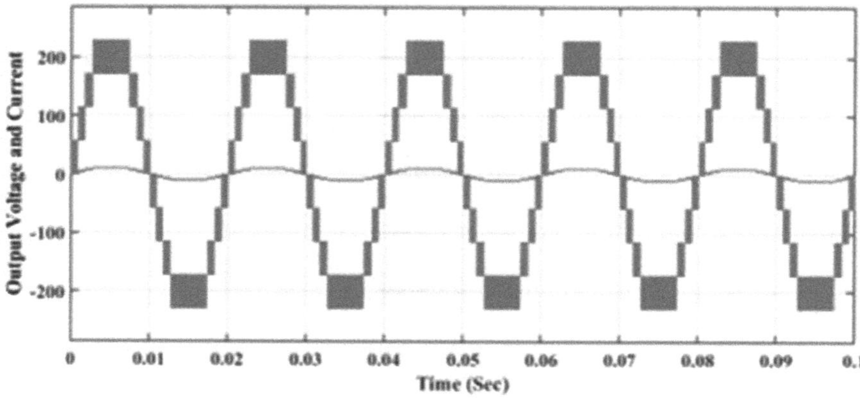

FIGURE 8.24 Nine-level output voltage.

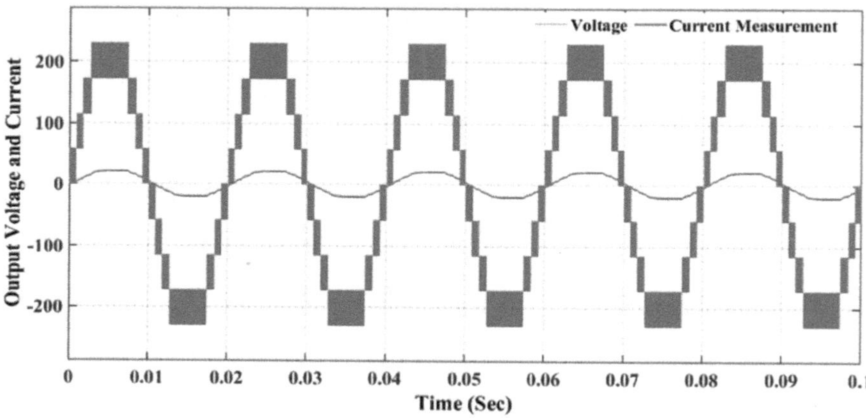

FIGURE 8.25 Performance characteristics of the nine-level inverter with output voltage, current, and capacitor voltage with resistive load.

signal modulates the width of the pulses. This technique allows for finer control over the output waveform, resulting in lower harmonic distortion and improved output quality. However, it requires more computational resources to generate and synchronize the carrier signal with the reference waveform. Both figures demonstrate how these PWM techniques influence the inverter's performance under varying load conditions and grid disturbances. The choice between triangular and carrier wave signal PWM depends on the desired trade-off between computational complexity and output waveform quality, highlighting the importance of optimizing control algorithms for efficient and reliable nine-level inverter operation, elucidating the generation of the output voltage waveform. Furthermore, Figures 8.25–8.27 delve into the performance characteristics of the inverter under different conditions, including resistive load, modulation index variations, and dynamic load changes, portraying its robustness and adaptability. Figure 8.28 conducts THD analysis on the output voltage

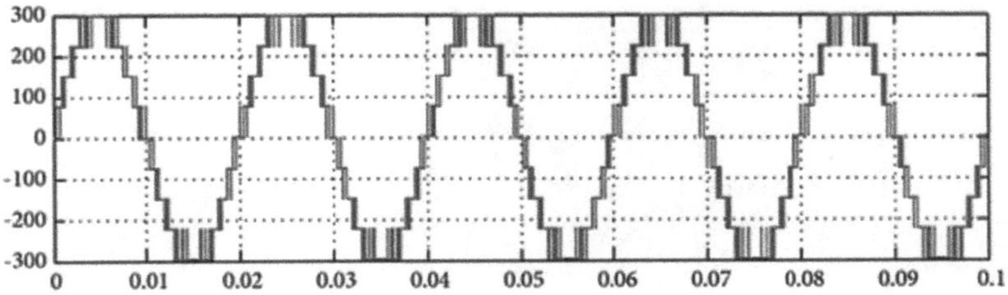

FIGURE 8.26 Performance characteristics of the nine-level inverter with output voltage, current, and capacitor voltage with resistive load for change in the modulation index.

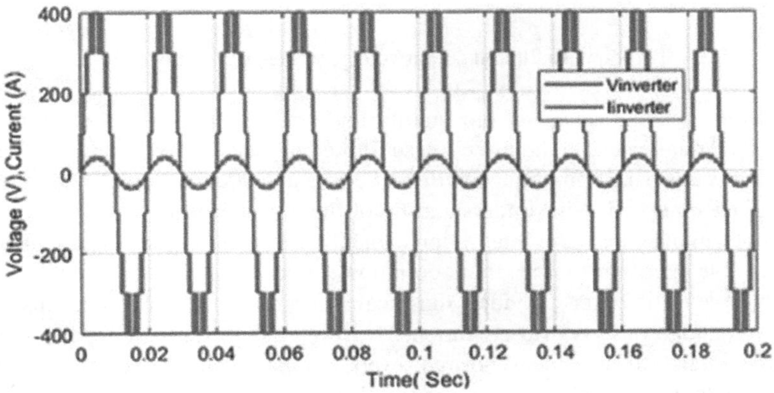

FIGURE 8.27 Performance characteristics of the nine-level inverter with output voltage, current, and capacitor voltage with dynamic load change conditions.

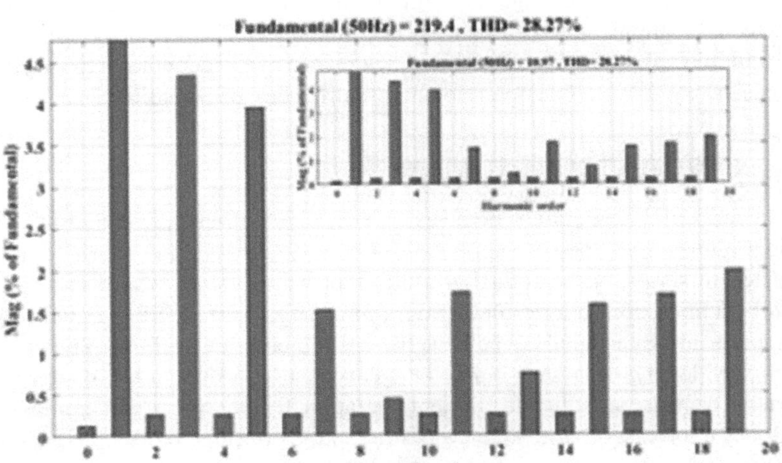

FIGURE 8.28 THD analysis output voltage and current.

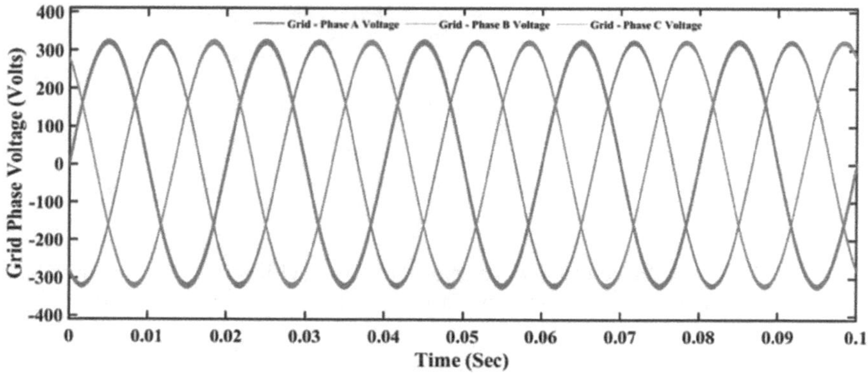

FIGURE 8.29 Three-phase filtered output voltage to the grid supply.

and current, evaluating waveform quality and grid integration suitability. Finally, Figure 8.29 demonstrates the three-phase filtered output voltage's compatibility with grid supply, ensuring compliance with grid codes and standards. Together, these figures offer a comprehensive understanding of the nine-level inverter system's operation, performance, and grid integration capabilities, essential for efficient utilization of renewable energy resources in modern power systems.

A nine-level inverter provides significant improvements in power quality and resilience under various grid conditions. During voltage sags, where the grid voltage momentarily drops below normal levels, a nine-level inverter can continue to supply stable and reliable power output without significant disruption. Its ability to generate higher-quality output waveforms with reduced harmonic distortion ensures compatibility with sensitive loads, maintaining operational efficiency even during grid disturbances. Similarly, in cases of voltage swells or over voltages, the nine-level inverter can efficiently manage and regulate the output voltage, preventing damage to connected equipment and ensuring grid compliance. The precise control afforded by its multiple voltage levels allows for quick response and adaptation to varying grid frequencies, ensuring stable operation across different grid conditions.

8.6 EXPERIMENTAL VALIDATION

The hardware implementation involves several key components, including multiple H-bridge units, a DC power supply, and a control circuit. The H-bridge units are connected in series, each producing a different voltage level. These units are controlled by a microcontroller or field programmable gate array (FPGA), which generates appropriate gate signals to achieve the desired output waveform. The DC power supply provides the necessary input voltage, typically through a set of capacitors or batteries. To ensure smooth and efficient operation, the control circuit employs PWM techniques to precisely switch the transistors in the H-bridge units. This modulation technique helps in minimizing harmonics and achieving a nearly sinusoidal output voltage. Figure 8.30 illustrates the three-phase filtered output voltage connected to the grid supply using a nine-level inverter with a digital storage oscilloscope. The system is comprised of various components, including a controller and driver

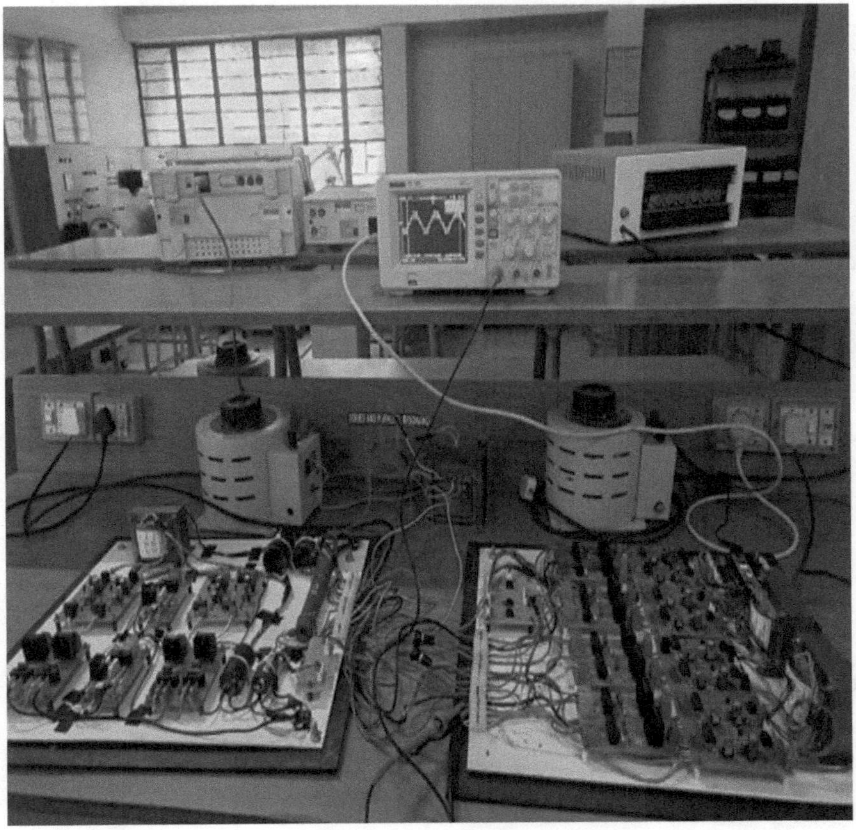

FIGURE 8.30 Three-phase filtered output voltage to the grid supply.

unit, specifically the TMS320F28445 microcontroller, which is responsible for managing the overall operation of the inverter. The inverter employs different power transistors, such as IPA60R480SA5, FA60R480S9, and IR60R4856, each with specific characteristics: a resistance (Rds(on)) of approximately 340–350 mΩ and varying capacitance values of 130pF, 110pF, and 100pF, respectively. These transistors are crucial for switching operations and efficient power conversion. The TLP230 is used as an optocoupler, providing electrical isolation between the high-power components and the low-power control circuitry. The inverter operates at a switching frequency of 2.5 kHz, which is appropriate for minimizing losses and ensuring efficient energy conversion. The output voltage is filtered to produce a cleaner sinusoidal waveform, suitable for grid integration at a standard frequency of 50 Hz. The hardware setup, labeled as FES8FT, highlights the integration and performance of these components, demonstrating effective control and conversion of DC to AC power. This ensures the stability and reliability of the power supplied to the grid, essential for both residential and industrial applications.

Figures 8.31–8.33 illustrate experimental waveforms for a nine-level inverter at modulation indices (Ma) of 0.5, 0.6, and 0.7, respectively, each showcasing three

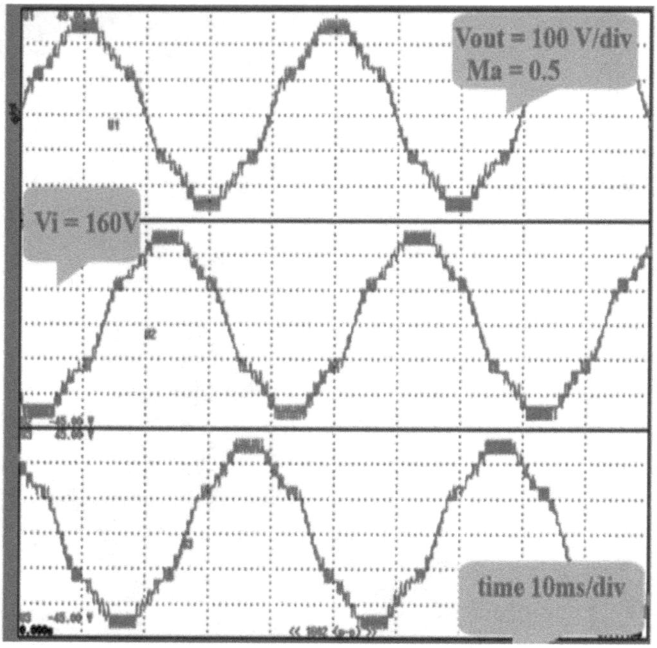

FIGURE 8.31 Experimental waveform Ma = 0.5 with three different phase modulations.

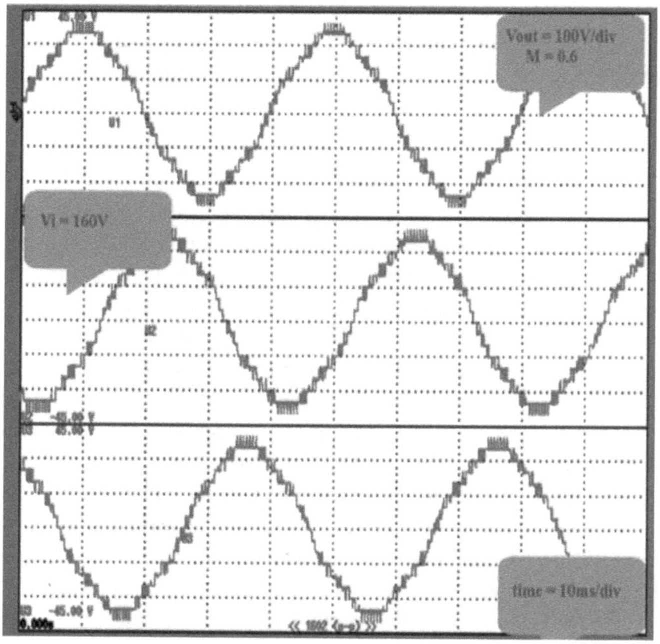

FIGURE 8.32 Experimental waveform Ma = 0.6 with three different phase modulations.

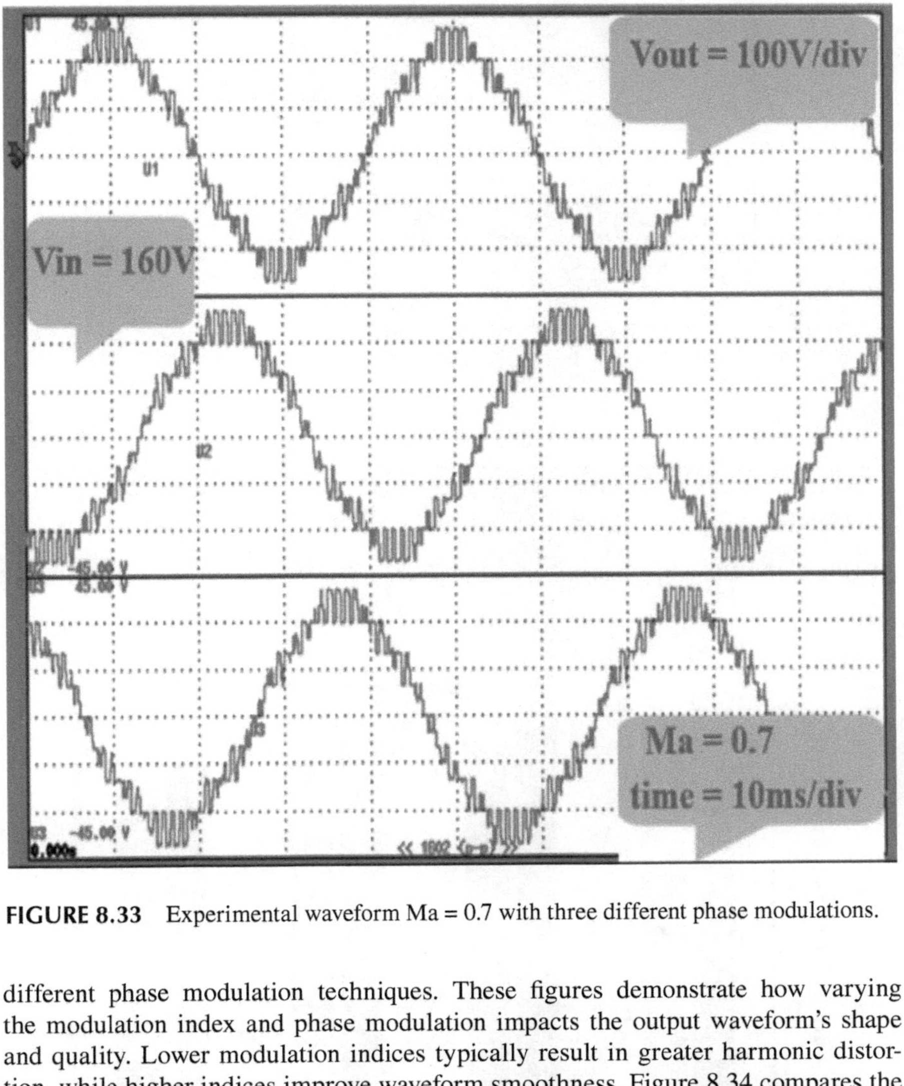

FIGURE 8.33 Experimental waveform Ma = 0.7 with three different phase modulations.

different phase modulation techniques. These figures demonstrate how varying the modulation index and phase modulation impacts the output waveform's shape and quality. Lower modulation indices typically result in greater harmonic distortion, while higher indices improve waveform smoothness. Figure 8.34 compares the efficiency of the inverter at a power output of 550W across different modulation techniques. This figure highlights which modulation method yields the highest efficiency under specified conditions. Figure 8.35 presents the THD comparisons for the same experimental setups, providing insights into the harmonic content of the output waveforms. Lower THD values indicate better waveform quality. Table 8.1 summarizes the experimental results and analysis of efficiency and THD, offering a comprehensive overview of how each phase modulation technique performs in terms of efficiency and harmonic distortion across different modulation indices. This table aids in identifying the optimal modulation strategy for achieving the best balance between efficiency and waveform quality.

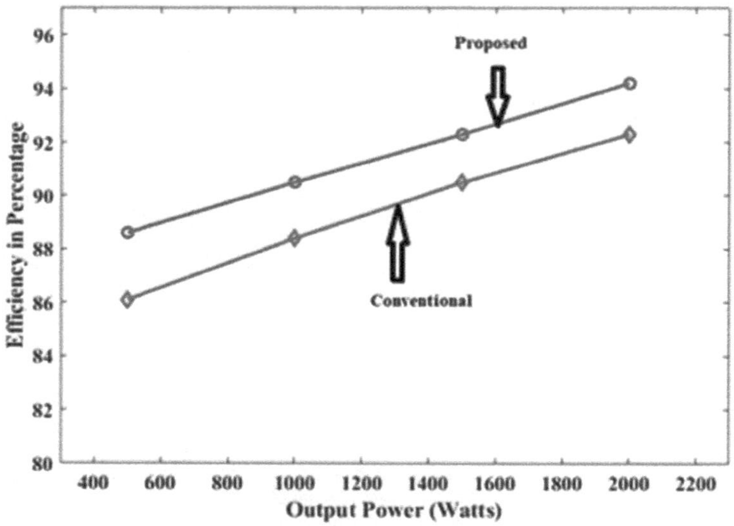

FIGURE 8.34 Efficiency comparisons when power is 550W.

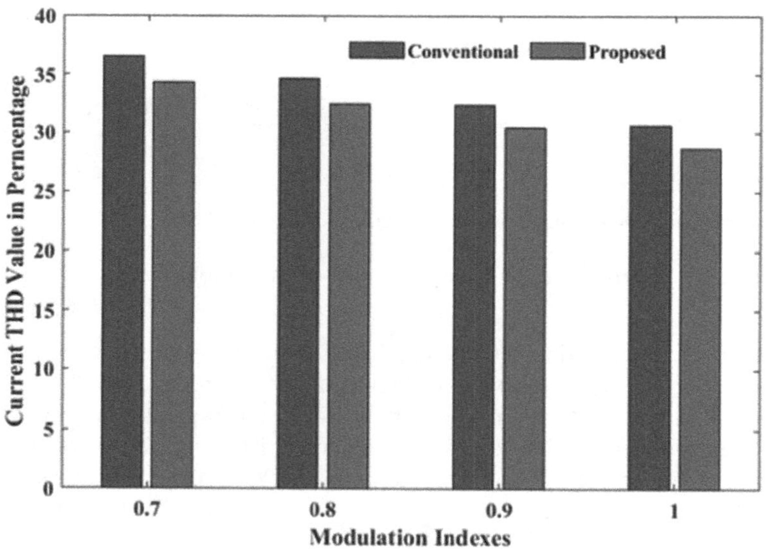

FIGURE 8.35 THD comparisons.

8.7 CONCLUSION

This chapter investigates advancements in power conversion solutions for three-phase grid-connected PV systems utilizing a nine-level inverter. Emphasizing THD, PWM technique, efficiency enhancement, and various inverter operation modes, it aims

TABLE 8.1
Experimental Results and Analysis

Parameters	Power/Load	Value
Efficiency	300W	(95.4–95.8) %
	550W	(96.3–96.9) %
	65W	(92.4–90.2) %
THD	RL load	1.5%

to optimize power quality and energy extraction from PV arrays. Integration of a boost converter augments efficiency by optimizing power flow. Evaluation of different inverter modes considers load variations and grid conditions. Through thorough analysis and simulation, this study provides insights into grid-connected PV systems with nine-level inverters, highlighting their potential to enhance power conversion efficiency and ensure high-quality power delivery to the grid. These findings contribute to advancing renewable energy integration into the grid, fostering sustainable energy practices, and promoting efficient utilization of solar resources. Additionally, investigating novel modulation techniques and hardware improvements can further enhance system efficiency and reliability. Integration with energy storage systems and hybrid renewable sources will also be investigated to address intermittency and grid stability concerns.

REFERENCES

1. M. Mohammadi, M. R. Banaei, and A. Yazdian, "Advanced Control Strategy for Nine-Level Inverter-Based Grid-Connected Photovoltaic Systems," *IEEE Transactions on Energy Conversion*, vol. 36, no. 2, pp. 778–789, 2021.
2. H. Zhang and Z. Xu, "Optimal Design and Analysis of a Nine-Level Inverter for Grid-Connected PV Systems," *IEEE Transactions on Industrial Electronics*, vol. 68, no. 8, pp. 6891–6900, 2021.
3. K. R. Cheepati, Ch. R. Reddy, M. M. Alqahtani, and M. Khalid, "Design of Triple Tuned Passive Harmonic Power Filter-A Novel Approach," *IEEE Canadian Journal of Electrical and Computer Engineering*, vol. 46, no. 4, pp. 270–277, 2023.
4. K. Liu and Y. Yang, "Efficiency Optimization of Nine-Level Inverter-Based PV Systems Using Genetic Algorithm," *IEEE Journal of Photovoltaics*, vol. 11, no. 7, pp. 1878–1885, 2021.
5. Y. Chen et al., "Model Predictive Control of Nine-Level Inverter for Grid-Connected PV Systems with Uncertainty," *IEEE Transactions on Control Systems Technology*, vol. 29, no. 1, pp. 324–335, 2022.
6. E. Parimalasundar, S. Jayakumar, T. Devaraju, and V. S. Chandrika, "Performance Analysis of a Seven-Level Multilevel Inverter in Grid-Connected Systems," *International Journal of Electrical and Electronics Engineering*, vol. 10, no. 6, pp. 9–22, 2023.
7. S. Li and J. Zhang, "Comparative Study of Nine-Level Inverter Topologies for Grid Integration of PV Systems," *IEEE Transactions on Power Delivery*, vol. 37, no. 3, pp. 1464–1473, 2022.
8. M. S. Sujatha, S. Sreelakshmi, and K. Suresh, "Mitigation of Harmonics for Five Level Multilevel Inverter with Fuzzy Logic Controller," *Electrical Engineering & Electromechanics*, vol. 01, no. 4, pp. 52–56, 2023.

9. B. Perumal, K. Suresh, and E. Parimalasundar, "Fault Analysis in the 5-Level Multilevel NCA DC–AC Converter," *Automatika*, vol. 64, no. 3, pp. 606–612, 2023.
10. R. Wang and Q. Zhou, "Voltage Regulation Strategy for Nine-Level Inverter in Grid-Connected PV Systems Considering Unbalanced Grid Conditions," *IEEE Transactions on Sustainable Energy*, vol. 13, no. 8, pp. 4639–4650, 2022.
11. H. Li et al., "Optimal Design of LCL Filter for Nine-Level Inverter in Grid-Connected PV Systems," *IEEE Transactions on Power Electronics*, vol. 37, no. 12, pp. 11092–11101, 2022.
12. K. Suresh and E. Parimalasundar, "IPWM Based IBMSC DC-AC Converter Using Solar Power for Wide Voltage Conversion System," *IEEE Canadian Journal of Electrical and Computer Engineering*, vol. 45, no. 4, pp. 394–400, 2022.
13. E. Parimalasundar and K. Suresh, "Fault Analysis and Compensation in a Five Level Multilevel DC-AC Converter," *El-Cezeri*, vol. 10, no. 1, 99–108, 2023.
14. C. Wu and W. Liu, "Dynamic Performance Analysis of Nine-Level Inverter in Grid-Connected PV Systems During Transient Events," *IEEE Transactions on Industry Applications*, vol. 58, no. 6, pp. 6752–6761, 2022.
15. B. N. Rao, Y. Suresh, B. S. Naik, and K. Aditya, "Implementation of Novel Toroidal Transformer-Based Single-Phase Multilevel Inverter," *Electrical Engineering,* vol. 106, no. 5, pp. 5291–5302, 2024.
16. H. Zhao et al., "Efficient Power Management Strategy for Hybrid Energy Systems with Nine-Level Inverter-Based PV Systems," *IEEE Transactions on Smart Grid*, vol. 13, no. 5, pp. 2871–2880, 2023.
17. M. Wang and X. Chen, "Optimization of Pulse Width Modulation Strategies for Nine-Level Inverter in Grid-Connected PV Systems," *IEEE Transactions on Industrial Informatics*, vol. 18, no. 1, pp. 101–110, 2025.
18. H. Zhou et al., "Experimental Validation of Nine-Level Inverter Performance in Grid-Connected PV Systems," *IEEE Transactions on Sustainable Energy*, vol. 15, no. 7, pp. 3812–3821, 2025.
19. K. Jena, K. K. Gupta, D. Kumar, and S. RanjanKabat, "Single-Stage Single-Phase 5-Level Switched-Capacitor Multilevel Inverter," Iranian Journal of Science and Technology Transactions of Electrical Engineering, vol. 48, no. 1, pp. 265–276, 2024.
20. E. Afshari et al., "Control Strategy for Three-Phase Grid-Connected PV Inverters Enabling Current Limitation under Unbalanced Faults," *IEEE Transactions on Industrial Electronics*, vol. 64, no. 11, pp. 8908–8918, 2017.
21. G. Zhang et al., "Real-Time Control and Monitoring System for Nine-Level Inverter-Based PV Systems," *IEEE Transactions on Power Delivery*, vol. 38, no. 9, pp. 3887–3896, 2023.
22. E. Parimalasundar, S. Jayakumar, R. Ravikumar, and K. Suresh, "Investigation Analysis of Open Circuit and Short Circuit Fault on Cascaded H-Bridged Multilevel Inverter Using Artificial Neural Network Approach," *International Journal of Electrical & Electronics Research*, vol. 10, no. 2, pp. 320–326, 2022.
23. J. Wang and Y. Zhang, "Dynamic Voltage Stability Analysis of Nine-Level Inverter-Based PV Systems with Energy Storage," *IEEE Transactions on Power Systems*, vol. 40, no. 2, pp. 1123–1132, 2022.
24. K. Suresh and E. Parimalasundar, "ITBC Controlled IPWM for Solar Based Wide Range Voltage Conversion System," *IETE Journal of Research*, vol. 70, no. 4, pp. 4278–4286, 2024.
25. N. Sandeep, J. S. M. Ali, U. R. Yaragatti, and K. Vijayakumar, "Switched-Capacitor-Based Quadruple-Boost Nine-Level inverter," *IEEE Transactions on Power Electronics*, vol. 34, no. 8, pp. 7147–7150, 2019.

26. Y. Zhang, Q. Wang, C. Hu, W. Shen, D. G. Holmes, and X. Yu, "A Nine-Level inverter for Low-Voltage Applications," *IEEE Transactions on Power Electronics*, vol. 35, no. 2, pp. 1659–1671, 2020.
27. K. Suresh and E. Parimalasundar, "Newly Designed Single-Stage Dual Leg DC-DC/AC Buck-Boost Converter for Grid Connected Solar System," *International Journal of Circuit Theory and Applications*, vol. 51, no. 11, pp. 5452–5469, 2023.
28. G. Chen, P. Lyu, J. Song, Z. Zhang, and L. Mo, "A Wide Input Voltage Range Switched-Capacitor Multilevel Inverter Based on Adjustable Number of Output Levels," *IEEE Transactions on Power Electronics*, vol. 38, no. 12, pp. 15000–15015, 2023.
29. S. P. Gautam, M. Jalhotra, L. K. Sahu, M. R. Kumar, and K. K. Gupta, "A Survey on Fault Tolerant and Diagnostic Techniques of Multilevel Inverter," *IEEE Access*, vol. 11, pp. 60866–60888, 2023.
30. A. Tirupathi, A. Kirubakaran, and S. V. Tirumala, "A New Hybrid Flying Capacitor–Based Single-Phase Nine-Level Inverter," *International Transactions on Electrical Energy Systems*, vol. 29, no. 12, pp. 01–15, 2019.

9 Multilevel Inverters for Electric Drives

Sathish Sathya Agila and Venkatraman Thiyagarajan

9.1 INTRODUCTION

The automobile industry is dedicating more and more of its research and development resources to electric vehicles (EVs) as a result of environmental degradation and the depletion of fossil fuels. Using EVs might end up in increased efficiency of energy, sharper acceleration, and less frequent maintenance, in alongside reducing or omitting the greenhouse emissions [1].

India may reduce its projected passenger road-based, mobility-related energy demand by 64% and its carbon emissions by 37% by 2030 if it promotes shared, electrically powered, and electronically connected transportation in the near future [2].

Electric vehicles can be classified into battery electric vehicles, hybrid electric vehicles, plug-in hybrid electric vehicle, and fuel cell electric vehicle. Battery electric vehicles are more efficient in comparison with other types of vehicles.

However, the biggest barriers to the swift development of transportation electrification include the prolonged charging times and constrained maximum driving distances of battery-powered electric vehicles before recharging [3].

On the other hand, the drivetrains of electric vehicles need electric machines that are vital. A significant number of traction drive systems have been trending toward including some kind of permanent magnet machine during the past several years.

The volume-specific power and propulsion efficiency of machines are constantly being improved, leading to innovative designs, advancements in fundamental machine configurations, and the introduction of new machine classes [4].

DC machines, induction machines, permanent magnet machines, switching reluctance machines, and synchronous reluctance machines are the fundamental machine types employed in electric vehicles. The numerous factors to be taken into consideration while picking the electric machine to be used in electric vehicles include torque ripple, vibration, control complexity, efficiency, power factor, cost, power density, and specific power [5].

Therefore, an effective controller is required for every electric vehicle with an electric motor, and multilevel inverters can be used to achieve this. Multilevel voltage source inverters often replace their two-level equivalents in medium- to high-voltage applications due to their ability to achieve reduced current and voltage total harmonic distortion (THD), lesser electro magnetic interference (EMI) and common mode voltage, improved effectiveness, and less electrical stress on devices.

 DOI: 10.1201/9781003496946-9

Numerous topologies that attempt to reduce switch count, improve modulation techniques, and be fault-tolerant have been developed by various researchers starting with traditional topologies [6–8].

Consequently, this article discusses the uses of multilevel inverters in electric drives and electric automobiles. It also discusses several topologies, modulation methods, and electric drive control ideas that have been proposed recently.

This article is divided into four sections. Section 9.2 enumerates the applications of multilevel inverters in electric drives, Section 9.3 illustrates the multilevel inverter applications in electric vehicles, and Section 9.4 deals with the conclusion.

9.2 MULTILEVEL INVERTERS IN ELECTRIC DRIVES AND APPLICATIONS

Numerous industrial and household applications, such as manufacturing plants, transit systems, garment factories, fans, pumps, motors, robots, and so on, heavily rely on the drive system. Electric motors, hydraulic engines, gas turbines, turbines that generate steam, and diesel or petrol engines are all powered by drives. This was created in 1838 in Russia by B.S. Iakobi, who experimented with a boat-propelled DC electric motor powered by a storage battery.

Today, electric drives can be encountered practically everywhere. Electric drives and applications greatly benefit from the use of multilevel inverters. Despite the fact that industrial applications can use a variety of drives, only synchronous and induction motor drives are covered in this review.

This research presents a multiple phase 11 level inverter that is tolerant to faults, and it was designed specifically for the pole-phase modulated multiphase induction machine [9], based on the double inverter idea. In this study, it is asserted that a topology with fewer switches and DC sources can withstand any fault scenario. Significant reductions were also made to the torque distortions.

A technique for connecting two separated load-commutated silicaon controlled rectifier (SCR)-based current source inverters in parallel mode to create a multilayer current waveform was presented in the work in [10].

The key element of this method is the introduction of a 30-degree phase difference between the two current source inverters. This phase shift enables the development of a multilevel current waveform. It offers substantial benefits in many applications, such as lower distortion levels and enhanced power delivery. Also, the induction motor drive can run at low speeds, providing full load torque, by employing the proposed design.

In [11], a suggested control method for adjusting the voltage proportion between a dual inverter's fixed capacitor and DC link was presented. In order to obtain twice the voltage utilization without compromising output quality, a ratio selection approach was offered.

The optimal voltage ratio under various load power factors was determined and discussed. It was shown that inverters with a ratio of 1:1 had a very high-voltage usage, whereas inverters with a ratio of 1:0.5 had the least amount of harmonic

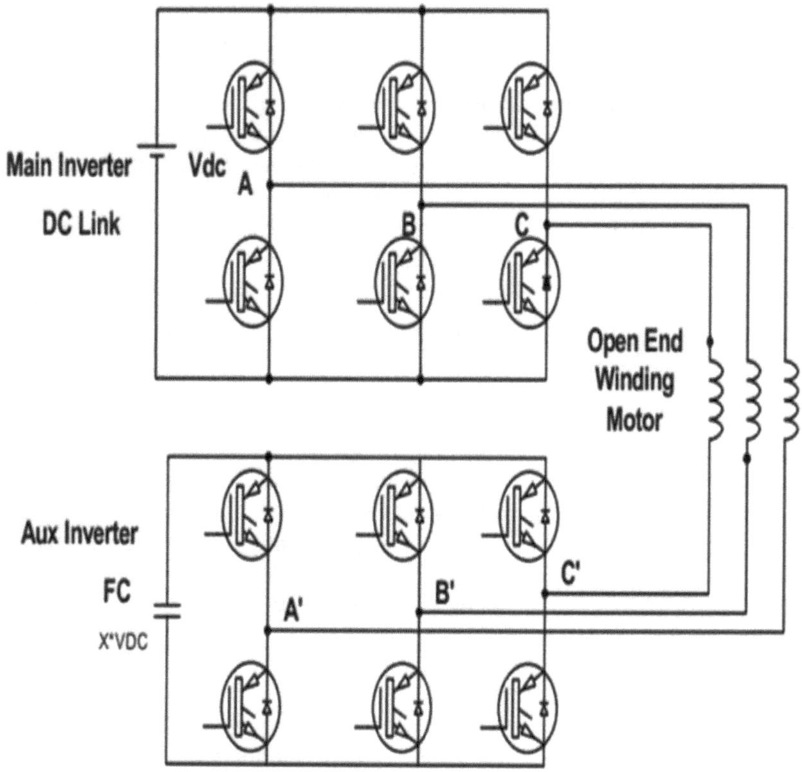

FIGURE 9.1 Schematic diagram of the dual inverter with FC [11].

distortion in the inverter output. The schematic diagram of the dual inverter with a fixed capacitor is shown in Figure 9.1.

For multilevel voltage generation, a nine-level topology with a single-DC-link and series-connected capacitors as stacked voltage sources is used in [12].

The motor currents in phase are used to balance any DC link capacitor voltage disturbance, independent of the power factor of the load or modulation indices. Without the use of additional self-balancing circuits, capacitor voltage self-balancing is achieved. For EV applications, these low-voltage capacitors can be replaced with batteries.

This paper proposed a five-level stacked cell technique with a lower component count, as depicted in Figure 9.2 [13]. Space vector modulation cannot be used with the suggested topology. Subsequently, an online common mode (CM) removal method and a loss steering technique were presented to enhance drive endurance and inverter dependability at lower speeds.

When operating at lower speeds, a significant value of capacitance is required for cascaded H-bridge (CHB) filters in a multilevel dodecagonal voltage space vector generating approach for variable-speed drive applications with single-DC-link operation.

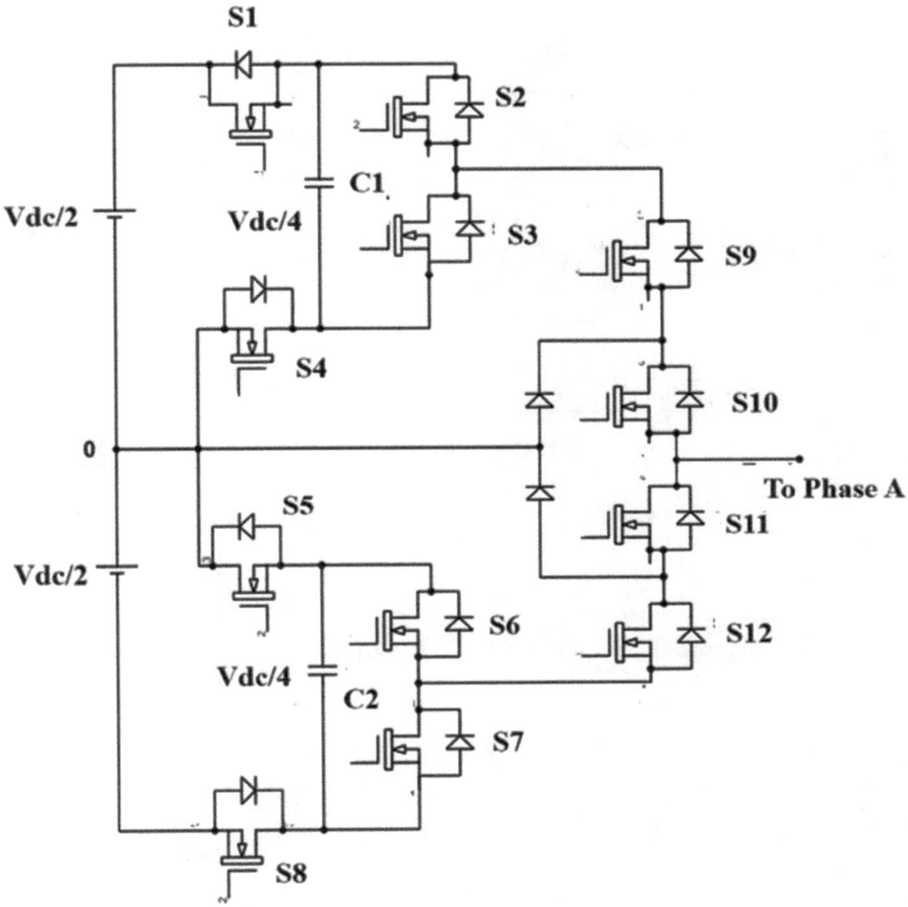

FIGURE 9.2 A five-level inverter topology proposed by [13].

A new technique for minimizing capacitance requirements for full-speed operation has been presented in [14], which involves constructing vector redundancies using modular and equal-voltage CHBs. Voltage ripple may be reduced by choosing the best vector redundancy selection by keeping an eye on instant capacitor voltages and phase currents. The suggested method lowered the capacitor size by 1/30th at 1 Hz. Figure 9.3 depicts the architecture of the inverter discussed in [14].

This paper [15] describes a medium voltage induction motor drive system fed by a 36-pulse converter and a five-level CHB multilevel inverter (CHB-MLI). The multiwinding transformer for a 36-pulse AC-DC converter is designed to use the fewest windings while improving power quality at the grid end. To regulate a five-level CHB-MLI-fed induction motor, modified carrier-level shifted pulse width modulation (PWM) with an intrinsic rotating switching pattern is employed. The performance of the drive system is assessed under several operating conditions, such as starting out, steady-state, speed, and load disturbances.

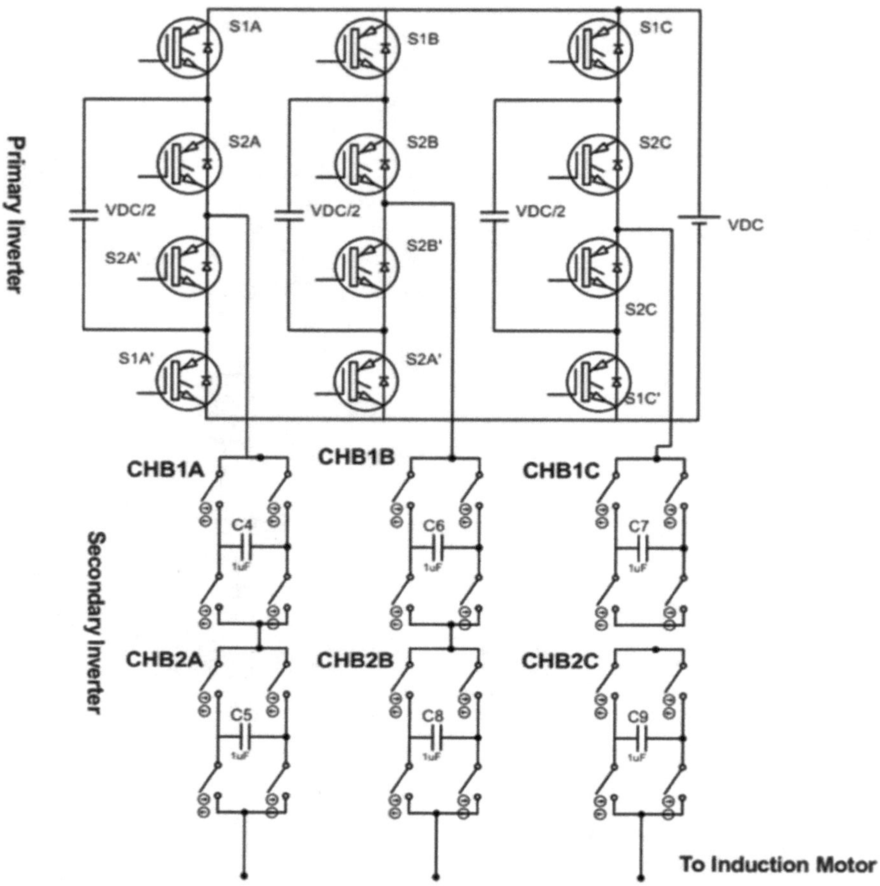

FIGURE 9.3 Inverter topology proposed in [14].

The concept of stacking three 17-level inverters to create a 49-level inverter appears in [16]. metal oxide semiconductor field effect transistors (MOSFETs) are employed in this design since it requires blocking very low voltages. The nearest level control is used to reduce switching losses.

This topology is ideal for electric vehicles because it requires less front-end DC voltage. The capacitor voltage balancing mechanism employed does not get impacted by any modulation index or load power factor. The topology proposed in [16] is shown in Figure 9.4.

A novel system consisting of a fuzzy logic controller, buck-boost converter, three-phase induction motor, and photovoltaic-based seven-level multilevel inverter is proposed by the research in [17], which also examines the speed of response and harmonics.

The current harmonics and the total harmonic distortion were reduced by using the fuzzy logic and proportional integral (PI) controller. The paper's shortcomings

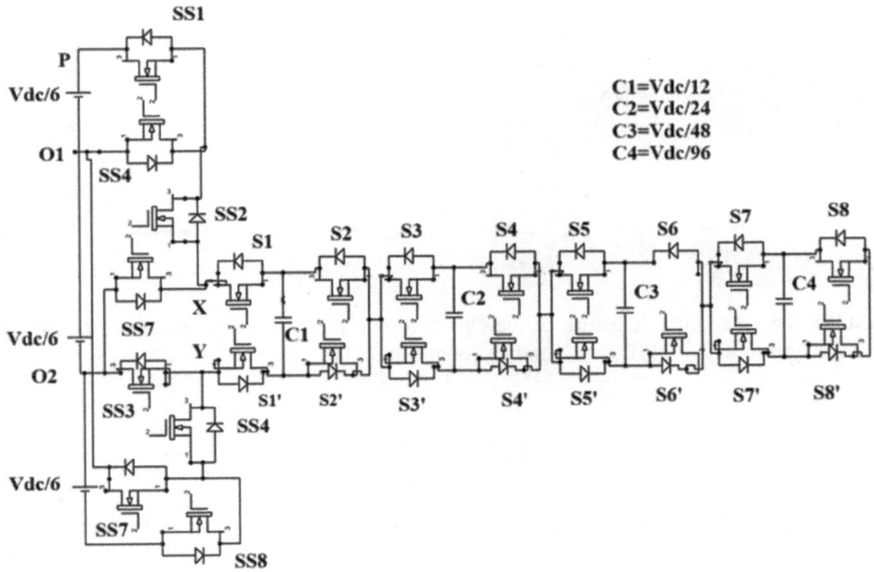

FIGURE 9.4 Topology proposed in [16].

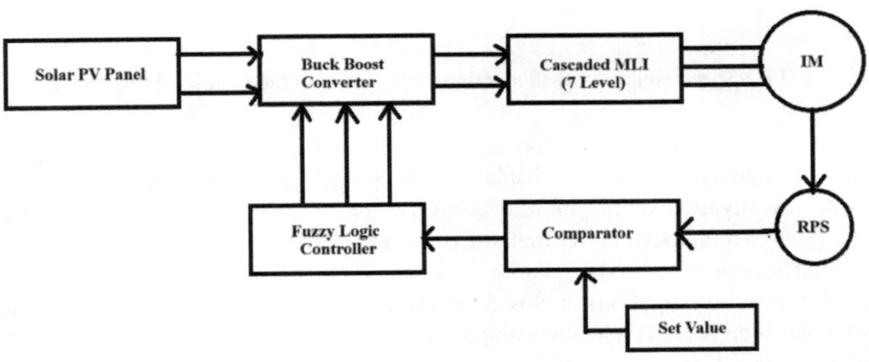

FIGURE 9.5 Block diagram of the proposed system in [17].

are related to synchronizing the system output with the grid and regulating the solar panel output. The overall block diagram of the proposed system is shown in Figure 9.5.

Narendran and Sureshkumar [18] propose a Landsman converter-based multilevel inverter in which the Landsman converter is positioned between the inverter and the DC source. The closed loop response of the Landsman converter-based multilevel inverter is compared with the Proportional Integral PI controller, fractional order Proportional Integral Derivative controller, proportional resonant, and hysteresis controller. The simulation results show that the dynamic response seemed to be improved using the hysteresis controller.

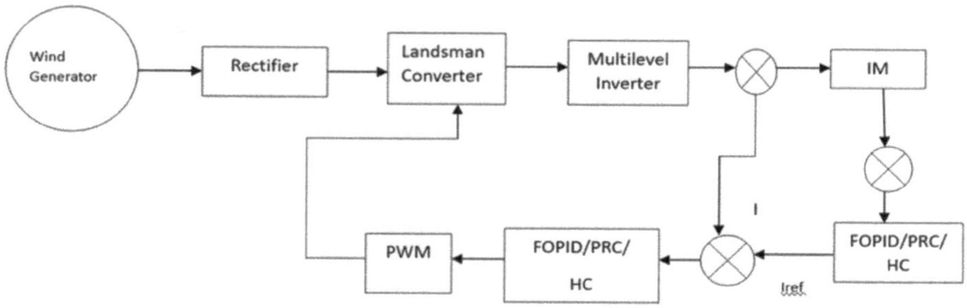

FIGURE 9.6 Modular structure of the closed loop Landsman converter-based multilevel inverter system with FOPID, PRC, and HC controllers [18].

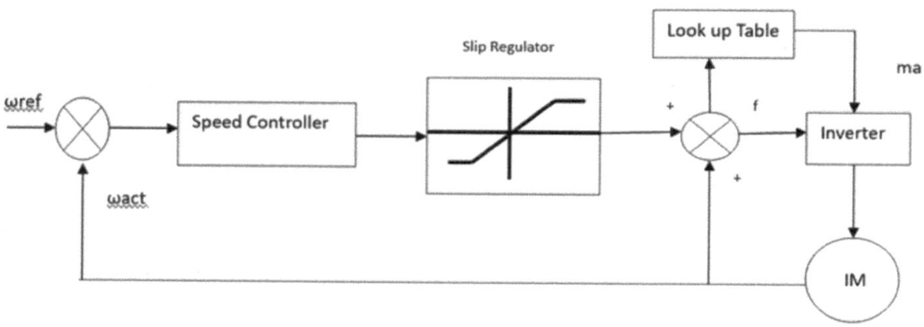

FIGURE 9.7 Speed control of the induction motor fed with the multilevel inverter [19].

The only drawback of the proposed controller is the circuit's extensive usage of passive components. Figure 9.6 displays the proposed system's modular structure.

The investigation of the multilevel-inverter-fed induction motor drive's closed loop model using constant *V/f* method is the key objective of the present research in [19]. Through slip speed regulation and steady *V/f*, induction motor speed control was obtained. Investigations of closed loop speed control utilizing proportional integral (controllers were conducted using a range of reference speeds and dynamic load disturbances.

Compared to VSI-fed induction motors, it was shown that multilevel-inverter-fed induction motors performed better in both steady-state and transient conditions. The sole flaw was that the hardware implementation wasn't employed to evaluate the simulation findings that were reported. The block diagram of induction motor speed control fed with the multilevel inverter is shown in Figure 9.7.

The research work between [9] and [19] is applicable for induction motor drives, whereas certain other recent studies involving permanent magnet synchronous motor drives are also discussed below.

In [20], using Park's transformation, a mathematical model of the permanent magnet synchronous motor (PMSM) is developed and examined for a trio of voltage supplies: a purely sinusoidal supply, a sinusoidal PWM inverter, and a multilevel inverter. When using a multilevel inverter, the torque ripples are significantly

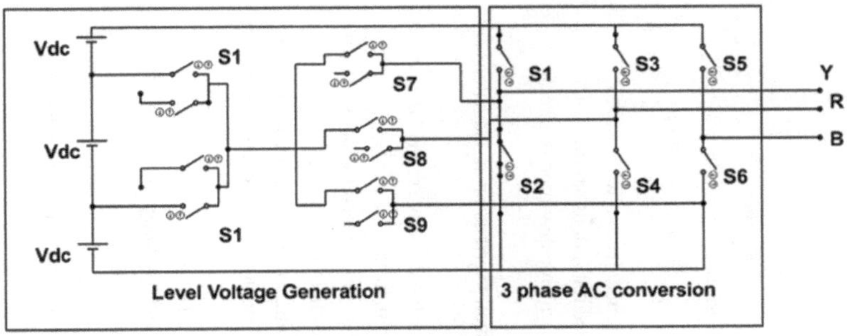

FIGURE 9.8 Inverter topology proposed in [20].

reduced as opposed to being completely absent in the pure sinusoidal case. This demonstrates why multilevel inverters are a wise choice for drive applications. The multilevel topology used is shown in Figure 9.8.

The majority of the research conducted between [21] and [22] is devoted to the speed control of PMSMs fed by multilevel inverters. A novel space vector PWM method that is feasible and not dependent on the levels was suggested in [21] for PMSM vector control.

In this method, the space vector plane is first decomposed, and the locations and duty cycles of the three closest vectors may be calculated immediately by finding the reference vector's location. This innovative approach reduces calculation complexity and does not call for a complex iterative algorithm. A five-level CHB inverter supplies power to this drive system.

It is established in [23] that an improved mathematical model of an interior permanent magnet synchronous motor exists and is being tested. This model accounts for the outcomes of saturation, cross-coupling, spatial harmonics, and iron loss.

In comparison to other traditional PWM techniques like sinusoidal phase shifted and switching frequency optimal phase shifted modulation techniques, it is demonstrated that the proposed harmonic mitigation algorithm improves the drive system's efficiency. However, high-speed and high-torque applications cannot use this tactic.

A predictive control strategy based on variable duty cycle was presented in [22] as a potential solution to the DC voltage unbalance of the CHB inverter. Every cell's DC supply may be completely used under imbalanced conditions, improving PMSM's steady-state performance. Reduced computations and low switching frequencies are two features of this control method.

The paper [24] analyzes the cause of the high neutral-point voltage (NPV) fluctuation in the three-level inverter using the traditional approach. It is suggested to use an enhanced NPV control technique based on the self-optimized sequence for switching to guarantee that the NPV can be fully handled with the least possible switching times. The results of the experiment show that it is possible to reduce the low-frequency harmonics in the output current and, at the lowest cost of an increase in switching loss, to reduce the NPV fluctuation of the recommended approach by a ratio of 46.3%–89.5%. Belt conveyor speed control is one use for this technique.

A brief summary of the multilevel inverter applications in the field of electric drives is mentioned in Table 9.1 below.

TABLE 9.1
Summary of Multilevel Inverter Applications in Electric Drives

Paper	Topology	Levels	DC Sources	Switches	Practical Implications	Results
9	Dual inverter	11 (normal) 5 (fault)	4	10 per phase	1. Improved reliability and wide torque-speed range. 2. Requires fewer DC sources and power electronic devices for implementation.	1. Improved performance of multiphase induction motor drive under faulty conditions. 2. Enhanced reliability and wide range of torque-speed for various applications.
10	New topology	Seven-level current waveform	–	10 per phase	1. Reduced harmonic distortion in motor current. 2. Regenerative braking capability demonstrated on an induction motor drive.	1. Reduced harmonic distortion of motor current. 2. Experimentally verified on an induction motor.
11	Dual inverter	7	1	12	1. Increase electric drive speed without flux weakening. 2. Optimize voltage ratios for inverter performance enhancement.	1. Identified optimal voltage ratios for inverter modulation indices. 2. Suggested method to increase output quality without sacrificing voltage utilization.
12	Multilevel inverter with the DC link	9	1	20	1. Voltage control using motor phase currents. 2. Suitable for medium voltage motor drives and EV applications.	1. DC link voltage deviation control algorithm verified at low power factor and unity power factor. 2. Experimental results presented for a nine-level inverter using a single DC link.
13	Stacked inverter	5	6	12 per phase	1. Reduced component count, volume, weight, and expenses. 2. Facilitates improved reliability, simple construction, and reduced electromagnetic interference.	1. Minimized component five-level inverter topology proposed. 2. Experimental results validate the feasibility of the proposed approach.

(Continued)

TABLE 9.1 (*Continued*)
Summary of Multilevel Inverter Applications in Electric Drives

Paper	Topology	Levels	DC Sources	Switches	Practical Implications	Results
14	Hybrid	3	1	12 per phase	1. Minimize capacitance rating for full-speed operation. 2. Optimize selection of vector redundancies to reduce voltage ripple.	1. Proposed algorithm optimizes vector redundancies to reduce voltage ripple.
15	CHB	5		12 per phase	1. Simplified design and manufacturing process for multiwinding transformer. 2. Enhanced efficiency of the induction motor drive system.	1. Multiwinding transformer design requires less isolated windings, simplifying the manufacturing process 2. Modified PWM technique enhances efficiency of the induction motor drive (IMD) system
16	Stacked inverter	49	3	24	1. Reduced switching losses due to nearest level control implementation. 2. Lower d_v/d_t during inverter operation for high-power drives.	1. Nearest level control used to reduce switching losses. 2. Experimental results show viability for high-power adjustable speed drives.
17	Cascaded	7	3	12	1. Enhances system performance through a fuzzy logic controller. 2. Improves power quality and system response in the renewable energy sector.	1. Improved system performance analyzed through MATLAB and Simulink simulations. 2. Fuzzy logic controller enhances response speed and system harmonic.

(*Continued*)

TABLE 9.1 (*Continued*)
Summary of Multilevel Inverter Applications in Electric Drives

Paper	Topology	Levels	DC Sources	Switches	Practical Implications	Results
18	Landsman converter between the DC source and inverter.	NA	NA	NA	1. A better time responsiveness and greater voltage gain. 2. Increased passive components are the only disadvantage.	1. Closed loop LC multilevel inverter systems systems with carrier based modulation control using proportional integral (CMC-PI), carrier-based modulation control using fractional order proportional-integral-derivative (CMC FOPID), and carrier-based modulation control using hysteresis control (CMC HC) are simulated using MATLAB. 2. Hysteresis controllers decrease steady-state error and settling time.
19	Cascaded inverter	5	2	8 per phase	1. The closed loop approach is evaluated for a range of baseline speeds as well as for changing loads and speed disturbances.	1. Torque and current ripples are reduced. 2. Reduced settling time.
20	Modified version of the full bridge	7	3	11	1. PMSMs offer improved performance, low torque ripple, and fault tolerance. 2. Multilevel inverter performance is evaluated and compared.	1. Mathematical model of PMSM tested with three voltage supplies. 2. Performance evaluation and comparative analysis of different voltage supplies.

(*Continued*)

TABLE 9.1 (*Continued*)
Summary of Multilevel Inverter Applications in Electric Drives

Paper	Topology	Levels	DC Sources	Switches	Practical Implications	Results
21	CHB cascaded hybrid clamped H-bridge (HCHB)	7	2	8	1. Simplifies reference voltage synthesis and reduces switching loss 2. Improves the performance of PMSM in the multilevel inverter motor drive.	1. Reducing computational complexity and improving PMSM performance. 2. Proposed space vector pulse width modulation scheme verified through simulation and experiment.
22	CHB	5	2	8 per phase	1. Torque ripple is not measured directly but as a function of current.	1. Motor efficiency improved by 15% with the hysteresis modulation modulation strategy.
23	CHB	Varying levels	1/cell/ phase	4 switches/ cell/phase	1. Improved PMSM performance under unbalanced conditions. 2. The duty cycle will be changed in real time based on each cell's actual DC voltage.	1. Reduced ripple under both balanced and unbalanced DC voltages. 2. Effective power control in the CHB inverter.

According to the literature review mentioned above, CHB inverters are the most recommended inverters for powering PMSM drives. Additionally, the majority of researchers appear interested in addressing the DC voltage imbalance while using CHB inverters, reducing torque ripples, and vector-controlling PMSM drives.

The summary of applications of multilevel inverters in the field of electric drives is presented in Table 9.1.

9.3 MULTILEVEL INVERTER APPLICATIONS IN ELECTRIC VEHICLES

Electric vehicles are a vital tool for reducing carbon emissions in transportation on roadways, and they account for around one-sixth of global emissions. Electric vehicles appeal to consumers because of their lower maintenance requirements, improved travel experiences, high efficiency, lack of noise, and reduced vibrations, among other features. Several topologies that have been recently proposed for electric vehicles have been discussed in this section.

Bidirectional multilevel inverters with fewer switches and capacitors are now available in a new design called T type multilevel inverters [25].

Some of the features of this novel topology include the absence of additional sensors, less complicated control loops, regular switching patterns, and the substitution of film capacitors for large electrolytic capacitors.

Two additional switches and one capacitor were used to address the primary issue with balancing the voltage across the capacitor. The cost of the power train is increased by the extra switches, but this is offset by a twofold increase in voltage gain. The proposed T type MLI topology is depicted in Figure 9.9.

In [26], it is suggested to use a cascaded multilevel inverter based on the master slave principle. Each phase consists of a master cell that is cascaded with 3N slave cells, where N cells were used for each phase, and they were made up of DC power supplies and H-bridge inverters.

In this research project, the phase disposition PWM method is used. The battery capacity is balanced using the bubble sort technique. The authors' claims that the suggested topology has good controllability, good security, good flexibility, and low complexity are based solely on simulation results, and the hardware prototype is unrecognized. The master slave topology for a single-phase system is shown in Figure 9.10.

The authors of [27] make a proposal for a nine-switched, three-level inverter for hybrid and all-electric vehicles. This work employs space vector PWM. Even so, there are fewer switches than in conventional inverters, there are more losses, and the overall harmonic distortion is about 22.85%.

Due to the fact that the voltage across a few power switches is about fifty percent of the voltage across the DC bus, this design is suitable for low-voltage applications. The fact that the simulated results are not confirmed by the hardware results is a further cause for concern.

By recommending a new topology called nested topology, the shortcomings of the conventional topologies were overcome [28]. The authors propose an inverter with four levels that uses sinusoidal level shift type PWM. With a lower modulation index, the harmonic levels rise. The fact that the results were not verified using the

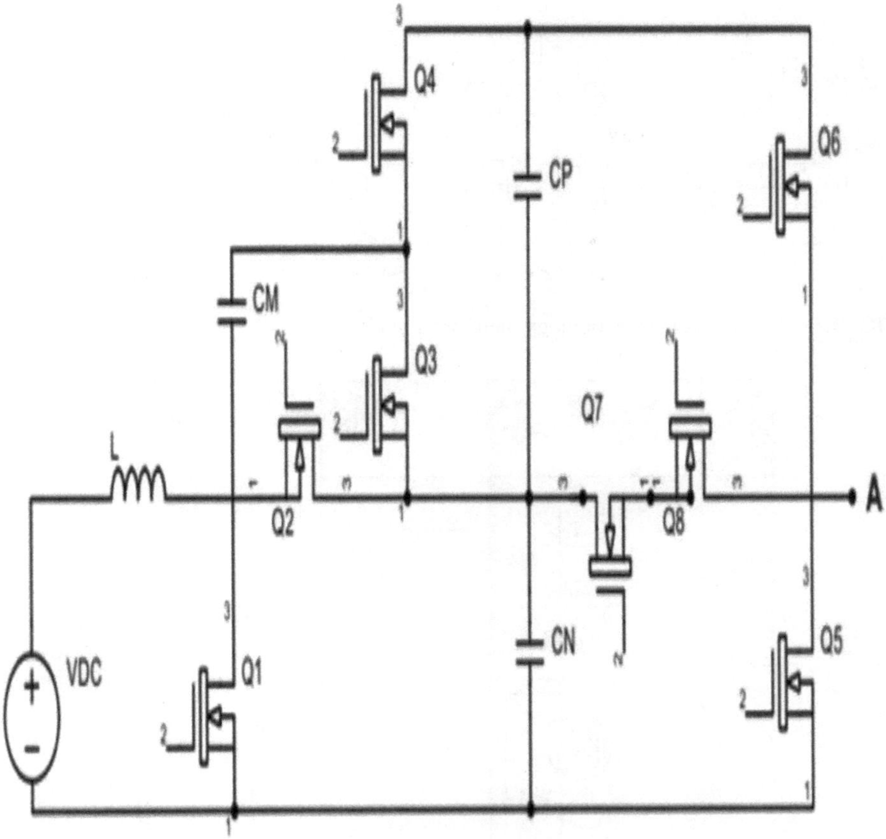

FIGURE 9.9 Proposed T type MLI topology in [25].

hardware configuration could be viewed as a backlog for this suggested topology. The suggested topology is shown in Figure 9.11.

By constructing multilevel converters one on top of each other with low-voltage space vector structures, which have many uses in electric vehicles, a higher number of levels was achieved using low-voltage devices [29].

In the suggested method, two five-level inverters are piled up to create a nine-level inverter. With this topology, the voltage across the capacitor can be stabilized for any modulation index and is unaffected by the load's power factor. To do this, the hysteresis algorithm is used. The nested topology is shown in Figure 9.12.

In [30], a different nested topology that incorporates a parallel enhanced commutation circuit with 17 levels was proposed. With the aid of this topology, the internal battery resistance of the system as a whole can be reduced by rearranging the individual battery cells. Due to the harmonic distortion being only in the range of 2.5%, this PECIN MLI enhances the electric vehicle's functionality.

Using no inductors, [31] proposes a 5-level CHB type inverter with boosting capability for electric vehicles and hybrid electric vehicles. Each leg of the typical

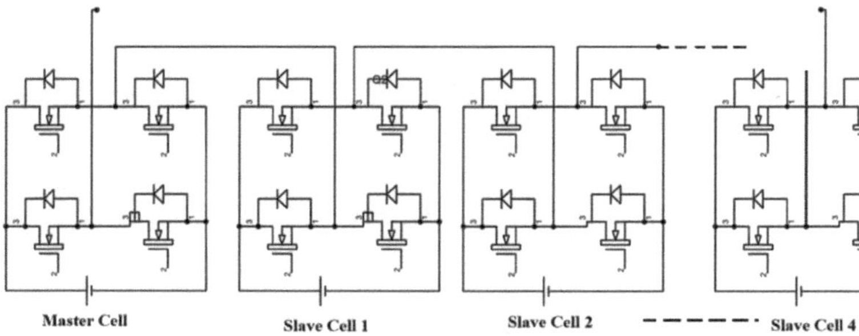

FIGURE 9.10 Master slave topology mentioned in [26].

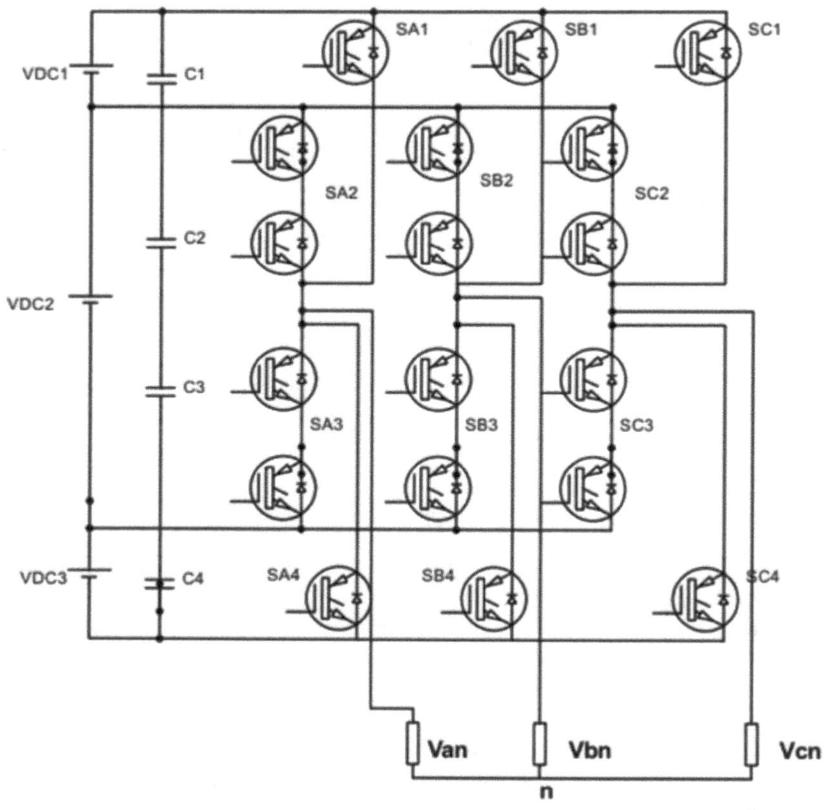

FIGURE 9.11 Nested topology discussed in [28].

three-phase inverter is linked in a sequence with a H-bridge that draws its DC power from a capacitor. By avoiding the large inductor, the power density is raised. Figure 9.13 shows the proposed topology in [31].

In [32], a 53-level inverter employing switching capacitors that reduces device count, cost, and space while still fulfilling IEEE requirements for total harmonic

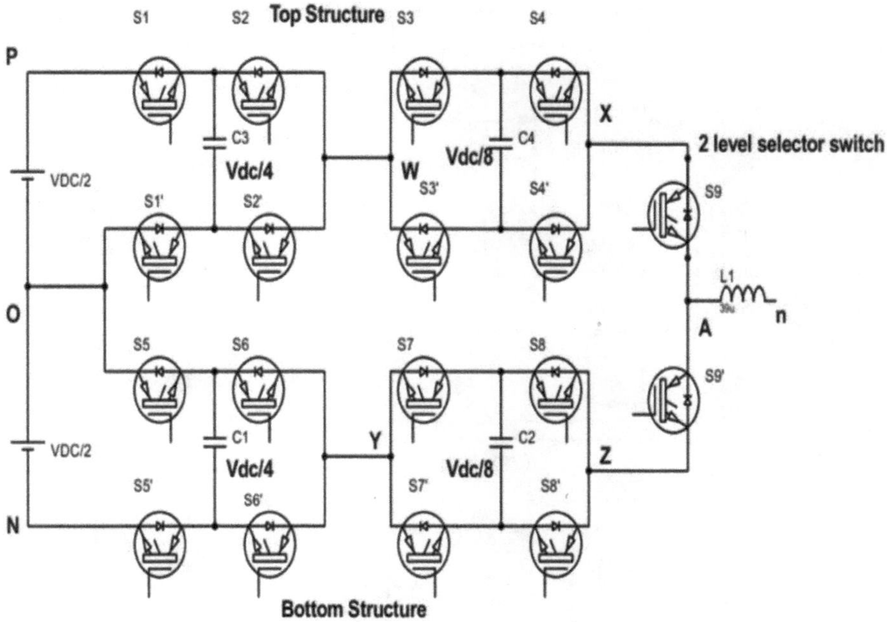

FIGURE 9.12 Stacked topology used in [29].

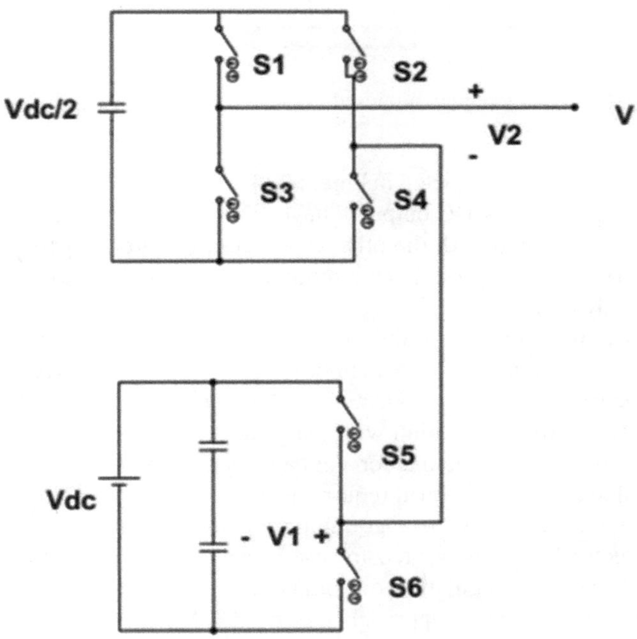

FIGURE 9.13 Topology proposed in [31].

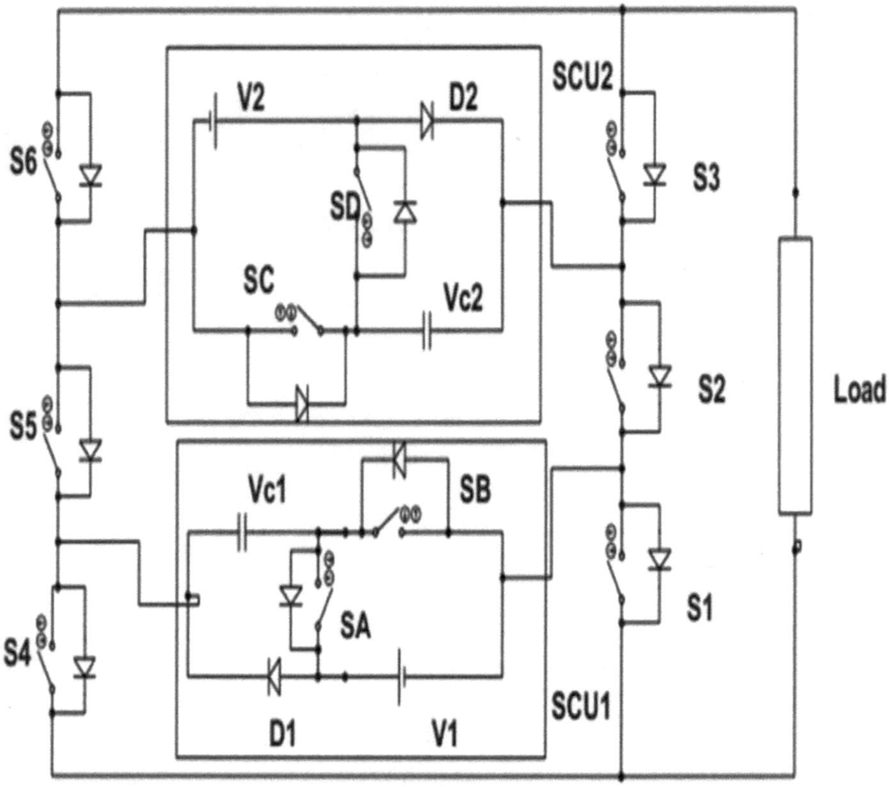

FIGURE 9.14 Design of topology proposed in [32].

distortion was designed as depicted in Figure 9.14. The perturb and observe technique was utilized to get a steady DC output voltage. The switching losses and conduction losses were not compared with the other topologies, despite the paper's promise of lower total harmonic distortion. Furthermore, the topic of capacitor voltage balancing was not addressed.

According to the study [33], multilevel inverters are important to combine the DC power of electric traction vehicles with microgrids and guarantee excellent AC power conversion for vehicle-to-grid applications. The work investigated MLI configurations for EV DC power integration with the grid. It also emphasized on increasing voltage and reducing switch counts for V2G applications. It was concluded that MLI must boost voltage to meet the grid requirements.

In [34], a novel sensorless voltage multiplier inverter that generates nine voltage levels effectively was presented for use in electric car applications, as shown in Figure 9.15. The self-regulation of voltage across the capacitors was achieved, and the nearest level control approach was used. This is appropriate for electric

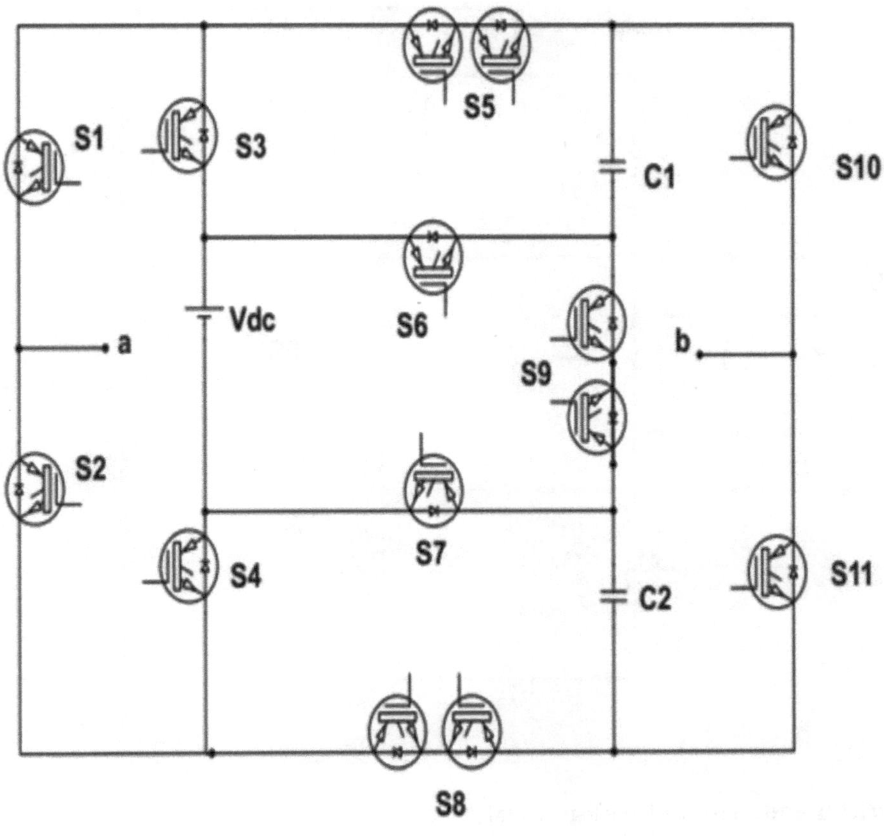

FIGURE 9.15 Topology proposed in [34].

cars since it achieves an efficiency of 94.74% under both static and dynamic load circumstances.

In order to provide improved power quality and decreased harmonic distortion, the study [35] presents updated multilevel and multistring inverters, shown in Figure 9.16. The recommended enhanced multilevel inverter has been constructed with fewer switches, and the novel multistring inverter model formulates an efficient combination of the conducting switches. More work is required to reduce the harmonic distortion further and enhance the robustness of the inverter model.

A summary of applications of multilevel inverters for electric vehicles is mentioned in Table 9.2.

In conclusion, multilevel inverters have significant uses in electric vehicles. In the research for electric vehicles, topologies like the CHB, nested topology, and stacked topology make significant contributions.

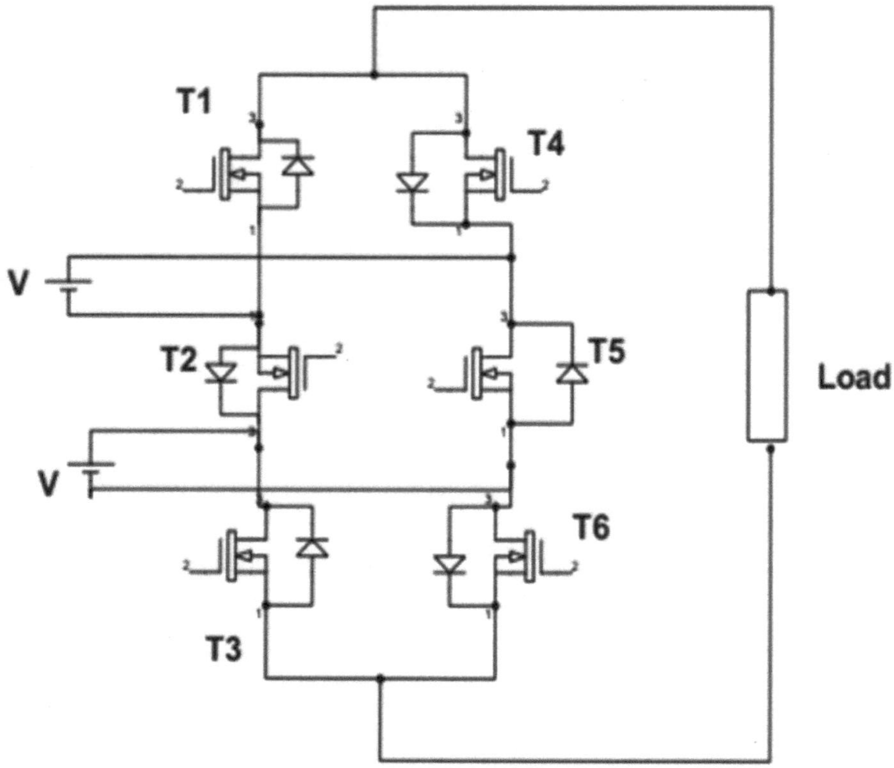

FIGURE 9.16 Proposed topology in [35].

9.4 CONCLUSION

An overview of multilayer inverters used in electric drives and electric cars is given in this study. In electric drives and electric cars, multilevel inverters are crucial components. When addressing multilevel inverters, complications of stabilizing voltage across capacitors, inhibiting circulating currents, and less stability due to the increased number of devices and capacitors have to be considered during their applications.

When considering drives, it appears that machine, drive, and thermal management integration into a single package is becoming a significant trend in research and development. Even though there have been a lot of research articles released in the last 10 years, this analysis has just looked at recent publications.

TABLE 9.2
Summary of Multilevel Inverter Applications in Electric Vehicles

Paper	Topology	Levels	DC Sources	Switches	Practical Implications	Results
24	T type inverter	5	1	8 per phase	1. Increased capacity for EV battery in chassis' space envelope. 2. Unique switching sequences, control circuits, and additional isolated sensors are not needed.	1. 20% size and weight reduction of the converter achieved. 2. Bulky electrolytic capacitors replaced with longer life film capacitors.
25	Master slave cascade multilevel inverter	9	5	20	1. Control permanent magnet synchronous switched reluctance motor in hybrid electric vehicles. 2. Validate the effectiveness and superiority of the proposed inverter through simulation.	1. Verification of the suggested master slave cascade inverter's excellence and efficacy. 2. Simulation results confirm control capability for PMSM in electric vehicles.
26		3	2	9	1. Improves system reliability. 2. Less cost.	1. Reduce d_v/d_t. 2. Reduced electromagnetic interference with a low switching frequency.
27	Nested MLI	4 level	3	6 per phase	1. No need for filters. 2. Reduced electromagnetic interference.	1. Less total harmonic distortion at modulation index = 1.0.
28	Stacked inverter	9 level	2	18	1. Improved reliability and reduced switching losses in inverters. 2. Higher efficiency and direct battery drive for electric vehicles.	1. Developed a nine-level inverter using two five-level inverters. 2. Ran an induction motor using the *V/f* control scheme.
29	Parallel enhanced commutation integrated nested	17 level	Based on the required output voltage.	4	1. Efficient battery system with minimal internal resistance. 2. Competitive multilevel inverter for battery electric vehicles.	1. The paper proposes a parallel enhanced commutation integrated nested multilevel inverter topology. 2. Measurements on a prototype show its function, performance, and efficiency.

(*Continued*)

TABLE 9.2 (*Continued*)
Summary of Multilevel Inverter Applications in Electric Vehicles

Paper	Topology	Levels	DC Sources	Switches	Practical Implications	Results
30	CHB	5 level	1	18	1. Eliminates bulky inductors, increases power density in inverters. 2. Wider modulation index range than traditional inverters for boosted AC voltage.	1. CHB inverter outputs boosted AC voltage without inductors. 2. Proposed design eliminates the need for bulky inductors in inverters.
31	Switched capacitor	53 level	3	14	1. Reduced number of devices, price, and volume of multilevel inverters 2. Stable DC voltage for electric vehicles under dynamic load variations.	1. Total harmonic distortion is less than 5%. 2. Tested the laboratory prototype under dynamic load variations with total harmonic distortion verification.
32	Switched capacitor	9	1	11	1. Suitable for electric vehicles and renewable energy sources. 2. Achieves 94.74% efficiency in intended operating point.	1. High efficiency. 2. Performance verified through simulation and experimental results.
33	Multistring inverter	3/5/7	2	6	1. Improved power supply for electric vehicles. 2. Enhanced power quality with reduced harmonic distortion.	1. Novel inverters show reduced switches and improved power quality. 2. Designed inverters outperform existing models in electric vehicle applications.
34	Multistring inverter	7	2	8	1. Combination of both multilevel and multistring inverters. 2. Reduced number of switches	More work is needed to increase the robustness and to decrease harmonics further.

REFERENCES

1. V. G. Madaram, P. K. Biswas, C. Sain, S. B. Thanikanti, P. K. Balachandran, Advancement of electric vehicle technologies, classification of charging methodologies, and optimization strategies for sustainable development - A comprehensive review, *Heliyon*, 10, 20, 2024.
2. S. Zhao, M. Guo, Electric vehicle power system in intelligent manufacturing based on soft computing optimization, *Heliyon*, 10, 21, 2024.
3. A. Poorfakhraei, M. Narimani, A. Emadi, A review of multilevel inverter topologies in electric vehicles: current status and future trends, *IEEE Open Journal of Power Electronics*, 2, 155–170, 2021.
4. Y. V. Sambhavi, V. Ramachandran, A technical review of modern traction inverter systems used in electric vehicle application, *Energy Reports*, 10, 3882–3907, 2023.
5. E. Agamloh, A. von Jouanne, A. Yokochi, An overview of electric machine trends in modern electric vehicles, *Machines*, 8, 20, 2020.
6. D. Ragul, V. Thiyagarajan, A novel fault-tolerant asymmetrical 21-level inverter topology with reduced components, *Rev Roumaine des Sciences Techniques-Series Electrotechnique et Energetique*, 68(2), 200–205, 2023.
7. V. Thiyagarajan, A new inverter module with minimum circuit components suitable for renewable energy applications, In *AIP Conference Proceedings*, Sikkim, India, vol. 2786, no. 1, pp. 1–9, 2023.
8. V. Thiyagarajan, A new 51-level asymmetrical inverter circuit with reduced number of components, *Lecture Notes in Electrical Engineering*, 973, 1–13, 2023.
9. B. P. Reddy, M. Rao, M. Sahoo, S. Keerthipati, A fault-tolerant multilevel inverter for improving the performance of a pole–phase modulated nine-phase induction motor drive, *IEEE Transactions on Industrial Electronics*, 65(2), 1107–1116, 2018.
10. S. C. Richu, P. P. Rajeevan, A load commutated multilevel current source inverter fed open-end winding induction motor drive with regeneration capability, *IEEE Transactions on Power Electronics*, 35(1), 816–825, 2020.
11. Z. Huang, T. Yang, P. Giangrande, S. Chowdhury, M. Galea, P. Wheeler, Enhanced performance of dual inverter with a floating capacitor for motor drive applications, *IEEE Transactions on Power Electronics*, 36(6), 6903–6916, 2021.
12. T. Debnath, K. Gopakumar, L. Umanand, D. Zielinski, K. Rajashekara, A nine-level inverter with single DC-link and low-voltage capacitors as stacked voltage sources with capacitor voltage control irrespective of load power factor, *IEEE Open Journal of the Industrial Electronics Society*, 3, 522–536, 2022.
13. A. Karthik, U. Loganathan, A reduced component count five-level inverter topology for high reliability electric drives, *IEEE Transactions on Power Electronics*, 35(1), 725–732, 2020.
14. M. Imthias, K. Raj, A. K. Yadav, K. Gopakumar, L. Umanand, C. Cecati, Minimization of switched capacitor voltage ripple in a multilevel dodecagonal voltage space vector structure for drives, *IEEE Transactions on Industrial Electronics*, 67(1), 126–135, 2020.
15. R. Kumar, P. Kant, B. Singh, Modified PWM technique for a multi-pulse converter-feed multilevel inverter-based IM drive, In *2020 IEEE International Conference on Power Electronics, Smart Grid, and Renewable Energy (PESGRE2020)* held in Cochin, India, pp. 1–6, 2020.
16. V. Nair, K. Gopakumar, L. G. Franquelo, A very high resolution stacked multilevel inverter topology for adjustable speed drives, *IEEE Transactions on Industrial Electronics*, 65(3), 2049–2056, 2018.
17. S. Chandrasekaran, S. Durairaj, S. Padmavathi, A performance evaluation of a fuzzy logic controller-based photovoltaic-fed multi-level inverter for a three-phase induction motor, *Journal of the Franklin Institute*, 358(15), 7394–7412, 2021.

18. A. Narendran, R. Sureshkumar, Hysteresis-controlled - landsman converter based multilevel inverter fed induction-motor system using PIC, *Microprocessors and Microsystems*, 76, 103099, 2020.
19. A Chitra, W. R. Sultana, J. Vanishree, S. Sreejith, S. Jose, A. J. Pulickan, Performance comparison of multilevel inverter topologies for closed loop v/f controlled induction motor drive, *Energy Procedia*, 117, 958–965, 2017.
20. M. Fahad, J. A. Lone, A. Sarwar, Modeling and performance analysis of a multilevel inverter fed PMSM open loop drive, In 2019 *International Conference on Electrical, Electronics and Computer Engineering (UPCON)*, Aligarh, India, pp. 1–6, 2019.
21. H. Yao, Y. Yan, T. Shi, G. Zhang, Z. Wang, C. Xia, A novel SVPWM scheme for field-oriented vector-controlled PMSM drive system fed by cascaded H-bridge inverter, *IEEE Transactions on Power Electronics*, 36(8), 8988–9000, 2021.
22. H. Yao, Y. Yan, Y. Cao, P. Song, T. Shi, A variable duty cycle-based predictive control for PMSM fed by cascaded H-bridge inverter, *IEEE Journal of Emerging and Selected Topics in Power Electronics*, 10(4), 4195–4206, 2022.
23. G. Schettino, C. Nevoloso, R. Miceli, A. O. D. Tommaso, F. Viola, Impact evaluation of innovative selective harmonic mitigation algorithm for cascaded H-bridge inverter on IPMSM drive application, *IEEE Open Journal of Industry Applications*, 2, 347–365, 2021.
24. X. Wu, C. Wang, J. Li, X. Zhang, G. Tan, Neutral point voltage control of the three-level inverter in permanent magnet synchronous motor based direct-drive belt conveyor, *Computers and Electrical Engineering*, 110, 108840, 2023.
25. A. Sheir, M. Z. Youssef, M. Orabi, A novel bidirectional T-type multilevel inverter for electric vehicle applications, *IEEE Transactions on Power Electronics*, 34(7), 6648–6658, 2019.
26. X. Pei, H. Li, Master-slave cascade multilevel inverter for motor drive control of electric vehicles, In *IECON 2019–45th Annual Conference of the IEEE Industrial Electronics Society*, Lisbon, Portugal, pp. 3210–3214, 2019.
27. S. Gokul, M. Karthik, R. N. Devi, R. Thangaraj, Three level inverters with nine switch topology for electric vehicle applications, *International Journal of Engineering Trends and Technology*, 44(2), 94–97, 2017.
28. N.K. Muthukuri, R.P. Narasipuram, S. Mopidevi, Performance analysis of nested multilevel inverter topology for 72V electric vehicle applications. *Journal Européen des Systèmes Automatisés*, 53(6), 925–930, 2020.
29. V. Nair, A. Rahul, R.S. Kaarthik, A. Kshirsagar, K. Gopakumar, Generation of higher number of voltage levels by stacking inverters of lower multilevel structures with low voltage devices for drives, *IEEE Transactions on Power Electronics*, 32(1), 52–59, 2017.
30. C. Terbrack, J. Stöttner, C. Endisch, Design and validation of the parallel enhanced commutation integrated nested multilevel inverter topology, *IEEE Transactions on Power Electronics*, 37(12), 15163–15174, 2022.
31. Z. Du, B. Ozpineci, L.M. Tolbert, J.N. Chiasson, DC–AC cascaded H-bridge multilevel boost inverter with no inductors for electric/hybrid electric vehicle applications, *IEEE Transactions on Industry Applications*, 45(3), 963–970, 2009.
32. C. Dhanamjayulu, S. Padmanaban, V.K. Ramachandaramurthy, J.B. Holm-Nielsen., F. Blaabjerg, Design and implementation of multilevel inverters for electric vehicles. *IEEE Access*, 9, 317–338, 2021.
33. Mohd Rizwan Khalid, Adil Sarwar, Ibrahim Alsaidan, Multi-level inverters interfacing electric vehicle charging stations with microgrid for vehicle-to-grid (V2G) applications, In Mohmmad Saad Alam, Reji Kumar Pillai, Murugesan Navaneetha Krishnan (eds), *Advances in Mechatronics and Mechanical Engineering (AMME) Book Series*, IGI Global Scientific Publishers, India, 2022.

34. A. Azimi, A. Khodaparast, M. Rasouli, J. Adabi, A novel sensorless voltage multiplier inverter for electric vehicle application, In *13th Power Electronics, Drive Systems, and Technologies Conference (PEDSTC)*, Tehran, Iran, pp. 609–614, 2022.
35. S. K. K. Sampathkumar, D. P. Kumar, Power quality assessment of novel multilevel and multistring inverters for electric vehicle applications, *Bulletin of Electrical Engineering and Informatics*, 11(4), 1818–1827, 2022.

10 Multilevel Inverter Solutions for Wind and Solar Hybrid Systems

Ramaprabha Ramabadran and Divya Thangaraj

10.1 INTRODUCTION

For many years, renewable energy sources have been considered a transformative solution to our growing global energy needs. The transition to renewable sources is not just a necessity but an opportunity to forge a more sustainable and resilient energy future. Investment and execution of energy policies and renewable technologies to accelerate this transition have become a key mission in countries worldwide. Increased electrification and the growth of clean electricity supply have meant that the consumption of clean electricity has grown by around 6 EJ, while modern biofuels grew by around 2.3 EJ [1]. The deployment of various renewable sources and biofuels is presented in Figure 10.1. Among them, the largest contribution comes from wind and solar in electricity sectors around the world.

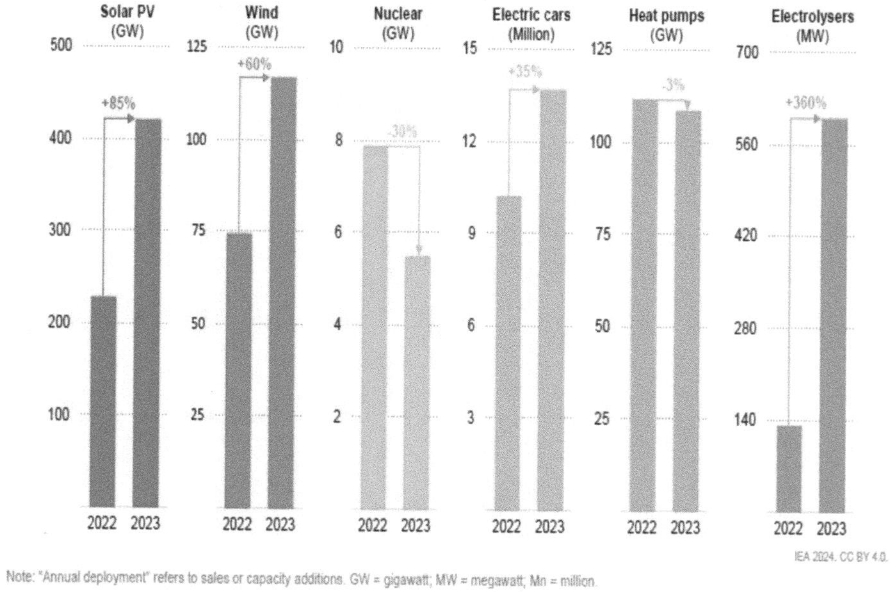

FIGURE 10.1 Annual deployment of major clean energy technologies, 2022 and 2023.

DOI: 10.1201/9781003496946-10

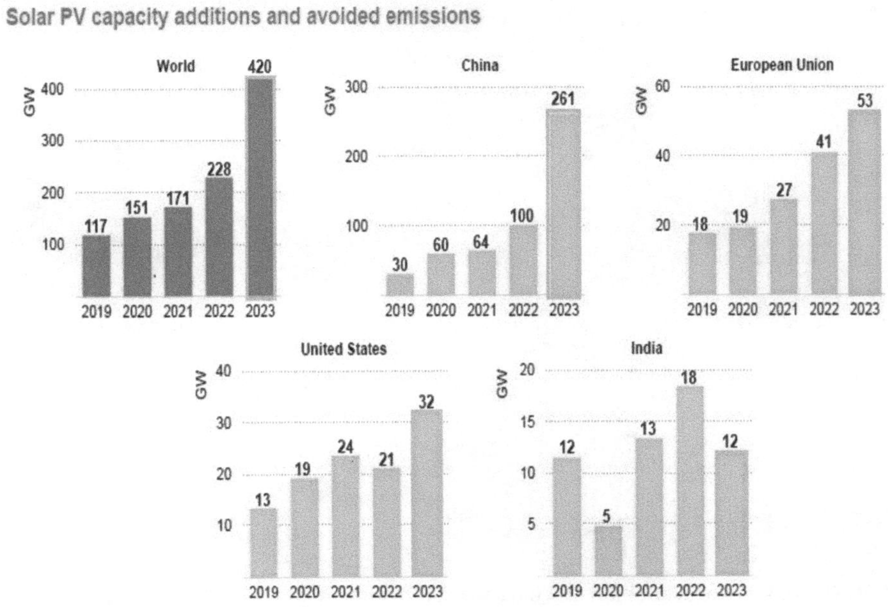

FIGURE 10.2 Global addition of solar capacity for year 2022–2023. (International Energy Agency.)

Global solar PV capacity additions increased by over 80% from 2022 to 2023 and broke a new record, reaching over 420 GW. Similarly, global wind capacity additions jumped almost 60% in 2023 to break the 2020 record. Onshore wind projects accounted for over 85% of global wind growth last year. The annual clean energy report from IEA (International Energy Agency) presented the growth of solar PV and wind energy capacity across the globe, shown in Figures 10.2 and 10.3.

Owing to this, in recent years, integrating nonconventional energy sources like wind and PV with the power grid has gained significant momentum [2]. However, energy from either PV or wind is intermittent and dependent mainly on climatic conditions. This has increased the demand for efficient power electronic converters to efficiently harness and integrate the power generated from these sources into the grid. Briefly discussing the hybrid system, the design considerations of the inverter for the wind–PV system are stated for reference.

Relating to this, in recent years, studies have observed that MLIs are more compatible with interfacing renewable energy sources with the grid. They are mainly employed in many industrial applications such as drives, filters, power factor compensators, etc. MLIs use several active and passive elements to give high output power from medium- or low-voltage sources. It gives a better voltage waveform with increased efficiency, and control is easy [3]. This chapter focuses on the design of an embedded source high-gain MLI for a wind–PV system. The said MLI embodies the characteristics of cascaded H-bridge type MLI and Z-source inverter types with its source embedded within the impedance network. The proposed MLI is viewed as

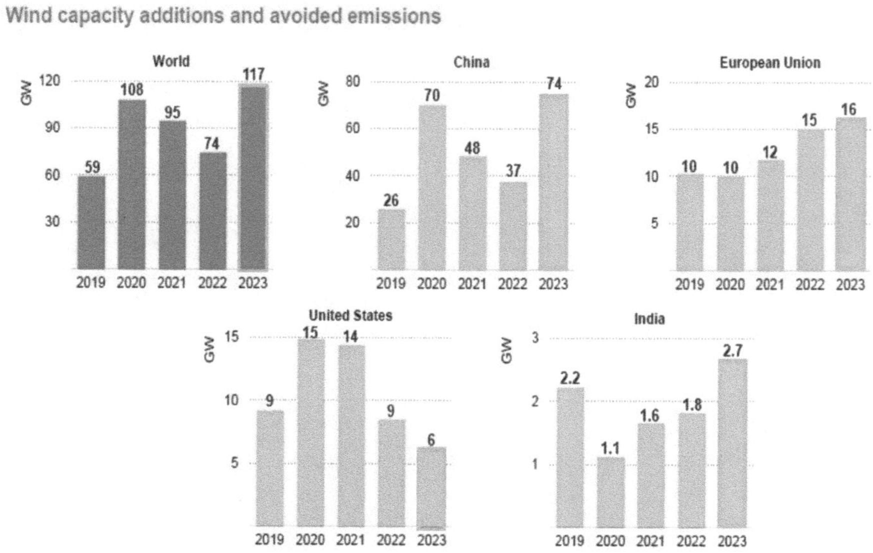

FIGURE 10.3 Global addition of wind capacity for year 2022–2023. (International Energy Agency.)

a promising inverter topology for all hybrid-related applications. Following this, a brief case study on the different design features for different components of wind–PV systems that need to be considered is presented in the following sections. This case study aims to provide a brief knowledge about various factors that contribute to the optimal design of the wind–PV hybrid system connected to the grid. Before concluding, this chapter also brushes upon a few emerging trends seen in this technology.

10.2 OVERVIEW OF THE WIND–PV SYSTEM

A wind–PV energy system, commonly known as a hybrid renewable energy system, combines the energy generated from PV solar panels and wind turbines. Integrating the two complementary sources that are observed to often peak during different periods ultimately boosts the overall system's reliability and energy output, as shown in Figure 10.4.

Initially, many hybrid systems were designed to be standalone systems that operate off-grid, as a means to reduce the total life cycle cost of the supplies in many off-grid situations. In recent years, the development and implementation of grid-connected hybrid systems have seen significant growth owing to the increasing demand for clean energy from renewable energy sources [4]. The growing market is promoted by various factors, including government support, stringent environmental regulations, and the need for reliable and uninterrupted power supply [5].

The wind–PV hybrid energy system mainly consists of a PV system, a wind turbine, an inverter, controllers, an energy storage system, which in many cases is batteries, charge controllers, and a control system to manage and optimize energy flow,

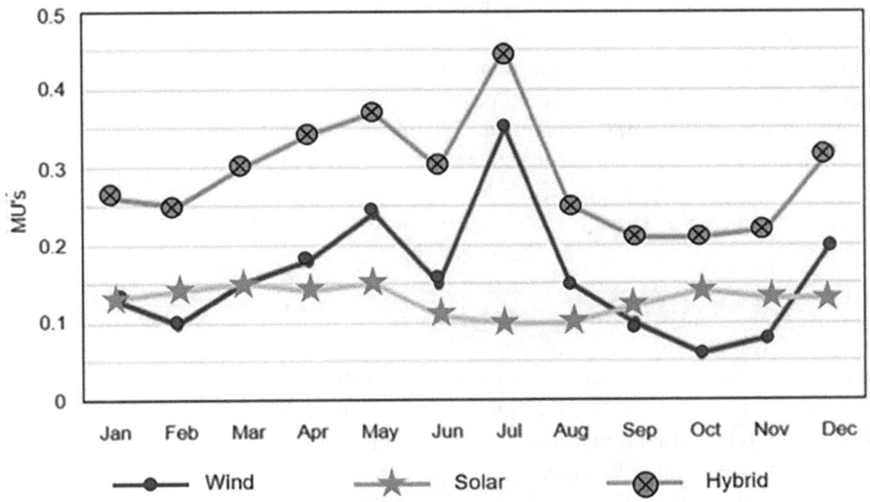

FIGURE 10.4 Monthly wind, solar, and hybrid generation profile.

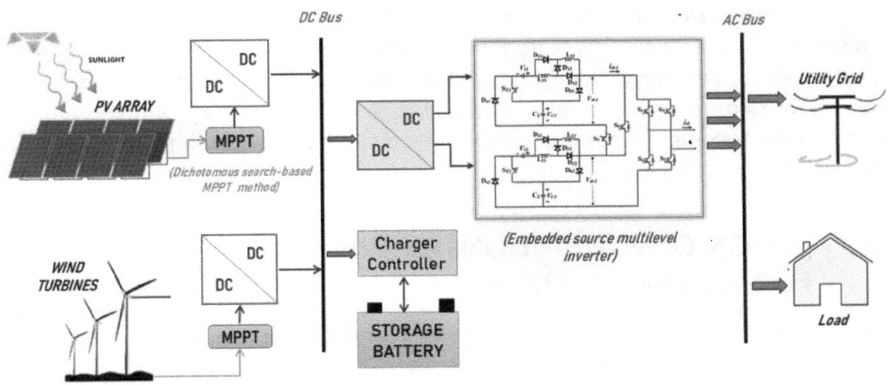

FIGURE 10.5 Grid-connected wind–PV system.

as shown in Figure 10.5. The foremost advantage of this hybrid system is its ability to take advantage of intermittent yet predictable and complementary aspects of the two renewable energy sources. The panels generate electricity during daylight hours, while turbines harness wind energy more consistently during different times of the day and seasons. Integration of such a complementary energy system can ensure a consistent generation of electricity throughout the day and across different weather conditions. Still, there may be instances when both the wind and the solar system fail to produce power. In such cases, most power systems rely on batteries and/or an engine generator, commonly powered by conventional fuels. The generator also recharges the batteries when they run low. However, an engine generator as a backup source increases the complexity of the overall system, but it can be efficiently controlled by modern electronic controllers. Likewise, it also reduces the size of the rest of the system.

In addition, the hybrid system enables increased energy generation potential as well as capacity utilization. The combined resources capture energy from two different sources at different times, resulting in maximizing the energy yield. Being complementary to each other, the electricity generation from one resource (like a wind turbine) is independent of the availability of the other resource (solar irradiance), and vice versa. This promotes better energy yield and a higher capacity factor. Eventually, energy consumption becomes optimized, reducing dependence on auxiliary power sources for backup. Furthermore, increased grid stability and reliability are another advantage of wind–PV hybrid systems [6,7]. By integrating with the grid, the hybrid system helps in balancing demand and supply. The surplus energy generated during peak solar or wind conditions can be fed back to the grid, thus making the grid more resilient. Also, the variability of renewable energy sources is eliminated through hybridization, reducing the risk of energy crises or blackouts.

In summary, wind–PV energy systems represent an innovative and sustainable approach to electricity generation. They allow reducing dependence on fossil fuels and associated greenhouse gas emissions and escalating energy independence. Furthermore, the systems can be placed in remote locations or regions with unreliable or no grid supply, which can serve as a reliable energy provider with minimal impact on the environment. Additionally, hybrid systems can be cost-effective, leveraging economies of scale and shared infrastructure to reduce installation and operational expenses compared to deploying standalone solar or wind projects. With the world transitioning toward a cleaner and more sustainable energy future, wind–PV energy systems combine the benefits of both technologies to offer a more sustainable approach to electricity production.

10.3 DESIGN CONSIDERATIONS OF INVERTERS FOR WIND–PV SYSTEMS

The inverter is the most vital component in a wind–PV or any hybrid system connected to the grid. They are in charge of converting the hybrid-generated DC power to AC power suitable for grid integration. Several critical factors should be considered while designing inverters for the wind–PV system to achieve optimal performance, reliability, and efficiency. In general, the primary functions of the inverters in the hybrid system can be classified as follows:

1. **Power Rating**: It is essential to determine a suitable power rating and the capacity limits for an inverter based on the combined capacity of the hybrid system. They should sustain the maximum power output from both sources along with factors like peak loads, intermittence in power generation, etc.
2. **Efficiency and Performance**: Almost always, the inverter design must focus on giving high efficiency conversion with minimal energy loss. The inverter must be equipped to handle different loads and environmental conditions.

3. **Parallel Input Sources**: Given that the hybrid system would be handling two different energy sources such as PV arrays and wind turbines, their inverter's design should include the capability of accepting parallel input. The two inputs should be integrated and run simultaneously for an uninterrupted and stable generation irrespective of fluctuations in the solar or wind conditions.
4. **Maximum Power Point Tracking (MPPT) Algorithms**: One of the essential components of the PV panels and wind turbines generating and harnessing systems is MPPT algorithms, for optimal energy extraction. The designed inverter must be feasible with advanced MPPT techniques designed for both solar and wind sources.
5. **Grid Integration and Standards Compliance**: Inverters in wind–PV systems must comply with grid integration standards and policies and integrate grid-tied operation. Features like anti-islanding protection, voltage regulation, and synchronization with utility grid parameters must be considered.
6. **Protection and Reliability**: The inverter design must be inclusive of features for protection against grid faults like overvoltage, overcurrent, short circuit, etc. It must have a durable design with reliable components for long-term operation.

There are other factors such as modularity, scalability, monitoring, and many environmental conditions that need to be considered for designing a proficient inverter for connecting the hybrid system to the grid. In conclusion, designing inverters for wind–PV systems should incorporate a broad set of aspects as discussed above. These design aspects need to be given careful consideration to develop inverters that optimize renewable energy utilization, support proper integration of solar and wind sources, and improve grid stability.

10.4 MULTILEVEL INVERTER FOR GRID APPLICATIONS

MLIs are essential for bridging renewable energy sources with the grid. With the inherent capability of output harmonic reduction, reduced blocking voltages for switching components, and reduced losses due to fewer commutation stresses, they make an excellent choice for high-power medium-voltage applications. With regards to grid systems, MLIs are a reliable means of transforming the DC output from renewable sources into suitable power for the grid. By utilizing several voltage levels, it is possible to obtain high-grade AC output with reduced harmonic distortion and increased efficiency compared to conventional two-level inverters. This process ensures stable power transmission and distribution which will improve the electricity network and the role played by renewable energy in it while also ensuring a more reliable and sustainable source of electricity. One of the most commercially preferred multilevel topologies is the cascaded H-bridge [8].

Cascaded H-bridge MLIs are made of series connections of several individually sourced single-phase inverters. This structure can reach medium output voltage levels using only standard low-voltage mature technology components. To increase the gain of the topology, each source is connected to its respective H-bridge using an impedance network to boost the output voltage. From literature [9], it was observed that an embedded switched inductor switched boost inverter (ESL-SBI) provides high boost capability and a reduced number of passive components, while displaying notably lower total harmonic distortion (THD), concerning hybrid system connecting to the grid. To further reduce the THD and provide a better output to feed the grid, a high-gain MLI using the ESL-SB inverter is proposed in the following section.

10.4.1 Embedded Source Switched Boost Multilevel Inverter

Employing the concept of the cascaded MLI, existing three-level ESL-SBIs are cascaded to derive an MLI. In this work, two ESL-SBIs are connected to get a five-level embedded type MLI, i.e., embedded switched inductor switched boost MLI. To reduce the overall number of switching devices in C-MLI topologies, a new switching network was proposed [10], which mainly consists of four switches making up the H-bridge network and two auxiliary switches to isolate the two sources. The embedded switched inductor switched boost MLI topology using the new switching network is shown in Figure 10.6 [11].

This inverter provides a buck/boost operation and dc/ac inversion in a single power stage using its impedance network for individual levels during shoot-through. The number of impedance networks consisting of inductors and capacitors depends on the level requirement. This network effectively uses the shoot-through stage to boost the DC-link voltage at each level while drawing continuous input current, owing to the placement of the source. The inverter modes of operation with reference to Figure 10.6, based on the load voltage and current polarity, are explained as follows:

- **Mode 1 (Shoot-Through Mode)**: All switches (S_1–S_4) are turned on, and the stored energy of the capacitor charges the inductor.
- **Mode 2 (Positive Power Mode)**: Only S_1 and S_4 are turned on, to obtain boosting mode. The inductor feeds back the capacitor's stored energy.
- **Mode 3 (Negative Power Mode)**: S_2 and S_3 are turned on, and the mode is similar to mode 2, except the inverter output is negative in polarity.

The basic characteristics design equations for the proposed MLI are given as follows:

- **Inductor Voltages**: $V_{L_n} = V_{i_n} + V_{C_n}$ (short-circuit mode); $V_{L_n} = \frac{V_{i_n} - V_{C_n}}{2}$ (power mode)
- **Capacitor and DC-Link Current**: $i_{\text{dc}_n} = 2I_{L_n} = -i_{C_n}$; $i_{C_n} = I_L - i_{\text{dc}_n}$
- **Output Voltage**: $v_o = V_{C1} + V_{C2} = V_{i1} + V_{i2}\left(\frac{1+D_S}{1-3D_S}\right)$, where D_S is the shoot-through duty cycle of the inverter.
- **Inductor Equation**: $L_n = \frac{RD_S(1-3D_S)\times 100}{f_s(1+D_S)x_L\%}$, where $x_L\% = \Delta i_{Ln}/I_L{*}\ 100$ (allowable current ripple).

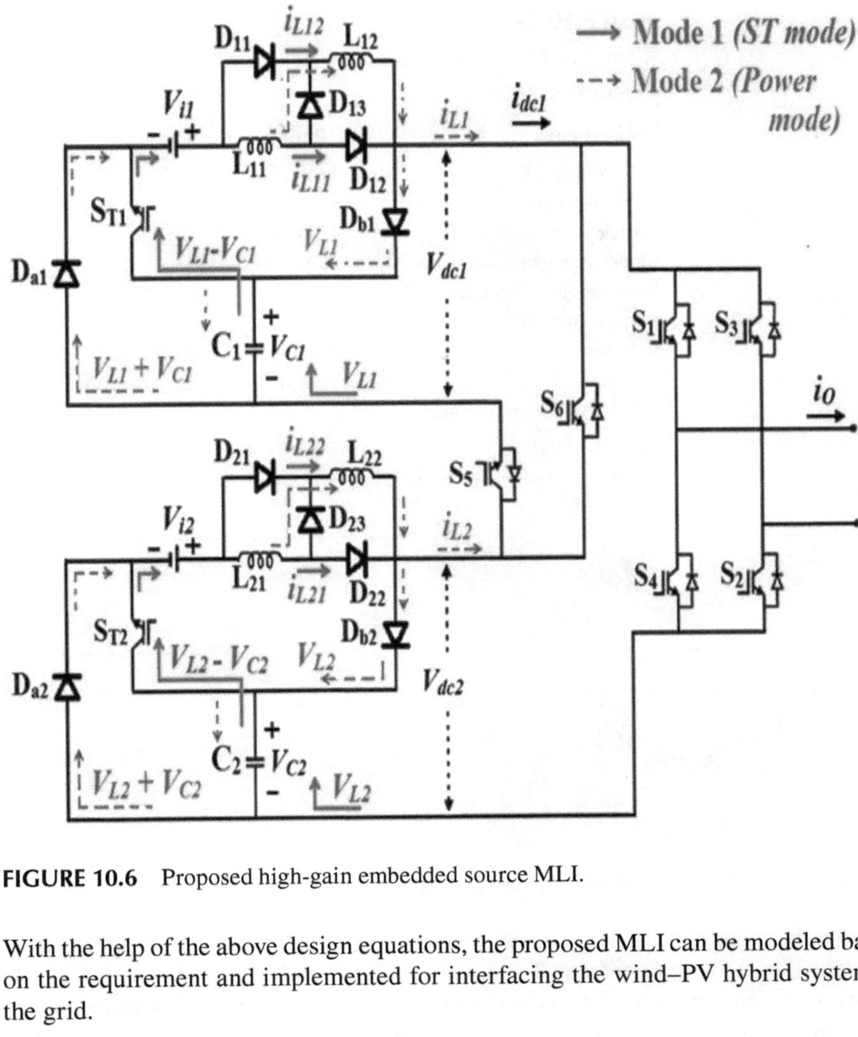

FIGURE 10.6 Proposed high-gain embedded source MLI.

With the help of the above design equations, the proposed MLI can be modeled based on the requirement and implemented for interfacing the wind–PV hybrid system to the grid.

10.4.2 Simulation and Hardware Study of the Proposed Multilevel Inverter Topology

In this section, the effectiveness of the proposed inverter is validated through simulations, and the obtained results will be supported by the hardware implementation of MLI. The simulation is carried out as a 2 kW system, whose parameter ratings are given in Table 10.1. Since the main idea of the work is to implement a high-gain MLI for the grid PV interface, the analysis is carried out for a five-level developed MLI.

With the values obtained in the simulation, a hardware model is developed to verify the results. Figures 10.7–10.9 show the simulation results of the proposed MLI, while Figures 10.10–10.14 show the hardware output of the proposed MLI.

TABLE 10.1
Ratings of the Proposed MLI Designed 2 kW

Components	Rating
Input voltage (V_{i1}/V_{i2})	44 V–76 V
Input current (I_i)	~30 A
Switching frequency (f_s)	25 kHz
Modulation index (M)	0.8–0.9
Shoot-through duty cycle (D_S)	0.2
Boost factor (B)	3
Inductors (L_1, L_2)	2.5 mH
Capacitor (C)	50 μF
Filter (C_f, L_f) – inverter side	1 μF, 1 mH
Filter (C_f, $L_{f1\text{-}1}$)	2 μF, 10 mH

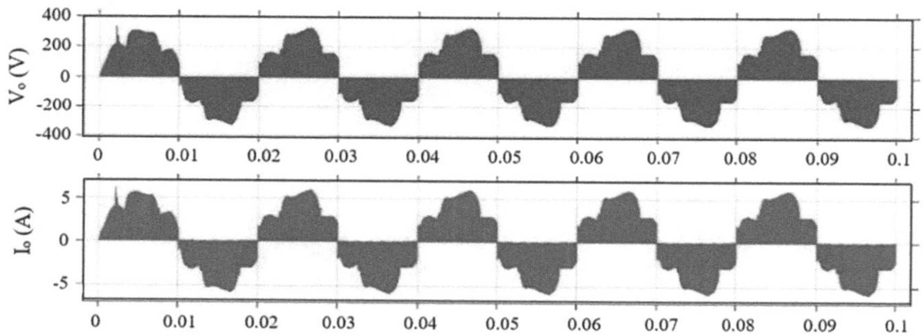

FIGURE 10.7 Staircase voltage and current output of the MLI.

From the work done, a comparison table of the simulation and hardware value of inverter parameters is presented in Table 10.2. This is to study the characteristics of different components in the module output from simulation and real-time hardware output.

As expected, the hardware realization of the developed high-gain MLI experiences additional losses across the components when compared with the simulation results.

10.5 HYBRID SYSTEM USING THE PROPOSED MULTILEVEL INVERTER: A DESIGN CASE STUDY

For the system integrating different renewable energy sources with varying characteristics, substantial focus needs to be given to the design of each component. The PV system, the wind turbine, the inverters, converters, and the control system should be tailored accordingly. In this section, the fundamental design requirements for designing and implementing an optimal 2 kW hybrid system are discussed.

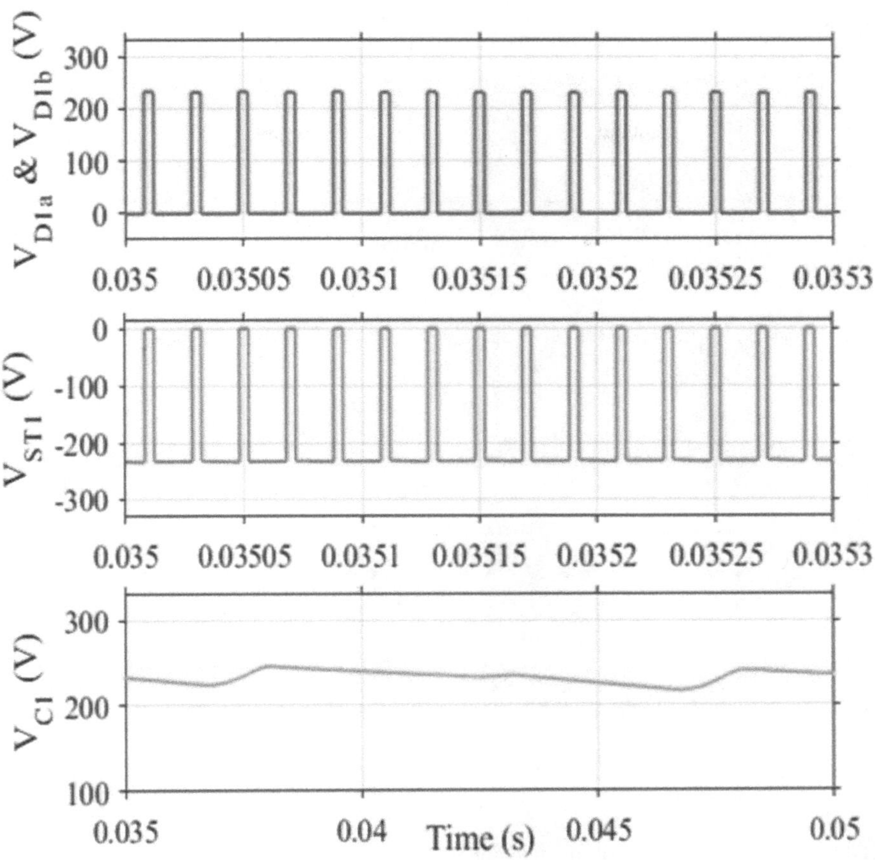

FIGURE 10.8 Voltage waveforms across the internal components.

10.5.1 Sizing of the PV Array with MPPT Control

Multiple solar panels that are made up of individual PV cells are electrically wired in series or parallel fashion to form a PV array. In this work, the modeling of the PV panel is done based on the one-diode model as shown in Figure 10.15 [12].

Based on this model, a single PV panel consists of 36 cells with an open circuit voltage (V_{OC}) of 21.24V, and a short-circuit current (I_{SC}) of 2.55 A. The output current of the one-diode PV model is given in equation (1). Solkar 36WP PV module is used in this work, whose design parameters under standard temperature and insolation ($T = 25°C = T_{ref}$ and $\lambda = 1{,}000\,W/m^2$) are given in Table 10.3.

$$I_{PV} = N_P \times I_{ph} - N_p \times I_o \left[\exp\left\{ \frac{q(V_{PV} + I_{PV} R_S)}{N_S A k} \right\} - 1 \right] - \frac{V_{PV} + I_{PV} R_S}{R_{sh}} \tag{10.1}$$

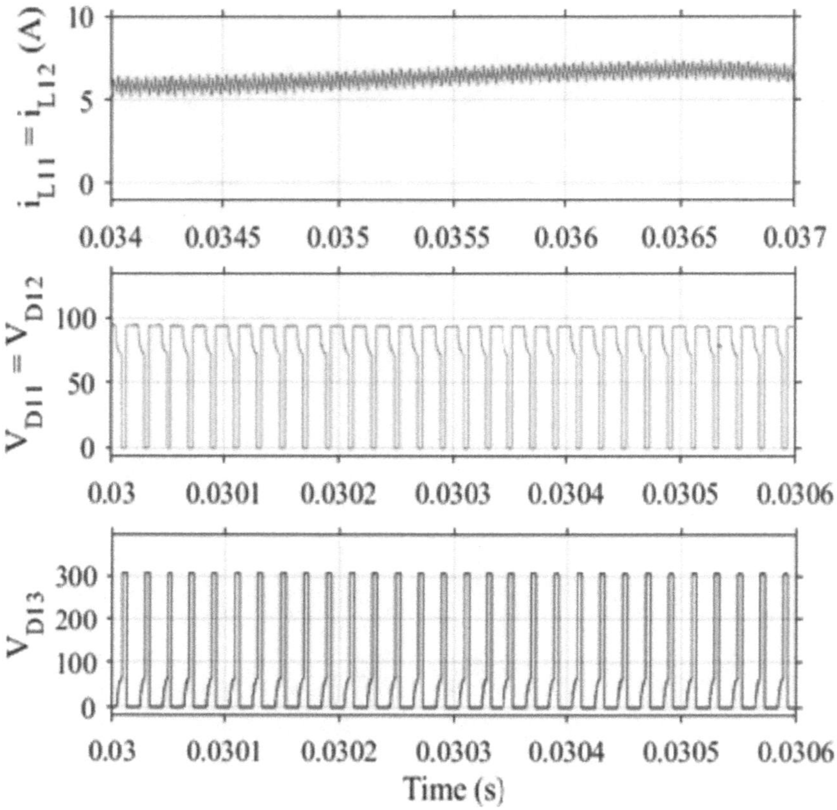

FIGURE 10.9 Waveforms of the inductor and the switched inductor cell diodes.

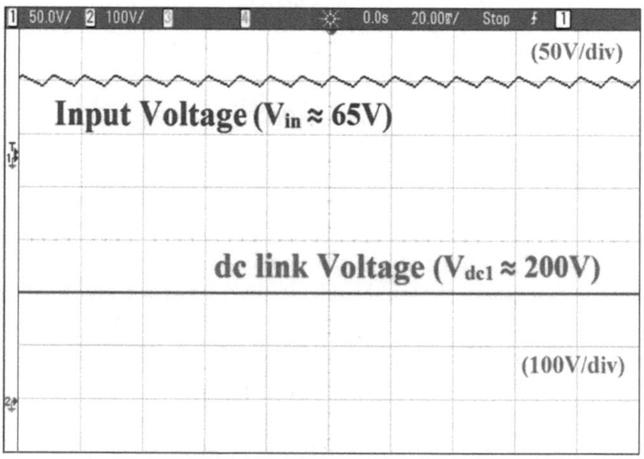

FIGURE 10.10 Input and DC-link measure of the hardware module (for $D_S = 0.2$).

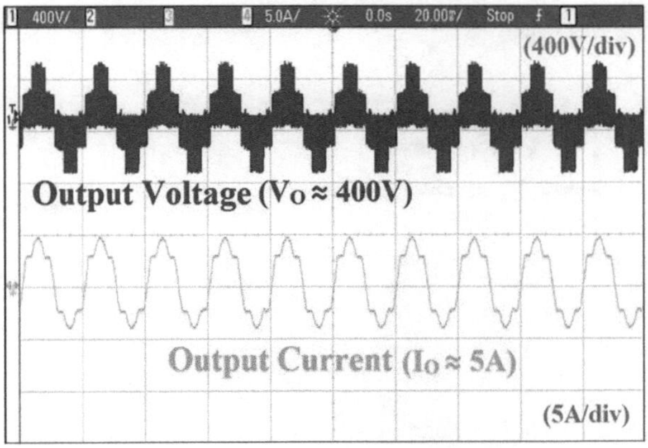

FIGURE 10.11 Output waveform of both the voltage and current from the module.

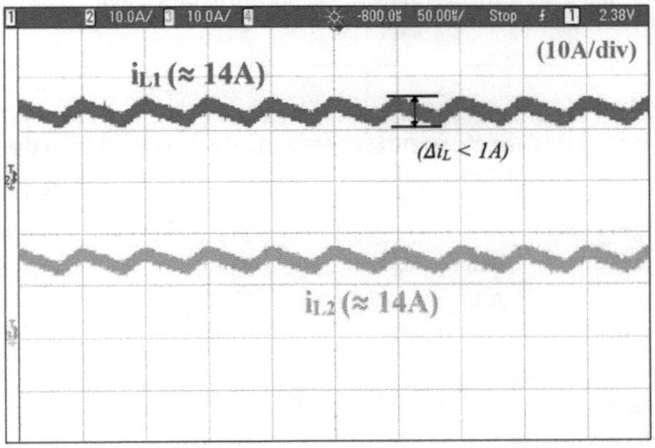

FIGURE 10.12 Inductor current waveform.

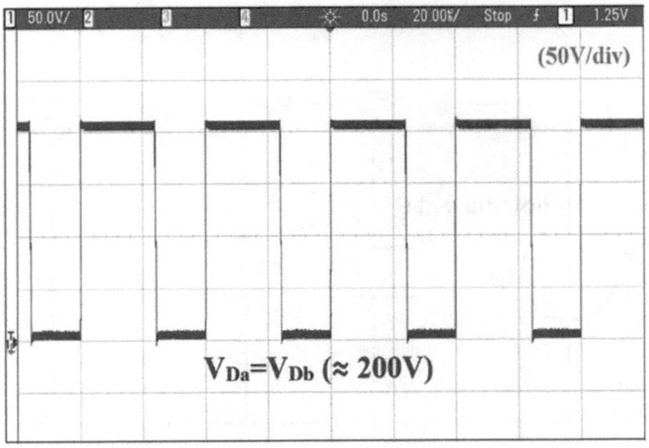

FIGURE 10.13 Auxiliary diode waveforms.

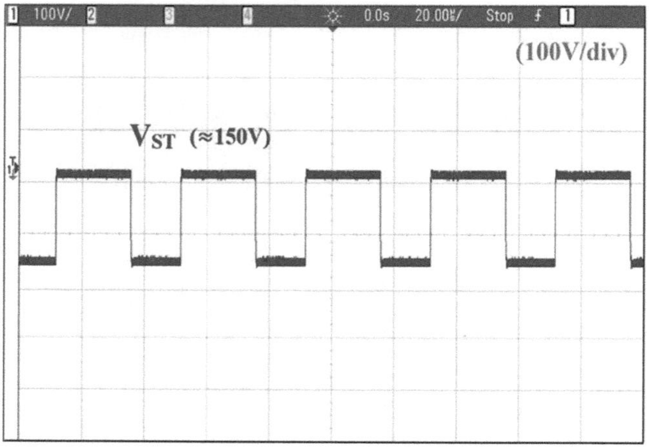

FIGURE 10.14 Voltage waveform across the shoot-through switch.

TABLE 10.2
Comparison of Parameters between Simulation and Hardware

Parameter		Simulation Value	Hardware Value
DC-link voltage (V_{dc})		217 V	≈200 V
Output voltage (V_O)	Before filter	410 V	≈400 V
	After filter	376 V	≈390 V
Output current (I_O)		≈5.4 A	≈5 A
Inductor current ($I_{L=}I_{L1=}I_{L2}$)		16.6 A	≈14 A
Inductor current ripple (Δi_L)		1.4 A	1 A
Shoot-through switch voltage (Capacitor voltage) ($V_{ST2=}V_{C2}$)		184 V	≈150 V

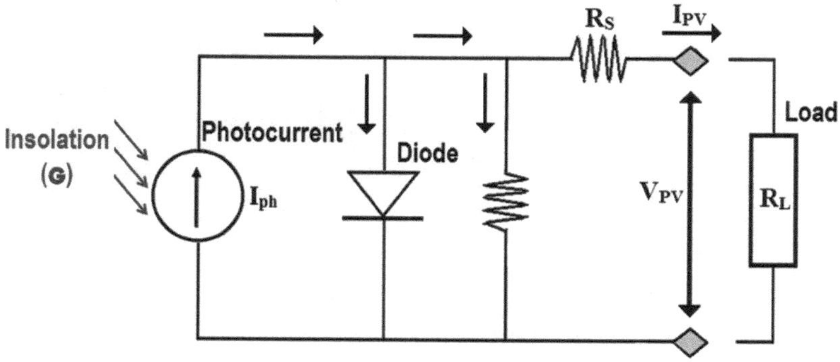

FIGURE 10.15 One-diode model of the PV panel.

TABLE 10.3
Parameters of 37 W Solkar PV Panel Given in the Datasheet

Parameter under STC (1,000W/m², 25°C)	Rating
P_{PV} (rated power)	37.08 W_p
V_{mp} (V at maximum power)	16.56V
I_{mp} (I at maximum power)	2.25 A
V_{OC} (at open circuit)	21.24V
I_{SC} (at short circuit)	2.55 A
N_S (number of cells in series)	36
N_p (number of cells in parallel)	1

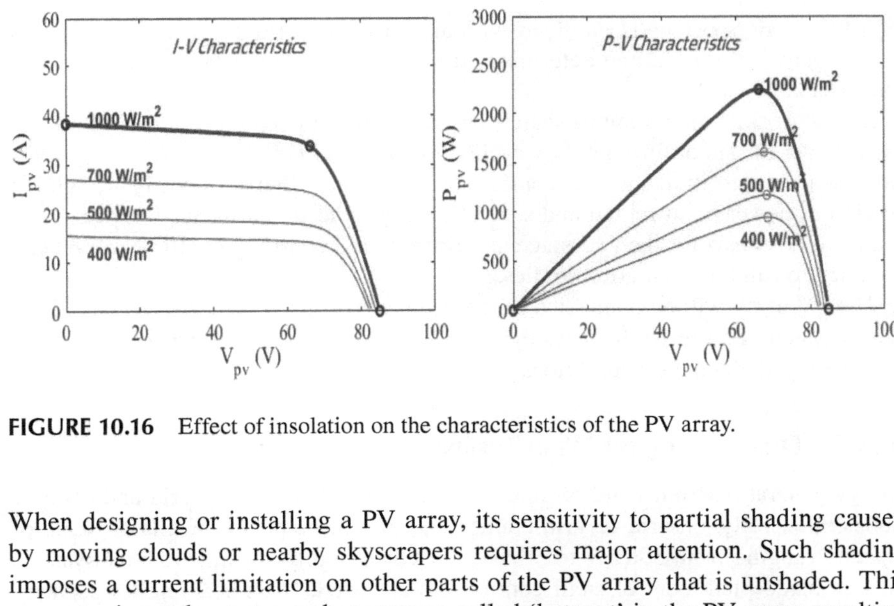

FIGURE 10.16 Effect of insolation on the characteristics of the PV array.

When designing or installing a PV array, its sensitivity to partial shading caused by moving clouds or nearby skyscrapers requires major attention. Such shading imposes a current limitation on other parts of the PV array that is unshaded. This current mismatch creates a phenomenon called 'hotspot' in the PV array, resulting in increased thermal stress on the shaded PV panel [13]. A substantial impact of partial shading is from the series–parallel connections of the panels ($m \times n$ panels), where series-connected (m) panels are more vulnerable to shading than the parallel-connected (n) ones. From literature [14], it is observed that a PV array with a shorter number of series-connected panels is expected to cause considerably less damage, as the high output current improves the current mismatch within the array. For a 2kW capacity, a 4×15 PV array is considered to be optimal for the hybrid system feeding the grid. The PV characteristics of the PV array are presented in Figures 10.16 and 10.17.

The PV module produces nonlinear output as ambient parameters vary, resulting in a complex relationship between temperature and total resistance. Therefore,

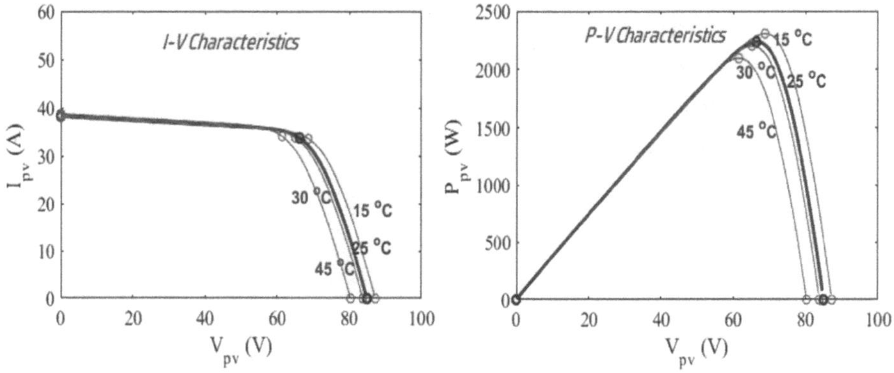

FIGURE 10.17 Effect of temperature on the characteristics of the PV array.

each PV array per stage is provided with an individual MPPT circuit that samples the PV output, tracks it, and determines the fitting modulation index for individual converters.

In this work, a dichotomous search-based algorithm is employed to have partial control over the shoot-through duty cycle of the inverter [15]. The modified dichotomous search algorithm method consists of two stages: first to identify the global maximum power point region and second to locate the actual maximum peak. The complete flowchart for the two-stage algorithm is shown in Figure 10.18, which can be coded to run the simulation studies.

The PV array will be connected to the dc/dc converter to feed the DC bus of the hybrid system. Figure 10.19 shows the voltage, current, and power waveforms of the converter, under standard conditions.

10.5.2 Designing of the Wind Turbine

Energy generation from wind turbines has been classified into on-grid and off-grid modes. Generally, on-grid or grid-connected wind turbines are of a large scale of capacity ranging from 500 kW to 5 MW, for which highly accurate control systems such as blade pitch control, stall control, etc are used. But for a smaller capacity (<100 kW), utilizing such complex control is not advisable, due to increased operational. In such cases, instead of fine-tuning to the required frequency directly, they produce DC power and store it in batteries.

To design a wind turbine energy system for a small capacity like 2 kW, several key factors must be deliberated. They are briefly discussed as follows:

1. **Turbine Type and Size**: The turbine chosen should efficiently capture the wind energy of the area (between the two types: horizontal axis and vertical axis). For such a small scale ranging from a few hundred watts to a few kilowatts in capacity, the *horizontal-axis wind turbine* is commonly preferred.

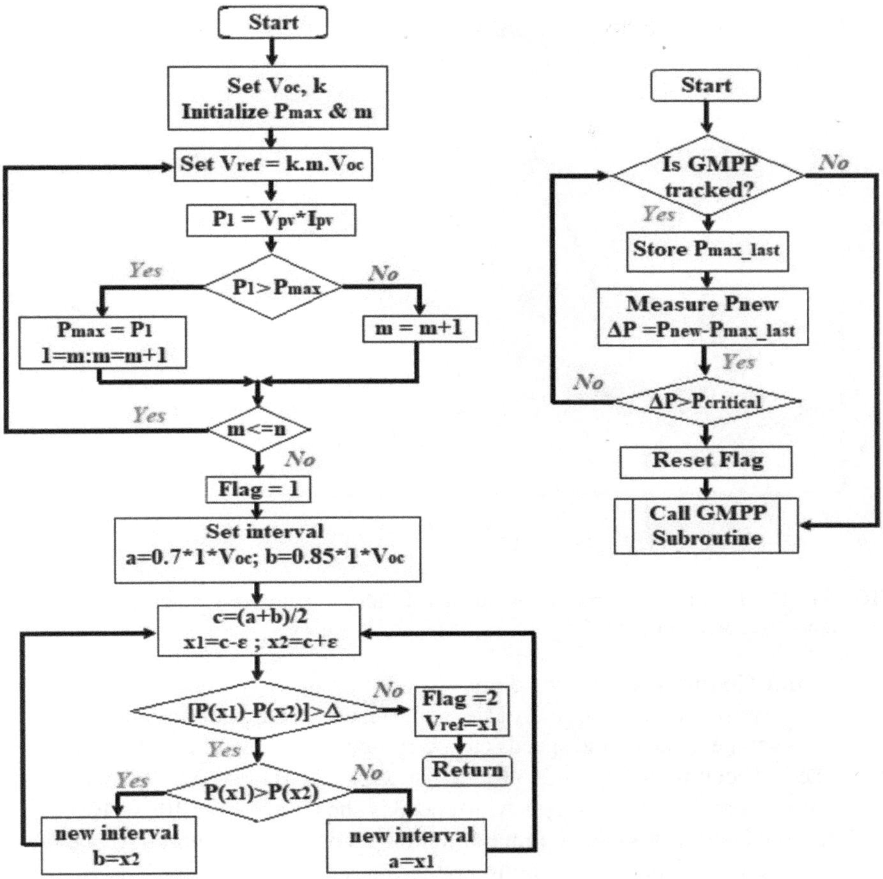

FIGURE 10.18 The working flowchart of the dichotomous search-based MPPT algorithm.

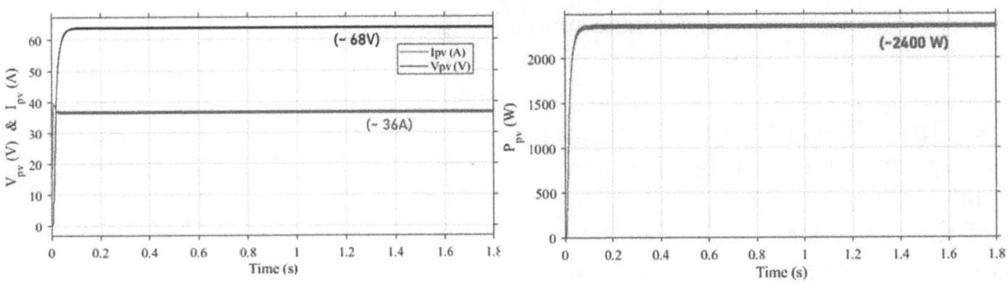

FIGURE 10.19 Voltage, current, and the power waveforms of the converter.

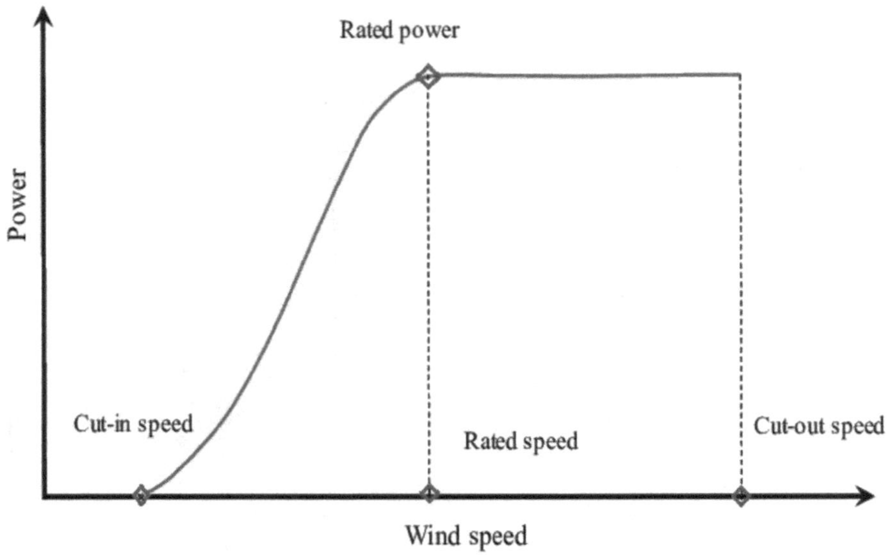

FIGURE 10.20 Power output vs wind speed (for a permanent magnet synchronous generator-based wind turbine).

2. **Wind Conditions**: The wind speed and its regularity must be measured at the site of the installation. By using wind resource maps and an anemometer, the average annual wind speeds can be computed, which will aid in deciding the perfect turbine size and tower height. A characteristic graph showing the turbine's output power versus wind speed is shown in Figure 10.20 [16].
3. **Tower Height**: An ideal optimal tower height needs to be chosen so that the turbine is positioned in the optimal wind stream. Taller towers usually capture superior and more regular winds. A tower that is 10–20 m high should be a reasonable choice for a 2kW turbine, depending on wind quality.
4. **Energy Production Estimation**: The use of the wind resource and the turbine's power curve to estimate annual energy production. For example, a 2 kW turbine at a good wind site can produce around 3,000–6,000 kWh per year.

Apart from these, permissions for installation, installation, maintenance, and monitoring performances all play an important role in designing and implementing a wind turbine system.

The design equations for a wind turbine can be referred to from [17]. For a capacity of 2 KW, the parameters for the wind turbine considered are given in Table 10.4. In this work, the wind turbine is designed using a permanent magnet synchronous generator, where permanent magnets are used instead of traditional field windings in the rotor. This facilitates turbine operation without a gearbox even at low speeds and helps maintain higher efficiency.

TABLE 10.4
Ratings of the 2 kW Wind Turbine

Parameters	Rating
Maximum power (P_{max})	2,400 W_p
Tip speed ratio (λ)	8.1
Turbine performance coefficient (C_p) (λ, β)	0.41
Pitch angle of the blade (β)	0°
Air density (ρ)	1.225 kg/m^3
Wind speed – (V_W)	12 m/s

10.5.3 Integration Challenges of Hybrid-Grid-Connected Power Systems

Integrating PV and wind energy systems into the existing power grid is essential for transitioning to a clean energy future, but this progression does not come without obstacles. The integration of these renewables presents a few technical, economic, and social challenges that need to be solved to guarantee steady power transmission. The following section discusses some factors, from synchronization issues to robust energy storage solutions. These challenges need understanding for the successful arrangement of PV and wind-based energy systems.

1. **Synchronization Issues**
   Synchronizing the output from PV and wind energy sources with the existing grid infrastructure is a significant challenge. Coordinating the operation of these systems with the existing grid infrastructure is complex due to differences in their generation patterns. The variability and uncertainty of renewable energy generation require exact real-time control over the frequency and voltage standards in smart grids. Furthermore, fluctuations in power generation as the weather changes can cause synchronization problems which makes it difficult for the grid to deliver a consistent output.
2. **Grid Stability and Reliability**
   Maintaining grid stability and reliability is a major concern when integrating PV and wind energy. Intermittent and variable power output from these sources will introduce frequency fluctuations, which can pose a major risk to the stability of the grid. Also, setting up high volumes of renewable energy, such as wind and solar, can undermine the security margins throughout power settings. This is particularly pronounced during peak generation periods when the supply may exceed demand, leading to potential overloading and system failures. Without proper management, these fluctuations can cause disruptions in power supply and can cause disturbances that affect both grid reliability and consumer power quality.
3. **Managing Intermittent Power Generation**
   One of the large barriers to integrating PV and wind is dealing with their intermittency. Being heavily dependent on weather conditions, the

predictability of energy production from these sources can be a little tricky. Weather forecasting models are necessary to predict energy generation, so being able to do this accurately is important. Furthermore, any energy fallout between the supply and demand can be controlled while providing timely and reliable solutions for grid balancing.

4. **Energy Storage Solutions**

   Energy storage devices are crucial to deal with the intermittency of PV and wind power. Battery and pumped hydro like storage systems can then store the excess energy to be used in periods with low production. Nevertheless, challenges that will arise with respect to the capacity and cost of these storage solutions are substantial. Additionally, the beautiful growth in these technologies clocked with PV and wind system integration demands a surge of energy storage infrastructure as well.

5. **Infrastructure and Grid Compatibility**

   Renewable resources must be distributed to the checkdown level, also requiring significant upgrades in the existing grid infrastructure and the transmission system, and enhancing substations and distribution networks to accommodate PV or wind systems. The image taken from local media ensuring that new renewable installations could work well with traditional grids and have a good energy transfer rate is essential to minimize possible losses.

6. **Economic and Financial Barriers**

   In solar and wind energy integration, very high installation and grid connection costs are required first. Due to unfavorable market structures and pricing mechanisms, renewable power generation faces certain economic hurdles. To create proper economic incentives and financial mechanisms, we need strong policy measures.

7. **Policy Hurdles and Environmental and Land Use Concerns**

   Enabling policy and regulatory systems is key to ensuring that PV and wind power are successfully integrated. Slow but lengthier and no less complex are the permitting processes associated with developing renewable energy projects. This will be key in removing the regulatory and policy impediments that stand between renewable resource integration.

   Acquiring land for large-scale solar and wind is particularly difficult. There are also environmental impacts, including habitat fragmentation and effects on wildlife, that must be taken into account in order to alleviate those factors as much as possible. The trade-offs between the benefits of renewable energy to reduce greenhouse gas emissions and more localized environmental impacts related to siting, design, construction, operation, or decommissioning need to be considered for long-term sustainable development.

8. **Community Acceptance and Social Impact**

   Public trust and local acceptance of renewable energy projects are key. This can help to facilitate the integration of PV and wind energy technology and establish positive relationships by resolving any local concerns involved in planning and development processes with communities.

To summarize, while combining PV and wind energy sources brings important advantages for the renewable power production, on a global scale, it raises numerous challenges. The synchronization problems, grid balancing challenges, integrated intermittent generation solutions, enabling storage technologies, supporting smart and robust market structures, and cost-adding infrastructure upgrades remain the critical areas to be addressed while integrating these renewable sources of energy together besides overcoming multiple economic and regulatory failures, environmental impacts, and accelerating social acceptance.

10.5.4 Design of an Electrical System

When designing an electrical system including the inverters, converters, and controllers for the hybrid energy system, one of the critical factors to consider is how to integrate solar PV and wind energy sources efficiently. For efficient operation of the hybrid systems, the following factors are to be considered for designing the electrical system.

- To decide on whether to use DC coupling, to combine the DC outputs of both the PV and wind sources before entering the inverter or AC coupling, and to convert the DC outputs of the two sources separately.
- To reduce the current harmonics injected into the grid within the standardized limit, a filtering network is required to act as a link between the inverter topology and the grid [18].
- The power conditioning equipment to manage system stability, including a proper synchronization system, to overcome the power quality issues that may be caused by some unpredictable disturbances. One of the most commonly used grid synchronization routines is the phase-locked loop, which is used in detecting the frequency and phase angle of the presumed input signal [19].
- To choose or design a proper energy storage system. Generally, a battery that is compatible with the inverter(s) is considered to be an efficient energy storage system to store the combined output from the hybrid system. In such cases, a charge controller should be used for battery charge and discharge to increase battery lifespan.
- Electrical protection to floor-protect anything that connects the source to the power. A surge protector is necessary in case the solar panel becomes faulty from lightning. Electrical grounding of the system is a must.
- A monitor system to oversee the performance of the solar and wind power devices, and system stability, to check for any fault conditions and implement corrective or preventive maintenance.

By considering the above aspects for the design of the overall electrical system, a sustainable and reliable renewable energy system that efficiently harnesses the benefits of solar and wind resources can be developed [20].

10.5.5 Economic Impacts of the System

When developing a converter/inverter topology, investigating its practical challenges is essential for several reasons. It aids in understanding the cost–benefit ratio, helping in analyzing the initial investments versus long-term savings. It helps in evaluating the possible savings in operation and maintenance costs (economic efficiency). Using MLIs with reduced switches in grid-connected systems can significantly impact the overall economic burden. They also improve the power quality and increase energy conversion efficiency, resulting in lower losses and very high system reliability.

In case of the proposed MLI, from a practical standpoint, one of the main attributes of the developed MLI is its easy scalability. Though the cascading switches are of a lower rating than the H-bridge switches, when the overall rating of the MLI is increased by adding more SL network, the rating of all the switches will go up, and hence, they will have to be replaced accordingly. On the other hand, when the overall power rating of the MLI is increased, by cascading another ESL-SBI structure to the MLI, only the main H-bridge network needs to be replaced according to the new power rating, while the cascading switches will be of the same rating (V_{in}). Hence, though scalability is easier than conventional type MLIs, the replacement of the H-bridge switches still poses an inconvenience.

Also, as the number of cascading networks increases, either in SL cells in each inverter or as a complete ESL-SBI network, the number of inductors increases. Though the rating of the inductor is small, it can still affect the overall output; hence, a proper filtering network is also required. Further study on the proposed MLI based on these economic factors and ways to improve its practical challenges will be conducted in the future work.

10.6 EMERGING TRENDS

In recent years, the energy industry has seen a notable surge in the renewable energy system, both in India and worldwide. Apart from the increasing demand for clean and renewable energy sources, technology advancements, policy support, and increasing environmental awareness have been contributing factors to this growth. One of its significant marks is the development of integrated wind–PV systems, designed to improve power generation from renewable resources and improve grid stability. The hybrid power system market is expected to continue its upward trajectory in the coming years, with a compound annual growth rate of over 6% during the forecast period. The integration of advanced technologies, growing investment in renewable energy projects, and the rise in off-grid electrification initiatives are expected to drive the market's future growth. In India, many utility-scale wind–PV hybrid projects are being proposed and implemented. As of 2023, India has successfully achieved 132.13 GW of renewable energy installed capacity and a total of 78,935.37 MW renewable energy capacity is under construction, among which 12,342.29 MW are of hybrid projects. The Kurnool ultramega solar park in Andra Pradesh is one of the largest solar parks with about 1,000 MW capacity [21]. Though it mainly focuses on solar energy generation, it also utilizes the region's wind resources. Adding a small capacity of wind energy in the park displays the benefits of such integrated systems to utilize the maximum potential. Such projects help in achieving India's target for renewable energy and support policies favoring hybrid projects [22].

The advancements in designs and technologies related to hybrid systems have gained traction worldwide. The development of control systems and inverters has facilitated the smooth operation of both PV and wind resources, aiding efficient power generation and grid integration.

The ensuing research directions are crucial in guiding the development of efficient, reliable, and sustainable inverter designs. Here, we present some of the key areas on which research can be carried out that would add to the capabilities and implementation of this system.

1. **Improved Control Algorithms**:

   There is a continuous pressing need for the development of advanced control strategies to improve the performance of MLIs in dynamic and fluctuating environments. Studies focused on model predictive control, adaptive control, and artificial intelligence-based systems will help in efficient reduction in response time as well as power quality optimization. Examination of the real-time data analytics integration as a means for predictive maintenance and fault detection could highly improve system reliability/longevity.
2. **Studies Based on Long-Term Performances**:

   It is necessary to evaluate the operational reliability and maintenance need of embedded MLI under practical work conditions in a comprehensive, long-term format. That research could also explore the impacts of aging on components, as well as thermal cycling and environmental factors affecting inverter performance over long periods.

   Test conditions are also something automotive test standards need to define. The need for standardized testing protocols for different operational scenarios will be pivotal for creating uniform performance metrics across multiple applications.
3. **Integration with Energy Management Systems**:

   Studies that investigate the synergetic effects of MLIs with smart grid technologies and energy management systems could be further developed to help optimize energy flow and storage. Investigating the communication and data exchange mechanisms between MLIs and other smart devices can lead to a more coordinated framework for energy distribution.
4. **Productivity and Cost Saving**:

   Ongoing research should also aim for new design topologies that enhance the efficiency of the MLIs while at the same time reducing manufacturing costs through modular structures executed in a scalable manner.
5. **Innovative Materials**:

   The search for novel semiconductors (e.g., silicon carbide, GaN) can open new avenues in thermal dissipation and switching speed aspects of multilevel inverters. In addition, research should address the need for novel dielectric materials in capacitors and insulators that can operate at higher voltage levels while increasing efficiency. In addition, studies into the sustainability aspect would potentially look at recyclable and eco-friendly materials to reduce the ecological footprint of inverter production.

The research areas reported above are important to further improve multilevel inverter performance within wind and PV systems, but are not limited to. These trends will not only aid in technological advancement but also promote sustainable energy for future energy systems.

Intelligent power management strategies, such as the use of dynamic load and battery integration, have also been embraced to enhance the overall system's flexibility and stability. Adding to this, the concept of "sector coupling" has emerged as a global movement, in which wind–PV systems are integrated with sectors such as transportation and industry through innovative methods like green hydrogen production. As is the case in countries like Germany and Denmark, where they have announced decarbonization of various sectors by utilizing the excess renewable energy, through this pioneering method.

To summarize, the last five years have been marked by the move toward more, integrated, complex, and advanced wind–PV systems, both in India and worldwide. This trend mainly accentuates the necessity of holistic planning when deploying renewable energy systems. It highlights the great potential hybrid systems especially wind–PV have in achieving the sustainable energy transition that helps combat climate change.

10.7 CONCLUSION

This chapter provided a comprehensive investigation of using an embedded source MLI for the integration of PV and wind systems with the grid. Starting with a brief overview of the wind–PV hybrid system, key factors for designing the inverters, such as power rating, efficiency, harmonic distortion, etc., are discussed. Based on this, MLIs, especially cascaded-type MLIs, are most suited for interfacing renewable sources with the grid. Hence, a type of cascaded H-bridge MLI is designed using a suitable high-gain converter (ESL-SBI) selected from the literature. The cascading is carried out through a modular H-bridge network that uses fewer switching components, increasing overall efficiency. This chapter further takes up a case study of designing a 2 kW hybrid system that focuses on the special criteria and factors associated with the PV array and wind system sizing. The case study also briefs on different components of the electrical system for the hybrid system. Additionally, emerging trends in wind–PV technology were discussed, highlighting the evolving landscape of renewable energy integration. The findings of these studies show the potential for this hybrid system with MLIs to enable sustainable and functional clean energy solutions. This chapter aims to highlight the significance of continued research and innovation in rising renewable energy technologies for a cleaner and more sustainable future.

REFERENCES

1. International Energy Agency. *Clean Energy Market Monitor* – March 2024. Website: www.iea.org.
2. Shingrakhia, H., Joshi, S., Shah, H. (2023) Review on optimum sizing of stand-alone hybrid renewable energy sources. In *IEEE IAS Global Conference on Renewable Energy and Hydrogen Technologies (GlobConHT)*, Male, Maldives, pp. 1–13.
3. Ahmad, S., Johari, S.H., Ahmad, A., Halim, M.F.M.A. (2015) Grid connected multilevel inverters for PV application. In *IEEE Conference on Energy Conversion (CENCON)*, Johor Bahru, Malaysia, pp. 181–186.
4. Khan, M.J. (2021) Review of recent trends in optimization techniques for hybrid renewable energy system. *Archives of Computational Methods in Engineering*, 28(3): 1459.
5. Fezzani, A., et al. (2023) Performances analysis of three grid-tied large-scale solar PV plants in varied climatic conditions: a case study in Algeria. *Sustainability*, 15(19): 14282.
6. Kumar, S., Sethuraman, C., Chandru, G. (2021) Design of control unit for off-grid and grid connected solar-wind hybrid system suitable for supplying power to both AC and DC loads. In *International Conference on Recent Trends on Electronics, Information, Communication & Technology (RTEICT)*, Bangalore, India, pp. 331–338.
7. Srividya, D., Gajula, U., Eragamreddy, G. (2021) Hybrid multilevel inverter fed isolated wind/pv system for the reduction of harmonics. *Journal of Chengdu University of Technology*, 26(8): 1–8.
8. Kumar, V., Kumari, P., Kumar, N. (2023) Comparison of different levels of cascaded H bridge multilevel inverter using PSPWM technique for EV applications. In *IEEE International Students' Conference on Electrical, Electronics and Computer Science (SCEECS)*, Bhopal, India, pp. 1–8.
9. Divya, T., Ramaprabha, R. (2020) Comparative topological study of embedded based switched boost inverter. In Jude Hemanth, D., Ambeth Kumar, V. D., Malathi S. (Eds.), *Advances in Parallel Computing* (IOS Press E-Books), vol. 37. Intelligent Systems and Computer Technology, pp. 37–44.
10. Kota, S., Muddineni, V., Dadiboina, A., Venna, G. (2016) Simulation and analysis of novel cascaded Z source inverter. *International Journal of Advances in Engineering & Technology*, 5: 207–215.
11. Divya, T., Ramaprabha, R. (2024) Embedded switched boost multilevel inverter for PV-fed single-phase grid. *Journal of Electrical Systems*, 20(1): 36–51.
12. Sutaya, I. W., Giriantari, I. A. D., Ariastina, W. G., Satya, I. N. (2023) Harnessing solar energy: enhancing maximum power output of parallel solar panels through external voltage analysis. In *International Conference on Smart-Green Technology in Electrical and Information Systems (ICSGTEIS)*, Kuta, Indonesia, pp. 184–189.
13. Vunnam, S., Vanithasri, M., Alla, R. (2023) An outline of solar photovoltaic systems impact on environment. *Bulletin of Electrical Engineering and Informatics*, 12(5): 2635–2642.
14. Divya, T., Ramaprabha, R. (2022) Embedded switched Z-source multilevel inverter for grid interfaced photovoltaic systems. *Advances in Electrical and Computer Engineering*, 22(3): 43–52.
15. Ramabadran, R., Sundaram, M. S. (2020) Dichotomous search based algorithm for tracking global peak in partial shaded photovoltaic array. *Revue Roumaine Des Sciences Techniques-Serie Electrotechnique Et Energetique*, 65(3–4): 211–216.
16. Shankara, H. M., Hegde, R., Gaonkar, R., Rai, N. (2023) A critical assessment of significant developments in wind turbine performance. *International Journal of Ambient Energy*, 45: 1–51.

17. Anjana Ethirajan, R. R. (2023). Mathematical modeling of PV and wind for 2.4 kW connected to a common DC bus. In *AIP Conference Proceedings*, vol. 2829, no. 1, p. 040003.
18. Kharade, J., Kabela, J., Joshi, P. (2023) Harmonics mitigation using DQ transform based solid state transformer (SST) for it and corporate buildings. *Proceedings on Engineering Sciences*, 5: 325–332.
19. Baid, S., Priyadarshi, A., Saha, R. (2023) A methodological approach of LCL filter design for industrial application using MATLAB script. In *2023 IEEE 3rd International Conference on Smart Technologies for Power, Energy and Control (STPEC)*, Bhubaneswar, India, pp. 1–5.
20. Ali, M. et al. (2023) Renewable energy sources-based hybrid microgrid system for off-grid electricity solution for rural communities. *Energy Science & Engineering*, 11: 1–14.
21. Sandhu, M., Thakur, T. (2020) Modified cascaded H-bridge multilevel inverter for hybrid renewable energy applications. *IETE Journal of Research*, 68: 3971–3983.
22. Pamuji, F. et al. (2023) Design and analysis of charging stations for battery electric bus with PV, wind turbine, and PLN grid. In *International Conference on Advanced Mechatronics, Intelligent Manufacture and Industrial Automation (ICAMIMIA)*, Surabaya, Indonesia pp. 1–6.

11 Electric Vehicle Charging Infrastructure
Current Technologies and Future Trends

Ponnupandian Aruna, Veeramani Vasan Prabhu, Palani Velmurugan, and Subbiah Sridharan

11.1 INTRODUCTION

One potential worldwide plan to reduce transportation-related carbon emissions is the shift to electric mobility. A small number of nations, including India, have joined the worldwide EV30@30 initiative, the goal of which is to sell 30% EVs by the year 2030 (EV30@30 campaign I Clean Energy Ministerial). The successful completion of this lofty shift depends on the availability and reliability of a widespread network of EV charging stations. The Indian government has put in place a number of laws that would facilitate the expansion of the charging infrastructure (Electric Vehicle I Government of India I Ministry of Power (powermin.gov.in)). Nevertheless, due to the unique qualities of this new infrastructure type, it must be tailored to the specific Indian transport ecosystem, and stakeholders must be empowered to back its actual expansion. In order to address local needs and integrate EV charging infrastructure into transportation and electrical networks, a contextual strategy is necessary to ensure its efficient and timely implementation.

EVs can be charged in a variety of ways, depending on location and requirement. Accordingly, charging infrastructure for EVs is of different types and designed for different applications. Specifications and standards for EV chargers, also known as electric vehicle supply equipment (EVSE), vary from one country to another, based on available EV models in the market and the characteristics of the electricity grid. EV charging infrastructure is crucial to the widespread adoption and practical utility of EVs. As we transition from fossil fuels to renewable energy sources, the development of an efficient and accessible charging network is essential for facilitating this shift and ensuring the sustainability of transportation systems.

The foundation of EV charging infrastructure encompasses various types of charging stations, differentiated primarily by their charging speed and installation location. Level 1 chargers, the most basic, use a standard household outlet and are suitable for overnight use at home. Level 2 chargers, which require a 240-V source, offer a faster charging speed and are commonly found in residential, commercial, and public settings. The fastest available, Level 3 chargers, also known as DC fast

DOI: 10.1201/9781003496946-11

chargers, can charge an EV battery to 80% capacity in under an hour, making them ideal for highway rest stops and urban areas where quick charging is necessary (Level 1 vs. Level 2 vs. Level 3 EV Chargers: Which is right for you? | Energy Sage). The strategic placement of these chargers in public locations, workplaces, and homes is vital for covering the diverse needs of EV users. Public charging stations are typically installed in accessible locations such as shopping centers, parking lots, and near major highways to ensure drivers can recharge during longer trips or while running errands. Workplace charging is an emerging trend that provides employees the convenience of charging their vehicles while at work. Moreover, the integration of smart technology into EV charging infrastructure is enhancing the user experience and efficiency of power usage. Smart chargers can adjust the charging rate based on grid demand, vehicle requirements, and user preferences, optimizing energy consumption and reducing costs (Best Smart EV Chargers—Clean Energy Reviews). As the number of EVs on the road continues to grow, the expansion and evolution of charging infrastructure must keep pace. This involves not only increasing the number of charging stations but also improving their technology and connectivity to support the next generation of EVs. The development of robust EV charging infrastructure is a pivotal component of the global strategy to reduce carbon emissions and promote sustainable transportation solutions.

This chapter delves into various aspects of EV charging infrastructure and technology, starting with the architecture and classification of EV chargers. It explores the use of industrial 3-phase Indian plugs and sockets for EV chargers, providing insight into the standards and configurations specific to the region. This chapter discusses how to choose the right charger by comparing slow and fast charging options, considering factors such as battery size and vehicle requirements. The complexities of fast charging for EV batteries are examined, highlighting the technical challenges and impacts on battery health. Additionally, this chapter introduces battery swapping technology, which offers an alternative to traditional charging by transforming battery ownership from a capital expenditure to an operational cost model. This includes battery leasing and swapping, enhancing EV convenience, and potentially improving financial viability. This chapter also addresses the cost structure and financial viability of battery leasing, as well as the security measures necessary for smart swappable batteries. The benefits of standardized battery systems for EVs are discussed, emphasizing the importance of evolution and standardization in EV charging protocols. This chapter outlines India's initiation into EV charging standards and future developments, such as AC-002 and DC-002 chargers, and examines the requirements for charging infrastructure and protocols to support EV adoption. The role of power electronics configuration in EV charging is highlighted, demonstrating its impact on efficiency and performance. Additionally, this chapter explores the impact of EV charging on the grid, balancing the positive aspects like grid stability through vehicle-to-grid (V2G) technology with challenges such as increased load during peak times.

This chapter further discusses companies involved in tie-ups for charging infrastructure development, showcasing collaborative efforts to expand the charging network. An introduction to wireless charging for EVs is provided, covering the principles and various types of wireless charging, followed by an examination of the challenges and future aspects of wireless power transfer (WPT) systems. Case studies of Tesla Superchargers and BMW's wireless charging technology offer real-world examples of advanced charging solutions. This chapter concludes with a comprehensive overview of the discussed topics, emphasizing the importance of continued innovation and standardization in the EV charging industry.

11.2 ARCHITECTURE OF EV CHARGERS

The architecture of an EV charger is shown in Figure 11.1. In the architecture of EV chargers, several critical components and functionalities ensure efficient and safe operation:

11.2.1 Rectifier in EV Chargers

A rectifier is used in EV chargers to convert alternating current (AC) from the grid into DC, which is required for charging the battery. The rectifier ensures a smooth and stable DC output, which is essential for the efficient charging of the EV battery.

11.2.2 Voltage and Current Control

Voltage and current control in EV chargers are vital for managing the power delivered to the vehicle's battery. Voltage control ensures that the correct voltage level is supplied to the battery, while current control regulates the amount of current flowing into the battery. These controls are necessary to match the battery's requirements, prevent overcharging, and enhance battery life.

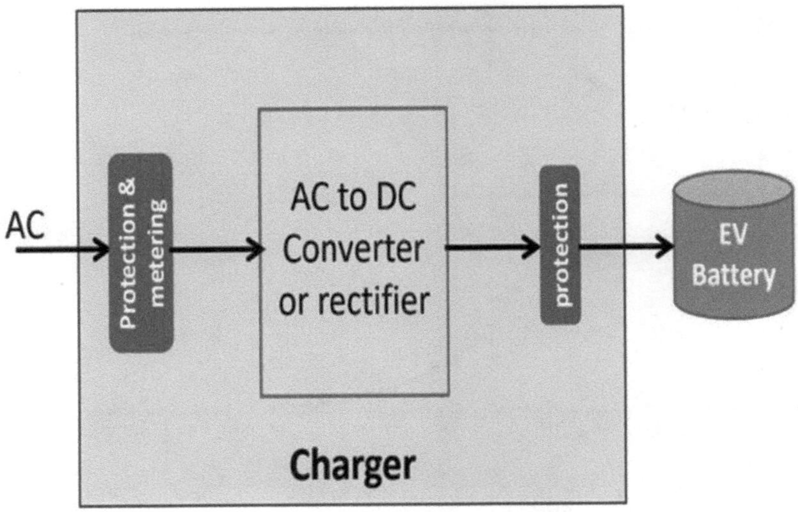

FIGURE 11.1 Architecture of an EV charger.

11.2.3 Protection and Metering at the Input Side

Protection mechanisms at the input side of an EV charger typically include surge protectors, circuit breakers, and residual current devices. These components safeguard the charger and connected vehicle from voltage spikes, overcurrent, and electrical faults. Metering devices are also incorporated to monitor the amount of energy consumed, providing data for billing and energy management purposes.

11.2.4 Protection at the Output Side

At the output side, protection includes features such as overcurrent protection, overvoltage protection, and short-circuit protection. These measures prevent damage to the vehicle's battery and the charger itself by ensuring that only safe levels of current and voltage are delivered during the charging process.

11.2.5 Methods of Charging

Constant current and constant voltage are two common charging methods for EV batteries, as shown in Figure 11.2:

Constant Current Charging: In this method, the charger supplies a constant current to the battery until it reaches a specified voltage. This approach is beneficial during the initial phase of charging, as it allows the battery to be charged quickly without overheating.

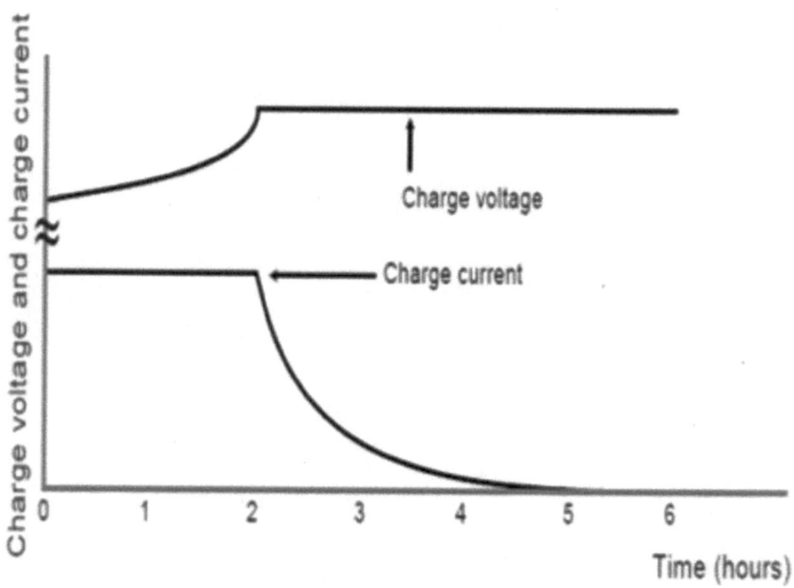

FIGURE 11.2 Constant current–constant voltage charging.

Constant Voltage Charging: Once the battery reaches the desired voltage, the charger switches to constant voltage mode. In this phase, the voltage is kept constant while the current gradually decreases as the battery becomes fully charged. This method helps to top off the battery safely and ensures that it reaches full capacity without overcharging. Combining these two methods allows for efficient and safe charging, maximizing battery life and performance. The charger typically starts with constant current charging and then transitions to constant voltage charging as the battery nears its full charge. This dual-phase approach is widely used in modern EV chargers to optimize the charging process.

11.3 CLASSIFICATION OF CHARGERS IN EV

Onboard and offboard chargers are pivotal in the EV ecosystem, facilitating the transition from ICEs to sustainable electric mobility. These chargers address the fundamental need for energy replenishment in EVs, each serving distinct roles to enhance the versatility and efficiency of the charging infrastructure (Liao et al., 2019; Zhang and Wang, 2018).

11.3.1 Onboard Chargers

Onboard chargers are integral components built into the EV. They are designed to convert AC from external power sources, such as household outlets or standard public charging stations, into the DC required to charge the vehicle's battery. The conversion process involves rectifying the AC input to a suitable DC output that matches the battery's charging requirements (Tate et al., 2009). Figure 11.3a and b shows Level 1 and Level 2 chargers. One of the primary advantages of onboard chargers is their convenience and ubiquity. Since they are built into the vehicle, drivers can charge their EVs using readily available electrical outlets, making it possible to charge at home, at work, or at any location with an accessible power supply. This capability is particularly beneficial for daily use and overnight charging scenarios, where the vehicle can be plugged in for extended periods without the need for specialized infrastructure (Liao et al., 2019). However, onboard chargers are typically limited in power due to size and thermal management constraints. Most onboard chargers are designed for Level 1 (120 V) or Level 2 (240 V) charging, which provides a slower charging rate compared to high-power options. Despite this, their versatility and ease of use make them a critical component for regular, everyday charging needs (Tate et al., 2009) (Table 11.1).

11.3.2 Offboard Chargers

Considering the excessive power consumption of Level 3, it is recommended to charge EV batteries at designated charging stations (Yilmaz and Krein 2012). Thus, offboard chargers shown in Figure 11.4 are the best way to describe Level 3 chargers. Offboard chargers, also known as external or DC fast chargers, are located outside the vehicle and are usually part of public or commercial charging infrastructure.

FIGURE 11.3 (a) Level 1 charger and (b) Level 2 charger.

TABLE 11.1
Specifications of Level 1 and Level 2 Chargers

Level 1	Level 2
Voltage: 120 V, 1-Phase AC	Voltage: 208 V or 240 V, 1-Phase AC
Amps: 12–16 A	Amps: 12–80 A
Charging load: 1.4–1.6 KW	Charging load: 2.5–19.2 KW
Charging time: 3–5 Miles per hour	Charging time: 12–60 Miles per hour

These chargers convert AC power from the grid to high-power DC, which is then delivered directly to the vehicle's battery. Offboard chargers bypass the vehicle's onboard charger, allowing for significantly higher power delivery and much faster charging times (Zhang and Wang, 2018). The primary advantage of offboard chargers is their ability to deliver rapid charging, often referred to as DC fast charging or rapid charging. These chargers can provide a substantial amount of energy in a short period, reducing the time needed to recharge an EV. This capability is essential for long-distance travel, commercial fleets, and scenarios where quick turnaround times are necessary (Tate et al., 2009). Offboard chargers are commonly found at highway rest stops, commercial areas, and dedicated EV charging stations. They support various power levels, with some providing up to 350 kW, which can add hundreds of miles of range in just a few minutes. This high-power charging is made possible by the external placement of the chargers, allowing for larger, more efficient conversion

FIGURE 11.4 Offboard charger.

TABLE 11.2
Specifications of Level 1 and Level 2 Chargers

Level 3
Voltage: 208 V or 480 V, 3-Phase AC Amps: >100 A
Charging load: 50–350 KW
Charging time: 10%–80% in 30 minutes

equipment that can handle the thermal and power demands of rapid charging (Zhang and Wang, 2018). Table 11.2 shows the specifications of Level 1 and Level 2 chargers.

11.3.3 Integration and Future Prospects

The integration of both onboard and offboard chargers in the EV ecosystem provides a comprehensive charging solution that addresses different user needs and scenarios. Onboard chargers offer flexibility and convenience for everyday use, while offboard chargers cater to high-power, rapid charging requirements for longer trips and commercial applications. Together, they form a robust charging infrastructure that supports the diverse demands of EV users (Liao et al., 2019). As the EV market continues to grow, advancements in charging technology are expected to further enhance the efficiency and accessibility of both onboard and offboard chargers. Innovations such as bidirectional charging, which allows vehicles to discharge power back to the grid or other devices, and improvements in battery technology will likely play significant roles in the evolution of EV charging infrastructure (Zhang and Wang, 2018). Furthermore, the development of standardized charging protocols and connectors, such as the combined charging system (CCS) and CHAdeMO, is crucial for ensuring compatibility across different vehicle models and charging networks. These standards help streamline the charging process and foster the widespread adoption of EVs (Tate et al., 2009).

11.3.4 Industrial 3-Phase Indian Plugs and Sockets of EV Chargers

Industrial 3-phase Indian plugs and sockets for EV chargers are designed to handle the high-power demands required for efficiently charging EVs. These connectors typically adhere to IS/IEC 60309 standards, which ensure their safety and compatibility across various applications. The 3-phase system allows for the distribution of power across three separate conductors, providing a more stable and efficient delivery of electricity. Figure 11.5 depicts industrial 3-phase Indian plugs and sockets of EV chargers.

This configuration is crucial for EV chargers, which require significant power to recharge batteries in a reasonable amount of time. The sockets and plugs are built to withstand higher currents and voltages, typically ranging from 16A to 125A and up to 690V. They are robustly constructed to endure harsh industrial environments, with features such as weatherproofing, resistance to dust and water ingress (often rated IP44 or higher), and durable materials that can handle the wear and tear of frequent use. These attributes make them well-suited for the demanding requirements of industrial EV charging stations, ensuring reliable and safe operation.

11.3.4.1 IP44 (2P+E)

IP44 (2P+E) refers to a type of electrical plug or socket designed to provide a reliable and safe connection in environments where protection against dust and water splashes is necessary. The "IP" stands for "Ingress Protection", with the first digit "4" indicating protection against solid objects larger than 1 mm, such as wires or small tools, and the second digit "4" signifying protection against water splashes from any direction. The "2P+E" denotes that the plug or socket has two active conductors (live and neutral) plus an earth (ground) connection, which is essential for safety by providing a path for electrical current to reduce the risk of electric shock. This configuration is commonly used in industrial settings, including for EV chargers, where robust and safe electrical connections are crucial.

11.3.4.2 IP67 (3P+N+E)

IP67 (3P+N+E) is a designation for a type of electrical plug or socket that offers high-level protection and a specific electrical configuration suitable for demanding environments. The "IP" stands for "Ingress Protection", with the first digit "6" indicating complete protection against dust ingress, ensuring no dust particles can enter the device. The second digit "7" signifies that the device is protected against the

FIGURE 11.5 Industrial 3-phase Indian plugs and sockets of EV chargers.

effects of temporary immersion in water up to 1 m depth for 30 minutes, making it highly suitable for outdoor and harsh industrial settings. The "3P+N+E" denotes that the plug or socket has three active conductors (3-phase power), a neutral conductor, and an earth (ground) connection. This configuration supports high-power loads and ensures electrical safety by providing a path for electrical current to prevent electric shock. Such connectors are essential in environments where robust, high-power electrical connections are required, such as in heavy industrial applications and advanced EV charging stations.

11.4 CHOOSING THE RIGHT CHARGER: HOW TO DECIDE BETWEEN SLOW AND FAST CHARGING

The driving range of an EV is determined by two key factors: the usable capacity of the battery and the energy consumption of the vehicle. The usable capacity, measured in kilowatt-hours (kWh), is calculated by multiplying the battery's total capacity by the State of Health (SoH) and the Depth of Discharge (DoD) that the battery can safely sustain. In formula terms, this is represented as

$$\text{Usable capacity} = \text{Battery capacity} \times \text{SoH} \times \text{DoD} \tag{11.1}$$

For optimal battery health and longevity, slow charging is recommended. This involves charging the battery at a rate of 0.1°C to 0.3°C, which translates to a charging time of approximately 3–10 hours to reach full capacity. This process is best conducted at moderate temperatures, ideally around 25°C, and often during nighttime hours when energy demand is lower.

However, there are circumstances where an EV needs to cover distances greater than its range on a single charge. In such cases, fast charging becomes a necessary option. Fast charging can replenish the battery's energy at a much higher rate, typically between 1°C and 3°C, allowing the vehicle to be charged in an hour or less—with the ideal scenario being a recharge time of just 10–15 minutes. This rapid charging enables longer distance travel without significant delays, providing greater flexibility and convenience for EV users.

11.4.1 C-Rate of a Battery

The C-rate of a battery is a measure that indicates the rate at which a battery is charged or discharged relative to its maximum capacity. A 1C rate means that the charge or discharge current will charge or discharge the entire battery in 1 hour. For example, charging a battery at 0.1°C–0.3°C is considered a slow charge, taking between 10 and 3.3 hours to fully charge the battery. Conversely, a lower DoD implies that less time is needed for charging.

Fast charging occurs at rates such as 1°C, 1.5°C, 2°C, or 3°C, where the battery can be charged from 20 minutes to 1 hour, assuming the battery is depleted to 100% DoD. The size of the battery and the power output of the charger determine whether the charging is considered fast or slow. For instance, a 5-kWh battery charged at 1 or 2 kW is undergoing a slow charge, while a 1 kWh battery charged at the same rate is considered a fast charge. This distinction is crucial because it illustrates that the charger's power output and the battery's size are collectively what define the charging speed, not the charger itself.

11.4.2 EV Charging: Matching Power to Battery Size

Traction power, which is the power required to move a vehicle, is a critical aspect when considering the changing needs of EVs. The battery size varies significantly across different types of EVs, influencing the charging speed when paired with chargers of various power outputs.

For instance, two-wheelers typically have a battery size ranging from 1 to 3 kWh, and e-rickshaws have batteries between 2.5 and 5 kWh. E-autos extend slightly higher with 3–6 kWh. As we move to four-wheelers, the battery sizes increase substantially; small cars have batteries ranging from 15 to 30 kWh, while large cars can have batteries between 30 and 75 kWh. The largest of the lot, buses, may have batteries anywhere from 100 to 400 kWh.

Given these battery sizes, a 10-kW charger would be considered fast for two-wheelers and three-wheelers as it can deliver a full charge relatively quickly. However, the same charger would be slow for all four-wheelers and buses due to their larger battery capacities. A 20-kW charger elevates the charging capability, being fast for two-wheelers, three-wheelers, and small four-wheelers, but still falls short for large four-wheelers and buses. For most vehicles, a 50 or 100 kW charger would provide a fast charging experience, although the largest buses and trucks would require even more powerful chargers to be considered fast.

This analysis underscores the importance of matching the charger's power output with the vehicle's battery capacity to determine the charging speed, which is essential for optimizing the charging infrastructure for various types of EVs.

11.4.3 The Complexities of Fast Charging for EV Batteries

Not all battery cells are capable of withstanding fast charging, and those that can require careful management to avoid significant impacts on battery life. Fast charging can accelerate battery degradation, necessitating that batteries are specifically designed and prepared for this process. These batteries often come at a higher cost, require extensive cooling systems, and need meticulous temperature management to handle the increased thermal loads during charging. Depending on the battery type, some may handle a fast charge rate of 1°C occasionally, while others can tolerate 2°C or even up to 3°C or 4°C without severe degradation. Discharge rates can generally be higher, with many batteries capable of 1°C discharge and short bursts at 2°C or 3°C. However, managing the temperature is critical to prevent overheating and ensure longevity (Wang et al., 2016; Yang et al., 2018). Effective temperature control remains a constant challenge, emphasizing the need for advanced thermal management systems in both charging and discharging scenarios.

11.5 BATTERY SWAPPING TECHNOLOGY

Battery swapping technology is an innovative approach designed to address key challenges associated with EV usage, particularly lengthy charging times and range anxiety. Instead of recharging the vehicle's battery, a depleted battery is exchanged for a fully charged one at specialized swapping stations, a process that can be completed in minutes, significantly reducing downtime compared to conventional charging methods. This technology offers several advantages: rapid battery exchange akin to refueling an internal combustion engine (ICE) vehicle, potential extension of battery life through centralized charging and maintenance, and lower upfront costs by decoupling the battery from the vehicle. Users can purchase EVs at a lower initial cost and lease the batteries, converting capital expenditure into operational expenditure. Additionally, with strategically located swapping stations, drivers can avoid long charging times, enhancing convenience and creating a more seamless driving experience. The technology of battery swapping is shown in Figure 11.6.

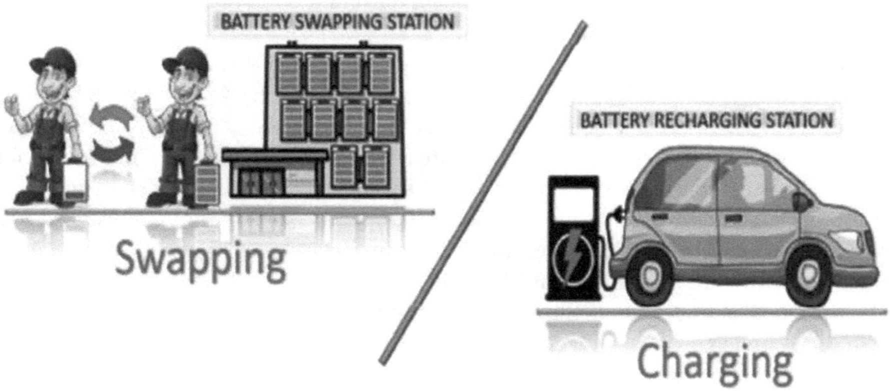

FIGURE 11.6 Technology of battery swapping.

This technology is particularly appealing for commercial fleets and urban areas where minimizing downtime is crucial. As the EV market continues to grow, battery swapping is emerging as a viable solution to complement traditional charging infrastructure, enhancing the overall feasibility and attractiveness of EVs (Doe, 2024; Green, 2024).

11.5.1 Transforming Battery Ownership: Converting Capital Costs into Operational

11.5.1.1 Costs for EVs

The capital cost of batteries, the primary energy container for EVs, is significantly higher compared to the cost of gasoline tanks in ICE vehicles. This disparity is not only evident in the initial capital cost but also in the life-cycle costs, which are influenced by various operational parameters such as the rate of charging/discharging (C-rate), temperature during charging/discharging, and the DoD (percentage of capacity used in each cycle). Additionally, interest rates play a crucial role in determining the operational costs of batteries, as they impact the overall financing expenses associated with battery procurement and maintenance.

To address these challenges, a novel mechanism is proposed wherein EV users lease charged batteries instead of purchasing them outright. This approach can effectively convert the high capital cost of batteries into operational costs, making EVs more financially accessible. Leasing batteries can mitigate the issue of range anxiety by ensuring that users have access to fully charged batteries when needed, thus extending the driving range of EVs without the need for significant upfront investment. Recent studies in 2024 have emphasized the importance of such innovative solutions in enhancing the adoption of EVs by reducing the financial burden on users and providing greater flexibility in battery management (Smith, 2024; Johnson, 2024). By transforming the battery ownership model, this leasing strategy not only lowers the barrier to entry for EV adoption but also aligns with broader economic trends and technological advancements aimed at sustainable transportation solutions.

11.5.2 Battery Leasing and Swapping: A New Model for EV Ownership

In an innovative shift from traditional EV ownership, customers can now purchase EVs without any batteries and lease charged batteries from outlets operated by an energy operator (EO). When the battery charge runs out, the customer visits an EO outlet, swaps the discharged battery for a charged one, and pays a usage charge. This model is akin to the LPG cylinder system in India, where the cylinder (battery) is not owned by the customer but leased. This system acknowledges that the battery is merely a container of electricity; it is the electricity that is consumed, not the container itself. Consequently, purchasing electricity as an energy source becomes very similar to buying petrol, with the added factor that customers also pay lease charges for the battery container. The EO is responsible for purchasing and owning the batteries, establishing charging stations, and setting up multiple swapping outlets. Customers pay usage charges for the charged battery during each swap. This approach not only reduces the initial cost of EVs for customers but also ensures a more efficient and convenient method of managing battery usage, addressing issues such as range anxiety and battery maintenance (Taylor, 2024; Brown, 2024).

11.5.3 Enhancing EV Convenience through Efficient Battery Swapping

The time required to swap discharged batteries with charged ones is just a few minutes, akin to the time taken to fill a petrol tank, eliminating waiting times for charging. This efficiency allows for the use of smaller batteries without inducing range anxiety, as the quick swap time ensures uninterrupted journeys. EOs provide multiple swapping outlets for user convenience, similar to the widespread availability of petrol stations. EV customers can sign up with an EO to regularly swap batteries at any of the EO's outlets. Charging can be done either at the swapping outlet itself or at centralized charging stations that support multiple nearby swapping outlets. These centralized stations utilize bulk chargers to simultaneously charge multiple batteries, ensuring that charged batteries are always available at each outlet for customer convenience. EOs manage the distribution of charged batteries, moving them between locations as necessary to maintain consistent availability. This system not only simplifies the battery management process but also enhances the overall convenience and practicality of EV ownership (Williams, 2024; Evans, 2024).

11.5.4 Cost Structure and Financial Viability of Battery Leasing for EVs

Battery leasing charges are determined based on the kilowatt-hours (kWh) used and the time of use, incorporating several cost factors. These include the cost of electricity to charge the batteries, depreciation, and interest costs for the batteries as detailed in Section 4.7. Additionally, the costs of bulk chargers, operational expenses for charging stations and swapping outlets, manpower, and the costs of moving batteries when necessary are factored in, along with a profit margin for the EO. The goal is to ensure that these leasing charges are competitive compared to the cost per kilometer for petrol in similar internal combustion vehicles. For optimal operations, an EO needs to purchase approximately 1.3–2 times the number of batteries relative to the number of vehicles served. This excess capacity ensures that there are always enough charged batteries available for swapping, thus maintaining seamless service for EV users and enhancing the overall feasibility and attractiveness of the battery leasing model (Smith, 2024; Johnson, 2024).

11.5.5 Strategic Investments and Data Management in Battery Leasing

EOs invest significantly in batteries, bulk chargers, and charging and swapping outlets, with bulk battery procurement driving down costs even for high-quality, long-life batteries. To maximize battery lifespan, EOs charge batteries using methods that optimize their performance, including the use of cooling systems for bulk chargers when necessary. Slow charging, over a period of 2–5 hours (0.2°C–0.5°C rate), is employed to further enhance battery life. Differential leasing charges may be applied based on the usage patterns of drivers, recognizing that regular fast driving and high current draw can negatively impact battery longevity. Central management systems (CMS) servers are implemented to record detailed data on cell and battery usage. These systems capture a variety of data during charge and discharge cycles to optimize both battery performance and financial returns. This comprehensive data collection and analysis enable EOs to fine-tune their operations, ensuring efficiency and cost-effectiveness while maintaining high service quality for EV users (Taylor, 2024; Brown, 2024).

11.5.6 Security Measures for Smart Swappable Batteries

Advanced locking mechanisms and security features are built into smart swappable batteries to reduce the risk of theft. Because these batteries can only be charged by approved chargers and can only power authorized vehicles, they are designed to be useless if stolen. Every battery has its own unique identifier, and authentication is made possible by encrypted key exchanges between the smart swappable battery and the charger or car. A valid authentication is required before the battery's internal relay can activate, making it impossible for illegal chargers or cars to switch it on. This security system ensures that even if a customer walks away with the battery, it remains an expensive but unusable resource without proper authorization. These measures not only protect the assets of the EOs but also provide peace of mind for users by ensuring that the batteries remain secure and functional only within the authorized network (Johnson, 2024; Evans, 2024).

11.6 BENEFITS OF STANDARDIZED BATTERY SYSTEMS FOR EVs

Standardizing battery systems across multiple brands of similar vehicles, such as small two-wheelers, can significantly reduce costs and improve product quality due to increased production volumes. For EOs, a larger market becomes accessible, enhancing the viability of their business model. EV customers benefit from a greater number of EO outlets and potentially have multiple EOs to choose from, increasing convenience and service accessibility. Without standardization, EOs would need to stock different types of batteries even for similar vehicles, leading to increased charged battery shelf-time and higher leasing costs. Standardized batteries streamline inventory management, reduce operational complexity, and lower costs, making the leasing model more attractive and feasible for both EOs and customers (Taylor, 2024; Green, 2024).

11.6.1 Evolution and Standardization of EV Charging Protocols Worldwide

The journey of standardizing EV charging protocols has been a dynamic process with different regions adopting and evolving standards over time. Japan was a pioneer in this field, introducing the CHAdeMO (derived from "let us go for a cup of tea") DC chargers as the first standardized protocol. China soon followed suit with the GuoBiao Tuijian (GBT) DC charger standards, which, although arriving later than CHAdeMO, featured more comprehensive protocols and updated capabilities. Europe entered the scene afterward, proposing even more advanced EV charging standards. The CCS was first standardized in Europe as a DC charger and later updated to CCS-2, which incorporated Grid to Battery Charging protocols, reflecting the latest technological advancements.

The United States adopted a mix of European and Japanese standards, coming in later in the standardization timeline. For AC chargers, standards were set by Europe, the USA, and Japan. The Level 1 AC charger, which does not communicate with the vehicle or the grid, was the simplest. The Level 2 AC charger allowed communication with the vehicle but did not specify grid communication. Level 3 chargers, being higher power DC chargers, enabled communication with both the vehicle and the grid. Today, all chargers use the same standard for communication with the grid, known as the open charge point protocol (OCPP), ensuring a unified and efficient charging infrastructure worldwide. The worldwide protocols of EV charging are given in Figure 11.7.

11.6.2 India's Initiation into EV Charging Standards

India began its journey into establishing EV charging standards in 2017, with a strong focus on affordability and practicality to suit the local market. The Department of Heavy Industry introduced the Bharat Chargers to create a standardized and cost-effective charging infrastructure. The AC-001 charger shown in Figure 11.8a was specified as the lowest power AC charger, delivering 10 kW at 230 V and without communication capabilities to either the grid or the vehicle. For DC charging, the DC-001 charger shown in Figure 11.8b was introduced as a low to medium power option, providing 15 kW at 48 V, 60 V, or 72 V.

This charger included OCPP communication to the grid and used a protocol based on the Chinese GBT standard for vehicle communication. While there have been discussions about the next iterations, AC-002 and DC-002, their specifications have yet to be finalized, reflecting ongoing efforts to adapt to evolving technological needs and market conditions.

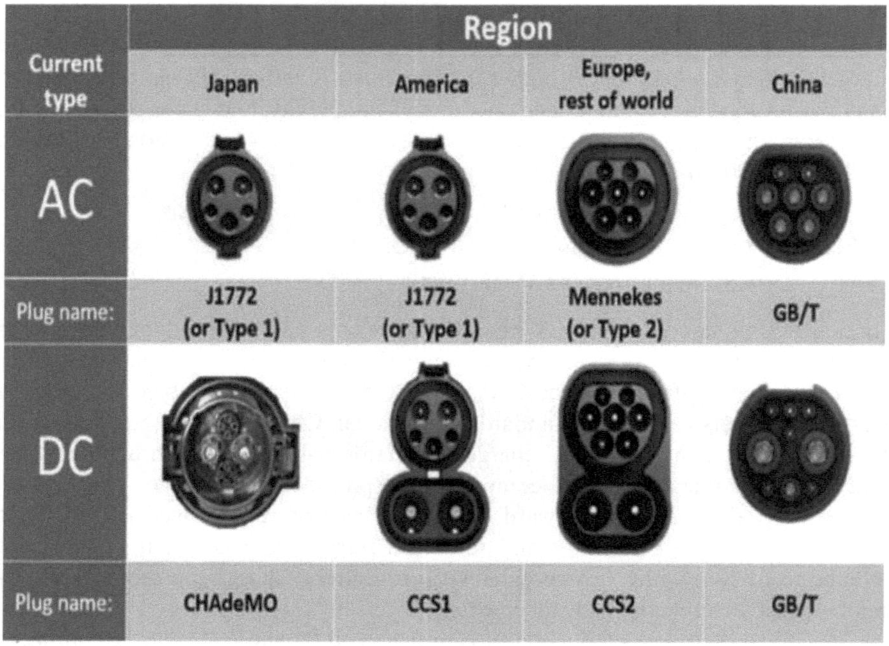

FIGURE 11.7 EV charging protocols worldwide.

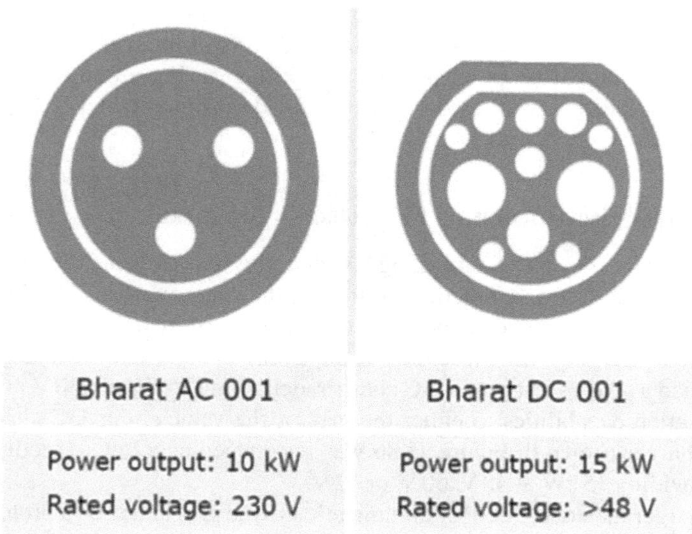

FIGURE 11.8 (a) Bharat AC 001 and (b) Bharat DC 001.

11.7 FUTURE DEVELOPMENTS: AC-002 AND DC-002 CHARGERS IN INDIA

India's ongoing efforts to enhance its EV charging infrastructure include the development of the AC-002 and DC-002 chargers. These chargers aim to build on the foundation set by the earlier AC-001 and DC-001 standards, addressing the need for higher power and more advanced communication capabilities to support the growing EV market in the country.

11.7.1 AC-002 Charger

India's ongoing efforts to enhance its EV charging infrastructure include the development of the AC-002 charger, a significant upgrade from the earlier AC-001 standard. The AC-002 chargers are designed to provide higher power levels than the AC-001, which is limited to less than 3 kW, allowing for faster charging times and greater convenience for EV owners.

Additionally, the AC-002 chargers shown in Figure 11.9 are anticipated to incorporate communication protocols with the vehicle and potentially with the grid. This advancement will enable better management of charging sessions and integration with smart grid technologies, supporting efficient energy use and grid stability. The AC-002 chargers are targeted for use in residential, commercial, and public charging locations such as apartment complexes, workplaces, and shopping centers, where faster and more reliable charging is essential. As India continues to expand its EV

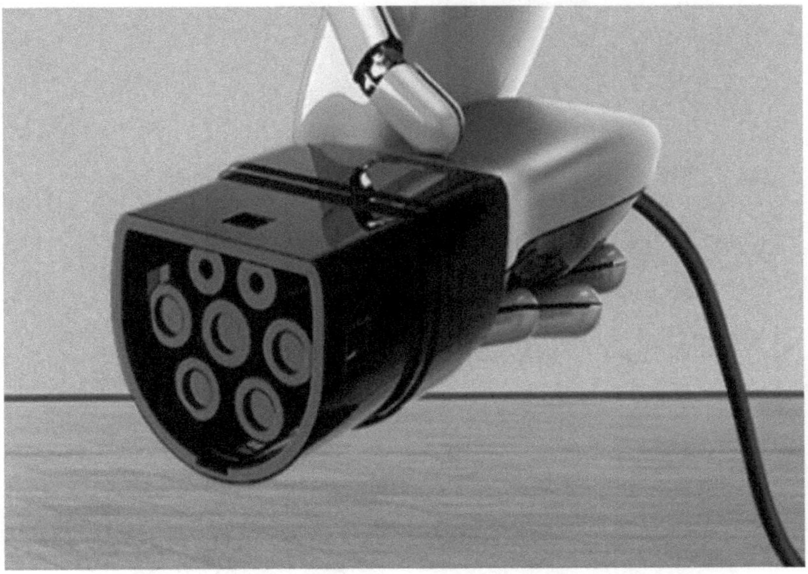

FIGURE 11.9 AC-002 charger.

infrastructure, the AC-002 chargers represent a crucial step toward a more robust and efficient charging network, facilitating the widespread adoption of EVs and contributing to sustainable transportation solutions.

11.7.2 DC-002 Charger

As part of India's continued efforts to improve its EV charging infrastructure, the development of the DC-002 charger represents a significant advancement over the existing DC-001 standard. The DC-002 chargers are expected to provide higher power outputs than the DC-001, which is limited to 15 kW at 48 V, 60 V, or 72 V, thereby enabling much faster DC fast charging. This enhancement will greatly reduce the time required to charge an EV, making it more convenient for users. Additionally, the DC-002 chargers will incorporate advanced communication protocols, including improved OCPP communication with the grid and enhanced vehicle-to-charger communication, ensuring better efficiency and safety. The specifications for DC-002 are designed to future-proof the infrastructure, ensuring compatibility with upcoming EV models that require higher power and advanced charging features. By focusing on higher power capacities and improved communication technologies, the DC-002 chargers will play a crucial role in supporting the widespread adoption of EVs in India, contributing to a sustainable and environmentally friendly transportation system. While the detailed specifications for the AC-002 and DC-002 chargers shown in Figure 11.10 are still under development, these chargers represent a significant

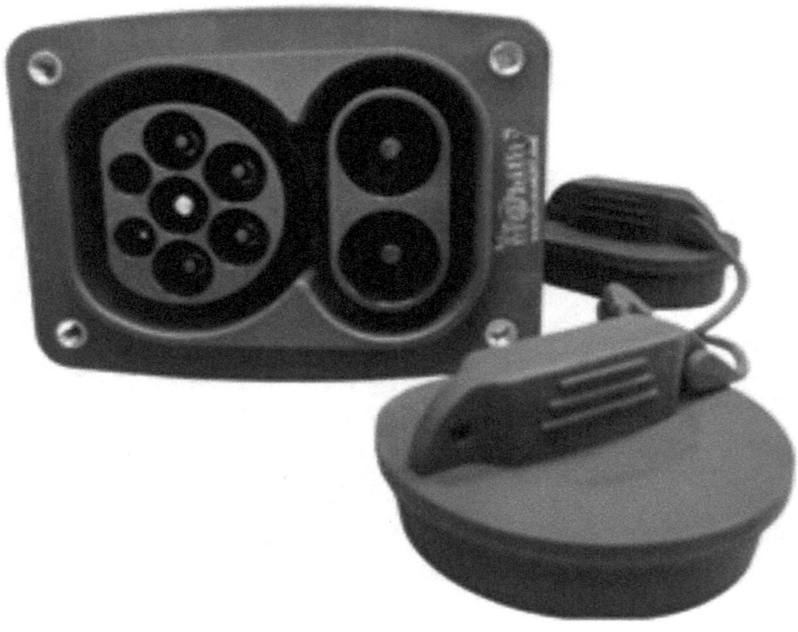

FIGURE 11.10 DC-002 charger.

TABLE 11.3
Cost–Benefit Comparison of the Chargers

Charging Type	Initial Investment	Operational Costs	Charging Speed	Convenience	Financial Viability
Level 1 (120V)	Low	Low	Slow	High	Best for low daily mileage, cost-sensitive users.
Level 2 (240V)	Moderate	Low to moderate	Moderate	High	Suitable for homeowners, workplaces, moderate to high daily mileage.
DC fast charging	High	High	Very fast	Moderate	Ideal for commercial use, high-traffic areas, long-distance travel.
Wireless	Very high	Moderate to high	Moderate	Very high	Best for niche markets, potential for future widespread use.

step forward in India's commitment to building a robust and efficient EV charging network. By focusing on higher power capacities and improved communication technologies, India aims to support the widespread adoption of EVs and contribute to a sustainable and environmentally friendly transportation system (Table 11.3).

The financial viability of EV charging technologies depends on the specific needs and circumstances of the user or business. Commercial applications and heavily populated regions are best served by DC fast charging and wireless technologies, while Level 2 and Level 1 chargers are more suited to home and office settings. Smart charging methods, like as wireless charging, may become popular as technology advances and prices fall.

11.8 CHARGING INFRASTRUCTURE AND PROTOCOLS FOR EVs

11.8.1 Charging Infrastructure

Charging infrastructure is crucial for the widespread adoption of EVs. It consists of various components that facilitate the efficient charging of EVs, including:

Electric Vehicle Supply Equipment (EVSE): Also known as charging stations, these are the hardware systems that deliver electrical energy from an electricity source to the EV. They can be categorized into three types based on the charging speed: Level 1 (slow), Level 2 (medium), and DC fast charging (fast). Figure 11.11 shows the charging infrastructure and protocols for EVs.

11.8.2 Central Management System (CMS)

This is a backend system that manages and controls multiple EVSEs. It provides functionalities such as user authentication, charging session management, payment processing, and real-time monitoring of the charging stations.

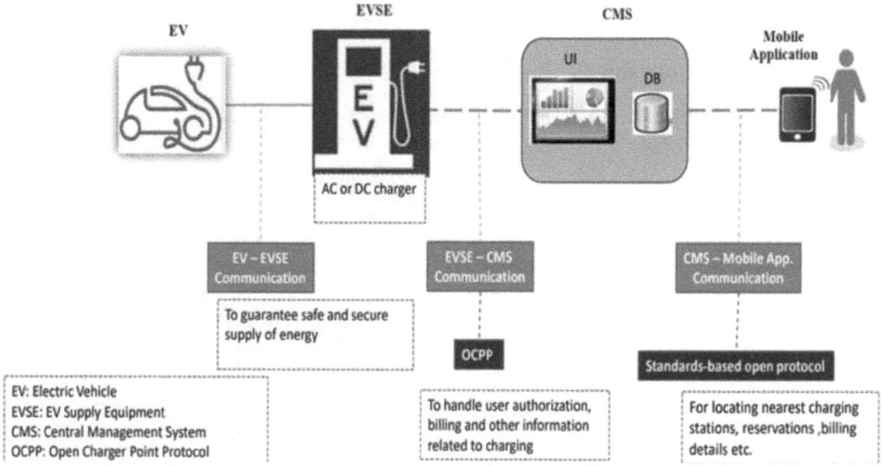

FIGURE 11.11 Charging infrastructure and protocols for EVs.

Mobile Applications: Apps that allow EV owners to locate charging stations, check their availability, initiate and terminate charging sessions, and make payments. These apps often integrate with CMS for a seamless user experience.

11.8.3 Communication Protocols

To ensure efficient and reliable charging, various communication protocols are used between different components of the charging infrastructure:

EV-EVSE Communication: This involves the exchange of information between the EV and the charging station to manage the charging process. Common protocols include:

CHAdeMO: A DC fast charging protocol developed in Japan, allowing for high-power charging.

Combined Charging System: A global standard supporting both AC and DC charging, providing a unified connector for all charging needs.

GB/T: The Chinese standard for EV charging, which includes both AC and DC charging specifications.

EV-CMS Communication: This involves the interaction between the EV and the CMS. Protocols used include:

ISO 15118: A standard for the communication between EVs and the EVSE, enabling features like Plug & Charge, where the vehicle and the charger communicate automatically for seamless charging and billing.

Mobile App Communications: These applications use internet-based communication to interact with the CMS. Application programming interfaces are often employed to provide services such as station location, status updates, user authentication, and payment processing.

Central Management System: This system uses various communication protocols to manage and control the charging network. It handles data from EVSEs, processes user transactions, monitors system health, and provides analytics.

11.8.4 Open Charge Point Protocol (OCPP)

The OCPP is an open standard protocol that enables the communication between EV charging stations and CMS. It ensures interoperability and flexibility within the charging network. Key features of OCPP include:

Interoperability: Allows charging stations from different manufacturers to communicate with a common central system, promoting a vendor-neutral environment.
Scalability: Supports the management of a large network of charging stations, accommodating future growth.
Remote Management: Enables remote monitoring, diagnostics, and maintenance of charging stations.
Smart Charging: Facilitates load management and dynamic adjustment of charging rates based on grid conditions and demand.

OCPP versions include:

OCPP 1.6: Widely adopted, supporting functionalities like smart charging and remote management.
OCPP 2.0: Introduces enhanced features such as improved security, transaction handling, and better support for advanced functionalities like V2G.

The charging infrastructure for EVs is a complex ecosystem involving various components and protocols to ensure efficient and reliable charging. Key elements include EVSE, CMS, mobile apps, and communication protocols like CHAdeMO, CCS, ISO 15118, and OCPP. These systems work together to provide seamless charging experiences, support the management of large charging networks, and enable advanced features like smart charging and Plug & Charge.

11.9 STANDARDIZATION REQUIREMENTS FOR EV CHARGERS

EV chargers require comprehensive standardization to ensure safety, interoperability, and efficiency across different manufacturers and regions. These standards cover various aspects, including electrical specifications, safety measures, communication protocols, and environmental considerations.

Electrical Specifications and Connectors: Standards define the electrical characteristics and connector types to ensure compatibility between EVs and charging stations. For AC charging, common standards include SAE J1772 (Type 1) for North America and IEC 62196 (Type 2) for Europe. DC fast charging is governed by standards like CHAdeMO, CCS, and GB/T in

China. These standards specify voltage levels, current ratings, and connector designs to facilitate safe and efficient power transfer.

Safety Standards: Ensuring safety during the charging process is paramount. Standards such as UL 2202 in North America and IEC 61851 internationally provide guidelines for the design and operation of EV chargers. These standards include requirements for overcurrent protection, ground fault protection, emergency shut-off mechanisms, and thermal management. Compliance with these standards helps prevent electrical hazards and ensures user safety.

Communication Protocols: Effective communication between the EV and the charger, as well as between the charger and the grid, is essential for efficient charging and grid integration. ISO 15118 is a key standard that enables vehicle-to-charger communication, supporting features like Plug & Charge, which allows for automated vehicle identification and billing. OCPP is widely used for communication between charging stations and CMS, enabling interoperability and remote management.

Power Quality and Efficiency: Standards like IEC 61000-3-2 address power quality issues, ensuring that chargers do not introduce significant harmonics into the grid. Power factor correction is often required to maintain a high-power factor, reducing reactive power and improving overall grid efficiency. These standards ensure that chargers operate efficiently and do not adversely affect the electrical grid.

Environmental and Mechanical Standards: Chargers must be designed to withstand various environmental conditions and mechanical stresses. Standards such as IEC 60529 define IP ratings, which indicate the level of protection against dust and water. Additionally, impact resistance and operational temperature ranges are specified to ensure that chargers can operate reliably in diverse environments.

Electromagnetic Compatibility: To prevent interference with other electronic devices, EV chargers must comply with electromagnetic compatibility standards such as CISPR 11. These standards set limits on electromagnetic emissions and susceptibility, ensuring that chargers do not cause or suffer from electromagnetic interference.

Global Harmonization and Regional Standards: While global standards are crucial, regional standards also play a significant role in addressing specific local requirements. For instance, the European Union's European Confirmity (EC) marking indicates compliance with regional safety, health, and environmental protection standards. Similarly, North American standards such as those from Underwriters Laboratories are tailored to local safety regulations.

11.10 STAGES OF CHARGING, PROTOCOLS, AND STANDARDS

11.10.1 CHAdeMO Protocol: Standards and Charging Stages

The CHAdeMO protocol is a widely adopted standard for DC fast charging, enabling efficient and safe energy transfer from the charger to the EV. This protocol is structured into several stages to ensure reliable communication and charging. The stages

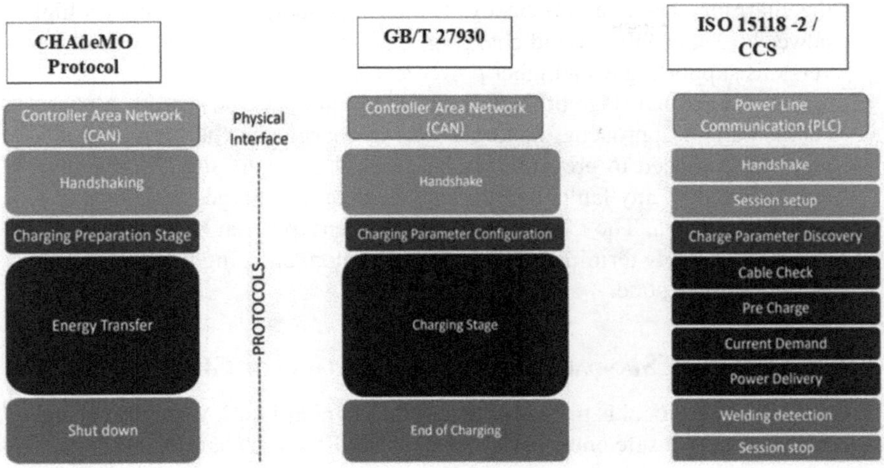

FIGURE 11.12 Stages of charging, protocols, and standards.

include controller area network (CAN) communication, handshaking, charging preparation, energy transfer, and shutdown, each with specific functions and standards. Figure 11.12 provides the stages of charging, protocols, and standards.

CAN Communication: The CHAdeMO protocol utilizes the CAN for communication between the EV and the charging station. This robust communication standard ensures reliable data exchange, allowing the EV and the charger to synchronize their operations. The CAN bus transmits critical information such as battery status, charging parameters, and fault conditions, ensuring a seamless charging process.

Handshaking: The charging process begins with the handshaking stage, where the EV and the charging station establish initial communication. During this stage, the EV identifies itself to the charger, and basic parameters such as voltage and current limits are exchanged. This step ensures that both the EV and the charger are compatible and ready for the subsequent stages of charging.

Charging Preparation Stage: Following successful handshaking, the charging preparation stage involves setting up the necessary conditions for energy transfer. The charger verifies the battery status and readiness of the EV, and safety checks are performed to ensure no faults are present. Parameters such as the initial charging voltage and current are also established during this stage.

Energy Transfer: Once the preparation is complete, the actual energy transfer stage begins. The charger delivers DC power to the EV's battery, with the charging parameters continuously monitored and adjusted as needed. The CAN communication ensures real-time data exchange, allowing the charger to adapt to the battery's state of charge and temperature, optimizing the charging speed and efficiency. The CHAdeMO protocol supports high power levels, enabling rapid charging, typically up to 62.5 kW, with newer versions supporting even higher power levels.

Shutdown: The final stage of the CHAdeMO protocol is the shutdown phase. As the battery approaches its target state of charge, the charging current is gradually reduced to prevent overcharging. Once the desired charge level is reached, or if any fault condition is detected, the charger initiates a controlled shutdown. The CAN communication ensures that both the EV and the charger safely terminate the charging session, ensuring that all systems return to a safe state.

11.10.2 Protocol Standards and Charging Stages of GB/T 27930

The GB/T 27930 protocol is the Chinese national standard for EV DC fast charging, ensuring efficient and safe energy transfer between EVs and charging stations. This protocol comprises multiple stages: CAN communication, handshaking, charging preparation configuration, charging stage, and end of charging, each governed by specific standards to ensure reliability and interoperability.

CAN Communication: The GB/T 27930 protocol utilizes the CAN for communication between the EV and the charging station. This communication backbone ensures robust and real-time data exchange, facilitating the coordination of charging parameters, battery status, and fault conditions. The CAN bus protocol is critical for maintaining synchronization and safety throughout the charging process.

Handshaking: The charging process initiates with the handshaking stage, where the EV and the charging station establish initial communication. During this stage, the EV identifies itself to the charger, and fundamental parameters such as voltage, current limits, and battery specifications are exchanged. This stage ensures compatibility and readiness for subsequent charging operations.

Charging Preparation Configuration: Following successful handshaking, the charging preparation configuration stage sets up the necessary conditions for energy transfer. This involves performing safety checks, verifying the EV's battery status, and configuring initial charging parameters such as target voltage and current. This stage is crucial for ensuring that both the EV and the charger are prepared for the actual energy transfer, minimizing risks and ensuring efficiency.

Charging Stage: The core of the protocol is the charging stage, where DC power is delivered from the charger to the EV's battery. During this stage, the charger continuously monitors and adjusts the charging parameters based on real-time data from the EV. The CAN communication facilitates this dynamic adjustment, optimizing the charging speed and ensuring safe battery management. The GB/T 27930 protocol supports high power levels, enabling rapid charging, which is essential for minimizing downtime for EV users.

End of Charging: The final stage of the protocol is the end of the charging phase. As the battery approaches its target state of charge, the charging current is gradually reduced to prevent overcharging. Upon reaching the desired charge level or if a fault is detected, the charger initiates a controlled shutdown. This phase ensures that the charging session is safely terminated, with the EV and the charger returning to a safe state. Proper communication during this phase is critical to ensure no residual currents or unexpected disconnections occur.

11.10.3 Protocol Standards and Charging Stages of ISO 15118-2/CCS

ISO 15118-2, part of the CCS, defines a comprehensive protocol for the communication between EVs and charging stations, ensuring secure and efficient charging sessions. This protocol leverages power line communication and includes several key stages: handshaking, session setup, charge parameter discovery, cable check, pre-charge, current demand, power delivery, welding detection, and session stop.

Power Line Communication: ISO 15118-2 utilizes power line communication for robust and secure data exchange over the charging cable. This communication method ensures that EVs and charging stations can coordinate charging processes, exchange necessary parameters, and ensure a high level of interoperability.

Handshaking: The charging session begins with the handshaking stage, where the EV and the charging station establish an initial connection. During this phase, the devices identify each other and exchange basic information, ensuring they are compatible and ready to proceed with the charging process.

Session Setup: Following a successful handshake, the session setup phase configures the communication link for the charging session. This involves the exchange of security certificates and the establishment of a secure communication channel. The session setup ensures that all data exchanged during the charging session is secure and authenticated.

Charge Parameter Discovery: In this stage, the EV and the charging station exchange detailed information about their capabilities and requirements. The EV communicates its battery status, preferred charging voltage, and current limits, while the charging station provides information on the available power and supported charging modes. This ensures that both the EV and the charger are aligned on the charging parameters.

Cable Check: Before initiating the actual charging process, a cable check is performed to verify the integrity and proper connection of the charging cable. This step is crucial to ensure there are no faults or issues that could disrupt the charging process or cause safety hazards.

Pre-Charge: Once the cable check is successful, the pre-charge phase prepares the EV's battery for charging. During this stage, the charger supplies a low current to the EV to confirm the battery's readiness and ensure all systems are functioning correctly before full power delivery begins.

Current Demand: In this phase, the EV communicates its current demand to the charging station. The charger adjusts the output to match the EV's requirements, ensuring an optimal charging rate. This dynamic adjustment is based on real-time data and continuous communication between the EV and the charger.

Power Delivery: The main charging phase, where the charger delivers the required power to the EV's battery. Throughout this stage, the charging parameters are continuously monitored and adjusted to maintain efficiency and safety. The EV and the charging station maintain a constant dialogue to ensure the charging process aligns with the battery's condition and state of charge.

Welding Detection: During the power delivery stage, welding detection mechanisms ensure that the contacts within the charging connector do not fuse together. This safety feature prevents potential hazards associated with welded contacts, ensuring the system can safely disconnect at the end of the session.

Session Stop: The final stage of the charging process involves safely terminating the charging session. The EV signals the charger to stop power delivery, and the communication link is closed. This phase ensures that all connections are safely disengaged, and the EV can be driven away without any residual currents or unexpected disconnections.

11.11 ROLE OF POWER ELECTRONICS CONVERTERS IN EV CHARGING

Power electronic converters are crucial in EV charging systems, enabling the efficient and safe transfer of electricity from the grid to the vehicle's battery. These converters, including AC/DC converters, DC/DC converters, and bidirectional converters, each play a specific role in the charging process. AC/DC converters, or rectifiers, convert the AC from the grid into DC necessary for charging the battery by using diodes and controlled rectifiers for rectification, capacitors and inductors for filtering, and voltage regulators for maintaining the correct output. DC/DC converters adjust the DC voltage to match the battery's requirements, utilizing high-speed switches, energy storage components, and feedback control systems for precise voltage and current regulation. Bidirectional converters facilitate two-way energy flow, enabling V2G functionality by operating in both charging and discharging modes. These converters are designed for high efficiency, precise control, flexibility in supporting various charging standards, and enabling V2G applications, which enhance grid stability and allow EVs to act as distributed energy resources. Recent advancements in power electronics, such as the use of silicon carbide (SiC) and gallium nitride (GaN) materials, further improve the performance and efficiency of these systems. The power electronic converters in EV are shown in Figure 11.13.

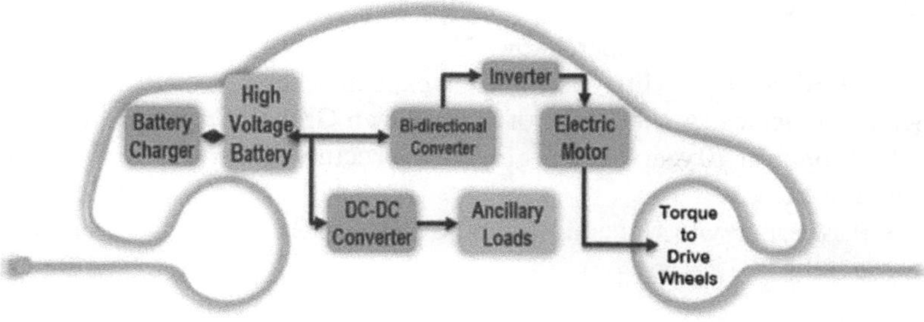

FIGURE 11.13 Power electronic converters in EV.

11.11.1 Policies Introduced to Develop Charging Infrastructure around the World

Governments worldwide have introduced various policies to promote the development of charging infrastructure for EVs. These policies aim to address the growing demand for EVs, reduce greenhouse gas emissions, and support sustainable transportation systems. In the United States, the National Electric Vehicle Infrastructure Program, part of the Bipartisan Infrastructure Law, allocates $5 billion over five years to help states create a network of EV charging stations (U.S. Department of Energy, 2021). Additionally, federal tax credits offer incentives for both individuals and businesses to install EV charging stations (IRS, 2021), while state-level initiatives like California's CALeVIP program provide funding for the installation of public charging stations (California Energy Commission, 2020). In the European Union, the Alternative Fuels Infrastructure Directive mandates member states to develop national policy frameworks for deploying EV charging infrastructure (European Commission, 2014). The EU's Green Deal aims to increase the number of charging points to at least 1 million by 2025 (European Commission, 2020), complemented by national initiatives such as Germany's subsidies for charging infrastructure under the "National Charging Infrastructure" funding guidelines (Federal Ministry for Economic Affairs and Energy, 2020). China has implemented the GB/T standards for charging and provides subsidies for constructing charging stations (Ministry of Industry and Information Technology, 2015), with its "Electric Vehicle Charging Infrastructure Development Guidelines" targeting the installation of 4.8 million charging points by 2025 (National Development and Reform Commission, 2020). In India, the Faster Adoption and Manufacturing of Hybrid and Electric Vehicles Scheme offers financial incentives for setting up charging infrastructure (Ministry of Heavy Industries, 2019), while the National Electric Mobility Mission Plan 2020 supports the development of a robust EV charging network (NITI Aayog, 2013).

11.11.2 Companies in Tie-Ups for Charging Infrastructure Development

Various companies and collaborations have emerged globally to support the expansion of EV charging networks. These tie-ups involve automakers, technology companies, and energy providers.

11.11.2.1 United States

Tesla: Operates its own Supercharger Network and has agreements with various retail and hospitality businesses to expand access.

Electrify America: A subsidiary of Volkswagen Group of America, investing $2 billion over 10 years in charging infrastructure as part of the Dieselgate settlement.

EVgo: Partnering with General Motors to add over 2,700 new fast chargers in the U.S. by 2025.

11.11.2.2 Europe

Ionity: A joint venture of BMW, Mercedes-Benz, Ford, Volkswagen Group, and Hyundai, aiming to build a network of high-power charging stations across Europe.

ChargePoint: Collaborating with companies like Daimler and BMW to expand their charging network across Europe.

BP and EnBW: Partnering to roll out fast charging points in Germany and other European countries.

11.11.2.3 China

State Grid Corporation of China: Partnering with automakers and technology firms to establish a comprehensive network of charging stations.

Tesla: Collaborating with local companies to expand its Supercharger Network within China.

NIO: Introducing innovative battery swap stations and working with Sinopec to expand the charging infrastructure.

11.11.2.4 India

Tata Power: Partnering with Tata Motors, MG Motors, and Jaguar Land Rover to develop a widespread charging network.

Fortum: Collaborating with various state governments and private entities to set up charging stations.

BPCL and Indian Oil Corporation Limited**:** Indian oil companies investing in EV charging infrastructure, partnering with private firms to establish charging points at fuel stations.

These policies and partnerships reflect a concerted global effort to build the necessary infrastructure to support the growing number of EVs and facilitate the transition to cleaner transportation.

11.12 INTRODUCTION TO WIRELESS CHARGING FOR EVs

Wireless charging for EVs represents a significant advancement in the automotive industry, offering a convenient and efficient alternative to traditional plug-in charging systems. This technology allows EVs to be charged without physical connectors, using electromagnetic fields to transfer energy between a charging pad on the ground and a receiver on the vehicle. Wireless charging enhances user convenience, reduces wear and tear on connectors, and can potentially be integrated into public infrastructure to support widespread EV adoption. With the growing demand for electric mobility, wireless charging is poised to play a critical role in the sustainable transportation ecosystem (Siqi et al., 2015).

EV owners would prefer WPT technology since it would eliminate the need for cumbersome charging. Charging becomes a breeze with wireless energy transfer to the EV. Drivers can simply park their cars and depart from a stationary WPT system. The EV could theoretically go on forever without stopping if it had a dynamic WPT system, which would allow it to be powered while moving. EVs that use wireless charging may also have a 20% or smaller reduction in battery capacity when compared to EVs that use conductive charging.

Figure 11.14 depicts a standard wireless EV charging setup. Charging an EV wirelessly involves multiple steps. An AC-to-DC converter with power factor adjustment is used to transform the utility's AC power into DC (Zhang et al., 2023).

A compensation network then drives the transmitting coil by converting the DC power to a high-frequency AC. A high-frequency isolated transformer can be used to add an extra layer of protection between the DC–AC inverter and the primary side coil in the event that the insulation on the former fails. An AC voltage is induced on the receiving coil by the high-frequency current in the transmitting coil. Both the efficiency and the amount of power delivered are enhanced when the secondary compensating network is under resonance.

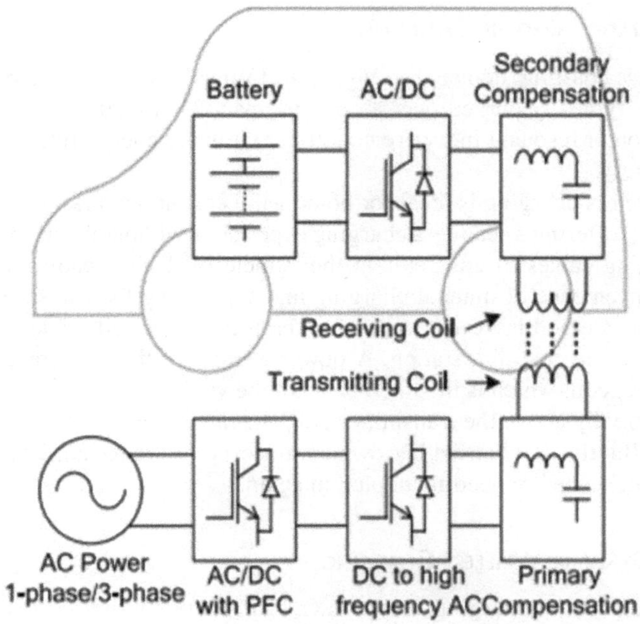

FIGURE 11.14 Standard wireless charging setup.

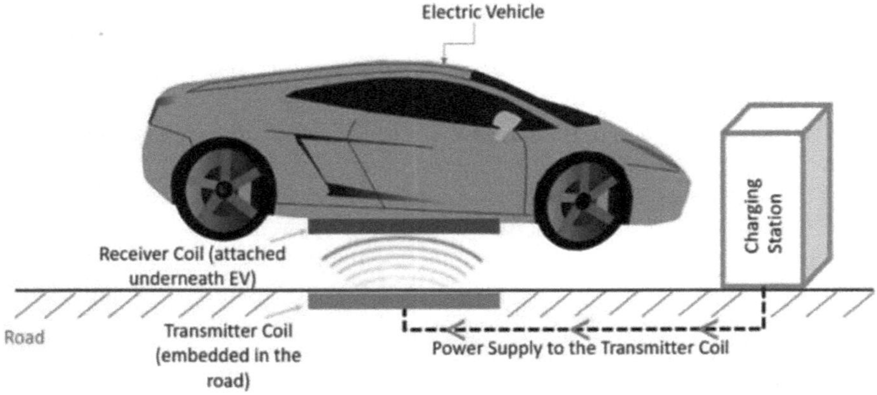

FIGURE 11.15 Static wireless charging.

Finally, in order to charge the battery, the AC is rectified. The following components make up a wireless EV charger: the sending and receiving coils, which are not connected but are physically separate, compensation network, the power electronics converters, and the coils, ferrite, and shielding. The wireless charging is classified into static wireless charging and dynamic wireless charging.

11.12.1 Static Wireless Charging

Static wireless charging depicted in Figure 11.15 involves charging the EV while it is stationary, typically parked over a charging pad. This method uses electromagnetic induction or resonant inductive coupling to transfer energy from the pad to the vehicle's battery.

Static wireless charging is ideal for home garages, public parking lots, and commercial fleets, offering a seamless charging experience without the need for plugging and unplugging cables. It ensures that the vehicle is always ready for use, reducing the inconvenience of manual charging and supporting the transition to electric mobility. This method involves connecting the transmitter coil, which is embedded in the road, to the charging station. A power source for the EV's reception coil is the transmitter coil, which is mounted beneath the vehicle. When the receiver coil is positioned directly above the transmitter coil, the intelligent wireless power transfer is executed. But this has caused EV owners to worry, or "range anxiety", about how far they can drive before needing to plug in again.

11.12.2 Dynamic Wireless Charging

The dynamic wireless charging, also known as in-motion wireless charging, allows EVs to charge while driving. This technology uses charging coils embedded in the road to transfer energy to the vehicle's receiver coils as it moves. The tracks of dynamic charging are shown in Figure 11.16.

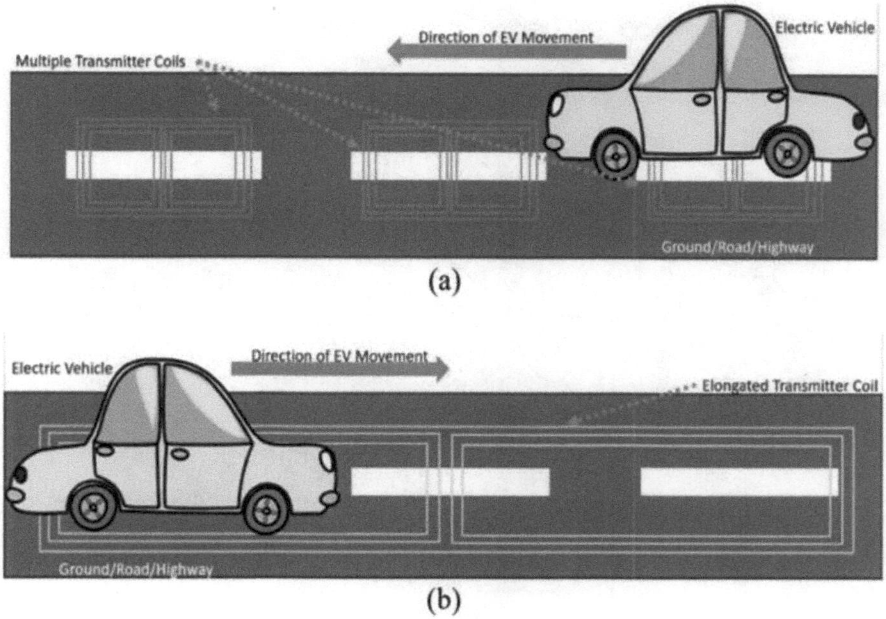

FIGURE 11.16 Tracks of dynamic wireless charging.

Dynamic wireless charging can significantly extend the driving range of EVs and reduce the need for large battery capacities, making EVs more efficient and practical for long-distance travel. This system requires substantial infrastructure investment but offers a visionary solution for continuous, on-the-go charging (Nanda et al., 2021). The transmitter coil for the DWC method can be in either an elongated or a lumped track, as shown in Figure 11.16a and b. When dealing with the lumped track, it's necessary to use numerous transmitter coils, each of which has its own compensating circuit. In contrast, the extended track has a length greater than that of the reception coil and is made up of a single transmitter coil. When comparing the two types of tracks, the one with lumps makes the charging coils and system circuits more complicated than the one with extensions. Because of this intricacy, the final cost of the charging system will be higher. But the lopped track is considerably simpler to use in order to turn on the active transmitter coil and turn off the idle transmitter coil. This capability has the potential to lessen power wastage during the DWC charging procedure. Thus, depending on the task at hand, each of them can be advantageous in its own way.

11.12.3 Inductive Charging

Inductive power transfer systems for EVs consist of several key components: primary compensation, secondary compensation, transmitter pad, receiver pad, AC/AC converter, AC/DC converter, controller, and battery energy storage systems. Each part plays a crucial role in the efficient transfer of energy from the charging infrastructure to the vehicle's battery. The inductive charging components of EV are shown in Figure 11.17.

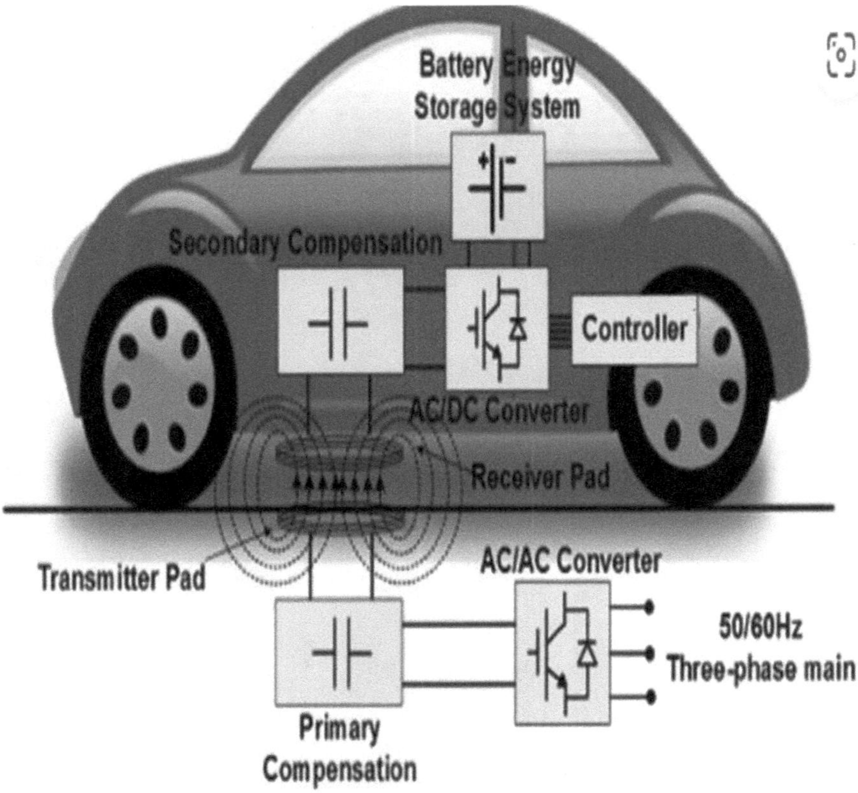

FIGURE 11.17 Inductive charging components in EV.

11.12.3.1 Primary Compensation

The primary compensation refers to the components and circuitry on the primary side (charging pad) that enhance the efficiency of power transfer. It typically includes capacitors and inductors arranged in series or parallel configurations to resonate at the operating frequency. This resonance minimizes reactive power and maximizes real power transfer, improving efficiency. The primary compensation circuit ensures that the energy generated by the AC/AC converter is optimally transferred to the transmitter pad, creating a strong and stable magnetic field.

11.12.3.2 Secondary Compensation

The secondary compensation is located on the vehicle side (receiver pad) and works in tandem with the primary compensation to maximize power transfer efficiency. It also involves capacitors and inductors that resonate at the same frequency as the primary side. This resonance ensures that the maximum amount of energy is captured by the receiver pad from the magnetic field. Secondary compensation reduces energy losses and stabilizes the power received before it is converted for storage in the battery.

11.12.3.3 Transmitter Pad

The transmitter pad, or primary coil, is embedded in the charging infrastructure (such as the ground or a parking lot). It generates an alternating magnetic field when an AC current passes through it. This magnetic field is crucial for inducing a voltage in the receiver pad. The design and positioning of the transmitter pad are critical for effective energy transfer, ensuring that the magnetic field is concentrated and aligned with the receiver pad.

11.12.3.4 Receiver Pad

The receiver pad, or secondary coil, is installed in the vehicle and captures the magnetic field generated by the transmitter pad. The alternating magnetic field induces an AC voltage in the receiver pad, which is then used to charge the vehicle's battery. The efficiency of the receiver pad depends on its alignment with the transmitter pad and the effectiveness of the secondary compensation circuit. Proper alignment and resonance ensure maximum energy capture and transfer.

11.12.3.5 AC/AC Converter

The AC/AC converter in the inductive power transfer system converts the grid AC power to a high-frequency AC power suitable for WPT. This conversion is necessary because high-frequency AC signals are more effective for creating a strong magnetic field for inductive charging. The AC/AC converter ensures that the power supplied to the transmitter pad is at the correct frequency and voltage for optimal performance.

11.12.3.6 AC/DC Converter

On the vehicle side, the AC/DC converter converts the high-frequency AC power received by the receiver pad into DC power. This DC power is required to charge the vehicle's battery. The AC/DC converter is designed to handle the varying power levels received from the inductive transfer process and provide a stable DC output to the battery energy storage system. Efficient conversion minimizes energy losses and ensures that the maximum amount of power is available for charging the battery.

11.12.3.7 Controller

The controller manages the entire inductive power transfer system, coordinating the operation of the AC/AC converter, primary and secondary compensation, and the AC/DC converter. It ensures that the system operates at the optimal frequency and power levels, adjusting parameters as needed to maintain efficiency and safety. The controller also handles communication between the vehicle and the charging infrastructure, ensuring proper alignment and initiating the charging process.

11.12.3.8 Battery Energy Storage System

The battery energy storage system in the vehicle stores the DC power converted from the received AC power. It consists of a battery management system that monitors and controls the charging process, ensuring that the battery is charged safely and efficiently. The battery management system optimizes the charging rate, manages thermal conditions, and protects the battery from overcharging or deep discharge, thereby extending the battery's lifespan and maintaining vehicle performance.

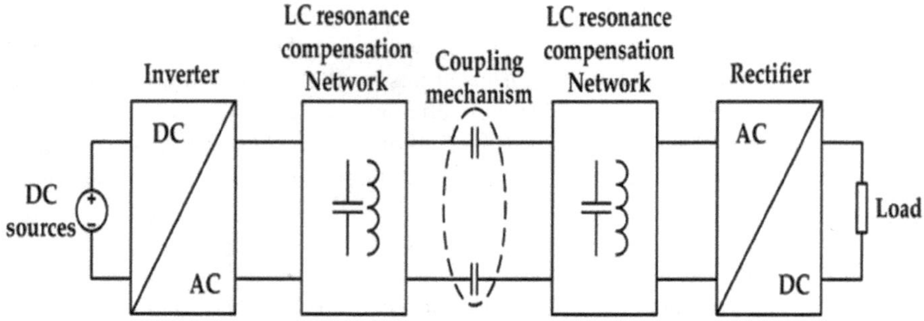

FIGURE 11.18 Capacitive charging components of EV.

11.12.4 Capacitive Charging

Capacitive wireless charging presented in Figure 11.18 uses electric fields instead of magnetic fields to transfer energy. It involves two conductive plates: one on the ground and one on the vehicle. When the vehicle is positioned correctly, an electric field is established between the plates, allowing energy to transfer and charge the battery (Zhang et al., 2023).

While capacitive charging is less common than inductive charging, it offers potential benefits such as lower cost and simpler integration with existing infrastructure. However, it generally provides lower power transfer efficiency and is more sensitive to alignment.

Capacitive wireless power transfer (CWPT) for EVs utilizes electric fields to transfer energy between two conductive plates. This system involves several critical components: DC source, inverter, Inductor-Capacitor (LC) resonance compensation networks at both input and output sides, a coupling mechanism, a rectifier at the output side, and the load. Each part plays a vital role in ensuring efficient energy transfer from the charging infrastructure to the vehicle's battery.

11.12.4.1 DC Source

The DC source in a CWPT system provides the initial power required to drive the system. This source typically comes from the electrical grid or a renewable energy system. The DC source ensures a stable and consistent supply of electrical energy, which is essential for the subsequent conversion and transfer processes. It acts as the primary power supply that initiates the energy transfer cycle.

11.12.4.2 Inverter

The inverter converts the DC power from the source into high-frequency AC power. This conversion is necessary because CWPT relies on high-frequency AC signals to create an effective electric field for power transfer. The inverter's output frequency is carefully controlled to match the resonant frequency of the LC compensation networks, optimizing the system's efficiency and power transfer capabilities.

11.12.4.3 LC Resonance Compensation Network at the Input Side

The LC resonance compensation network at the input side consists of inductors (*L*) and capacitors (*C*) arranged to form a resonant circuit. This network is designed to resonate at the operating frequency of the inverter, minimizing reactive power and maximizing real power transfer. By tuning the LC network to the correct frequency, the system can achieve high efficiency, reducing energy losses and enhancing the strength of the electric field generated between the plates.

11.12.4.4 Coupling Mechanism

The coupling mechanism in CWPT involves two conductive plates: one on the ground (transmitter plate) and one on the vehicle (receiver plate). When the vehicle is positioned correctly, an electric field is established between these plates. The high-frequency AC power from the inverter creates oscillating electric fields between the plates, allowing energy to be transferred across the air gap. The efficiency of this coupling mechanism depends on factors like plate alignment, distance, and environmental conditions.

11.12.4.5 LC Resonance Compensation Network at the Output Side

The LC resonance compensation network at the output side is similar to the input side but is positioned on the vehicle. This network also resonates at the operating frequency, optimizing the reception of the transferred energy. It ensures that the maximum possible power is captured from the electric field and converted into a form suitable for charging the vehicle's battery. Proper tuning of this network is crucial for maintaining high efficiency and stable power transfer.

11.12.4.6 Rectifier at the Output Side

The rectifier converts the high-frequency AC power received by the vehicle into DC power, which is required for charging the battery. The rectifier typically uses diodes or other semiconductor devices to perform this conversion efficiently. The DC output from the rectifier is then regulated to match the charging requirements of the battery, ensuring a stable and safe charging process.

11.12.4.7 Load

The load in a CWPT system is the battery energy storage system of the EV. The DC power from the rectifier is used to charge the battery, which stores energy for later use in powering the vehicle's electric motor. The battery management system oversees the charging process, optimizing the charge rate and ensuring the battery operates within safe limits to maximize its lifespan and performance.

11.12.5 Challenges and Future Aspects of WPT Systems

WPT systems face several challenges that must be addressed to realize their full potential. Key issues include efficiency and energy losses due to misalignment and distance between the transmitter and receiver, which can significantly impact the effectiveness of power transfer. Proper alignment is crucial, especially in dynamic charging scenarios where the vehicle is in motion. Heat dissipation is another concern, as WPT systems can generate substantial heat that needs to be managed to prevent damage. Additionally, electromagnetic interference poses a challenge, requiring careful design to avoid disruption to other electronic devices and ensuring safety for human exposure. The lack of standardization and compatibility across different devices and manufacturers also hinders widespread adoption. Cost remains a barrier, as the technology and components required for WPT systems can be expensive.

Despite these challenges, the future of WPT systems is promising. Advances in materials science, circuit design, and resonant frequency tuning are expected to improve efficiency and reduce energy losses.

Research into new methods such as magnetic resonance and metamaterials could extend the range of WPT systems, enabling practical applications like dynamic wireless charging for EVs. Integration with public infrastructure, such as roads and parking lots, will provide seamless charging experiences and support broader EV adoption. Smart charging systems that integrate with smart grids and internet of things (IoT) will offer optimized and automated charging based on grid demand, energy prices, and user preferences. Efforts to develop universal standards and improve interoperability will facilitate greater compatibility between different devices and manufacturers. Cost reduction through technological advancements and economies of scale will make WPT systems more accessible. Overall, WPT systems have the potential to contribute significantly to sustainable and efficient energy transfer solutions in the future.

11.13 IMPACT OF EV CHARGING IN THE GRID

The integration of EVs into the grid can be viewed from two perspectives: positive and negative impacts. These are elaborated in Tables 11.13.1 and 11.13.2. Power utilities face substantial challenges due to EVs. Over-integration of EVs into the distribution network can destabilize the grid. This instability arises from alterations in load profiles, imbalances in voltage and frequency, excessive injection of harmonics, and power losses. The over-saturation of EVs in the grid can lead to degradation of power quality, peak loads, and issues in power regulation. However, these issues can be mitigated using sophisticated power management techniques (Azadfar et al., 2015). The beneficial effects of EV integration are encapsulated in Figure 11.19.

11.13.1 Negative Impacts of EV–Grid Integration

Effect on power quality	The impact of an electric grid on a V2G network is primarily evaluated by examining the power supply's characteristics (Khalid et al. 2019; Durante et al., 2017).
	Uncertainties in EV charging can lead to issues such as grid overvoltage, power quality degradation, increased line damage, and higher current faults (Godina et al., 2019).
	High-frequency converters, which convert AC power to DC for EV charging, introduce harmonics into the grid, resulting in substandard power quality (Karmaker et al., 2019).
	Harmonics overload distribution transformers, reducing their lifespan (Meyer et al., 2011).

(Continued)

(*Continued*)

Effect on voltage stability	Nonlinear EV loads drawing large power over short durations cause instability in the power system (Dharmakeerthi et al., 2013). EVs increase the chances of power system disruptions and prolong the return to a steady state (Ul-Haq et al., 2015). A well-managed electric vehicle grid integration (EVGI) system can enhance grid reliability (Ma et al., 2017).
Effect on the stability of power grids	New EV loads need to be incorporated into the system, necessitating an evaluation of the IEEE 3 bus system with and without charging stations. A stability analysis must be conducted before adding new EV loads to an unstable electrical system (Onar and Khaligh, 2010). The electric network is analyzed for small-signal stability. An EV is considered as a constant load and impedance (Das and Aliprantis, 2008). EVs modeled as constant power loads were found unstable, necessitating the use of intelligent control algorithms to control EVs modeled as constant impedance loads (Yan et al., 2010).
Effects on the supply and demand balance in the power grid	EV charging stations introduce harmonics into the grid system due to the power electronics converter. Therefore, the charging architecture influences the variations in total harmonic distortion (Habib et al., 2015). Harmonic distortion variability ranges from 7% to 99% for EV chargers (Mahalik et al., 2010). Researchers have found that EV penetration levels can be increased based on capacitor banks without causing voltage harmonic violations (McCarthy and Wolfs, 2010).
The overloading of components	The large number of EVGIs generates or transmits an additional load demand (Yong et al., 2015). Some existing power system components are not designed to handle the additional loads, leading to overloaded components and shortened transformer lifespans (Yan and Kezunovic, 2012).
Increased load demand	Up to 1,000 TWh of additional loads can be added, a 25% increase over current levels (Bowermaster et al., 2017). Charges during peak times increase due to uncontrolled EV charging, significantly burdening utilities (Hadley and Tsvetkova, 2009).

11.13.2 Positive Impacts of EV–Grid Integration

Reduction of peak loads	The use of time-of-use tariffs and coordinated charging strategies makes the grid more energy-efficient without adding more generating capacity (Kristoffersen et al., 2011). Several bidirectional grid-connected EVs are used as part of the industrial power grid. Battery management systems may impose limitations on the type and extent of usage-driven charge and discharge signals. This adversely affects the penetration of EVs into the grid system (Hofmann et al., 2015). In practical circumstances, 96% of the peak demand on the power system can be reduced. When the EV battery has a large capacity, it can significantly contribute to reducing peak domestic and grid demand, leading to better utilization of the power system (Garwa and Niazi, 2019).

(*Continued*)

(Continued)

Provision of ancillary services	Ancillary services to power production ensure reliability, demand, supply, and grid stability (White and Zhang, 2011).
	The use of V2G technology can provide ancillary services, improving the stability of the electric grid (Guille and Gross, 2009).
Support for renewable energy	Operating EVs as energy storage can reduce the uncertainty of renewable energy (Saber and Venayagamoorthy, 2010).
	Using EVs as a buffer for renewable energy reduces emissions and can save money (Domínguez-Navarro et al., 2019).
Frequency regulation	Frequency regulation corrects grid frequency deviation (Garwa and Niazi, 2019).
	Through V2G technology, the load can be balanced by regulating grid frequency (Ur Rehman and Riaz, 2017).
	EV integration can also help minimize frequency fluctuations caused by sudden changes in load when EVs are integrated with the grid (Tian et al., 2012).
Voltage regulation	Voltage regulation is crucial to ensure the stability of reactive power supply and demand. EVs can react quickly to regulatory signals as they are autonomously controlled by the vehicle (Qiangqiang et al., 2012).
	EVs autonomously process regulation signals, allowing them to react quickly to safety signals. EV charging is halted when the system voltage drops below a certain level and resumes when the system voltage reaches a certain level (Donadee and Ilić, 2012).

The integration of EV charging into the grid has both positive and negative impacts. On the positive side, EVs can help stabilize the grid through V2G technology, where parked EVs can discharge power back to the grid during peak demand, thereby balancing load and enhancing grid resilience. Additionally, the growth of EV charging infrastructure can drive investments in renewable energy and smart grid technologies, promoting a cleaner and more efficient energy system. However, the increasing demand for electricity from EVs can strain the grid, especially during peak charging times, potentially leading to overloads and requiring significant upgrades to the existing infrastructure. The uneven distribution of charging stations can also create localized demand spikes, challenging grid management. Therefore, strategic planning, advanced grid management, and investment in renewable energy sources are essential to mitigate these impacts and harness the benefits of EV integration.

11.13.3 EV Adoption: Insights into Consumer Preferences and Infrastructure Development

Understanding the intricacies of consumer preferences, charging habits, and the barriers to EV adoption is essential for developing effective strategies to promote EV usage. Consumer decisions are influenced by a range of factors, including vehicle range, charging speed, cost, brand trust, and environmental impact. Simultaneously, charging habits, such as the preference for home, public, or workplace charging, play a critical role in shaping the EV experience. Despite the growing interest in EVs, several barriers continue to impede their mass adoption. High initial costs, range

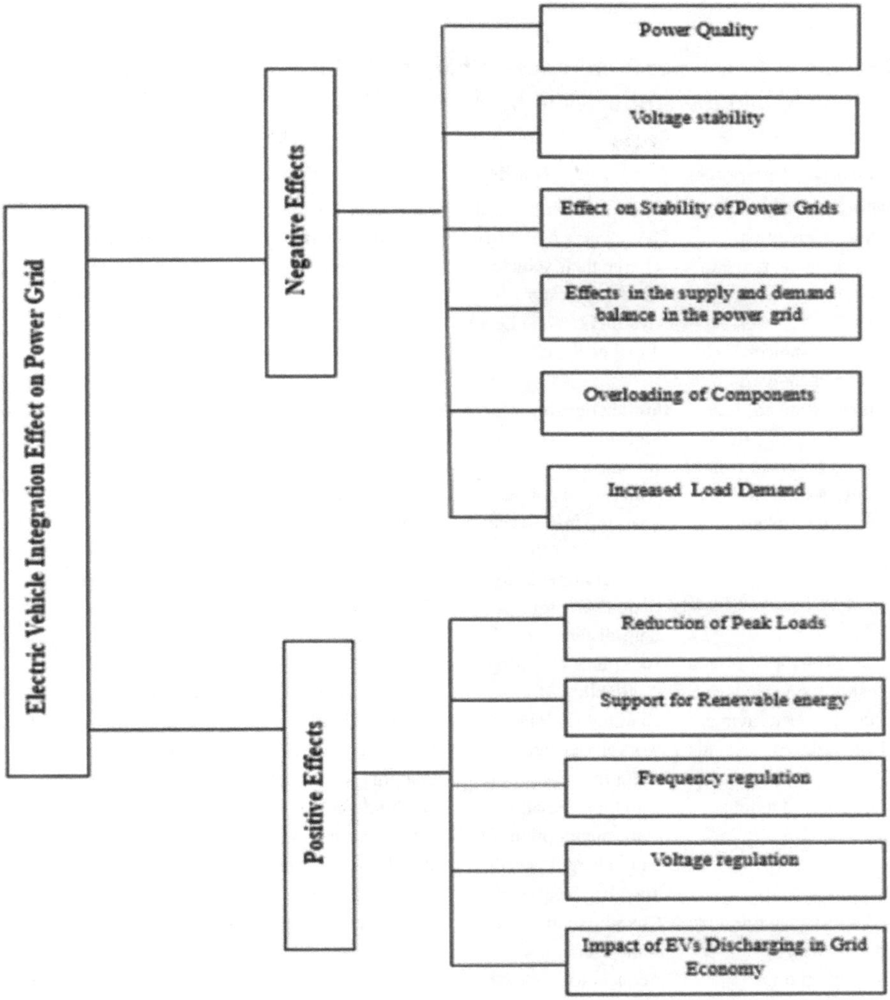

FIGURE 11.19 Impact of EV on the power grid.

anxiety, inadequate charging infrastructure, concerns about battery life, and limited model availability are among the primary challenges that need to be addressed. Overcoming these barriers requires a comprehensive approach involving government incentives, investments in infrastructure, advancements in battery technology, and educational initiatives to raise awareness and dispel misconceptions. Table 11.4 provides insights into consumer preferences, charging habits, barriers to EV adoption, and the strategies to be practiced to overcome the barriers.

The transition to electric mobility is not merely a technological shift but a comprehensive transformation involving economic, social, and environmental dimensions. By understanding and addressing the intricate dynamics of consumer behavior,

TABLE 11.4
Insights of Consumer Preferences, Charging Habits, Barriers to the EV Adoption, and the Strategies to be Practiced to Overcome the Barriers

Consumer Preferences	Charging Habits	Barriers to Adoption	Strategies to Overcome Barriers
Vehicle range: Consumers often prioritize the range of an EV, preferring models that offer longer distances between charges.	**Home charging**: Most EV owners prefer to charge their vehicles at home, typically overnight, using Level 1 or Level 2 chargers.	**Initial cost**: Despite long-term savings, the higher upfront cost of EVs compared to ICE vehicles can deter potential buyers.	**Incentives and subsidies**: Government incentives, such as tax credits and rebates, can lower the effective cost of EVs.
Charging speed: Fast charging capabilities are a significant factor, as users want to minimize downtime.	**Public charging stations**: Access to reliable and widespread public charging infrastructure is crucial for those without home charging options and for long-distance travel.	**Range anxiety**: Fear of running out of battery without access to a charging station remains a significant concern.	**Investment in infrastructure**: Expanding and enhancing public charging networks can reduce range anxiety and improve convenience.
Cost: Both the upfront cost of the vehicle and the long-term savings from reduced fuel and maintenance costs influence adoption.	**Workplace charging**: Availability of charging stations at workplaces can encourage EV adoption, providing a convenient option for daily charging needs.	**Charging infrastructure**: Insufficient or unreliable public charging infrastructure can hinder adoption, especially in areas with fewer stations.	**Education and awareness**: Increasing consumer knowledge about the benefits of EVs and the realities of charging and maintenance can alleviate concerns.
Brand trust: Established automotive brands with a reputation for reliability can sway consumer choices.	**Charging frequency**: Charging habits vary; some users prefer frequent top-ups, while others may charge only when necessary, similar to refueling a traditional car.	**Battery life and replacement costs**: Concerns about battery degradation and the high cost of replacement can affect purchasing decisions.	**Improving battery technology**: Advances in battery technology that increase range, reduce costs, and enhance longevity can boost consumer confidence.
Environmental impact: Increasing awareness and concern about climate change drive some consumers to choose EVs for their lower environmental footprint.		**Limited model availability**: A limited selection of EV models, especially in certain vehicle categories (e.g., trucks, SUVs), can restrict consumer choices.	**Diverse model availability**: Introducing more EV models across various segments can cater to a wider range of consumer needs and preferences.

(*Continued*)

TABLE 11.4 (*Continued*)
Insights of Consumer Preferences, Charging Habits, Barriers to the EV Adoption, and the Strategies to be Practiced to Overcome the Barriers

Consumer Preferences	Charging Habits	Barriers to Adoption	Strategies to Overcome Barriers
		Technological uncertainty: Rapid advancements in EV technology might make consumers hesitant to invest, fearing their vehicle could become obsolete quickly.	

charging habits, and adoption barriers, stakeholders can pave the way for a sustainable future in transportation. The collective efforts of governments, manufacturers, and consumers are crucial in accelerating this transition, ensuring that EVs become a mainstream choice that contributes to a cleaner, greener, and more efficient transportation ecosystem. Through concerted actions and strategic investments, the potential for EVs to revolutionize the transportation sector can be fully realized, leading to significant environmental benefits and a more sustainable world.

11.13.4 Transforming EVs from Passive Loads to Active Grid Stabilizers

Once viewed merely as a passive load on energy systems, EVs are now playing an active role in enhancing grid stability and load management. With the integration of advanced artificial intelligence and machine learning algorithms, an increasing number of smart charging EVs can monitor real-time grid conditions. These technologies help EV owners to avoid charging during high-demand periods, such as evenings or extreme weather conditions like heat waves and cold snaps. By strategically reducing energy consumption during these peak times, EVs can actively participate in peak shaving, thereby aiding in the prevention of power fluctuations and outages. Smart charging technology continually evolves to support renewable energy sources such as solar, wind, and hydropower. By aligning EV charging times with high energy output cycles, we can maximize clean energy utilization and reduce reliance on fossil fuels. Achieving precise synchronization, however, necessitates advanced artificial intelligence forecasting and seamless communication between EVs, charging stations, and power companies. This exchange of information relies on vehicle-to-everything and V2G technologies as shown in Figure 11.20.

V2G (Vehicle-to-Grid) allows EVs to function as mobile energy storage and distribution units. During off-peak periods, they draw power, and during demand spikes, they supply stored energy back to the grid, enhancing grid

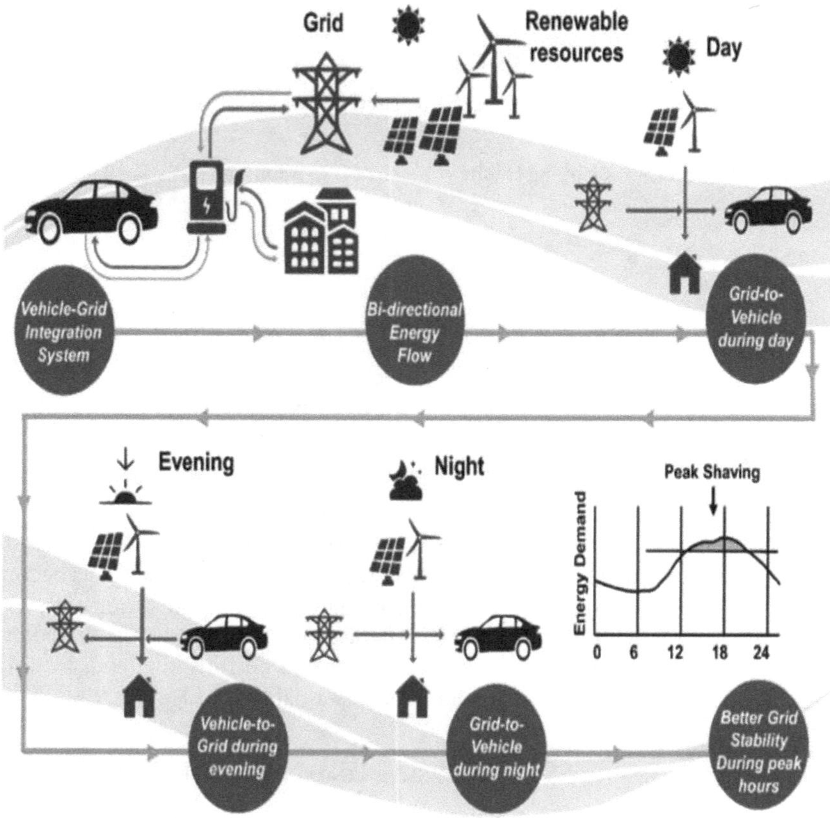

FIGURE 11.20 Visualization of vehicle-to-grid (V2G) dynamics across daylight hours.

resilience. V2G also addresses renewable energy variability by efficiently storing excess energy during peak generation and redistributing it when production drops.

V2X (Vehicle-to-Everything) extends beyond grid interfacing, transforming EVs into decentralized endpoints. It supplies power to homes and businesses during blackouts or voltage drops, ensuring a stable energy supply. Additionally, V2X integrates EVs into smart home systems, allowing owners to monetize their vehicle's energy storage capabilities and helping power companies optimize real-time energy distribution.

11.13.5 Role of Smart Grids in Mitigating the Potential Issues of EV Integration into the Conventional Grid

The integration of EVs into the power grid presents both opportunities and challenges. As the adoption of EVs accelerates, the demand for electricity surges, which can strain the existing grid infrastructure, leading to potential issues such as grid congestion, voltage instability, and increased peak loads. Smart grid technologies offer a comprehensive solution to these challenges by enhancing the flexibility, resilience, and efficiency of the power grid. Through advanced data analytics, real-time monitoring, and bidirectional energy flow, smart grids facilitate the seamless integration of EVs while optimizing energy distribution and utilization.

Smart grid technologies play a crucial role in mitigating potential issues associated with integrating EVs into the electrical grid. They enable better management of energy demand and supply, allowing for effective load balancing and peak shaving, which are essential to prevent grid congestion and undervoltage issues (Secchi et al., 2023). By incorporating advanced V2G systems, smart grids facilitate bidirectional energy flow, enabling EVs to act as distributed energy storage units, thus enhancing grid stability and resilience (Mastoi et al., 2023). Moreover, smart grids integrate renewable energy sources and optimize charging strategies to reduce the strain on the grid, utilizing algorithms like the slime mold algorithm to coordinate EV charging and mitigate the adverse effects of uncoordinated EV integration (Abid et al., 2023; Barman et al., 2023). These technologies also support real-time monitoring and control through IoT-based solutions, improving the efficiency and reliability of power distribution networks (Ullah et al., 2023). Overall, smart grids are instrumental in addressing the challenges of EV integration, ensuring a sustainable and reliable energy system.

Figure 11.21 presents the integration of smart grid technologies with EVs and renewable energy sources. Smart grids leverage advanced data analytics and machine learning algorithms to predict and manage energy demand more accurately, allowing for dynamic adjustment of EV charging schedules to avoid overloading the grid during peak hours (Li and Mi, 2015). The integration of distributed energy resources such as solar panels and wind turbines with smart grid systems also enables the effective use of renewable energy for EV charging, reducing dependency on traditional power sources and lowering carbon emissions (Ntombela et al., 2023). Furthermore, smart grids support the deployment of decentralized energy storage solutions, where EV batteries can store excess energy during low-demand periods and release it during high-demand times, thus providing additional grid support and enhancing energy efficiency (Waseem et al., 2023). These systems also incorporate robust cybersecurity measures to protect against potential threats and ensure the integrity and reliability of grid operations. By fostering improved communication and coordination between different grid components, smart grids facilitate a more resilient and flexible energy infrastructure that can accommodate the growing number of EVs and the diverse energy needs of modern societies (Hossain et al., 2023).

Thus, smart grid technologies play a pivotal role in mitigating the potential issues associated with the integration of EVs into the power grid. By leveraging sophisticated algorithms, distributed energy resources, and robust control systems, smart grids ensure balanced energy supply and demand, reduce the risk of grid overload, and enhance the overall stability of the electrical network. The integration of renewable energy sources further complements these efforts, promoting sustainable and efficient energy use. As the adoption of EVs continues to rise, the deployment of smart grid technologies will be crucial in supporting a resilient and adaptive energy infrastructure, ultimately facilitating a smooth transition toward a cleaner and more sustainable future.

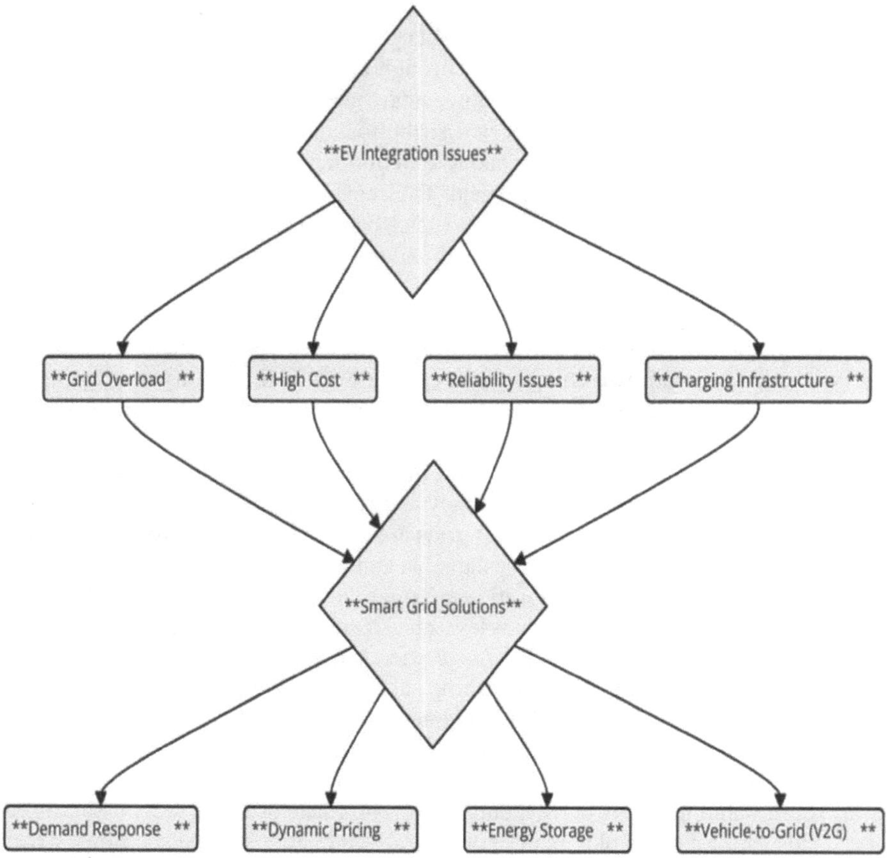

FIGURE 11.21 Integration of smart grid technologies with EVs and renewable energy sources.

11.14 CASE STUDIES

11.14.1 A Brief Overview of Tesla

Tesla, an American company specializing in EVs and clean energy, was established in 2003 by Elon Musk (Figure 11.22). The company's mission is to expedite the global shift toward sustainable energy. Tesla offers a variety of EVs, energy storage solutions, and solar products. The Tesla Supercharger Network, a system of high-speed charging stations for Tesla EVs, was introduced. Since its inception in 2012, the network has grown to encompass over 45,000 Superchargers globally. The Supercharger Network is engineered to alleviate range anxiety and augment user convenience by delivering a fast and reliable charging infrastructure.

FIGURE 11.22 Tesla Supercharger.

11.14.2 Strategic Placement of Tesla's Supercharger Stations

Tesla has tactically deployed Supercharger stations worldwide to enable Tesla owners to journey over extended distances without the concern of depleting their charge. The Supercharger Network is strategically situated along key highways and routes, facilitating long-distance travel for Tesla owners. To accommodate the requirements of urban residents, Tesla has also established Supercharger stations in city locales.

11.14.3 Engineering That Powers the Tesla Supercharger Network

The Tesla Supercharger Networks shown in Figure 11.23 utilize high-speed DC rapid charging technology to provide a quick and reliable charging solution. The Supercharger Network has the capacity to charge a Tesla vehicle to approximately 80% in just 30 minutes. Tesla integrates its proprietary technology to ensure the Supercharger Network is fast, reliable, and secure.

FIGURE 11.23 Tesla Supercharger Network.

11.14.4 Issues and Difficulties

Tesla encountered numerous obstacles in constructing the Supercharger Network, encompassing regulatory barriers, elevated expenses, and technical difficulties.

In addition, Tesla confronted issues while broadening the Supercharger Network to new territories due to variances in rules and infrastructure.

11.14.5 How Do You Think Tesla Handled These Challenges?

11.14.5.1 Comprehensive Approaches

11.14.5.1.1 Negotiations with Regulators

a. Participated in forward-thinking discussions with regulatory authorities to simplify the process of obtaining permits.
b. Worked in tandem with governmental bodies to formulate policies that are beneficial for the infrastructure of EV charging.

11.14.5.1.2 Minimizing the Expenses

a. Pioneered proprietary charging solutions to enhance efficiency and curtail the total costs of infrastructure.

b. Established strategic alliances to distribute infrastructure costs and expedite implementation.

Advancements in Technology: (a) Allocated resources to research and development to tackle technical obstacles, guaranteeing uninterrupted charging experiences for users. (b) Capitalized on breakthroughs in battery technology to boost charging velocity and dependability.

Standardization on a Global Scale: (a) Promoted universal charging standards, aiding in the implementation of compatible connectors. (b) Worked in conjunction with global partners to establish a consistent, uniform charging experience.

Localized Modification: (a) Executed comprehensive market analysis to comprehend local requirements and customize infrastructure suitably. (b)Modified Supercharger blueprints to cater to a variety of grid and power supply conditions.

11.14.6 Highlights of Tesla Supercharger

The Tesla Supercharger Network serves as a unique selling point for Tesla, aiding in its rise as a frontrunner in the EV industry. By strategically placing Supercharger stations and employing high-speed DC fast charging technology, Tesla has effectively mitigated range concerns and improved user experience.

The Supercharger Network's speed, reliability, and safety are guaranteed by Tesla's exclusive technology.

11.15 WIRELESS CHARGING IMPLEMENTATION – BMW

11.15.1 Important Obstacles in Wireless Charging of BMW

Efficient Operation: The charging system should not squander energy.

Convenience: The charging mechanism needed to be user-friendly and suitable for regular use.

11.15.2 Triumphing Over Obstacles

As far as efficiency goes, the BMW Wireless Charging method is about 85% efficient.

Simple operation: By positioning the car over a base pad, the BMW Wireless Charging system allows high-voltage batteries to receive electric energy directly from the mains supply (Figure 11.24).

11.15.3 Highlights of Wireless Charging in BMW

Innovation: New and improved EV usability and convenience are the results of innovative charging technology like BMW Wireless Charging.

Efficiency: The efficiency rate of the BMW Wireless Charging system is approximately 85%, making it exceptionally efficient.

FIGURE 11.24 Wireless charging of BMW.

Ease of Use: Benefiting from its intuitive design, the BMW Wireless Charging system is ideal for regular usage.

Future: It will soon be even easier to charge the battery of an EV than it is to fill up a gas tank, thanks to innovations like BMW Wireless Charging.

11.16 OTHER CHARGING INFRASTRUCTURES AROUND THE WORLD

11.16.1 ChargePoint's Public Charging Network

Businesses and individuals alike can benefit from its smart-energy management solutions as they make the switch from gas to electricity (Figure 11.25). With more than 25,000 charging sites nationwide and 100 more stations planned for "express charging corridors" along the Atlantic and Pacific coasts, vehicles will have no trouble connecting to a network of EV chargers, hastening the departure from fossil fuels (https://unfccc.int/).

11.16.2 IONITY's High-Power Charging Network

IONITY is a joint venture of major automakers (BMW, Daimler, Ford, and Volkswagen Group) to create a high-power charging network across Europe (Figure 11.26). The high-power charging network for EVs spans 24 European countries, featuring charging stations with multiple charging points per location to ensure accessibility and convenience. These stations are equipped with chargers offering

FIGURE 11.25 Charge point's public charging network.

up to 350kW charging capacity, enabling ultrafast charging for minimal downtime. Utilizing the European charging standard CCS, the network ensures the widest compatibility across various EV models. Furthermore, the network is powered entirely by renewable energy, promoting carbon-neutral driving and contributing to a sustainable and eco-friendly transportation system (Home - Home (ionity.eu)).

11.16.3 BP Pulse Charging Hubs

BP Pulse is a leading provider of EV charging solutions, offering a comprehensive network of charging stations across the UK. As part of BP's commitment to a low-carbon future, BP Pulse provides a range of charging options, including rapid and ultrafast chargers, to meet the diverse needs of EV drivers (Figure 11.27).

The network is strategically located at convenient points such as fuel stations, retail sites, and other public locations, ensuring ease of access for users. BP Pulse also emphasizes reliability and user experience, with a focus on maintaining high operational standards and providing customer support. By integrating renewable

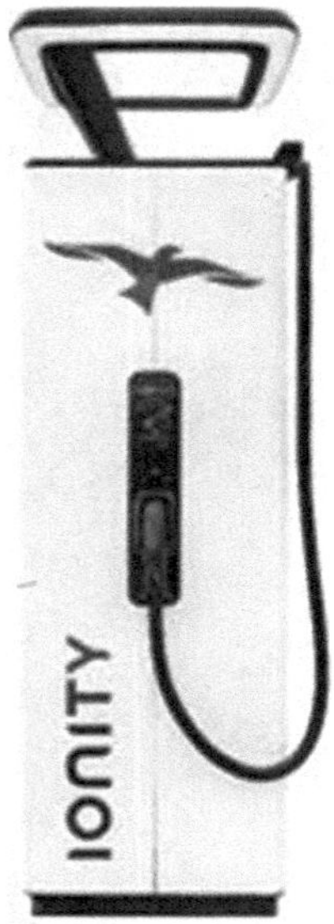

FIGURE 11.26 Ionity's high-power charging network.

FIGURE 11.27 BP pulse charging hubs.

energy sources where possible, BP Pulse supports sustainable transportation and contributes to reducing the overall carbon footprint of EVs (BP Pulse I EV Charging Solutions).

11.16.4 Greenlots Charging Solutions in Singapore

Greenlots, a member of the Shell Group, is a prominent provider of EV charging solutions in Singapore, focusing on integrating smart charging technologies with the national grid. Greenlots offers a variety of charging options, including both AC and DC fast chargers, catering to the diverse needs of EV drivers across residential, commercial, and public locations. The company's charging stations are designed to support efficient energy management, leveraging advanced software to optimize charging schedules and reduce costs. By incorporating renewable energy sources where feasible, Greenlots aims to promote sustainability and reduce carbon emissions. The company's initiatives in Singapore are part of a broader effort to enhance the city's EV infrastructure, making electric mobility more accessible and reliable for residents and businesses alike (Singapore (greenlots.com)) (Figure 11.28).

FIGURE 11.28 Greenlots charging solutions in Singapore.

11.17 CONCLUSION

The exploration of EV charging infrastructure reveals a complex and evolving landscape characterized by diverse charging methods, architectures, and standards. The architecture of EV chargers ranges from basic slow chargers to advanced fast charging systems, with battery swapping technology offering an alternative approach to extend vehicle range and reduce downtime. Standardized battery systems and charging protocols, such as the emerging AC-002 and DC-002 standards in India, are crucial for ensuring compatibility and efficiency across different EV models and charging networks. The role of power electronics converters in managing charging processes and the introduction of wireless charging technologies represent significant advancements in the field. However, the impact of EV charging on the grid necessitates careful consideration, with smart grid technologies playing a key role in addressing potential challenges and integrating EVs into the existing infrastructure. Case studies, such as Tesla's Supercharger Network, provide valuable insights into the practical application and performance of various charging solutions globally. Overall, the future of EV charging infrastructure will depend on continued innovation, standardization, and the effective management of grid interactions to support the growing adoption of EVs.

REFERENCES

Abid, M.S., Apon, H.J., Alavi, A., & Hossain, M.A. (2023). Mitigating the effect of electric vehicle integration in distribution grid using slime mould algorithm. *Alexandria Engineering Journal*, 64, 785–800.

Barman, P., Dutta, L., Bordoloi, S., & Kalita, A. (2023). Renewable energy integration with electric vehicle technology: A review of the existing smart charging approaches. *Renewable and Sustainable Energy Reviews*, 183, 113518.

Bowermaster, D., Alexander, M., & Duvall, M. (2017). The need for charging: Evaluating utility infrastructures for electric vehicles while providing customer support. *IEEE Electrification Magazine*, 5(1), 59–67.

Brown, D. (2024). Battery swapping and leasing: Transforming electric vehicle infrastructure. *Electric Vehicle Research and Development*, 16(3), 200–220.

Das, T., & Aliprantis, D.C. (2008). Small-signal stability analysis of power system integrated with PHEVs. In: *2008 IEEE Energy 2030 Conference*. IEEE, Atlanta, GA, pp. 1–4.

Dharmakeerthi, C.H., Mithulananthan, N., Saha, T.K., 2013. Impact of electric vehicle load on power system oscillatory stability. In: *2013 Australasian Universities Power Engineering Conference (AUPEC)*. Hobart, TAS, Australia, pp. 1–6.

Doe, J. (2024). Battery swapping: A game-changer for electric vehicles. *Journal of Sustainable Transportation*, 33(1), 85–102.

Domínguez-Navarro, J.A., Dufo-López, R., Yusta-Loyo, J.M., et al. (2019). Design of an electric vehicle fast-charging station with integration of renewable energy and storage systems. *International Journal of Electrical Power & Energy Systems*, 105, 46–58.

Donadee, J., & Ilić, M. (2012). Stochastic co-optimization of charging and frequency regulation by electric vehicles. In: *2012 North American Power Symposium (NAPS)*. IEEE, Champaign, IL, pp. 1–6.

Durante, L., Nielsen, M., & Ghosh, P. (2017). Analysis of non-sinusoidal wave generation during electric vehicle charging and their impacts on the power system. *International Journal of Process Systems Engineering*, 4(2–3), 138–150.

Evans, K. (2024). Centralized charging and battery swapping: Improving EV infrastructure. *Sustainable Energy Solutions*, 18(2), 110–130.

Garwa, N., & Niazi, K.R. (2019). Impact of EV on integration with grid system–A review. In: *2019 8th International Conference on Power Systems (ICPS)*. IEEE, Jaipur, India, pp. 1–6.

Godina, R., Rodrigues, E.M.G., Paterakis, N.G., et al. (2016). Innovative impact assessment of electric vehicles charging loads on distribution transformers using real data. *Energy Conversion and Management,* 120, 206–216.

Green, L. (2024). The future of EV charging: The role of battery swapping stations. *Electric Vehicle Research and Development*, 16(2), 120–140.

Guille, C., & Gross, G. (2009). A conceptual framework for the vehicle-to-grid (V2G) implementation. *Energy Policy,* 37(11), 4379–4390.

Habib, S., Kamran, M., & Rashid, U. (2015). Impact analysis of vehicle-to-grid technology and charging strategies of electric vehicles on distribution networks–a review. *Journal of Power Sources*, 277, 205–214.

Hadley, S.W., & Tsvetkova, A.A. (2009). Potential impacts of plug-in hybrid electric vehicles on regional power generation. *The Electricity Journal*, 22(10), 56–68.

Hofmann, M., Raab, S., Schaefer, M., et al. (2015). Measurements on vehicle to grid application in industrial power grid for peak load reduction: Robust bidirectional charger for series production electric and plug-in hybrid vehicles. In: *2015 IEEE 6th International Symposium on Power Electronics for Distributed Generation Systems (PEDG)*. IEEE, Aachen, Germany, pp. 1–5.

Hossain, S., Rokonuzzaman, M., & Rahman, K. S. (2023). Grid-vehicle-grid (G2V2G) efficient power transmission: An overview of concept, operations, benefits, concerns, and future challenges. *Sustainability*, 15, 5782.

Johnson, A. (2024). Mitigating range anxiety through battery leasing. *Electric Vehicle Research and Development*, 15(1), 45–60.

Karmaker, A.K., Roy, S., & Ahmed, M.R. (2019). Analysis of the impact of electric vehicle charging station on power quality issues. In: *2019 International Conference on Electrical, Computer and Communication Engineering (ECCE)*. IEEE, Cox's Bazar, Bangladesh, pp. 1–6.

Khalid, M.R., Alam, M.S., Sarwar, A., et al. (2019). A comprehensive review on electric vehicles charging infrastructures and their impacts on power-quality of the utility grid. *ETransportation,* 1, 100006.

Kristoffersen, T.K., Capion, K., & Meibom, P. (2011). Optimal charging of electric drive vehicles in a market environment. *Applied Energy*, 88(5), 1940–1948.

Li, S., & Mi, C.C. (2015). Wireless power transfer for electric vehicle applications. *IEEE Journal of Emerging and Selected Topics in Power Electronics*, 3(1), 4–17.

Liao, F., Molin, E., & van Wee, B. (2019). Consumer preferences for electric vehicles: A literature review. *Transport Reviews*, 39(3), 252–275.

Ma, G., Jiang, L., Chen, Y., et al. (2017). Study on the impact of electric vehicle charging load on nodal voltage deviation. *Archives of Electrical Engineering*, 66 (3).

Mahalik, M., Poch, L., Botterud, A., et al. (2010). Impacts of plug-in hybrid electric vehicles on the electric power system in illinois. In: *2010 IEEE Conference on Innovative Technologies for an Efficient and Reliable Electricity Supply*. IEEE, Waltham, MA, pp. 341–348.

Mastoi, M.S., Zhuang, S., Munir, H.M., Haris, M., & Hassan, M. (2023). A study of charging-dispatch strategies and vehicle-to-grid technologies for electric vehicles in distribution networks. *Energy Reports*, 9, 1777–1806.

McCarthy, D., & Wolfs, P. (2010). The HV system impacts of large-scale electric vehicle deployments in a metropolitan area. In: *2010 20th Australasian Universities Power Engineering Conference*. IEEE, Christchurch, New Zealand, pp. 1–6.

Meyer, J., Hähle, S., Schegner, P., et al. (2011). Impact of electrical car charging on unbalance in public low voltage grids. In: *11th International Conference on Electrical Power Quality and Utilisation*. IEEE, Lisbon, Portugal, pp. 1–6.

Nanda, N.N., Hanifah, M.S.A., Yusoff, S.H., et al. (2021). In-depth perception of dynamic inductive wireless power transfer development: A review. *International Journal of Power Electronics and Drive Systems (IJPEDS)*, 12(3), 1459–1471.

Ntombela, M., Musasa, K., & Moloi, K. (2023). A comprehensive review of the incorporation of electric vehicles and renewable energy distributed generation regarding smart grids. *World Electric Vehicle Journal*, 14, 176.

Onar, O.C., & Khaligh, A. (2010). Grid interactions and stability analysis of distribution power network with high penetration of plug-in hybrid electric vehicles. In: *2010 Twenty-Fifth Annual IEEE Applied Power Electronics Conference and Exposition (APEC)*. IEEE, Palm Springs, CA, pp. 1755–1762.

Qiangqiang, L., Guoding, Z., Honghua, G.E., et al. (2012). Technical and economic analysis of vehicle-to-grid support peak electricity. *Electric Power*, 4, 92–95.

Saber, A.Y., & Venayagamoorthy, G.K. (2010). Plug-in vehicles and renewable energy sources for cost and emission reductions. *IEEE Transactions on Industrial Electronics*, 58(4), 1229–1238.

Secchi, M., Barchi, G., Macii, D., & Petri, D. (2023). Smart electric vehicles charging with centralised vehicle-to-grid capability for net-load variance minimisation under increasing EV and PV penetration levels. *Sustainable Energy, Grids and Networks*, 35, 101120.

Smith, J. (2024). Innovative financial models for EV adoption: A review. *Journal of Sustainable Transportation*, 32(2), 120–135.

Tate, E. D., Harpster, M. O., & Savagian, P. J. (2009). The electrification of the automobile: From conventional hybrid, to plug-in hybrids, to extended-range electric vehicles. *SAE International Journal of Passenger Cars-Electronic and Electrical Systems*, 1(1), 156–166.

Taylor, R. (2024). Revolutionizing EV ownership: The battery leasing model. *Journal of Sustainable Transportation*, 33(2), 145–160.

Tian, W., He, J., Niu, L., et al. (2012). Simulation of vehicle-to-grid (V2G) on power system frequency control. In: *IEEE PES Innovative Smart Grid Technologies*. IEEE, Tianjin, China, pp. 1–3.

Ul-Haq, A., Cecati, C., Strunz, K., et al. (2015). Impact of electric vehicle charging on voltage unbalance in an urban distribution network. *Intelligent Industrial Systems,* 1(1), 51–60.

Ullah, Z., Rehman, A.U., Wang, S., Hasanien, H.M., & Luo, P. (2023). IoT-based monitoring and control of substations and smart grids with renewables and electric vehicles integration. *Energy*, 282, 128924.

UrRehman, U., & Riaz, M. (2017). Vehicle to grid system for load and frequency management in smart grid. In: *2017 International Conference on Open-Source Systems & Technologies (ICOSST)*. IEEE, pp. 73–78.

Wang, Q., Jiang, B., Li, B., & Yan, Y. (2016). A critical review of thermal management models and solutions of lithium-ion batteries for the development of pure electric vehicles. *Renewable and Sustainable Energy Reviews*, 64, 106–128.

Waseem, M., Khan, M. A., Goudarzi, A., & Fahad, S. (2023). Incorporation of blockchain technology for different smart grid applications: Architecture, prospects, and challenges. *Energies*, 16, 820.

White, C.D., & Zhang, K.M. (2011). Using vehicle-to-grid technology for frequency regulation and peak-load reduction. *Journal of Power Sources*, 196(8), 3972–3980.

Williams, P. (2024). Streamlining EV usage: The impact of battery swapping. *Journal of Electric Mobility*, 34(1), 75–90.

Yan, J., Xu, G., Qian, H., et al. (2010). Battery fast charging strategy based on model predictive control. In: *2010 IEEE 72nd Vehicular Technology Conference-Fall*. IEEE, Ottawa, Canada, pp. 1–8.

Yan, Q., & Kezunovic, M. (2012). Impact analysis of electric vehicle charging on distribution system. In: *2012 North American Power Symposium (NAPS)*. IEEE, Champaign, IL, pp. 1–6.

Yang, X., Zhang, G., Liu, C., & Li, S. (2018). A review of lithium-ion battery thermal management system and application in vehicles. *Renewable and Sustainable Energy Reviews*, 82, 3880–3895.

Yilmaz, M., & Krein, P. T. (2012). Review of battery charger topologies, charging power levels, and infrastructure for plug-in electric and hybrid vehicles. *IEEE transactions on Power Electronics*, 28(5), 2151–2169.

Yong, J.Y., Ramachandaramurthy, V.K., Tan, K.M., et al. (2015). A review on the state-of-the-art technologies of electric vehicle, its impacts and prospects. *Renewable and Sustainable Energy Reviews*, 49, 365–385.

Zhang, C., & Wang, Z. (2018). A review of charging infrastructure for electric vehicles in China. *Energy Procedia*, 144, 90–97.

Zhang, J. et al. (2023). A review of capacitive power transfer technology for electric vehicle applications. *Electronics,* 12(16), 3534.

12 Electric Vehicle Charging Infrastructure

Challenges, Innovations, and Policy Recommendations

Jawahar Marimuthu and Edward Rajan Samuel Nadar

12.1 INTRODUCTION

One of the main ways to lessen greenhouse gas emissions and the transportation sector's reliance on fossil fuels is through the increasing usage of EVs. According to Chen et al. (2014), EVs provide a long-term way to lessen dependency on fossil fuels and cut greenhouse gas emissions. Nonetheless, the distribution network has difficulties when it comes to charging a sizable fleet of EVs dispersed throughout residential areas. Ramachandran et al. (2015) discuss the inclusion of low-carbon alternatives such as electricity, hydrogen, and biofuels in the transport sector to reduce greenhouse gas emissions and provide a sustainable future.

In order to optimize the placement and sizing of EV charging stations, Islam et al. (2015) review various optimization techniques employed over the last decade. Girardi et al. (2015) compare the environmental performances of EVs and internal combustion engine vehicles, focusing on the production of electricity used to charge EVs. Tang et al. (2016) propose a probabilistic modeling approach to estimate the nodal EV load at fast charging stations, emphasizing the importance of EVs in reducing greenhouse gas emissions and fossil fuel dependence. Sun et al. (2017) introduce a software-defined green framework for hybrid EVcharging networks, highlighting the potential of EVs to reduce greenhouse gas emissions and dependence on fossil fuels. Liu et al. (2018) analyze the energy consumption and greenhouse gas emissions in China's road transport sector, emphasizing the role of EVs in reducing reliance on fossil fuels in the long term. Singh et al. (2019) study the environmental feasibility of incorporating electric taxis in South Africa, emphasizing the importance of clean electricity in reducing environmental impact. The potential of plug-in automobiles to minimize road transport emissions is highlighted by García-Afonso et al. (2021), who undertake a simulation research study to assess energy consumption and Well-to-Wheels greenhouse gas emissions of several light-duty powertrain architectures. Range anxiety is a problem that EV users have, and Shrestha et al. (2022) address it, stressing how important it is to find a solution in order to promote

 DOI: 10.1201/9781003496946-12

the widespread adoption of EVs and further cut greenhouse gas emissions. Overall, the examined literature emphasizes how crucial EVs are to cutting greenhouse gas emissions and the transportation sector's reliance on fossil fuels.

The importance of EVCI for widespread EV adoption is a critical factor in overcoming barriers to the mass adoption of EVs. Madina et al. (2016) highlight the high investment costs associated with EVs and the necessary infrastructure, presenting a methodology for assessing business models for charging stations. Ferguson et al. (2018) discuss the optimal planning of workplace EVCI to take advantage of smart charging opportunities while considering investment costs. Mouhrim et al. (2018) emphasize the need for an optimal location of wireless charging infrastructure along transport routes to facilitate EV adoption. Mobarak and Bauman (2019) introduce vehicle-directed smart charging strategies to mitigate the impact of long-range EV charging on distribution transformer aging, showcasing the importance of intelligent charging methods. Heredia et al. (2020) evaluate smart charging for EV-to-building integration, stressing the importance of demand flexibility for efficient grid operation with increased EV adoption. Jones et al. (2020) forecast a significant increase in EV demand by 2030 and highlight the critical raw materials needed for EV production, underscoring the importance of infrastructure for mass EV adoption. Asensio et al. (2021) address interoperability issues in EV infrastructure and present a high-resolution dataset of real-time EV charging transactions at a corporate campus, emphasizing the need for data sharing and analysis. Roy and Law (2022) use machine learning to examine spatial disparities in EV charging station placements, highlighting inequities that hinder widespread EV adoption. Andrenacci et al. (2022) analyze charge profiles of EVs in different cities to optimize EVCI design and maximize benefits for users and managers. In conclusion, the literature reviewed underscores the critical role of EVCI in facilitating widespread EV adoption by addressing investment costs, optimizing infrastructure planning, implementing smart charging strategies, and ensuring equitable access to charging stations. These studies collectively emphasize the importance of EVCI in overcoming barriers to mass EV adoption and promoting sustainable transportation.

12.2 EV CHARGING LEVELS AND TECHNOLOGIES

Infrastructure for convenient and effective charging of EVs is essential to their mass adoption. The three most common EV charging levels – Level 1, Level 2, and DC Fast Charging – are thoroughly covered in this chapter. We delve into their power ratings, charging speeds, and ideal use cases to empower researchers, policymakers, and EV users with a clear understanding of this critical aspect of EV ownership. The transportation sector is undergoing a paradigm shift toward electric mobility. However, range anxiety, fueled by concerns over charging infrastructure and time, remains a significant barrier for EV adoption. Understanding the different levels of EV charging is crucial to alleviating these concerns and optimizing the EV ownership experience. EV charging levels are categorized based on the voltage and current delivered, directly impacting charging speed. There are three primary levels, and their specifications are listed in Table 12.1.

TABLE 12.1
Charging Levels and Their Specifications

Level	Voltage	Current	Power Rating (kW)	Charging Time (BEV, Depleted to 80%)	Typical Use Case
Level 1	120V AC	12A	1.4-3.6	40–50+ hours	Overnight charging at home
Level 2	240V AC	32A	7.2–19.2	4–8 hours	Home charging, public charging
DC Fast Charging	400–900V DC	Up to 500A	Up to 350+	15–45 minutes	Long-distance travel, public charging stations

12.2.1 Level 1 (120V AC)

This is the slowest option for charging, and it usually uses a regular household plug. Power ratings range from 1.4 to 3.6 kW, resulting in charging times of 40–50+ hours for a depleted battery in a Battery Electric Vehicle (BEV). It is best suited for overnight charging at home for vehicles with short daily commutes.

12.2.2 Level 2 (240V AC)

This is the most common charging option for home and public settings. Power ratings typically range from 7.2 to 19.2 kW, significantly reducing charging times compared to Level 1. A full charge for a BEV can be achieved in 4–8 hours, making it suitable for daily commutes and longer trips with planned charging stops.

12.2.3 DC Fast Charging (Level 3)

This is the fastest charging option, utilizing high-voltage DC to bypass the car's on-board AC/DC converter. Power ratings can reach up to 350 kW or even higher, enabling rapid charging in 15–45 minutes for a typical BEV. DC Fast Charging is ideal for long-distance travel and public charging stations where extended dwell times are undesirable.

It is important to remember that the charging timeframes listed above are estimates that can vary depending on a number of factors:

- **Battery Capacity:** Larger batteries naturally take longer to charge.
- **Vehicle's Charging Acceptance Rate:** This refers to the maximum power an EV can handle.
- **Environmental Conditions:** Extreme temperatures can impact charging efficiency.
- **Charger Availability and Power Output:** Not all Level 2 or DC Fast Charging stations offer the maximum power ratings.

Users can plan their trips and choose the appropriate charging infrastructure by having a clear understanding of the various EV charging levels, as seen in Figure 12.1.

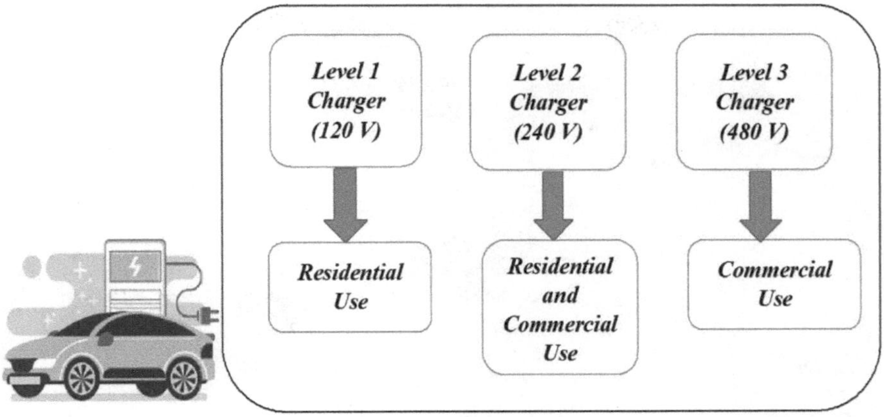

FIGURE 12.1 Types of e-vehicles charging levels.

DC Fast Charging gives a quick fix for long-distance travel, Level 2 offers a balance for regular commuting, and Level 1 charging takes care of nighttime top-ups. Charge times are predicted to continue getting shorter as EV infrastructure and technology advance, which will help reduce range anxiety and hasten the shift to more environmentally friendly transportation.

12.3 POWER CONVERTER TECHNOLOGIES FOR EV CHARGING STATIONS

The widespread adoption of EVs necessitates the development of robust and efficient charging infrastructure. This chapterexplores the critical role of power converter technologies in EV charging stations. We delve into three key converter types: AC/DC converters for grid connection, DC/DC converters for battery interfacing, and bi-directional converters for potential Vehicle-To-Grid (V2G) applications. We discuss the operating principles, advantages, and considerations for each converter type, highlighting their contributions to optimal EV charging and grid integration.

The burgeoning EV revolution necessitates a robust charging infrastructure capable of efficiently transferring energy between the grid and EVs. Power converters play a pivotal role in this process, transforming electrical energy to meet the specific voltage and current requirements of EVs. This chapter explores the three fundamental converter technologies prevalent in EV charging stations as either on-board or off-board charges as shown in Figures 12.2 and 12.3.

12.3.1 On-Board and Off-Board Battery Charging

In the realm of EVs, the terms "on-board" and "off-board" typically refer to the charging infrastructure rather than the battery itself. However, for clarity, we'll delve into both the battery and charging aspects.

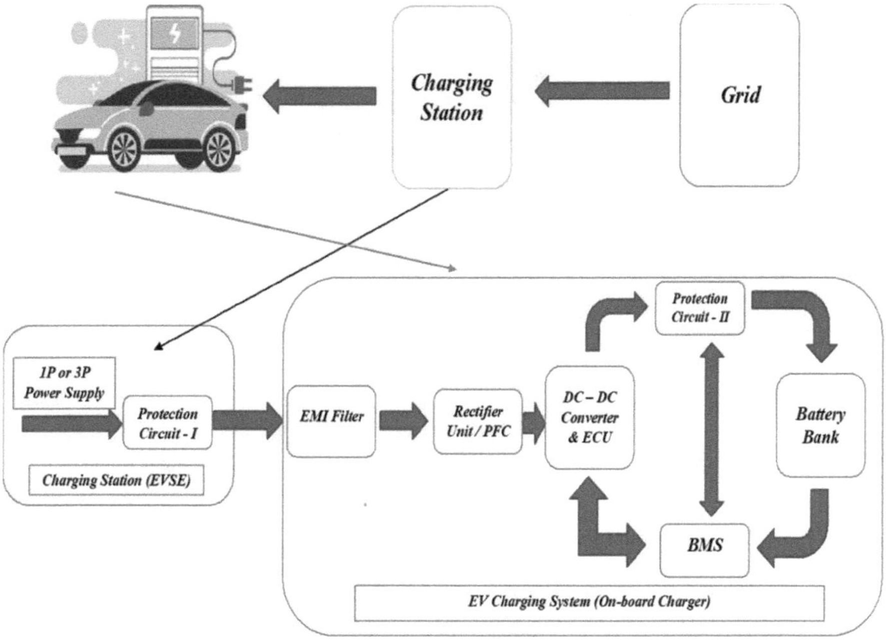

FIGURE 12.2 On-board charger.

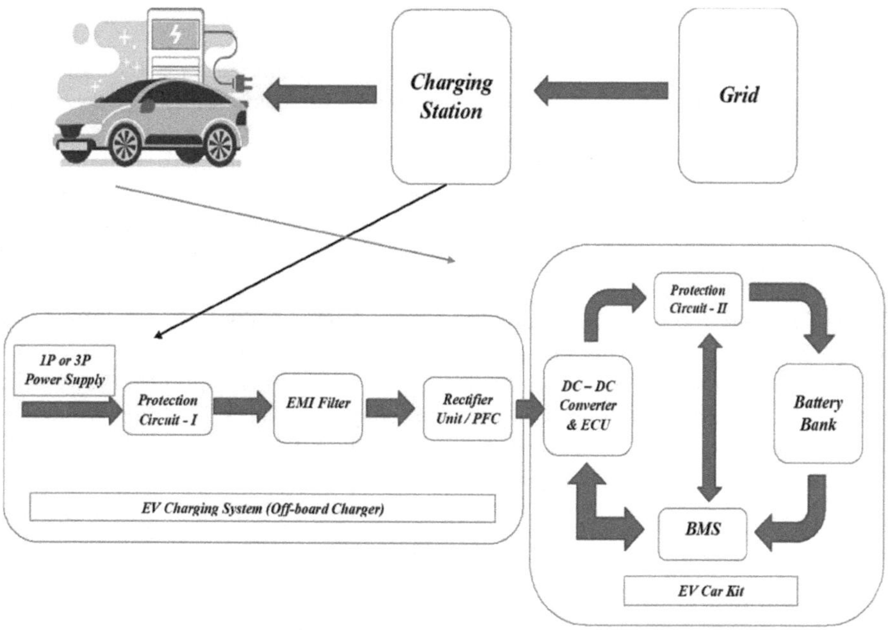

FIGURE 12.3 Off-board charger.

An on-board battery is physically located within the vehicle. It's the primary power source for the electric motor, providing the energy needed for propulsion. These batteries are typically lithium-ion based, known for their high energy density and relatively low weight. The size and capacity of the on-board battery determine the vehicle's driving range.

12.3.1.1 On-board Charging

The process of transforming external AC power into DC power appropriate for charging the vehicle's battery is known as "on-board charging". The EV undergoes this process inside. On-board chargers, also known as Level 1 and Level 2 chargers, are usually utilized for slower charging speeds. On-board chargers are useful, but their charging power is limited, which might affect how long it takes to charge an item.

12.3.1.2 Off-board Charging

Off-board charging involves external charging stations that deliver DC power directly to the vehicle's battery. This method bypasses the on-board charger, enabling much faster charging speeds, commonly known as DC Fast Charging or Level 3 charging. Off-board chargers are typically found in public charging stations or commercial locations. They require specialized equipment and infrastructure but offer significant advantages in terms of charging time.

12.3.1.3 Key Differences and Considerations

The primary distinction between on-board and off-board charging lies in the charging speed and power delivery. On-board charging is slower but more convenient, while off-board charging is faster but requires dedicated infrastructure. The choice between the two often depends on factors such as the location of charging, available duration, and the desired charging speed.It's important to note that while the term "off-board battery" might be technically correct, it's not commonly used in the EV industry. The focus is primarily on the charging infrastructure, with on-board and off-board referring to the location of the charging equipment rather than the battery itself.

- **AC/DC Converters:** Interface the charging station with the AC power grid, rectifying the incoming AC voltage to a regulated DC voltage suitable for DC Fast Charging or further conversion within the station.
- **DC/DC Converters:** Bridge the voltage gap between the high-voltage DC output from the AC/DC converter and the lower voltage DC battery pack of the EV, enabling safe and efficient charging.
- **Bi-directional Converters:** Hold the potential to revolutionize the grid by facilitating V2G integration. These converters not only enable charging but also allow EVs to feed excess stored energy back into the grid during peak demand periods.

12.3.2 AC/DC Converters

AC/DC converters are the workhorses of EV charging stations, responsible for converting the incoming AC grid voltage (typically 400–800V) into a high-voltage DC output suitable for DC Fast Charging applications. Two primary topologies dominate this space:

- **Vienna Rectifier:** This high-efficiency converter utilizes synchronized switches to achieve near-unity power factor (ratio of real power to apparent power), minimizing grid current distortion and maximizing energy transfer.
- **Two-Stage Converter:**This method uses a DC-DC converter after the Power Factor Correction stage. A high power factor is guaranteed by the Power Factor Correction stage, and the output voltage for charging is regulated by the DC-DC converter. The optimal output voltage range, cost, and efficiency are some of the criteria that influence the decision between various topologies.

12.3.3 DC/DC Converters

When connecting the higher voltage DC output of the AC/DC converter to the EV's lower voltage DC battery pack, DC/DC converters are essential components. Usually, they are buck converters, which adjust the voltage and current to meet the needs of the battery during charging. Galvanic isolation between the grid and the battery is provided by isolated DC/DC converters, which improves safety.

12.3.4 Bi-directional Converters for V2G Integration

V2G technology allows EVs to act as distributed energy storage units, potentially stabilizing the grid by feeding excess stored energy back during peak demand periods. This necessitates bi-directional converters that can operate in both charging (DC-to-DC boost) and discharging (DC-to-AC inversion) modes. Popular bi-directional converter topologies for V2G applications include:

- **Vienna Rectifier with Integrated Boost Converter:** This single-stage converter offers a compact and efficient solution for V2G integration.
- **Isolated Bi-directional Converter:** This topology utilizes a transformer to provide galvanic isolation between the grid and the battery, enhancing safety for V2G operation.

The development of efficient and cost-effective bi-directional converters is crucial for realizing the full potential of V2G and promoting a more sustainable and resilient electric grid.

Power converter technologies are the backbone of efficient and reliable EV charging infrastructure. AC/DC converters ensure seamless grid connection, DC/DC converters facilitate safe and optimized battery charging, and bi-directional converters pave the way for V2G integration. Continued advancements in converter design and control strategies will be instrumental in supporting the widespread adoption of EVs and fostering a smarter and more sustainable electric grid.This chapter has explored the fundamental converter technologies for EV charging stations, as shown in Figure 12.4. Future research directions include:

- Highlighting the impact of converter efficiency on overall charging station performance and grid integration.

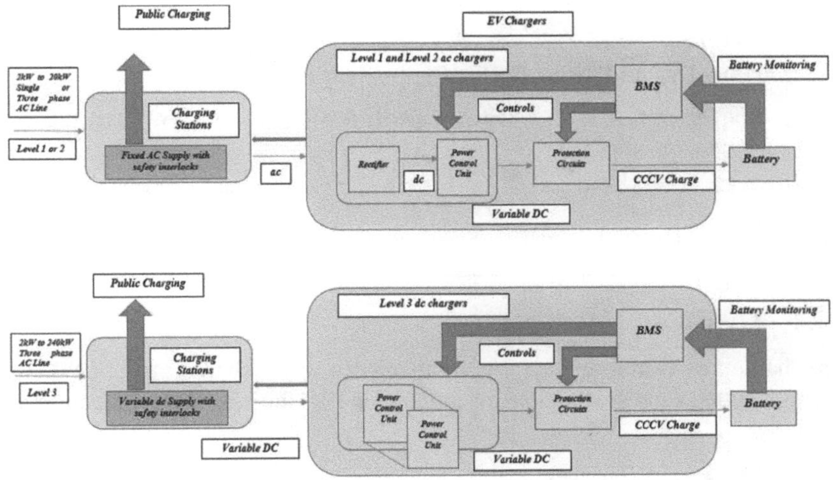

FIGURE 12.4 Block diagram of EV charging types (Tahir et al., 2021).

- Exploring the role of intelligent control algorithms in optimizing converter operation and maximizing energy transfer efficiency.
- Discussing the challenges and opportunities associated with integrating renewable energy sources with EV charging infrastructure using advanced power converter technologies.

By addressing these considerations, we can ensure the continued development of robust and efficient power converter solutions that empower the future of electric mobility and sustainabletransportation.

12.4 DIVERSITY IN EV CHARGING INFRASTRUCTURE: PUBLIC, PRIVATE, AND FLEET OPTIONS

The growing market for EVs requires a stable and diverse infrastructure for charging in order to meet the diverse needs of drivers. The three main categories of EV charging stations are covered in this chapter: fleet charging stations, private charging stations at homes and offices, and public charging stations. Every category has a specific function and is essential to the general EV adoption process. Table 12.2 describes the different e-vehicle segments and their battery characteristics.

12.4.1 Public Charging Stations

Public charging stations provide essential on-the-go charging for EV owners, fostering long-distance travel and spontaneous trips. These stations are typically located at convenient public spaces like shopping centers, gas stations, and along highways. Public stations offer a range of charging speeds, with Level 2 AC chargers being the most prevalent. However, DC fast chargers are becoming increasingly common, enabling rapid top-ups for extended journeys.

TABLE 12.2
Various EVs and Their Battery Specifications

Vehicle Segment	Battery Capacity (KWh)	Battery Voltage (V)
Electric two wheelers	1.2–3.3	48–72
Electric three wheelers	3.6–8	48–60
Electric cars (2nd Gen)	21	72
Electric cars (3rd Gen)	30–80	350–500

12.4.2 Level 2 AC Chargers

These chargers deliver faster charging compared to Level 1 chargers (commonly found household outlets) and are suitable for topping up EVs during errands or commutes.

12.4.3 DC Fast Chargers

EVCI stands for Electric Vehicle Charging Infrastructure. DC fast chargers are a critical component of this infrastructure. They are devices that deliver DC electricity to EVs, rapidly replenishing their batteries. Unlike slower AC chargers, DC fast chargers bypass the on-board AC-to-DC converter in the EV, allowing for much quicker charging times. The charging process begins when an EV is plugged into a DC fast charger. The charger then delivers high-voltage DC electricity directly to the EV's battery. This direct power transfer eliminates the need for the vehicle's internal charging system to convert AC to DC, significantly reducing charging time.

12.4.3.1 Benefits of DC Fast Chargers

Fast chargers for DCs have various benefits. First of all, when compared to AC chargers, they significantly shorten the charging period. This is essential for long-distance EV travel since it allows drivers to easily refuel and resume their trips. Furthermore, by easing range anxiety– a major worry among prospective EV buyers– DC fast charges can aid in the wider adoption of EVs. The power output of DC fast chargers ranges from 50 to 350 kW. Quicker charging times are correlated with higher power outputs. For instance, an EV's battery that is compatible with a 350 kW charger can be charged to 80% capacity in 15–30 minutes.

12.4.3.2 Charging Standards and Compatibility

DC Fast Charging standards have been developed to ensure compatibility between various EVs and chargers. AC and DC charging are supported by the widely used Combined Charging System standard. Tesla's unique connector and CHAdeMO are examples of other standards. Compatibility is a crucial factor to take into account when selecting a DC fast charger.

12.4.3.3 Charging Infrastructure and Network Development

The expansion of the EV market depends on the installation of DC fast chargers. To satisfy the growing demand for rapid charging choices, governments, charging network providers, and private companies are making significant investments in the construction of charging infrastructure. These networks of charging stations give EV users easy access to rapid chargers at a variety of sites, including public parking lots, retail centers, and rest stops along highways. The electrical grid may be impacted if DC fast chargers are widely used. Grid operators must make investments in smart charging technologies and infrastructure upgrades to handle the rising demand for electricity. Reducing the environmental impact of EV charging can also be achieved by including renewable energy sources in the infrastructure for charging.

12.4.3.4 Future Trends and Innovations

The environment for EV charging is changing quickly. It is anticipated that improvements in battery technology and infrastructure for charging will result in even quicker charging rates and greater charging capacity. Bi-directional charging technology will also enable EVs to supply grid support services by allowing them to not only consume electricity but also send it back into the system. DC rapid chargers are necessary for the widespread use of electric cars. One of the main concerns of potential EV purchasers is addressed by offering quick and easy charging choices. The future of electric transportation will be greatly influenced by DC fast chargers as the infrastructure for charging grows and technology advances.

12.4.4 Private Charging Stations at Homes and Workplaces

Home and workplace charging stations are the backbone of everyday EV use. These stations offer convenient overnight or workday charging, ensuring EVs are readily available for daily commutes. Level 2 AC chargers are most commonly installed in these settings due to their affordability and ability to fully charge most EVs overnight.

12.4.5 Level 2 AC Charger Installations

- **Home Charging:** Homeowners can install Level 2 chargers in their garages or carports, providing a seamless charging experience.
- **Workplace Charging:** Businesses can install charging stations at workplaces, incentivizing employee adoption of EVs and promoting sustainability initiatives.

12.4.6 Fleet Charging Stations

Fleet charging stations are specifically designed to replenish the power of EV fleet batteries. Similar to how traditional vehicles require refueling, EVs need to recharge to maintain operability. These stations are typically installed in commercial or industrial locations and feature multiple charge points, allowing for simultaneous charging of several vehicles. Fleet charging, also known as fleet electrification, is crucial for businesses and organizations operating large numbers of EVs, ensuring their vehicles remain ready for operation.

To facilitate efficient fleet charging, a robust charging infrastructure is essential. This involves the installation of charging stations capable of drawing power from the grid and accommodating various EV models. The infrastructure can be designed with or without existing electrical systems, depending on the specific needs of the fleet. While transitioning to EVs offers environmental and cost benefits, fleet owners must carefully consider factors such as demand charges and energy costs associated with EV infrastructure. The expansion of EV adoption is driving the growth of fleet charging stations. These stations are becoming increasingly common in public areas, commercial premises, and dedicated charging depots. This growing infrastructure ensures convenient access to charging facilities for EV fleets, enabling smooth operations and reducing range anxiety for drivers. As the number of EVs continues to rise, the availability of fleet charging stations will be vital in supporting the widespread adoption of sustainable transportation.

Fleet charging stations offer various benefits, including reduced operating costs compared to traditional fuel, decreased environmental impact, and improved brand image for eco-conscious businesses. Additionally, advancements in charging technology and the development of intelligent charging management systems are optimizing charging efficiency and maximizing the utilization of charging infrastructure. By investing in fleet charging stations, businesses can contribute to a cleaner and more sustainable future while enhancing their fleet operations.

12.4.7 Smart Charging Integration

Smart charging systems can optimize energy usage and minimize grid strain during peak hours. They can also schedule charging based on vehicle usage patterns and electricity prices.The diversification of EV charging infrastructure, encompassing public, private, and fleet options, is crucial for facilitating a smooth transition toward widespread EV adoption. Public stations provide essential accessibility for long-distance travel, while private stations at homes and workplaces cater to daily commutes. Fleet charging stations ensure the efficient operation of commercial EVs. Continued innovation in charging technology, coupled with strategic infrastructure development, will be paramount in creating a robust and user-friendly EV charging ecosystem.

12.5 THE GEOGRAPHIC DIVIDE: DISTRIBUTION PATTERNS OF EV CHARGING STATIONS AND THEIR IMPACT ON EV ADOPTION

The distribution of EV charging stations plays a critical role in shaping EV ownership and usage patterns. This chapter analyzes the current landscape of charging station deployment and its impact on consumer behavior.

12.5.1 Uneven Distribution and Range Anxiety

A significant challenge lies in the uneven distribution of charging stations. Urban areas often boast a higher concentration of public stations compared to rural or suburban regions. This disparity creates "charging deserts", areas with limited access to charging infrastructure, which can exacerbate range anxiety – the fear of running out of power before reaching a charging station. This anxiety disproportionately affects potential EV adopters residing outside major cities, hindering their willingness to switch from gasoline-powered vehicles.

12.5.2 Public vs. Private Charging

The current focus on public charging infrastructure development can overshadow the importance of private charging options. Widespread availability of Level 2 AC chargers at homes and workplaces would significantly alleviate range anxiety and encourage EV ownership, particularly for those with predictable driving patterns.

12.5.3 Impact on Usage Patterns

The distribution of charging stations influences how people utilize their EVs. Residents with access to home charging are more likely to rely on their EVs for daily commutes, leading to increased overall EV usage. Conversely, those dependent on public charging stations may restrict their EV use for longer trips or those requiring spontaneous charging needs.

12.5.4 Socioeconomic Considerations

The uneven distribution of charging infrastructure can exacerbate socioeconomic disparities. Residents in low-income neighborhoods often lack access to private parking or the resources to install home charging stations. Additionally, public charging costs can be a barrier for low-income EV owners. This necessitates targeted government policies and initiatives to ensure equitable access to charging infrastructure.

12.5.5 The Power of Data

Data-driven approaches are crucial for optimizing charging station deployment. Analyzing anonymized EV usage data can pinpoint areas with high demand, enabling more targeted infrastructure development. Additionally, studying charging patterns can inform strategies for increasing utilization of existing stations.Bridging the geographic divide in charging station availability is essential for fostering widespread EV adoption. Policies incentivizing home and workplace charging installations, coupled with strategic public station placement informed by data analysis, are key strategies. Furthermore, ensuring equitable access through targeted programs for low-income communities will be paramount in creating an inclusive EV ecosystem. The distribution patterns of EV charging stations significantly impact EV ownership and usage trends. Addressing the uneven distribution, promoting private charging options, and prioritizing data-driven infrastructure development are critical steps toward a future dominated by sustainable electric mobility.

12.6 ASSESSING CHARGING DEMAND AND SETTING TARGETS

Accurately assessing charging demand is paramount for the successful deployment of EV charging infrastructure. It provides a foundation for informed decision-making, enabling stakeholders to make strategic investments in the right locations, with the appropriate capacity, and at the optimal time. By anticipating charging needs, infrastructure providers can ensure a seamless charging experience for EV drivers, thereby accelerating EV adoption.

12.6.1 Key Factors Influencing Charging Demand

Several factors contribute to charging demand, including the number of EVs in operation, driving patterns, charging behaviors, and the availability of alternative charging options. Analyzing these elements is crucial for generating reliable demand forecasts. Geographic variations in EV penetration, population density, and charging infrastructure availability also impact demand patterns, necessitating a localized approach to assessment.

12.6.2 Data Collection and Analysis

Comprehensive data collection is essential for accurate demand assessment. This involves gathering information on EV ownership, charging station usage, traffic patterns, and demographic data. Advanced data analytics techniques can then be employed to identify trends, correlations, and patterns in charging behavior. By leveraging data-driven insights, stakeholders can gain a deeper understanding of charging demand dynamics.

12.6.3 Demand Forecasting Methodologies

Various forecasting methodologies can be applied to estimate future charging demand. These include statistical models, machine learning algorithms, and simulation-based approaches. Selecting the appropriate method depends on the available data, desired level of accuracy, and forecasting horizon. It is often beneficial to combine multiple methodologies to enhance forecast reliability.

12.6.4 Setting Charging Targets

Based on the assessed charging demand, appropriate charging targets can be established. These targets should align with broader sustainability goals, such as reducing greenhouse gas emissions and improving air quality. They should also consider the desired level of charging infrastructure penetration, taking into account factors like EV adoption rates, grid capacity, and available investment.

12.6.5 Balancing Supply and Demand

Achieving a balance between charging supply and demand is critical for optimal infrastructure utilization. Overprovisioning can lead to idle charging stations and increased costs, while underprovisioning can result in long wait times and user dissatisfaction. Dynamic pricing strategies and demand management measures can be implemented to optimize resource allocation and ensure efficient charging infrastructure utilization.

12.6.6 Incorporating Stakeholder Perspectives

Effective demand assessment and target setting require collaboration with various stakeholders, including EV manufacturers, charging point operators, grid operators, policymakers, and endusers. Engaging with these stakeholders helps to identify their needs, expectations, and challenges, leading to more informed and inclusive decision-making.

12.6.7 Continuous Monitoring and Adaptation

The EV landscape is rapidly evolving, necessitating ongoing monitoring and adaptation of charging demand assessments and targets. As EV adoption rates increase and charging technologies advance, demand patterns will change. By regularly reassessing the situation and making adjustments as needed, stakeholders can ensure that charging infrastructure remains aligned with evolving needs.By following these steps and fostering collaboration among stakeholders, it is possible to effectively assess charging demand and set ambitious yet achievable targets for the development of a robust and sustainable EV charging infrastructure.

12.7 LOCATION PLANNING AND ALLOCATION FOR EVCI

Effective location planning and land allocation are crucial for the successful deployment of EVCI. Key considerations include identifying strategic locations with high EV traffic, accessibility, and proximity to urban centers. Land requirements vary based on charging station capacity and ancillary facilities. Factors such as land ownership, zoning regulations, and grid capacity must be evaluated. Public–private partnerships can facilitate land acquisition and infrastructure development. Additionally, cost-benefit analysis should be conducted to optimize land utilization and investment returns. Leveraging existing infrastructure like parking lots, gas stations, and commercial buildings can streamline the deployment process. Ultimately, a well-planned EVCI network is essential for accelerating EV adoption and reducing range anxiety.

12.7.1 Location Selection Criteria for EVCI

The effective uptake of electric cars depends on the location of EV charging facilities. Important considerations for choosing a location include things like accessibility, traffic patterns, population density, and distance from residential and commercial areas. EVCI implementation is most appropriate in high-traffic corridors, public transit hubs, and places with a high population of electric car owners. Optimizing the efficiency and sustainability of the charging network also requires taking into account the possibility for integration of renewable energy sources and the availability of grid infrastructure.

12.7.2 Location Allocation for EVCI

Once potential locations have been identified, a rigorous allocation process is necessary to determine the optimal placement of EV charging stations. This involves analyzing factors like charging demand, competition from existing charging infrastructure, and the cost of land and installation. Geographic Information Systems can be employed to visualize and analyze spatial data, aiding in identifying areas with the highest charging demand and potential return on investment. A balanced approach, considering both urban and rural areas, is crucial to ensure equitable access to charging facilities.

12.7.3 Design Considerations for EVCI

The design of EV charging stations should prioritize user convenience and experience. Factors such as charging speed, connector types, payment options, and lighting should be carefully considered. The physical layout of the charging station, including the number of charging points, parking spaces, and accessibility for people with disabilities, is also essential. Additionally, the integration of smart charging technologies can optimize grid utilization and reduce charging costs. Robust and reliable power supply infrastructure is paramount to prevent charging interruptions and ensure the smooth operation of the EVCI.

12.7.4 EVCI and Smart Grid Integration

Infrastructure for EV charging has a significant potential impact on the advancement of smart grids. Electric cars can function as energy storage devices by combining bi-directional charging capabilities, supporting the grid during periods of high demand. Furthermore, incorporating renewable energy sources into the EVCI can support sustainable transportation and lower carbon emissions. Advanced metering infrastructure and real-time data analytics are necessary to track charge trends, enhance grid performance, and spot demand response program prospects.

12.8 ACHIEVING EFFECTIVE EV–GRID INTEGRATION

The integration of EVs into the power grid is a complex challenge that requires careful planning and coordination. As EV adoption surges, the grid faces increasing strain due to the additional electricity demand for charging. To ensure a seamless transition, it's crucial to balance the needs of EV owners with the grid's capacity and stability.

12.8.1 The Role of Smart Charging

Smart charging is a cornerstone of effective EV–grid integration. By optimizing charging times based on grid conditions, electricity costs, and driver preferences, smart charging helps to distribute the load more evenly. This reduces the risk of grid overloads during peak demand periods, enhancing grid reliability.

12.8.2 Leveraging V2G Technology

The technology known as V2G is a paradigm change in the energy sector. With V2G, bi-directional energy flow between the vehicle and the power grid is made possible by converting electric cars (EVs) into mobile energy storage units. This novel idea has enormous potential to improve grid stability, include renewable energy, and reward EV owners monetarily. Fundamentally, V2G technology enables EVs to feed extra energy back into the grid during times of high demand in addition to using it to take electricity from it for charging. Bi-directional chargers, which aid in the transmission of energy, enable this. Grid operators can boost overall grid dependability, optimize energy distribution, and lower peak load by efficiently controlling the charging and discharging processes.The potential of V2G to facilitate the incorporation of renewable energy sources is one of its biggest benefits. The intermittent nature of solar and wind power generation causes variations in the supply of electricity on the grid. When there is an excess of renewable energy produced, V2G-enabled EVs may store it and release it when demand is high, stabilizing the grid and optimizing the use of clean energy. Additionally, V2G technology can offer EV owners significant financial advantages. EV owners can make money by selling electricity back to the grid by taking part in V2G initiatives. This extra source of income can help defray the cost of car ownership and encourage the use of EVs. Additionally, V2G can lower electricity costs by enabling smart charging, which allows EVs to be charged off-peak hours when energy prices are lower.

Nonetheless, there are a number of obstacles in the way of V2G technology's broad implementation. It is vital to ensure that EV batteries are safe and long-lasting by charging and discharging them frequently. To further enable energy trading between EVs and the grid, strong market mechanisms and communication infrastructure must be developed. There is no denying that V2G has potential advantages despite these difficulties. In order to create a sustainable and resilient energy future, V2G is expected to be essential as EV adoption and grid modernization advance. We can improve grid efficiency, lower carbon emissions, and generate new revenue streams by utilizing EVs as distributed energy resources. A potential answer to the problems with incorporating renewable energy and controlling grid fluctuations is provided by V2G technology. V2G can accelerate the shift to a cleaner and more efficient energy system by enabling EV owners to actively engage in the energy market. Future developments in technology and market dynamics should contribute to the growing significance of V2G in the energy landscape.

12.8.3 Grid Infrastructure Upgrades

To accommodate the growing EV charging demand, grid infrastructure upgrades are essential. This involves expanding the distribution network, installing new substations, and strengthening grid components. By investing in grid modernization, utilities can ensure a robust and reliable power supply for both EV charging and other electricity consumers.

12.8.4 Demand Response Programs

Demand response programs encourage EV owners to modify their charging habits during periods of high demand. Utilities may lessen grid stress by encouraging EV users to move charging to off-peak hours by providing cash prizes or other incentives.

12.8.5 Renewable Energy Integration

Combining EV charging with renewable energy sources can start a positive feedback loop for sustainability. Transportation has a far less carbon footprint when EVs are charged with renewable power. Additionally, EVs can store extra renewable energy produced when demand is low, which improves grid stability even more.

12.8.6 Data Management and Analytics

Data plays a crucial role in effective EV–grid integration. By collecting and analyzing data on EV charging patterns, grid conditions, and electricity prices, utilities can make informed decisions about grid management and infrastructure planning. Advanced data analytics can also identify opportunities for optimizing charging schedules and maximizing the benefits of V2G technology.

12.8.7 Policy and Regulatory Framework

A supportive policy and regulatory environment is essential for accelerating EV adoption and grid integration. Governments can play a vital role by providing incentives for EV purchases, investing in charging infrastructure, and establishing clear guidelines for grid operators and EV manufacturers.

12.8.8 Public Awareness and Education

Fostering widespread adoption of EV–grid integration requires increasing public understanding of its benefits. Providing EV owners with information on demand response programs, smart charging, and V2G technologies can enable them to take an active role in grid management. By putting these plans into practice and encouraging cooperation between utilities, EV producers, legislators, and end users, we can make it possible for EVs and the grid to coexist peacefully in the future, enabling clean energy and sustainable mobility.

12.9 EVCI MODEL PLANNING AND IMPLEMENTATION

EVCI model planning involves a comprehensive assessment of a region's EV adoption potential, charging infrastructure needs, and available resources. Key steps include conducting a thorough demand analysis to identify charging hotspots, evaluating grid capacity and integration requirements, and developing a strategic roadmap for infrastructure deployment. Stakeholder engagement is crucial, encompassing collaboration with government agencies, utilities, private sector players, and the public to align interests and secure necessary permits and approvals. Financial planning, including identifying funding sources and developing sustainable business models, is essential for the long-term viability of the EVCI.

12.9.1 EVCI Model Implementation

EVCI model implementation focuses on the physical establishment of charging stations. This phase involves site selection based on demand analysis, grid capacity, and accessibility. Installation and commissioning of charging equipment follow, ensuring compatibility with different EV models and adherence to safety standards. Grid upgrades may be necessary to accommodate increased electricity demand. Smart charging technologies can be integrated to optimize grid utilization and reduce charging costs. Ongoing monitoring and maintenance are crucial to ensure optimal performance and user satisfaction. Public awareness campaigns can promote EV adoption and educate users about charging etiquette.

12.9.2 EVCI Model Case Studies

Case studies offer valuable insights into successful EVCI models. Cities like Oslo, Norway, and Amsterdam, Netherlands, have implemented extensive charging networks through public–private partnerships, combining government incentives with private investments. These cities have prioritized convenient charging locations, integrated smart charging technologies, and offered additional EV benefits to encourage adoption. On the other hand, challenges faced by certain regions highlight the importance of addressing factors like grid limitations, charging costs, and user concerns. Analyzing these case studies helps identify best practices and potential pitfalls for future EVCI deployments.

12.9.3 EVCI Model Evaluation and Improvement

Continuous evaluation is essential to assess the effectiveness of the EVCI model. Key performance indicators include charging station utilization, user satisfaction, grid integration performance, and economic viability. Data analysis helps identify areas for improvement, such as optimizing charging station locations, adjusting pricing strategies, and enhancing user experience. Feedback from EV users and stakeholders is valuable for refining the model. As EV technology and market dynamics evolve, the EVCI model must adapt accordingly to stay relevant and support the transition to sustainable transportation.

12.10 CHALLENGES AND LIMITATIONS OF CURRENT EVCI

There is great potential for lowering reliance on fossil fuels and mitigating climate change from the shift to EVs. Inadequate EVCI stands in the way of a widespread adoption of EVs, nevertheless. This study examines three major issues brought on by insufficient EVCI: geographical variations in station availability, charging speed restrictions, and range anxiety.

12.10.1 Range Anxiety: A Psychological Hurdle

For prospective EV adopters, range anxiety– the worry of running out of battery power before reaching a charging station– is a major obstacle. A small public charging network and a car's shorter driving range than a gasoline-powered vehicle may put off buyers, especially those who are used to the convenience of petrol stations. Those who live in places with poor charging infrastructure or frequently travel long distances are disproportionately affected by this fear.

12.10.2 The Charging Speed Dilemma: Balancing Convenience and Efficiency

While Level 2 AC chargers are readily available and suitable for overnight charging at homes or workplaces, their charging times can be lengthy, typically ranging from several hours to overnight. This can be inconvenient for individuals requiring a quick top-up during commutes or spontaneous trips. Conversely, DC fast chargers offer significantly faster charging, but their current limitations include:

- **Lower Ubiquity:** Compared to Level 2 chargers, DC fast chargers are less prevalent, particularly along highways and in rural areas.
- **Battery Compatibility Issues:** Not all EVs are compatible with the highest charging speeds offered by DC fast chargers.
- **Potential Battery Degradation:** Frequent use of DC Fast Charging can potentially accelerate battery degradation, impacting an EV's overall lifespan.

12.10.3 The Geographic Divide: Urban Oasis, Rural Desolation

The current distribution of EV charging stations exhibits a significant geographical disparity. Urban areas tend to boast a higher concentration of public charging stations, catering to a denser population and higher EV adoption rates. However, rural and remote regions often face a dearth of charging infrastructure, creating "charging deserts". This uneven distribution exacerbates range anxiety for potential EV owners residing in these underserved areas, hindering their transition to electric mobility.

Insufficient EVCI poses a multifaceted challenge to widespread EV adoption. Addressing range anxiety, optimizing charging speeds, and ensuring equitable geographical distribution of charging stations are crucial steps toward a robust and user-friendly EV charging ecosystem. By implementing targeted policies, fostering technological advancements, and promoting data-driven infrastructure development, we can close the EVCI gap and open the door to a future of environmentally friendly transportation.

12.11 ADVANCEMENTS IN EV CHARGING TECHNOLOGY

The EV revolution hinges on a user-friendly charging experience. While current infrastructure offers some solutions, limitations like charging times and range anxiety remain. This chapter explores two emerging technologies with the potential to revolutionize the EV charging landscape: ultra-fast charging and battery swapping stations.

12.11.1 Ultra-Fast Charging: Powering Up in Minutes, Not Hours

The goal of ultra-fast charging methods is to shorten charging times considerably as compared to the quickest DC fast chargers. Some of the prototypes of these systems have reached an astounding 1 MW (1,000 kW), operating at power levels above 350 kW. This means that big battery top-ups may need as little as 15 minutes to charge, which would be just as convenient as filling up a car with gasoline.

12.11.2 Benefits of Ultra-Fast Charging

- **Reduced Range Anxiety:** By enabling rapid charging, ultra-fast technology can significantly alleviate range anxiety, making EVs more appealing for long-distance travel and spontaneous trips.
- **Increased Station Utilization:** Faster charging times can lead to increased utilization of existing charging stations, as shorter charging sessions free up stations for more EVs.
- **Enhanced Grid Integration:** Advancements in battery management systems and smart charging technologies can ensure efficient grid integration during ultra-fast charging processes.

12.11.3 Challenges and Considerations

- **Infrastructure Investment:** Developing and deploying ultra-fast charging infrastructure requires significant investment in high-power equipment and grid upgrades.
- **Battery Compatibility:** Current battery technologies may not be fully optimized for ultra-fast charging speeds. Collaborations between EV manufacturers and charging infrastructure providers are crucial for ensuring compatibility.
- **Thermal Management:** Ultra-fast charging generates significant heat, necessitating robust thermal management systems to prevent battery degradation.

12.11.4 Battery Swapping Stations: A Speedy Alternative

Globally, the shift towardEVs is gathering pace; nonetheless, a significant obstacle to their extensive implementation has been the duration of charging. Even with the improvements in infrastructure, charging still takes a lot longer than refueling a conventional gasoline car. Battery switching stations provide a viable way to overcome this difficulty. A battery switching station is essentially a location where EV owners can easily swap out a depleted battery for a fully charged one. This procedure cuts down on charging time dramatically and may be finished in a matter of minutes, much like filling up a gas-powered automobile. For business fleets, such as delivery services or taxis, where vehicle uptime is critical, this quick interchange is especially advantageous. Range anxiety is a prevalent issue among prospective EV purchasers, and one of the main benefits of battery switching is its potential to alleviate it.

By offering a network of switching stations, drivers may travel farther with confidence and assurance that they won't run out of fuel. Furthermore, since the batteries in the swapping pool may be better maintained and monitored, battery switching can assist in maximizing battery life.Battery switching stations have complicated technology. Accurate alignment, safe connections, and effective battery management systems are necessary. The stations must follow strict guidelines to guarantee safety in order to avoid problems like battery fires. In addition, the quantity of swapping stations, the cost of batteries, and the service's price structure all affect how profitable battery swapping stations may be. Battery changing has a lot of possibilities, but it also has drawbacks. Interoperability between various EV models and battery types depends on standardization. Building and maintaining a network of exchanging stations may also be quite expensive. However, the infrastructure for battery swapping is expected to grow due to ongoing technical developments and rising EV demand. For electric cars, battery swapping stations provide a strong substitute for conventional charging. They can speed up the adoption of EVs by solving range anxiety and drastically cutting down on charging times. Notwithstanding the challenges, this technology has great promise for the development of electric transportation in the future.

12.11.5 Benefits of Battery Swapping Stations

- **Unparalleled Convenience:** Swapping batteries offers the fastest turnaround time, ideal for busy individuals or long-distance journeys.
- **Standardization Potential:** Standardized battery formats could facilitate interoperability between different EV models and swapping station providers.
- **Battery Health Optimization:** By decoupling charging from the vehicle, battery swapping stations can potentially optimize battery health and lifespan.

12.11.6 Challenges and Considerations

- **Standardization Hurdles:** Achieving widespread adoption hinges on standardization of battery sizes and connectors across different EV manufacturers.
- **Logistics and Costs:** Deploying and maintaining a network of swapping stations requires extensive logistical planning and significant initial investment.
- **Second-Life Battery Management:** Strategies for effectively managing and redeploying used batteries from swapped EVs are crucial for sustainability considerations.

Ultra-fast charging and battery swapping stations hold immense promise for revolutionizing the EV charging experience. While each technology presents unique challenges, their potential to address range anxiety and enhance user convenience is undeniable. Continued research, collaboration between stakeholders, and strategic infrastructure development will determine the pace and extent of their integration into the EV ecosystem. As these technologies evolve, the future of EV charging appears increasingly bright, paving the way for a more sustainable and convenient transportation landscape.

12.12 THE ROLE OF GOVERNMENT AND POLICY

A reliable and easily accessible EVCI is necessary for the mass deployment of EVs. Incentives and policies from the government are essential for advancing EVCI development and addressing current obstacles. This study investigates the ways in which financial incentives, legal requirements, and R&D expenditures might hasten the adoption of EVCIs.

12.12.1 Financial Incentives: Electrifying Consumer Choice

Financial subsidies can act as a powerful catalyst for EVCI development by influencing consumer behavior and encouraging private investment. Here are two key strategies:

- **EV Purchase Subsidies:** Government subsidies that reduce the upfront cost of EVs can incentivize consumers to switch from gasoline vehicles, thereby increasing demand for charging infrastructure.
- **Charging Station Installation Subsidies:** Financial assistance for installing charging stations at homes, workplaces, and public locations can stimulate private investment in EVCI development. This can encompass rebates for home charger installations, tax credits for businesses installing workplace chargers, and grants for public charging station development projects.

12.12.2 Regulatory Frameworks: A Nudge in the Right Direction

Regulations and mandates can create a supportive environment for EVCI development by placing the onus on relevant stakeholders:

- **Utility Regulations:** Regulatory frameworks can incentivize electric utilities to invest in grid modernization and expand charging infrastructure. This could involve mandating utilities to allocate a specific percentage of their profits toward EVCI development or setting targets for the number of charging stations to be installed within their service areas.
- **Business Mandates:** Regulations requiring businesses with large parking lots (shopping malls, office buildings) to allocate a certain percentage of parking spaces for EV charging can significantly expand charging availability in convenient locations.

12.12.3 Investing in Innovation: Powering the Future of EVCI

Government investment in R&D holds immense potential for accelerating the development of next-generation charging technologies:

- **Ultra-Fast Charging Technologies:** Supporting research on ultra-fast charging technologies that can significantly reduce charging times is crucial for mitigating range anxiety and enhancing user convenience.
- **Battery Swapping Infrastructure:**Investing in R&D for standardized battery swapping systems can foster innovation and expedite the deployment of this potentially transformative technology.
- **Smart Charging Technologies:** Government funding for research on smart charging technologies can optimize grid integration during charging processes, ensure efficient energy usage, and minimize peak demand.

Government policies and incentives play a pivotal role in accelerating EVCI development. By implementing a combination of financial subsidies, regulations, and R&D investments, governments can create an environment conducive to private sector participation and technological innovation. This collaborative approach is essential for overcoming existing challenges and fostering a future dominated by sustainable electric mobility.

12.13 FUTURE DIRECTIONS FOR EVCI

The burgeoning EV market necessitates a future-proof EVCI capable of supporting widespread adoption and catering to diverse user needs. This chapter explores key areas demanding focus, including standardization, grid modernization, innovative solutions for urban environments, and strategies for facilitating long-distance travel.

12.13.1 Standardization: The Cornerstone of Interoperability

Standardization of charging protocols and connector types is paramount for ensuring seamless user experience and maximizing EVCI utilization. A lack of standardization creates compatibility issues, potentially deterring consumers due to concerns about limited charging options. Collaborative efforts between governments, automakers, and charging infrastructure providers are crucial for establishing universal standards.

- **Global Harmonization:** International cooperation can pave the way for globally recognized charging standards, simplifying EV travel across borders.
- **Plug-and-Play Convenience:** Standardized connectors should enable effortless charging across different EV models and charging stations.

12.13.2 Grid Modernization: Powering the Future of Mobility

Existing electrical grids will unavoidably be strained by the widespread use of EVs. In order to meet the growing demand and provide a consistent power supply for charging stations, grid modernization initiatives are crucial.

- **Smart Grid Integration:**A smart grid is an updated electrical system that makes use of cutting-edge technology to increase electricity delivery's sustainability, dependability, and efficiency. Fundamentally, it entails the integration of automation, control, and digital communication technologies throughout the whole value chain of electricity, from production to consumption. The capacity of a smart grid to enable two-way communication between utilities and customers is its fundamental component. This makes it possible to monitor power flow, demand variations, and system performance in real time. Sophisticated sensors and energy meters gather information on energy consumption, enabling detailed examination and enhancement. Furthermore, this data is processed by clever algorithms that use it to determine the best course of action for the production, distribution, and use of electricity. The ability of smart grid integration to handle the growing share of renewable energy sources is a major benefit. The grid becomes more robust and ecologically beneficial when solar, wind, and other renewable energy sources are integrated efficiently. Furthermore, smart grids enable users to actively participate in energy management by implementing demand response initiatives. Consumers might possibly lower their energy bills and help to system stability by managing their patterns of power usage. Moreover, smart grid technologies improve resilience and dependability of the system. Modern fault isolation and detection systems allow for quick outage response, reducing consumer impact. The frequency and length of blackouts can be decreased by automatically rerouting electricity around damaged grid segments thanks to self-healing capabilities. In the end, smart grid integration opens the door to an electrical system that is more consumer-focused, sustainable, and efficient. Smart grids are revolutionizing the production, distribution, and consumption of electricity by utilizing digital technology. This paradigm shift is necessary to provide a safe energy future, mitigate climate change, and fulfill the world's expanding energy demands.

- **Investment in Upgradation:** Significant investment is required in grid infrastructure expansion and upgrading transmission lines to handle the increased power needs of EVs.

12.13.3 Urban Innovation: Charging Solutions for Dense Environments

Urban environments pose unique challenges for EVCI development. Limited space necessitates innovative charging solutions that optimize space utilization and cater to a high concentration of EVs.

- **On-Street Lamppost Charging:** Integrating charging points into existing lampposts can provide a convenient and space-saving solution for urban areas.
- **Wireless Charging Technologies:** Wireless charging pads embedded in parking spaces offer a contactless and user-friendly charging experience, potentially revolutionizing urban EV charging.

12.13.4 Long-Distance Travel: Addressing Range Anxiety on the Road

Facilitating long-distance travel with EVs requires a well-distributed network of high-powered charging stations along highways and major travel corridors. Additionally, exploring alternative solutions can further enhance user confidence and range capabilities.

- **Ultra-Fast Charging Infrastructure:** Deployment of ultra-fast charging stations along key routes can significantly reduce charging times, making long journeys more feasible.
- **Battery Swapping Stations:** Strategic placement of battery swapping stations can offer a rapid alternative for long-distance travel, reducing reliance on lengthy charging sessions.

12.14 CONCLUSION

The future of EVCI necessitates a multi-pronged approach. Standardization will ensure user convenience, grid modernization is crucial for supporting widespread adoption, urban solutions will cater to dense environments, and long-distance travel considerations will alleviate range anxiety. By continuously innovating and implementing strategic policies, we can create a robust and user-friendly EVCI ecosystem that propels us toward a sustainable transportation future. This future necessitates collaboration between stakeholders, continued research and development, and a commitment to building a charging infrastructure that empowers seamless EV adoption across diverse driving needs and geographies.

REFERENCES

Andrenacci, N., Karagulian, F. and Genovese, A., Modelling charge profiles of electric vehicles based on charges data, *Open Research Europe*, 1, 156, 2022.

Asensio, O. I., Cade Lawson, M. and Apablaza, C. Z., Electric vehicle charging stations in the workplace with high-resolution data from casual and habitual users, *Scientific Data*, 8, 168, 2021.

Chen, N., Tan, C. W. and Quek, T. Q. S., Electric vehicle charging in smart grid: Optimality and valley-filling algorithms, *IEEE Journal of Selected Topics in Signal Processing*, 8, 1073–1083, 2014.

Ferguson, B., Nagaraj, V., Kara, E.C. and Alizadeh, M., Optimal planning of workplace electric vehicle charging infrastructure with smart charging opportunities, *21st International Conference on Intelligent Systems*, Maui, HI, 2018.

García-Afonso, O., Santana-Méndez, I., Delgado-Torres, A. M. and González-Díaz, B., On the road to a sustainable transport mobility in isolated power systems: The role of light-duty powertrain electrification, *Journal of Cleaner Production*, 320, 128646, 2021.

Girardi, P., Gargiulo, A. and Brambilla, P. C., A comparative LCA of an electric vehicle and an internal combustion engine vehicle using the appropriate power mix: The Italian Case Study, *The International Journal of Life Cycle Assessment*, 20, 1127–1142, 2015.

Heredia, W. B., Chaudhari, K., Meintz, A., Jun, M. and Pless, S., Evaluation of smart charging for electric vehicle-to-building integration: A case study, *Applied Energy*, 266, 114803, 2020.

Islam, M., Shareef, H. and Mohamed, A., A review of techniques for optimal placement and sizing of electric vehicle charging stations, *Przegląd Elektrotechniczny*, 91, 122–126, 2015.

Jones, B., Elliott, R. J. R. and Nguyen-Tien, V., The EV revolution: The road ahead for critical raw materials demand, *Applied Energy*, 280, 115072, 2020.

Liu, F., Zhao, F., Liu, Z. and Hao, H., China's electric vehicle deployment: Energy and greenhouse gas emission impacts, *Energies*, 11, 3353, 2018.

Madina, C., Zamora, I. and Zabala, E., Methodology for assessing electric vehicle charging infrastructure business models, *Energy Policy*, 89, 284–293, 2016.

Mobarak, M. H. and Bauman, J., Vehicle-directed smart charging strategies to mitigate the effect of long-range EV charging on distribution transformer aging, *IEEE Transactions on Transportation Electrification*,5, 1097–1111, 2019.

Mouhrim, N., Alaoui, A. E. H.and Boukachour, J., Pareto efficient allocation of an in-motion wireless charging infrastructure for electric vehicles in a multipath network, *International Journal of Sustainable Transportation*, 13, 419–432, 2018.

Ramachandran, S. and Stimming, U., Well to wheel analysis of low carbon alternatives for road traffic, *Energy and Environmental Science*, 8, 3313–3324, 2015.

Roy, A. and Law, M., Examining spatial disparities in electric vehicle charging station placements using machine learning, *SSRN Electronic Journal*, 83, 103978,2022.

Shrestha, S., Baral, B., Shah, M., Chitrakar, S. and Shrestha, B., Measures to resolve range anxiety in electric vehicle users, *International Journal of Low-Carbon Technologies*,17, 1186–1206, 2022.

Singh, R., Kumar, A., Singh, A. R., Naidoo, R., Bansal, R. C. and Kumar, P., Environmental feasibility of incorporation of electric taxis in South Africa, *The Journal of Engineering*, 2019, 5078–5084, 2019.

Sun, Y., Hu, X., Liu, X., He, X. and Wang, K., A software-defined green framework for hybrid EV-charging networks. *IEEE Communiation Magazine*, 55, 62–69, 2017.

Tahir, Y., et al., State-of-the-art review on topologies and control techniques of solid-state transformers for electric vehicle extreme fast charging, *IET Power Electron*. 14, 1560–1576, 2021.

Tang, D., Wang, P. and Wu, Q., Probabilistic modeling of nodal electric vehicle load due to fast charging stations, *International Conference on Probabilistic Methods*, Beijing, China, 2016.

13 Compact Electric Tractor Charging at the Farm

N. Subramani, S. Allirani, D. Logesh, K. Shobika, J. Madhan Kumar, and B. Kaviya

13.1 INTRODUCTION

The increasing demand for emission reduction and energy efficiency improvements around the globe has highlighted the importance of EV technology in reducing the environmental effect of transportation. Although EVs hold great potential for a cleaner and more sustainable future, a major obstacle to their broad acceptance is the scarcity of EV charging stations. This problem is especially serious in the agriculture industry, where switching to electric tractors has the potential to significantly reduce reliance on fossil fuels. With a focus on the special requirements of electric tractors in agricultural settings, this project intends to solve the urgent problem of EV charging infrastructure. The suggested remedy focuses on the creation of a novel Electric Tractor Charging Architecture that makes use of a well-balanced combination of Conventional and Renewable energy sources. This strategy aims to guarantee a steady and effective power supply for electric tractor charging stations by fusing the sustainability of renewable energy sources with the dependability of traditional energy sources. The incorporation of solar or wind energy into the infrastructure for charging is in line with the objectives of global sustainability and helps to lower the carbon footprint associated with agriculture.

13.2 FIELD SURVEY

A comprehensive field survey was conducted to gather feedback from a broad spectrum of stakeholders within the agricultural domain, including farmers, agriculture-based machinery and automobile manufacturers, agricultural farm fields at various areas across Tamil Nadu such as Coimbatore and Theni, and the Tamil Nadu Agricultural University Agro-Machinery Department (shown in Figures 13.1 and 13.2).

This deliberate endeavor sought to compile many viewpoints regarding the use of agricultural machinery, difficulties farmers encounter, and areas open to innovation. Farmers shared their perspectives on the pricing, upkeep, usability, and performance of both conventional and next-generation automotive machinery used on farms. Manufacturers of automobiles and agricultural machines provided insights on the state of technology, new trends, and development obstacles. The Tamil Nadu Agricultural University Agro-Machinery Department and Thudiyalur Co-operative Agricultural Services Ltd. consultations added scholarly knowledge and research perspectives to the survey like tractor usage of different types of farmers, operating time and fuel consumption, and different costs associated. Stakeholders may

DOI: 10.1201/9781003496946-13

FIGURE 13.1 Field survey at Tamil Nadu Agricultural University.

FIGURE 13.2 Outlooks of the field survey.

prioritize research projects, pinpoint areas for improvement, and encourage innovation in agricultural EV solutions by combining these varied points of view. This cooperative project emphasizes how crucial inclusive stakeholder interaction is to bringing about constructive change in the agriculture industry.

13.3 STAKEHOLDER FEEDBACK

The survey uncovered a number of urgent problems that stakeholders in the farm machinery industry were facing. Among the main issues mentioned were high fuel expenses, a short operating range, frequent troubleshooting needs, the demand for frequent oil changes, the need for mechanical part replacement, and the general lack of affordability (Figure 13.3).

Stakeholders also stated that they wanted rental services to help with financial concerns. Regarding expectations, stakeholders stressed the need for machinery to have a long lifespan and be durable, as well as the necessity of lowering range anxiety and guaranteeing reliability, all of which are included in Figure 13.4. Similar to the adaptability provided by USB "C" connectors, they likewise emphasized the significance of adaptability to different instruments and settings. Stakeholders also pushed for outcome-based research to produce workable solutions that give farmers real value.

They emphasized how crucial it is to support fresh ideas that raise agricultural output and raise the standard of living in farming communities. The field survey yielded priceless information that will direct future efforts in the farm machinery sector to meet stakeholder expectations and tackle these issues.

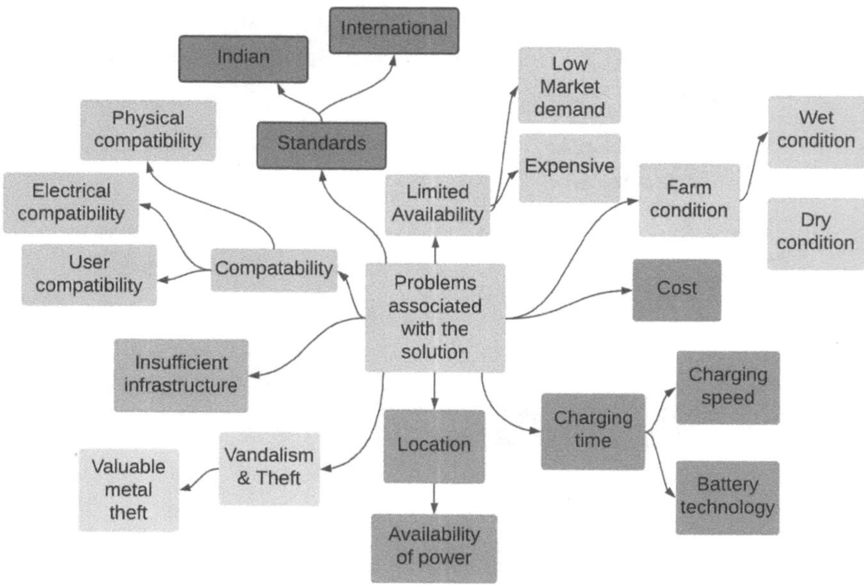

FIGURE 13.3 Problems associated with the solution development.

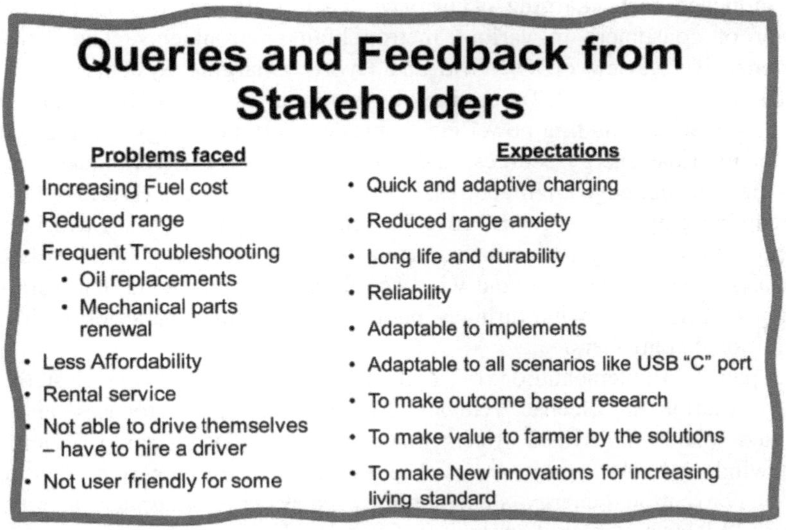

FIGURE 13.4 Queries and feedback from stakeholders.

13.4 LITERATURE SURVEY

The study "A Hybrid Energy Hub Investigation with Renewables and Electric Vehicle in a Smart Microgrid Lab" (Riccardo et al., 2021) highlights how microgrids can improve energy efficiency by encouraging decentralized energy distribution. The paper "Electric Tractors: Survey of Challenges and Opportunities in India" (Malik and Kohli, 2020) provides an overview of the Indian tractor market and highlights the steps taken by the government to reduce air pollution caused by farm equipment. It also suggests that adding electric drives to tractors and other agricultural equipment has a number of benefits, such as increased functionality and improved energy economy. The paper "An Introduction to Electric Tractor" (Abdul Lateef Dar, 2022) describes how the design of electric tractors that run on direct current (DC) batteries has revolutionized agricultural methods. Farmers' farming operations become simpler with the removal of fuel-related complications, such as the logistics of refueling and engine maintenance. Farmers find that using an electric tractor makes farming their property easier and more cost-effective. The study "An In-Depth Analysis of Electric Vehicle Charging Station Infrastructure, Policy Implications, and Future Trends. Energy Reports" (Mastoi et al., 2022) examines efforts to standardize the infrastructure, issues associated with it, and research and development linked to charging station infrastructure in order to improve research in the future.

The forecasting strategy that the study "Electric Vehicle Charging Hub Power Forecasting: A Statistical and Machine Learning Based Approach. Energies" (Lo Franco et al., 2023) attempts to take into account all the major aspects that could influence each charging occurrence. Their statistical pattern predicts the behavior of consumers in various metropolitan environments. The article "A Comprehensive Review on Off-Grid and Hybrid Charging Systems for Electric Vehicles" (Rituraj et al., 2022) examines the many subsystems of off-grid and hybrid charging systems, including power electronics converters, energy storage systems, single or multiple energy sources, and energy management techniques, and discusses their cutting-edge architectures. In the work "Design of a Hybrid Solar-Wind Powered Charging Station for Electric Vehicles" (Li et al., 2013), a hybrid solar–wind-powered charging station was created to supply electricity for EVs based on Tangshan's coastal areas' sun and wind conditions. Key elements that were taken into consideration were wind turbines, photovoltaic modules, batteries, an inverter, and additional controllers.

The paper "Implementation of a Grid-connected Solar-Wind System with Charging Station for Electric Vehicles" (Yadav et al., 2022) focuses on a grid-connected solar–wind hybrid system that includes an EV charging station. Grid power, wind, and solar energy are used to power the charging system. Our dependency on conventional energy is decreased by the system's integrated operation. In the work "Electric Vehicle Community Charging Hubs in Multi-Unit Dwellings: Scheduling and Techno-Economic Assessment" (Zhang et al., 2023), they present the idea of communal charging facilities for multi-unit dwellings, where chargers can be shared. Their design reduces the overall waiting time at the charging station. In Chicago, IL, New York City, NY, and Los Angeles, CA, they conduct a techno-economic study to gauge the charging hub's effectiveness and analyze the levelized cost of charging. In the study, "Smart Charging Strategy for Electric Vehicle Charging Stations" (Moghaddam et al., 2017), they describe a smart charging strategy in this study for a plug-in vehicle network that provides several charging alternatives, such as battery swapping capabilities at charging stations, DC rapid charging, and alternating current (AC) level 2 charging. When a plug-in vehicle needs access to charging facilities, the problem of locating the best charging station is modeled as a multi-objective optimization problem. The objective is to choose a station that minimizes the costs associated with charging, trip time, and charging time. The work "Modeling and Control of Hybrid Electric Vehicles: A Case Study for Agricultural Tractors" (Jia et al., 2018) introduces a high-fidelity forward model for series hybrid electric agricultural tractors. Two benchmarking rule-based power management strategies, thermostat and power follower control, are designed and investigated to address the control issue brought on by the mixed power loads and complex tractor-implement operations.

The study "Novel Grid-Connected Solar/Wind Powered Electric Vehicle Charging Station with Vehicle-To-Grid Technology" (Fathabadi, 2017) describes the design and construction of a new vehicle-to-grid, grid-connected solar and wind-powered EV charging station. A photovoltaic array, a wind energy conversion system, two unidirectional (DC)/(DC) converters connected to the photovoltaic array and wind energy conversion system, a unified maximum power point tracking controller, fifteen bidirectional (DC)/(DC) converters connected to the fifteen charging stations allocated for EV charging, and a three-phase bidirectional (DC)/(AC) inverter connected to the grid make up the solar/wind-powered EV charging station. In the work "Modeling of Hybrid Wind Photovoltaic System with Energy Storage for Rural Application" (Manojkumar et al., 2014), the hybrid wind–photovoltaic standalone system simulation utilized for distant applications as standalone systems is listed. Power demand, voltage, and current are the factors that determine the best way to transfer power from the sources. Setting the DC bus voltage to a set value yields the best overall system control. MATLAB®/Simulink® was used to model the suggested hybrid system, and the outcomes were displayed.

An updated assessment of the use of electric tractors is presented in the work "Conception of an Electric Propulsion System for a 9-kW Electric Tractor Suitable for Family Farming" (Melo et al., 2019). Also presented and discussed is a bespoke drive system for the idea of a unique low-power electric tractor ideal for family farms. Aspects of energy generation, transmission, conversion, storage, use, conservation, and management are all covered by the system that has been introduced, along with sustainability-related issues. A 9-kW prototype is given, consisting of two lead-acid battery banks, two independent inverters, and three-phase induction motors. The article "Integration of PV and Wind Energy with Grid and to Charge Electric Vehicles Battery" (Bind and Gupta, 2023) suggests an unconventional hybrid energy system for charging the batteries of EVs that combines solar and wind power. A 1.5 kW solar PV system is paired with a self-excited, 2.5 kW squirrel cage induction generator powering the wind system. The paper "Electrification of Agricultural Machinery: A Review" (Scolaro et al., 2021) provides a thorough analysis of the electrification of agricultural machinery, concentrating on hybrid electric tractors and their accessories. By examining the key conclusions from the literature and taking into account cutting-edge technology, the introduction of electric drives in farming tractors is covered in detail. Economic analyses, upcoming trends, and manufacturer proposals and prototypes are also examined. Solar Charged Electric Farming Tractors, a masters' thesis (2015), by Akriditis, deals with utilizing solar energy-based charging and economic analysis with technical aspects for the European region.

13.5 RENEWABLE ENERGY RESOURCE ASSESSMENT DATA FOR THE TYPICAL FARM AREAS IN WHICH FIELD SURVEY IS CONDUCTED

The farm areas where field surveys were taken were analyzed for daily and yearlong renewable data. Since it required advanced equipment for finding the renewable data, the team utilized the satellite data from the web firms – Global Solar Atlas, Global Wind Atlas, which are supported by the World Bank and ESMAP and used by many researchers. The data available were validated and found that they are nearly accurate, by consulting field experts. The different farm areas were selected, and the data from the firm were used for consideration (Figures 13.5 and 13.6).

13.5.1 Farm Location: at Theni Area (10.053286, 77.512663), Southwestern District in Tamil Nadu State of India

13.5.1.1 Solar Resource Data

Map data (Per day)

Specific photovoltaic power output	PVOUT specific	4.316	kWh/kWp per day
Direct normal irradiation	DNI	3.828	kWh/m² per day
Global horizontal irradiation	GHI	5.424	kWh/m² per day
Diffuse horizontal irradiation	DIF	2.540	kWh/m² per day
Global tilted irradiation at optimum angle	GTI opta	5.524	kWh/m² per day
Optimum tilt of PV modules	OPTA	12 / 180	°
Air temperature	TEMP	26.1	°C
Terrain elevation	ELE	N/A	

Map data (Per year)

Specific photovoltaic power output	PVOUT specific	1575.5	kWh/kWp
Direct normal irradiation	DNI	1397.1	kWh/m²
Global horizontal irradiation	GHI	1979.8	kWh/m²
Diffuse horizontal irradiation	DIF	927.1	kWh/m²
Global tilted irradiation at optimum angle	GTI opta	2016.4	kWh/m²
Optimum tilt of PV modules	OPTA	12 / 180	°
Air temperature	TEMP	26.1	°C
Terrain elevation	ELE	N/A	

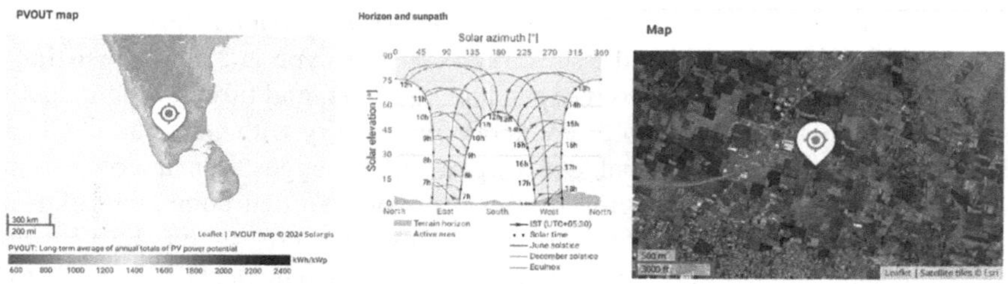

FIGURE 13.5 Solar radiation data (/day and /year), Horizon – Sun path for the farm location.

13.5.1.2 Wind Resource Data

Almost a major portion of the farms surveyed were capable of renewable (either/both wind–solar) or/and conventional grid supply. This is possible because Indian geography is tropical, which makes it have abundant solar energy and a good wind profile (Figures 13.7 and 13.8).

13.6 System Design Brainstorming for Solutions and Conceptualization

Farmers who use electric tractors in the current system encounter practical difficulties because there are few charging facilities in rural areas. Due to the lack of a specialized charging infrastructure, farms must make lengthy excursions to remote charging locations, which adds to operational inefficiencies and may prevent electric tractors from being widely adopted in the agriculture industry. Figure 13.9 tabulates the viability of different types of solution architecture brainstormed by the team.

Integrating a solar-panel-based charging arrangement right in the field is one creative strategy. This approach gives electric tractors an independent and renewable power source by incorporating an additional battery with the same specs. By utilizing solar energy, this not only lessens reliance on traditional energy networks but also advances sustainability and is in line with international environmental goals.

FIGURE 13.6 Annual averages, monthly averages, average hourly profiles, and PV power output for solar data.

In addition, a grid-supplied charging board installation at the farm has been contemplated, taking subsidies into account. This strategy encourages farmers to switch to electric tractors by utilizing the current power infrastructure and adding financial incentives. The project is to improve the overall accessibility of charging stations, meeting the demands of farmers in rural areas and encouraging the broad adoption of electric tractors by merging grid supply with strategically positioned charge boards on farms. To overcome the logistical difficulties involved with on-farm charging, a centralized pole configuration has been put in place.

This setup is intended to transmit power to the tractor through a sturdy wire that retracts like a measuring tape. Moreover, a pantograph-like charging system that makes use of the tractor's community power path has been investigated. For effective charging, this method uses a central charging station with an extending pantograph arm that connects to the tractor.

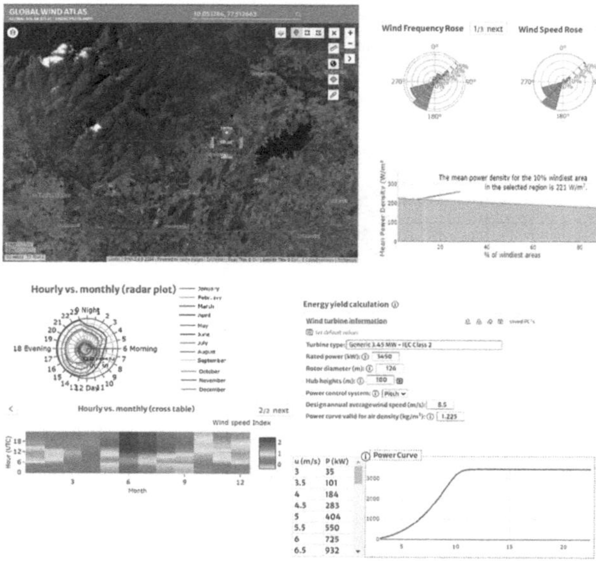

FIGURE 13.7 Wind speed, frequency, power data for a given farm location, and mean speed with variability for the farm location (monthly, hourly).

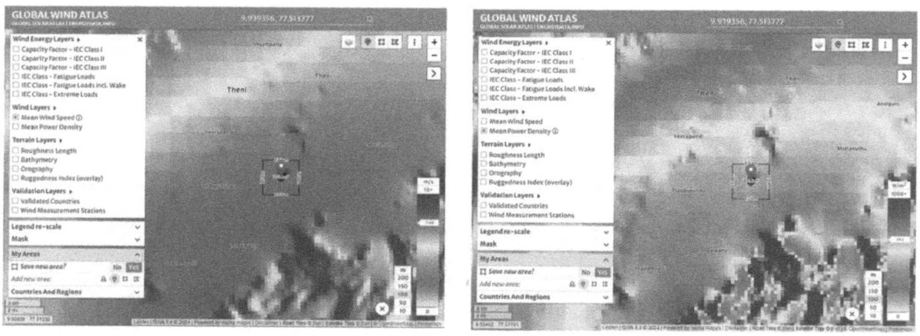

FIGURE 13.8 Mean wind speed and mean power density contour maps of the surveyed farm location area.

By enabling numerous tractors to be charged concurrently, the community power path concept promotes cooperative charging practices, which maximizes resource efficiency and shortens total charging periods. An ethanol engine-coupled alternator-based hybrid onboard charging system has been included to expand the charging choices. With this hybrid solution, an alternator and onboard ethanol engine enable electric tractors to provide their own electricity. The shortcomings of the current system are evident in how little accessible and long-lasting the charging choices for electric tractors are.

The team brainstormed all these ideas and analyzed the pros and cons of each of them. Finally, the team concluded with a best-case solution that has significant

Sl.No	Idea	Feasibility level
1.	Solar panel based charging setup at field with another auxiliary battery of same specifications.	✓
2.	Charging board installation at farm using the Grid supply with subsidy considerations	✓
3.	Centralised pole setup giving supply to tractor through a rugged wire which doesn't touches the land retracting like a measuring tape	✓
4.	Pantograph like charging using a community power path for the tractor	✓
5.	Ethanol engine coupled alternator based hybrid on board charging	✓

FIGURE 13.9 Viability of different types of solutions.

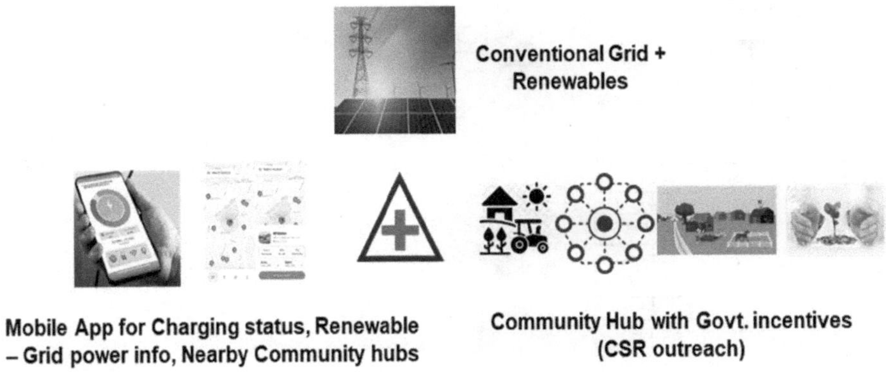

FIGURE 13.10 Proposed solution framework representation.

advantages, better implementation options, feasibility, and better outcomes. The table in Figure 13.9 summarizes all the ideas brainstormed and their feasibility level with the shaded market. Finally, the team proceeds with the idea of a solar-panel-based charging setup in the field with an auxiliary battery. The team later collaborates and makes changes to enhance the outcomes of the idea.

We simulate the complex dynamics of power generation using MATLAB, integrating solar and wind energy sources with the choice of traditional grid supplies. The simulation calculates the exact amount of electricity produced from renewable sources and assesses if it is adequate to charge the battery of the electric tractor.

13.6.1 Proposed System Architecture

Two methods are used to implement the suggested Electric Tractor Charging Architecture: a physical prototype and MATLAB simulation. First, the model is developed as per the requirements and modeled in MATLAB Simulink. Later, a small-scale PoC is developed for physical validation and testing (Figures 13.10 and 13.11).

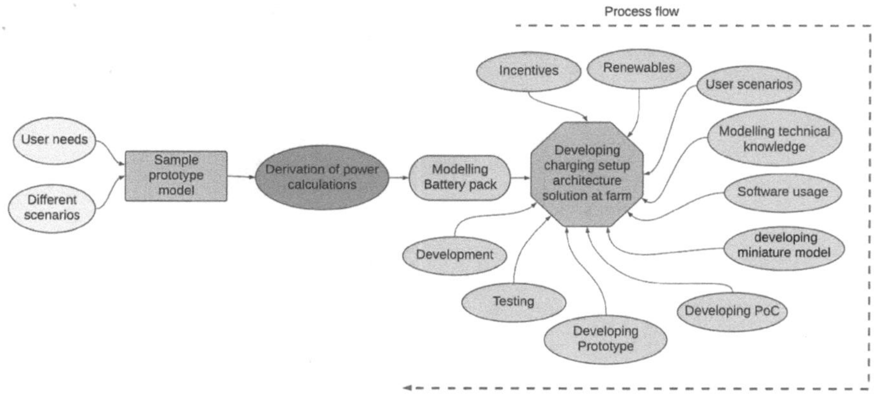

FIGURE 13.11 Process flow for the development of the solution architecture.

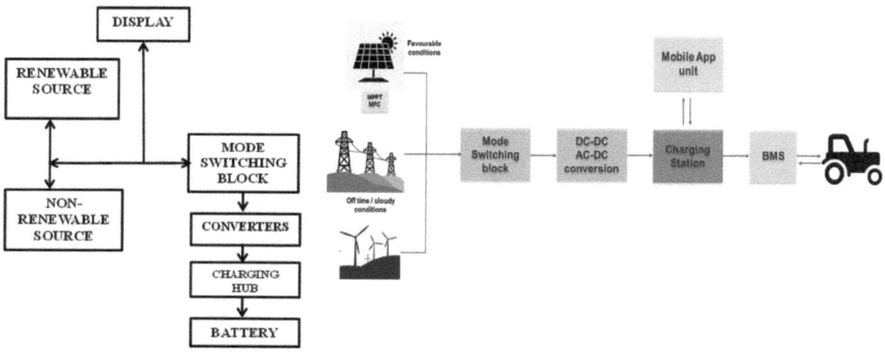

FIGURE 13.12 Block diagram.

Before diving into the model-based simulation, a methodology flow is developed to have a first-hand view of the future work. Also, a process flow diagram is also developed by considering the steps involved in the development of the solution framework. Developing these methodologies and process flow diagrams offer the team to visualize the path ahead in the development of the solution framework (Figure 13.12).

The system is made to smoothly transition to non-renewable power sources in situations where the output of renewable energy is insufficient, guaranteeing an uninterrupted and dependable charging procedure for the electric tractor. In order to generate renewable energy, the physical prototype integrates wind turbines and solar panels at the same time as the MATLAB model. Measurement tools are incorporated into the prototype to assess the real-time output from these renewable sources.

The system makes a dynamic decision on whether to pull power from the traditional grid based on this measurement. As demonstrated in the block diagram below, this smooth transition between renewable and non-renewable sources guarantees an uninterrupted and effective charging process for the electric tractor, in line with the project's sustainability aims.

Additional choice of community charging hub can be implemented as like as in fleet charging in a common place of a village where a majority of the electric tractors/vehicles are present. Also, a company providing electric tractors can implement the communal charging hub as a CSR activity or infrastructure too (Figure 13.13).

13.6.2 Model-Based System – Software Simulation

Before the actual prototype is fully completed, the MATLAB simulation is a vital testing ground for optimizing system performance and fine-tuning algorithms. With this combined technique, the suggested Electric Tractor Charging Architecture may be thoroughly examined, guaranteeing its efficacy, dependability, and suitability for a variety of real-world situations.

The Electric Tractor Charging Architecture's Simulink model includes all the necessary parts to replicate how traditional and renewable energy sources are integrated. The model contains battery charging systems, power management logic to dynamically switch between energy sources based on measured output, and power measurement blocks for real-time monitoring of solar panels and wind turbines. Extra energy is controlled by a bidirectional power flow block, and load simulation evaluates how the system reacts to changing tractor power requirements. To ensure a comprehensive assessment, the model includes environmental factors and blocks for performance metrics visualization. Before a physical prototype is implemented, the Simulink model provides a reliable testing environment for optimizing performance, fine-tuning algorithms, and verifying the architecture's resilience. This is made possible by the integration of external components (Figures 13.14 and 13.15).

FIGURE 13.13 Methodology for the development of the solution architecture.

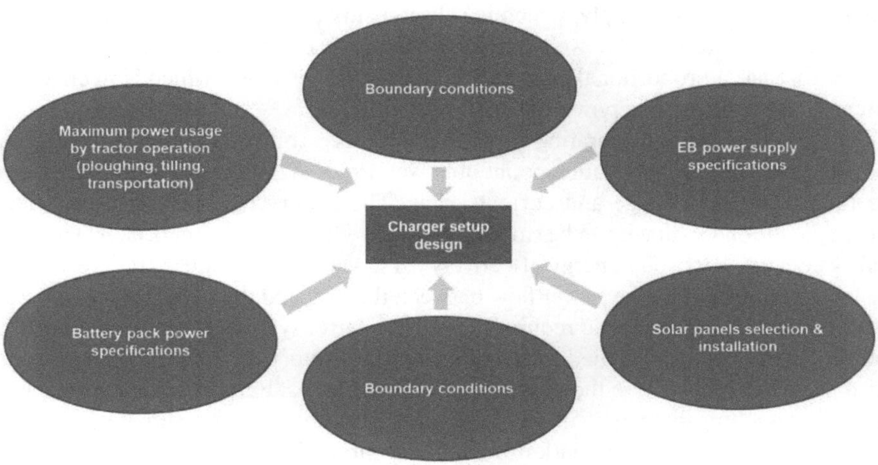

FIGURE 13.14 Factors involved in the development of the solution architecture.

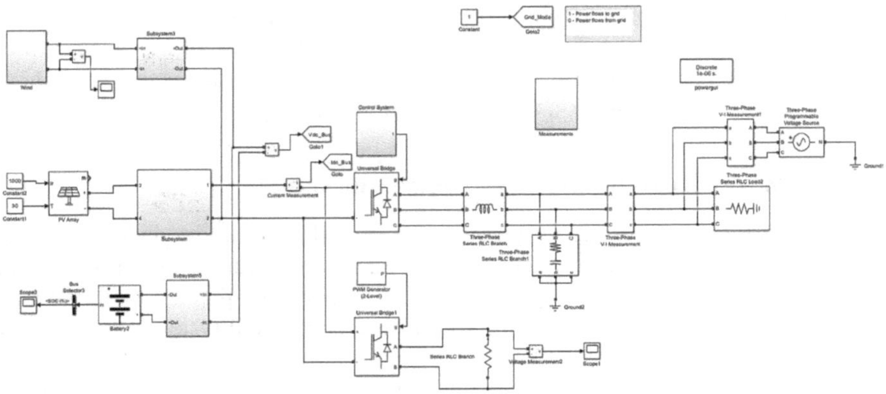

FIGURE 13.15 Complete system model.

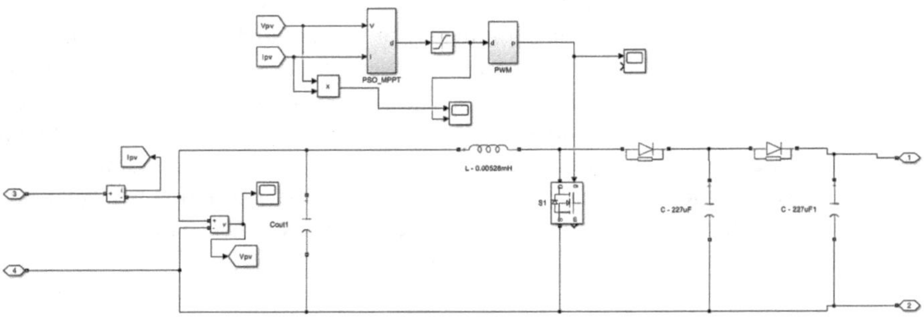

FIGURE 13.16 Subsystem 1 (solar).

To maximize its performance, the solar subsystem in Figure 13.16 has predetermined reference settings for temperature and irradiance. Solar panels provide the subsystem with a DC supply, which is subsequently sent via a bridge rectifier to be converted into useful DC power. The DC power is then routed into a DC-to-DC converter that has a proportional-integral (PI) controller installed, which is in charge of managing the maximum power point tracking algorithm.

By modifying their operating point appropriately, this guarantees that the solar panels produce their optimum amount of power. Pulse Width Modulation is also used to further control voltage and current levels. The battery receives the transformed energy at the end so that it can be stored and used to charge electric cars later. The solar subsystem converts solar energy effectively using various techniques (Figure 13.17).

In order to regulate the power flow between the grid and the battery, or vice versa, depending on the linked load requirements, the control system block is essential. The power supply can be switched from the grid to the battery or from the battery to the grid more easily thanks to the pulses this block produces. Furthermore, it guarantees smooth switching between single-phase and three-phase power supplies based on the attached load. The simulation model in Figure 13.18 shows this functionality. By optimizing power distribution and use, the control system block improves the charging infrastructure's efficiency and dependability.

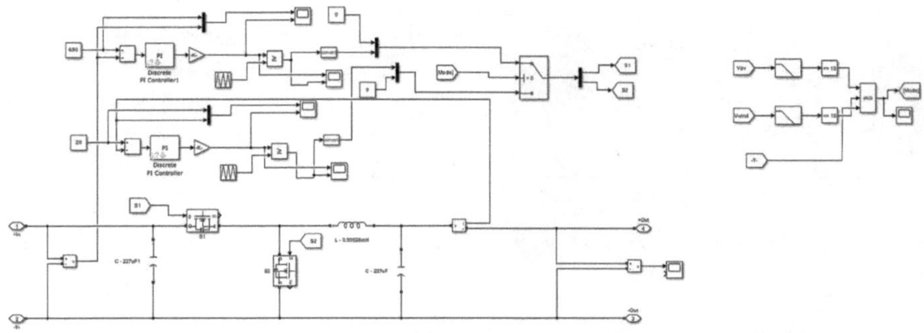

FIGURE 13.17 Subsystem 2 (swapping backup hub).

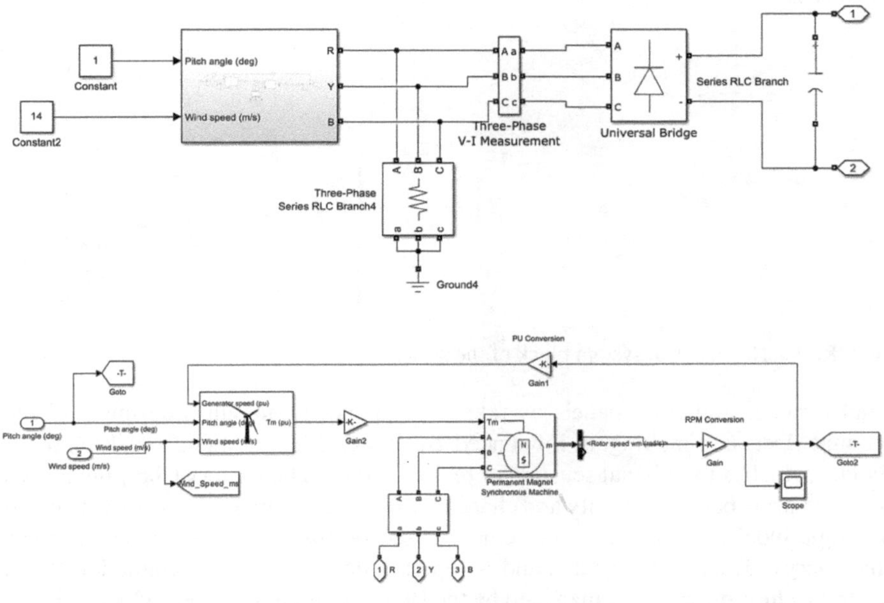

FIGURE 13.18 Subsystem 3 (wind), wind power generation system model.

The wind subsystem in the simulation block shown in Figure 13.19 is made up of various essential parts that are needed to effectively utilize wind energy. First, a steady reference voltage that is set to a predefined value serves as a benchmark for the system's performance. The wind energy production system supplies the wind subsystem with AC, which is then sent to a bridge rectifier to change it into DC. This DC output is then sent to a boost converter, where a PI controller optimizes energy conversion by controlling the voltage and current levels. Lastly, the battery receives the transformed energy for storage and subsequent usage in charging electric cars. By using this configuration, the wind subsystem efficiently gathers and transforms wind energy into a form that can be used, enhancing the charging infrastructure's overall sustainability and efficiency.

Several Simulink blocks, each with a critical function, replicate the Electric Tractor Charging Architecture. The Renewable Energy Sources Block replicates

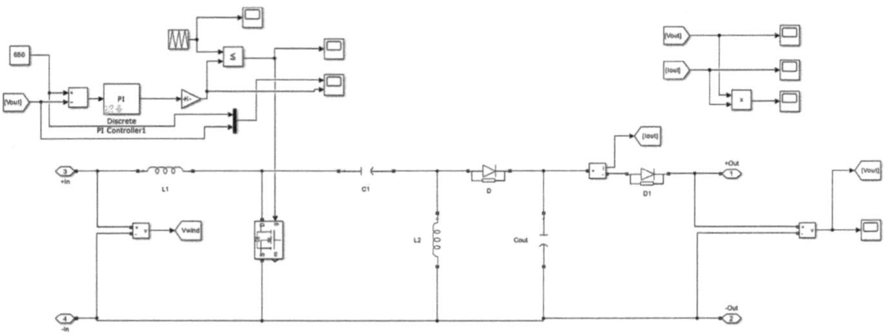

FIGURE 13.19 Subsystem 3 (wind), wind power conditioning system for the wind subsystem.

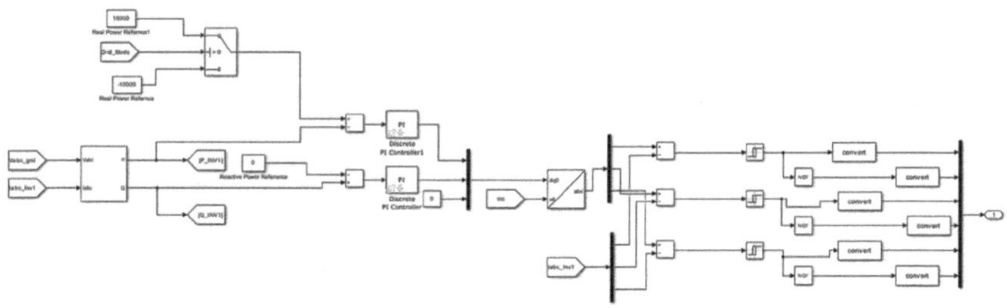

FIGURE 13.20 Control system block of the system.

wind turbines and solar panels by taking wind speed and sunshine intensity into account. Decision-making is influenced by Power Measurement and Monitoring Blocks, which act as virtual sensors for real-time data. The Battery Charging System Block controls battery capacity and charging rates. Decisions are made by the switching logic blocks and power management based on the output of measured renewable energy. Tractor energy demand is represented by the Load Simulation Block, while surplus power flow is managed by the Bidirectional Power Flow Control Block. Measurement and simulation of circumstances are done via the System Performance Metrics and Environmental Circumstances Blocks. The Integration with External Factors Block guarantees adaptation to real-world settings, while the Visualization Blocks create visual representations that facilitate learning. The model guarantees the system's performance through iterative refinement in Simulink, offering a strong basis for the physical prototype implementation by directing algorithm fine-tuning and improving dependability for practical deployment (Figures 13.20–13.22).

13.6.3 RESULTS AND DISCUSSION

Developing a decentralized infrastructure for charging directly on farms is the main idea behind the suggested solution. With limited availability of charging stations, especially in rural regions, our Compact Charging Framework seeks to address that

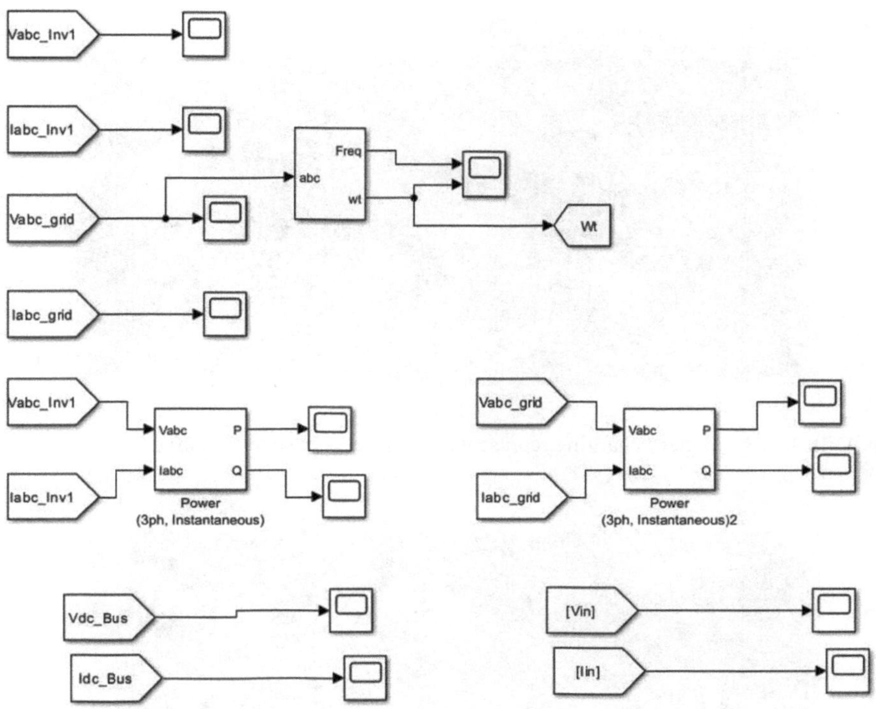

FIGURE 13.21 Measurement system block.

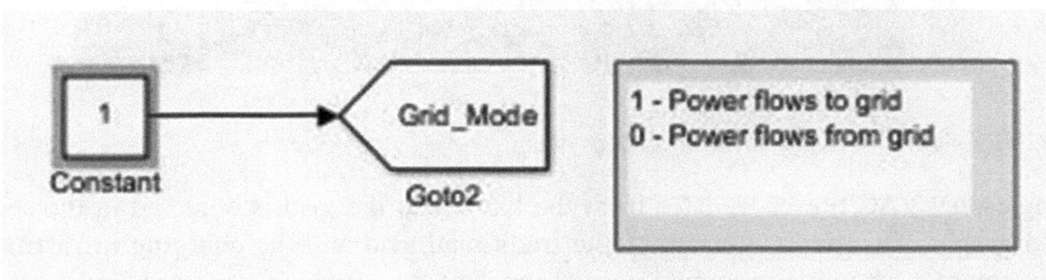

FIGURE 13.22 Mode switching system block.

issue. With regard to the system's performance, the following figures give a visual depiction of the simulation results. While Figure 13.25 depicts the battery's power exchange with the grid, Figure 13.23 offers a summary of the dynamics of battery charging, in terms of state of charge (Figure 13.24).

Two essential components of the Electric Tractor Charging Architecture are illustrated by the simulation results: grid interaction and battery charging. The battery's state of charge, shown as a percentage, is depicted over time in the first graph. This graph sheds light on the charging process by illustrating how the battery's energy level changes as it gets power from the grid or renewable sources. On the other hand,

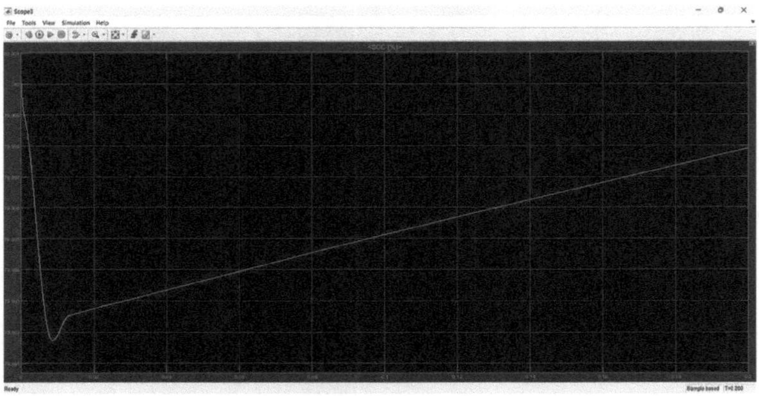

FIGURE 13.23 Battery charging representation in terms of state of charge.

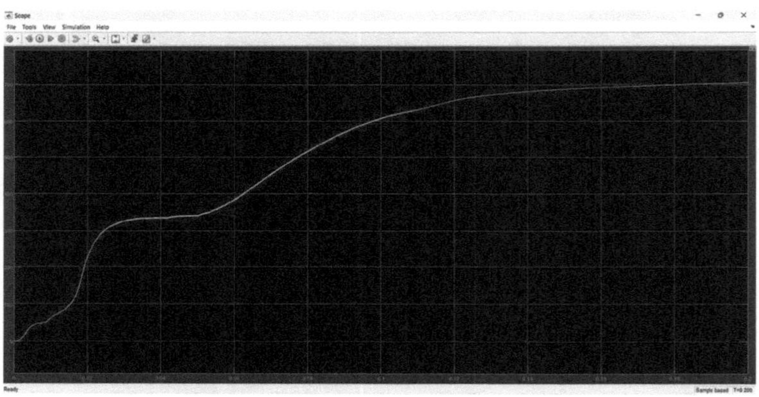

FIGURE 13.24 Output of the wind subsystem.

single-phase AC power transfer from the battery to the grid is depicted in the second graph. This graph shows how the traditional grid and the charging infrastructure interact, showing moments when extra battery energy is exported to the grid. In order to maximize energy efficiency and maintain grid stability, bidirectional power flow is necessary. This enables surplus energy produced by renewable sources to be dispersed effectively (Figure 13.25).

When combined, these graphs provide a thorough understanding of how the charging infrastructure functions, emphasizing the energy transfer dynamics between the grid, renewable energy sources, and batteries. Insights into system performance, efficiency, and the overall efficacy of the Electric Tractor Charging Architecture in satisfying energy demands while advancing sustainability and grid integration can be gained from analyzing these data (Figures 13.26–13.31).

It is found that at both 1 & 0 Grid_mode scenarios, the Voltage & Current output were found similar. But there are some significant observable variations in Real and Reactive parts of the power in both states (Figures 13.32–13.35).

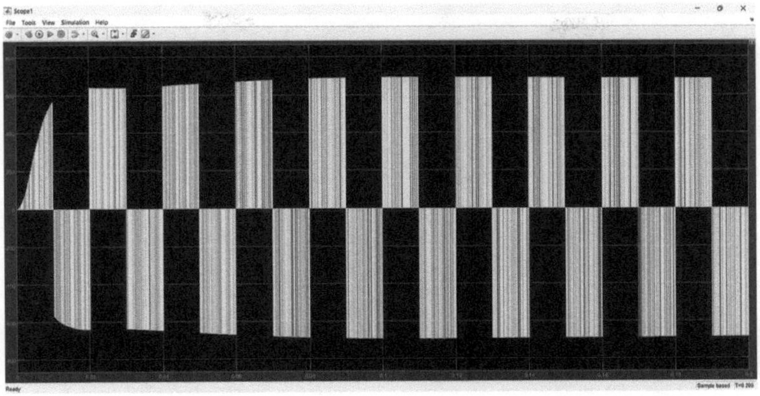

FIGURE 13.25 Battery to grid supply voltage.

FIGURE 13.26 Voltage in the DC bus.

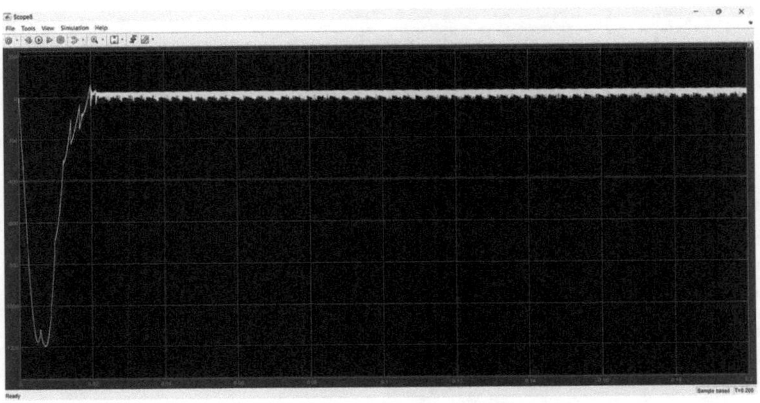

FIGURE 13.27 Current in the DC bus.

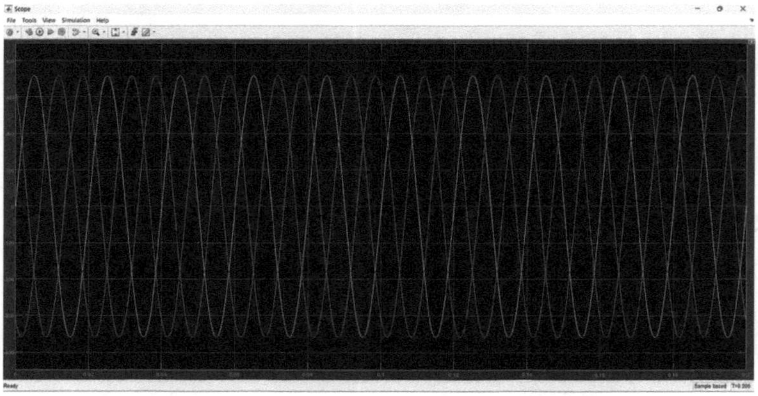

FIGURE 13.28 Voltage output from the inverter.

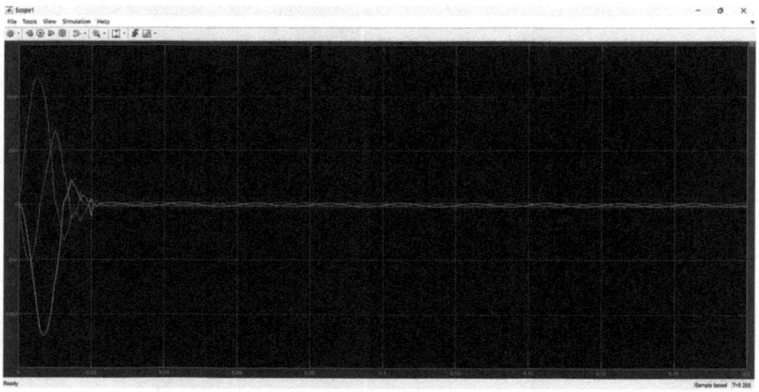

FIGURE 13.29 Current output from the inverter.

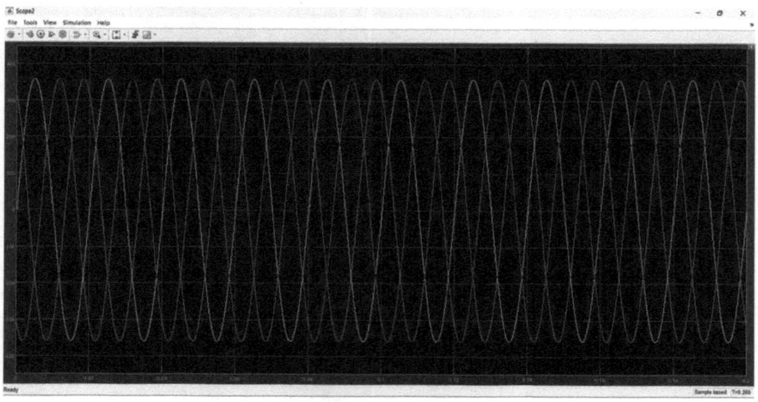

FIGURE 13.30 Voltage output from the grid.

FIGURE 13.31 Current output from the grid.

FIGURE 13.32 Real and reactive power of inverter three-phase output (P & Q) and grid three-phase output from grid (P & Q) [power flowing to grid scenario – 1 state].

FIGURE 13.33 Real and reactive power of inverter three-phase output (P & Q) and grid three-phase output from grid (P & Q) [power flowing from grid scenario – 0 state].

FIGURE 13.34 Small-level PoC - hardware setup.

The bidirectional charging network prototype is shown in Figure 13.34. It is made up of Arduino microcontroller development board, 12 V 20 W solar panel, 3 Li-ion cell series battery pack, mode switching module, grid-based 12 V charger, and small battery management system for controlled charging. This introductory level prototype was tested under lighting conditions at the first step in the indoor lab for about 2 hours, and it was found that the switching to renewable occurred when the light intensity was increased artificially by flashlights. The prototype is being tested at the miniature level, and it needs to be developed further. The results were obtained for different test environments and discussed in the next section of the

FIGURE 13.35 Preparation of the PoC for testing.

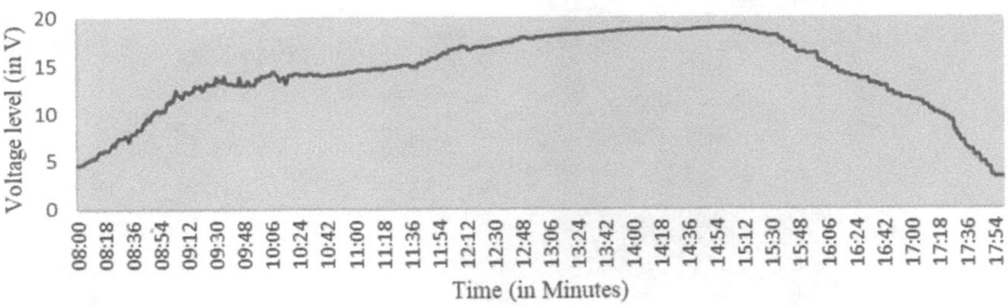

GRAPH 13.1 Generated voltage vs time.

paper. Later, after fabrication of medium-level PoC with a simulator environment, a real-time testing is planned to be done at the pilot scale, and it will be discussed in future work.

13.7 TESTING OF THE SMALL-LEVEL POC AND DATA ANALYSIS

The small-level PoC was tested under sunlight conditions as a real-time solar panel does (PoC mimics the large-level system on a small scale) (Graph 13.1 and Table 13.1). The PoC had a 12 V 20 W solar panel as a renewable part. The Arduino microcontroller is connected with a host PC using an RS232 cable and logs the data in the serial monitor, at a sampling rate of 1 minute from 08:00 AM to 17:59 PM. The results showed that between the timeline of 11 AM and 16:30 PM (~5:30 hours), the output voltage of the panel is found to be more than the rated voltage of the panel, i.e., >12 V. The charging mode switching block also made charging with respect to the panel output. But sometimes the mode switching block went to the conventional grid supply-based charging whenever it is interfered with opaque objects or a very high overload is given in the charging side. This can be avoided by correct practices, load balancing circuit topologies, or intelligent management systems.

TABLE 13.1
Testing Details of the PoC Developed

Solar panel used	12 V, 20 W
Sampling rate	1 minute
Data logging platform	Arduino Serial monitor in host PC
Total duration of data logged	600 minutes
Peak recorded voltage level	18.93 V
Minimum recorded voltage level	3.2 V
Average voltage level	14.2239 V
Time available for better charging	461 minutes
Efficient time percent over total time	76.83%

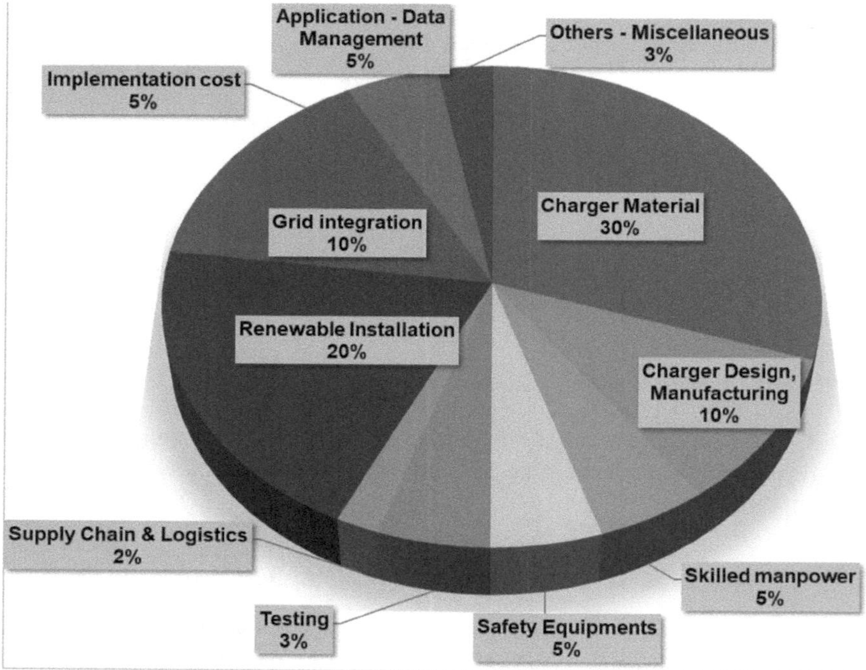

FIGURE 13.36 Estimated cost and effort split-up for real-time implementation of the solution framework.

A base level cost and effort were analyzed for real-time implementation and each sector were waited accordingly. Factors considered are as follows: charger design and manufacturing, skilled manpower, safety requirements, testing, supply chain & logistics, renewable installation, grid integration, implementation cost, mobile or web application – data management and other miscellaneous factors. A corresponding pie chart is developed and is shown in Figure 13.36.

13.8 TECHNICAL ANALYSIS OF THE SOLUTION ARCHITECTURE

The following detailed economic analyses explain the cost associated with the solution architecture. It was calculated for the one common hub integrated majorly with wind mills and ten electric tractors connected with the hub. A current electric tractor available in market – Sonalika Tiger has following technical specifications (Figure 13.37 and Table 13.2).

The charging voltage and current for the battery pack are calculated analytically. They would be approximately 120–230 V (sometimes 400 V for fast charging) and 8.5–12 A (In the case of fast charging). For the common Indian domestic household charging, the charging household voltage is 230 V and charging current is <= 10 A (if charged for 10 hours (overnight charging)) and >15 A, if fast charging is used (4 hours).

$$P = \text{charging time}/E = 4\text{ hours}/25{,}500\text{ Wh} = 6{,}375\text{ W}$$

$$I = P/V = 6{,}375\text{ W}/400\text{ V} = 15.9\text{ A}(3P)$$

$$I = P/V = 6{,}375\text{ W}/230\text{ V} = 27.71\text{ A}(1P)$$

$$P = \text{charging time}/E = 10\text{ hours}/25{,}500\text{ Wh} = 2{,}550\text{ W}$$

$$I = P/V = 2{,}550\text{ W}/230\text{ V} = 11\text{ A}$$

FIGURE 13.37 Sonalika Tiger Tractor model at a farm with the implement.

TABLE 13.2
Specifications of the Sonalika Tiger Electric Tractor

Tractor power	15 hp/11 kW
Battery type	Lithium-ion
Battery capacity (A)	250/270/350
Induction motor - capacity max/peak	10.5/21.3
Transmission – forward speed range (km/h)	2.12–21.3
Battery capacity	25.5 kWh
Charging time	10 hours
Fast charging time	4 hours
Top speed	24.9 km/h

FIGURE 13.38 Solar-based pump irrigation found in the field visit and electrical characteristics data of a panel.

While going through field visits, we found that typical solar panels used in agricultural farms are of 300 W type. The following picture also shows the panel specifications and arrangements which are used for daily pump irrigation purposes. On an average, the 300 W panels are ten in total number and they are connected in such a way that five panels in series connection provide 220–230 V. Another set of five panels in similar series connection are connected to the previous set in parallel to provide around 9–10 A. This panel arrangement provides enough power output of 1.9–2.5 kW from morning 10 AM to evening 3 PM, thus providing 9.5–11.5 kW in maximum manner total per day for 5 hours. This is enough for charging the above said E-Tractor battery pack in a small-level hub arrangement (Figure 13.38).

The panels can be increased as the tractor demand increases. The hub is not solely relying on the solar energy. The area like Nagalapuram area of Theni district (which the author field surveyed) has a large number of wind mills which produce more energy than the hub's energy requirement. This makes the system reliable during night time. Also, the common conventional grid is easily available. This can be used for worst-case scenarios (heavy raining, night time) where renewable energy is not sufficient. The hub's specifications can be easily scaled up if the demand rises in future.

13.9 ECONOMIC VIABILITY ANALYSIS OF THE SOLUTION ARCHITECTURE

For purely grid-based charging without renewable dependence, it takes 2.5 kWh/10 hours normal overnight charging, 6 kWh/4 hours fast charging. This leads to 20–25 units of energy consumption. Currently, the Tamil Nadu state of India has zero billing for agriculture. It allocated agricultural power consumption under full subsidy. A 16-hour three-phase power supply for farmers is provided under that. Normally, the loads of Agri-Electricity connections are small lights and majorly water pumps for irrigation. They require 5 kWh of energy consumption per hour. On average, daily energy consumption would be 5–15 kWh for the agricultural practices (Graph 13.2). Mostly, the pattern of electricity usage for farming varies over a year. Some particular months like those in which crops are grown require electricity at a significant level (Figures 13.39 and 13.40).

Over a year, nearly 2–4 months require major electricity demand. Other months require minimum or null electricity, because of the agricultural patterns. Even if the tractor is used daily for initial and terminal agricultural practices like tilling, sowing, and harvesting, the electricity usage is accountable for a maximum of nearly 3 months. Thus, the tractor is going to consume 800 $<= x <=$ 1,200 units of electricity. The government can also subsidize this if possible. If it made to come under domestic usage, the cost associated per day usage is ₹ 100–150, if 1 unit is billed around ₹ 5.5. We also had a discussion with a local farmer who owns 15 HP compact

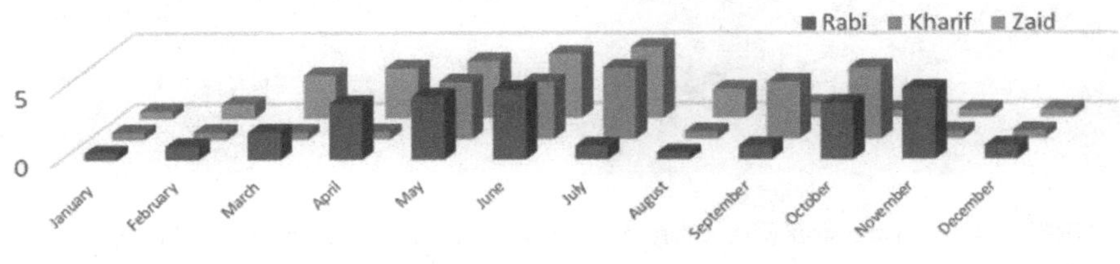

GRAPH 13.2 Tractor usage over a year during three different seasons in India.

Name	Sowing	Harvesting	Major crops
Sornavari (chittirai pattam)	April-May	August-September	Millets and cotton
Samba (Adipattam)	July-August	January-February	Paddy and sugarcane
Navarai	November -December	February-March	Fruits, vegetables, cucumber and watermelon

FIGURE 13.39 Different types of agricultural seasons of the Tamil Nadu state in India.

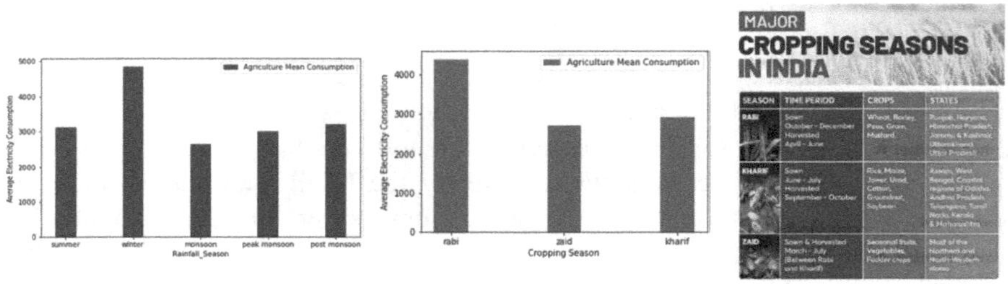

FIGURE 13.40 Different seasons in India and months with crop details with electricity consumption.

FIGURE 13.41 Discussion with a farmer who owns 15 hp compact and 45 hp large diesel tractors.

and 45 HP large Diesel tractors. We discussed about their use cases, challenges faced and about our solution framework (Figure 13.41).

The price of a 15 HP diesel engine tractor costs around ₹3–4 lakhs (Mahindra Yuvraj 215 NXT – ₹3.20–3.30 lakh, Swaraj 717 – ₹2.50–3.49 lakh). The normal consumption for the 15 hp diesel tractor typically ranges from approximately 0.75–1.2 L/h under moderate load conditions. For daily operation with 5 hours of combined practices (tilling, transportation, others), the diesel costs around ₹500.

The farmer has to expend >= ₹300 minimally per day for the basic agricultural activities minimally throughout the usage months over a year. On average, the diesel cost is around ₹400. If he uses it for 3 months (90 days) over a year cumulatively, he spends >= ₹36,000.

For maintenance like oil change, parts repair (bearing, gears) and replacing, etc., he uses around ₹20,000 per year in a cumulative manner. Also, if the farmer wants to do a new agricultural practice, he needs to buy additional implements like a new rotovator, power take-off, etc. This also adds up the grand total to nearly ₹5 lakhs (Table 13.3).

TABLE 13.3
Expenses Associated with the Diesel Tractor of 15 hp Power

Type of the Expense	Name of the Cost	Cost in INR ₹
Initial investment (one-time)	Initial cost of the diesel tractor (off-road price)	3,20,000
	Other initial costs (RTO, road tax, insurance, others)	20,000
	Additional cost for implements	50,000
Operating costs	Fuel cost over a year of usage (cumulative of all days)	36,000
	Maintenance cost	20,000
	Miscellaneous	5,000
Total		4,51,000

TABLE 13.4
Expenses Associated with the Electric Tractor of 15 hp Power

Type of the Expense	Name of the Cost	Cost in INR ₹
Initial investment (with consideration of the EV incentives) (one-time)	Initial cost of the electric tractor (off-road price)	5,00,000
	Other initial costs (RTO, road tax, insurance, others)	10,000
	Additional cost for implements	50,000
Operating costs	Electricity cost over a year of usage (cumulative of all days)	10,000
	Maintenance cost	5,000
	Miscellaneous	2,000
Total		5,77,000

The performance degradation of the diesel tractor is around 7–10 years. After that, the parts start to degrade because of the use in dusty, muddy, and water areas. After that, the frequency of parts repair increases every year. This can only be compensated if the income generated is increased.

Now let's consider the electric tractor segment. A 15 hp electric tractor in India – Sonalika Tiger (whose technical specifications are considered above) is around ₹ 6 lakhs. There is also a subsidy by the government for the E-Tractor under several schemes for the next decades. This also reduces the total cost of purchase by ₹ 5 lakhs. Also, there is a reduction and nullification of taxes associated with the purchase. The electricity bills may be subsidized in future, as in the current scenario. Currently, everyday electricity costs around ₹ 100 and cumulatively over ₹ 10,000 (Table 13.4).

13.10 COMPARATIVE COST ANALYSIS OF THE DIESEL AND ELECTRIC TRACTORS OVER 10 YEARS PERSPECTIVE

Let's consider specified annual increases in fuel and electricity prices of 4.5% and 0.5% (as per the last 5-year trends).

Let's calculate the total cost and Return on Investments (ROI) for both the diesel and electric tractors over a 7-year period, considering the new annual increases in fuel (4.5%) and electricity (0.5%) prices.

Diesel Tractor:

Costs:

1. **Initial Investment:**
   - Initial cost of the Diesel tractor: ₹ 3,20,000
   - Other initial costs: ₹ 20,000
   - Additional cost for implements: ₹ 50,000

     Total Initial Investment:

     Total Initial Investment} = ₹ 3,20,000 + ₹ 20,000 + ₹ 50,000 = ₹ 3,90,000
2. **Operating Costs:**
   - Fuel cost over a year of usage (Year 1): ₹ 36,000
   - Maintenance cost: ₹ 20,000 per year
   - Miscellaneous: ₹ 5,000 per year
3. **Fuel Cost Increase:**
   - Yearly Fuel Cost Calculation:

     Fuel Cost Year 1 = ₹ 36,000

     Fuel Cost Year 2 = ₹ 36,000 × 1.045 = ₹ 37,620

     Fuel Cost Year 3 = ₹ 37,620 × 1.045 = ₹ 39,313

     Fuel Cost Year 4 = ₹ 39,313 × 1.045 = ₹ 41,086

     Fuel Cost Year 5 = ₹ 41,086 × 1.045 = ₹ 42,944

     Fuel Cost Year 6 = ₹ 42,944 × 1.045 = ₹ 44,886

     Fuel Cost Year 7 = ₹ 44,886 × 1.045 = ₹ 46,912

     Total Fuel Cost over 7 Years:

     Total Fuel Cost} = ₹ 36,000 + ₹ 37,620 + ₹ 39,313 + ₹ 41,086 + ₹ 42,944 + ₹ 44,886 + ₹ 46,912 = ₹ 2,88,761
4. **Other Operating Costs:**

   Total Maintenance Cost} = ₹ 20,000 × 7 = ₹ 1,40,000

   Total Miscellaneous Cost} = ₹ 5,000 × 7 = ₹ 35,000
5. **Total Operating Costs over 7 Years:**

   Total Operating Costs} = ₹ 2,88,761 + ₹ 1,40,000 + ₹ 35,000 = ₹ 4,63,761
6. **Total Costs over 7 Years:**

   Total Costs = {Total Initial Investment} + {Total Operating Costs}

   Total Costs = ₹ 3,90,000 + ₹ 4,63,761 = ₹ 8,53,761

Electric Tractor:

Costs:

1. **Initial Investment:**
   - Initial cost of the Electric tractor: ₹ 5,00,000
   - Other initial costs: ₹ 10,000
   - Additional cost for implements: ₹ 50,000

     Total Initial Investment:

     Total Initial Investment} = ₹ 5,00,000 + ₹ 10,000 + ₹ 50,000 = ₹ 5,60,000

2. **Operating Costs:**
   - Electricity cost over a year of usage (Year 1): ₹ 10,000
   - Maintenance cost: ₹ 5,000 per year
   - Miscellaneous: ₹ 2,000 per year
3. **Electricity Cost Increase:**
   - Yearly Electricity Cost Calculation:

     Electricity Cost Year 1 = ₹ 10,000
     Electricity Cost Year 2 = ₹ 10,000 × 1.005 = ₹ 10,050
     Electricity Cost Year 3 = ₹ 10,050 × 1.005 = ₹ 10,100
     Electricity Cost Year 4 = ₹ 10,100 × 1.005 = ₹ 10,151
     Electricity Cost Year 5 = ₹ 10,151 × 1.005 = ₹ 10,202
     Electricity Cost Year 6 = ₹ 10,202 × 1.005 = ₹ 10,253
     Electricity Cost Year 7 = ₹ 10,253 × 1.005 = ₹ 10,304

   Total Electricity Cost over 7 years:

   Total Electricity Cost = ₹ 10,000 + ₹ 10,050 + ₹ 10,100 + ₹ 10,151 + ₹ 10,202 + ₹ 10,253 + ₹ 10,304 = ₹ 71,060
4. **Other Operating Costs:**
   Total Maintenance Cost} = ₹ 5,000 × 7 = ₹ 35,000
   Total Miscellaneous Cost} = ₹ 2,000 × 7 = ₹ 14,000
5. **Total Operating Costs over 7 Years:**
   Total Operating Costs = ₹ 71,060 + ₹ 35,000 + ₹ 14,000 = ₹ 1,20,060
6. **Total Costs over 7 Years:**
   Total Costs} = {Total Initial Investment} + {Total Operating Costs}
   Total Costs} = ₹ 5,60,000 + ₹ 1,20,060 = ₹ 6,80,060

Revenue and ROI Comparison:

Assume the tractor generates an annual revenue of ₹ 2,00,000 for both tractors:

Diesel Tractor:

Total Revenue over 7 years} = 7 × ₹ 2,00,000 = ₹ 14,00,000
ROI = [{₹ 14,00,000 − ₹ 8,53,761}/{₹ 8,53,761}] × 100
ROI ~= 64.0%

Electric Tractor:

Total Revenue over 7 years} = 7 × ₹ 2,00,000 = ₹ 14,00,000
ROI = [{₹ 14,00,000 − ₹ 6,80,060}/{₹ 6,80,060}] × 100
ROI ~= 105.9%

Conclusion:

- **Initial Investment:** The electric tractor has a higher initial investment (₹ 5,60,000) compared to the diesel tractor (₹ 3,90,000).
- **Operating Costs:** The electric tractor has significantly lower operating costs (₹ 1,20,060) compared to the diesel tractor (₹ 4,63,761) over 7 years.
- **ROI:** The electric tractor has a higher ROI (105.9%) compared to the diesel tractor (64.0%).

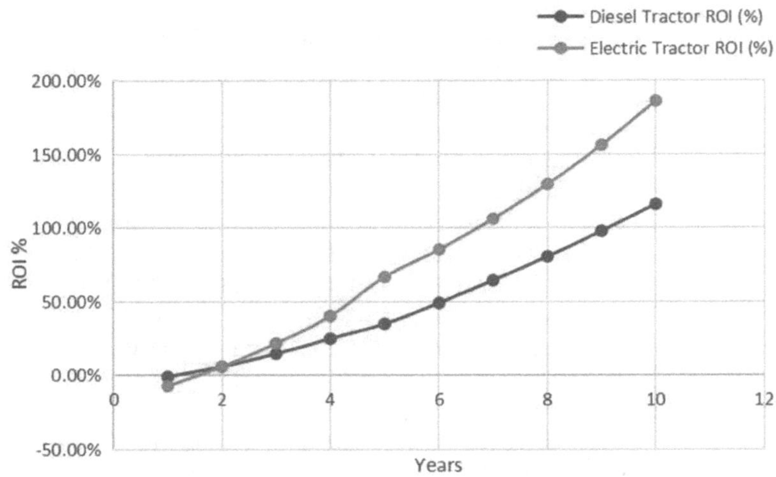

GRAPH 13.3 ROI % of diesel and electric tractors over 10 years.

The electric tractor offers a better ROI over a 7-year period, even with the incremental increase in electricity prices. The lower operating costs and the rising fuel prices contribute to its higher ROI (Graph 13.3).

The electric tractor, despite a higher initial investment, offers a better ROI over 7 years due to its lower operating costs and minimal increase in electricity prices. It offers a higher ROI over 10 years, mainly due to its significantly lower operating costs. This difference between electric tractors and diesel tractors can be further increased up to nearly 25%, if the renewables are integrated/government subsidies are increased/new technological innovations in EVs.

Now let us calculate the ROI for combined charging at the hub which uses renewables, where three electric tractors are charged. Consider a solar panel of 340 W. It costs around ₹9000. Ten panels cost around ₹90,000. The other accessories like connectors and controllers cost up to ₹15,000. On average, the complete ten panel setups cost around ₹1,10,000. There is very little maintenance cost associated with the solar panels. The average life of a solar panel ranges from 20–30 years. After that, the production capacity is reduced to 80%. One wind turbine of 5 kw rating costs around ₹2.5 lakhs. (Assuming maintenance costs for solar panels as ₹5,000 and ₹15,000 yearly for a wind turbine.)

To calculate the power output per day and ROI for the setup using a combination of solar panels and a wind turbine, let's break down the details:

Power Output Calculation:

Solar Panels:

Number of panels: 10

Power per panel: 340 W

Total power from solar panels: 10×340 W = 3400 W = 3.4 kW}

Effective daily power output (5 hours of peak sunlight): 3.4 kW × 5 hours = 17 kWh/day}

Wind Turbine:

Power rating: 5 kW

Assumed average daily power output (considering 24-hour operation at average 30% efficiency): 5 kW × 24 hours × 0.3 = 36 kWh/day}

Combined Power Output:

Total daily power output (50% solar and 50% wind): 17 kWh/day × 0.5 + 36 kWh/day × 0.5 = 8.5 kWh/day + 18 kWh/day = 26.5 kWh/day

Cost Calculation:

Initial Investment:

Solar panels setup: ₹ 1,10,000

Wind turbine: ₹ 2,50,000

Total initial investment: ₹ 1,10,000 + ₹ 2,50,000 = ₹ 3,60,000

Annual Maintenance Costs:

Solar panels: ₹ 5,000

Wind turbine: ₹ 15,000

Total annual maintenance cost: ₹ 5,000 + ₹ 15,000 = ₹ 20,000

ROI Calculation:

Annual Electricity Savings:

Total annual power output: 26.5 kWh/day × 365 days = 9,672.5 kWh/year

Cost of electricity (assuming ₹ 10/kWh): 9,672.5 kWh/year × ₹ 10/kWh = ₹ 96,725/year

Total Savings over 7 Years:

Electricity cost savings: ₹ 96,725 × 7 = ₹ 6,77,075

Total maintenance cost over 7 years: ₹ 20,000 × 7 = ₹ 1,40,000

Net savings (excluding initial investment): ₹ 6,77,075 – ₹ 1,40,000 = ₹ 5,37,075

ROI:

ROI calculation over 7 years:

ROI = ₹ 5,37,075/₹ 3,60,000 × 100 ~= 149.2%

The combination of solar panels and a wind turbine offers a substantial ROI of approximately 149.2% over 7 years, making it a highly viable option for sustainable power generation compared to diesel and electric tractors alone (Graph 13.4 and Table 13.5).

The potential financial benefits for farmers from switching to renewable energy sources like solar and wind power are numerous. Here are some key benefits:

Reduction in Operating Costs: Solar panels and wind turbines can significantly reduce or even eliminate electricity bills. Over time, this leads to substantial savings, as evidenced by the 7-year ROI calculation showing significant savings and returns. Unlike diesel, whose prices can fluctuate, the cost of maintaining solar and wind energy systems is relatively stable, helping farmers better predict and manage their expenses.

Increased Profit Margins: As shown in the table, the ROI for a combination of solar and wind energy setups significantly surpasses that of diesel tractors, leading to better profit margins. With the longevity of solar panels

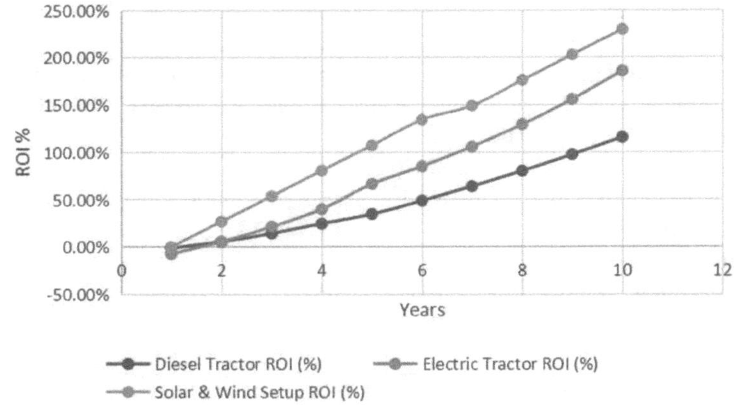

GRAPH 13.4 Comparison plot of ROI % of diesel, electric and renewables over 10 years.

TABLE 13.5
ROI % of Diesel, Electric and Renewables over 10 Years

Year	Diesel Tractor ROI (%)	Electric Tractor ROI (%)	Solar and Wind Setup ROI (%)
1	−1.20	−7.40	0.00
2	5.70	5.50	26.80
3	14.40	21.40	53.70
4	24.60	39.90	80.50
5	34.70	66.40	107.30
6	48.80	85.00	134.20
7	64.00	105.90	149.20
8	80.20	129.40	176.00
9	97.50	155.90	202.90
10	115.80	185.80	229.70

(20–30 years) and wind turbines, the initial investment pays off over the long run, resulting in higher cumulative savings.

Government Incentives and Subsidies: Many governments offer tax credits and deductions for renewable energy investments, reducing the initial capital burden. Farmers can often benefit from grants and subsidies aimed at promoting renewable energy adoption, further reducing upfront costs.

Environmental and Social Benefits: Using renewable energy reduces the carbon footprint, contributing to environmental sustainability. Eco-friendly farming practices can attract consumers and businesses interested in sustainable products, potentially opening up new markets and increasing sales.

Energy Independence: By generating their own power, farmers become less dependent on external fuel sources, which can be subject to geopolitical and supply chain disruptions. Renewable energy provides a hedge against the volatility of fuel prices, as seen with diesel, where prices can increase annually by significant percentages.

Additional Revenue Streams: Farmers with renewable energy setups can sell excess power back to the grid, creating an additional revenue stream. Farmers can lease land to energy companies for the installation of solar panels or wind turbines, providing a steady income.

Reduced Maintenance Costs: Solar panels generally have lower maintenance costs compared to diesel tractors, contributing to overall savings. Solar panels and wind turbines are built to last, requiring fewer repairs and replacements over time.

A practical example of these benefits can be seen in the calculated ROI table, where the solar and wind setup shows an ROI of 149.2% over 7 years, compared to 64.0% for diesel and 105.9% for electric tractors. This highlights the financial viability and long-term benefits of investing in renewable energy for farming operations.

By transitioning to renewable energy, farmers can enjoy significant financial benefits, enhance sustainability, and position themselves as leaders in the agricultural industry.

If the hub is made as a CSR-based infrastructure in a large scale in a village, definitely the profile of costs changes significantly. Normally, a mid (e.g. Exide batteries, Waree renewables, etc.)–large cap (Bosch, Mahindra, etc.) company like spends around ₹ 10 lakhs to ₹ 100 crores. If they and government subsidies were concentrated toward the initial investments in the solution architecture, then the ROI can be achieved within the span of 3–5 years. The farmers can also easily adapt to the infrastructure, and they turn toward electric mobility.

The long-term sustainability and potential environmental impact of the proposed charging infrastructure for electric tractors using a combination of solar panels and wind turbines are significant and multifaceted.

Resource Efficiency: Solar and wind power are renewable resources, ensuring a consistent energy supply without depleting natural resources. Solar panels and wind turbines convert sunlight and wind into electricity without significant environmental degradation.

Longevity of Equipment: Solar panels typically have a lifespan of 20–30 years, while wind turbines can last 20–25 years. This long lifespan reduces the frequency of replacements and associated environmental impacts.

Energy Independence: By generating electricity on-site, farmers reduce their dependence on fossil fuels, mitigating the impact of fuel price volatility and supply disruptions. Energy production minimizes transmission losses and infrastructure costs associated with centralized power grids.

Economic Viability: Over time, renewable energy systems reduce operational costs, providing a financial cushion against rising electricity prices.

Incentives and Subsidies: Government incentives for renewable energy adoption further enhance the economic viability of the proposed infrastructure.

13.11 POTENTIAL ENVIRONMENTAL IMPACT

Reduction in Greenhouse Gas Emissions: Solar and wind energy systems produce no direct emissions, significantly reducing the carbon footprint compared to diesel-powered tractors. Over time, this can lead to substantial reductions in greenhouse gas emissions, contributing to climate change mitigation. Even when considering the entire lifecycle of solar panels and wind turbines (including manufacturing, transportation, installation, and disposal), the emissions are significantly lower than those of fossil-fuel-based energy systems.

Biodiversity and Land Use: Solar panels can be installed on rooftops or marginal lands, minimizing competition with agricultural land. Wind turbines require more space, but can often coexist with agricultural activities. While wind turbines can pose risks to bird and bat populations, careful site selection and technological innovations (like bird-safe turbines) can mitigate these impacts.

Water Use: Unlike coal or nuclear power plants, which require significant water for cooling, solar panels and wind turbines require negligible water, conserving this vital resource.

Waste and Recycling: The disposal and recycling of solar panels and wind turbines are critical issues. Current efforts focus on improving recycling technologies to recover valuable materials like silicon, silver, and rare earth metals from solar panels, and steel and rare earth magnets from wind turbines.

Governments are increasingly implementing regulations to ensure proper disposal and recycling of renewable energy equipment, reducing environmental impacts.

13.12 Potential Socio-Economic Impact

Job Creation: The renewable energy sector creates jobs in manufacturing, installation, maintenance, and research and development. These opportunities can boost local economies and provide stable employment for communities.

Community Empowerment: Decentralized renewable energy systems can improve energy access in remote and rural areas, empowering communities and supporting local development initiatives.

Educational and Training Programs: Renewable energy projects often include educational components, providing training and capacity-building opportunities for local populations.

13.13 Challenges and Mitigation Strategies

Intermittency of Renewable Sources: To address the intermittency of solar and wind power, energy storage systems (like batteries) can store excess energy during peak production times for use during periods of low production. Developing efficient, cost-effective storage solutions is critical for the long-term sustainability of renewable energy systems. Combining solar and wind energy with other renewable sources or backup systems (like bioenergy) can ensure a reliable energy supply.

Initial Investment Costs: Access to affordable financing is crucial for the widespread adoption of renewable energy. Governments and financial institutions can support through grants, low-interest loans, and other financial instruments. Technological advancements and economies of scale are driving down the costs of solar panels and wind turbines, making them increasingly affordable for farmers and other small-scale users.

13.14 CALCULATION OF EMISSIONS FROM DIESEL AND ELECTRIC TRACTORS

To calculate the reduction in carbon and harmful emissions by switching from diesel to electric tractors powered by solar and wind energy, we need to consider the emissions produced by diesel tractors and the lack of emissions from renewable energy sources (Graphs 13.5 and 13.6).

Step-by-Step Calculation

1. **Emissions from Diesel Tractors:**
   Diesel Consumption: 36,000 L/year
   CO_2 Emissions per Liter of Diesel: Approximately 2.68 kg CO_2 per liter
   Annual CO_2 Emissions = 36,000 L/year × 2.68 kg CO_2/L = 96,480 kg CO_2/year
2. **Emissions from Electric Tractors:**
   Electricity Source: Solar and Wind (zero emissions during operation)
   Grid Electricity Emissions (for Comparison): Approximately 0.5 kg CO_2 per kWh (if grid electricity is partially used)

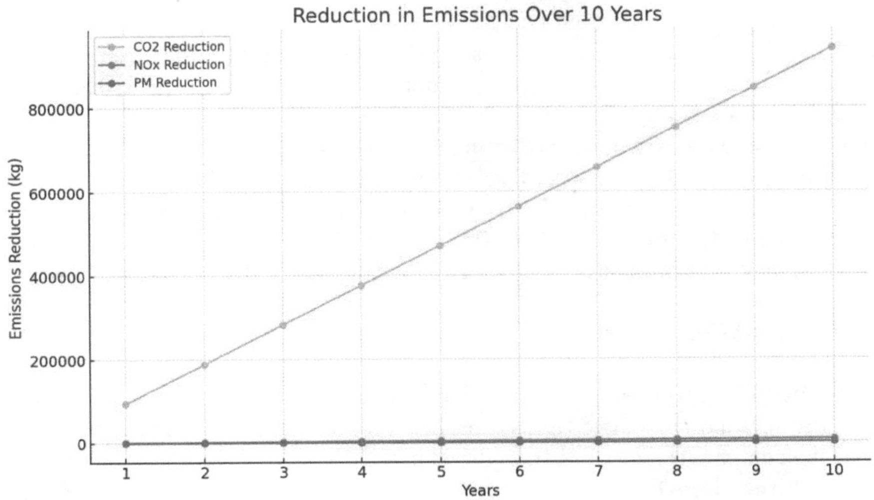

GRAPH 13.5 Reduction of emissions plot over 10 years.

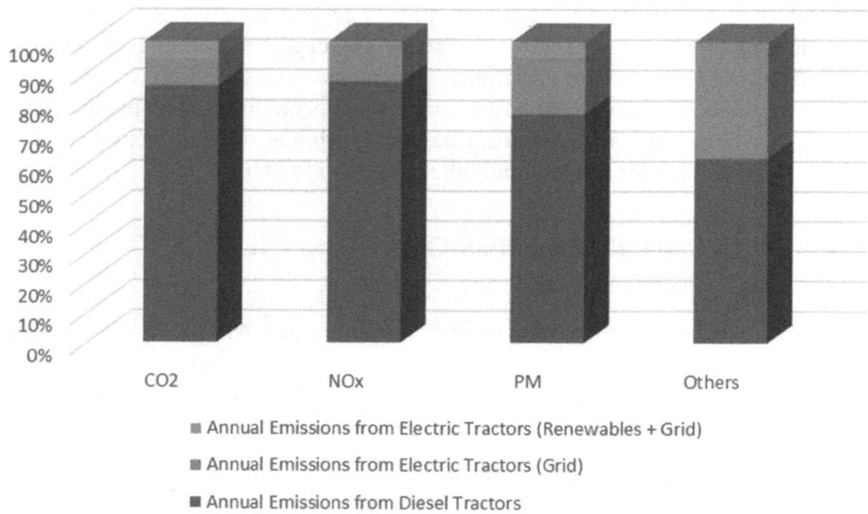

GRAPH 13.6 Different pollutants % over emissions of three different topologies.

3. **Emissions Reduction Calculation:**
   Energy Consumption of Electric Tractors:
   Annual energy requirement: 10,000 kWh/year (as previously stated)
   Emissions from Grid Electricity (if used for 50% of the time):
   Annual Emissions from Grid = 10,000 kWh/year × 0.5 × 0.5 kg CO_2/kWh = 2.50 kg CO_2/year
   Net Emissions Reduction:
   Reduction in CO_2 Emissions = 96,480 kg CO_2/year (diesel) −2,500 kg CO_2/year (grid) = 93,980 kg CO_2/year
   Harmful Emissions from Diesel:
   Diesel engines also emit harmful pollutants such as nitrogen oxides (NO_x) and particulate matter (PM). These can be calculated similarly.
4. **Emissions of NO_x and PM from Diesel Tractors:**
   NO_x Emissions per Liter of Diesel: Approximately 20 g (0.02 kg)
   PM Emissions per Liter of Diesel: Approximately 0.4 g (0.0004 kg)
   Annual NO_x Emissions = 36,000 L/year × 0.02 kg NO_x/L = 720 kg NO_x/year
   Annual PM Emissions = 36,000 L/year × 0.0004 kg PM/L = 14.4 kg PM/year

Additional Considerations

1. **Lifecycle Emissions:** Manufacturing and recycling solar panels and wind turbines have associated emissions, but these are generally much lower than the continuous emissions from diesel use.
2. **Indirect Benefits:** Reducing local air pollution improves public health, potentially reducing healthcare costs and increasing quality of life.

3. **Sustainability:** Using renewable energy sources helps conserve finite fossil fuels and reduces environmental degradation associated with extraction and combustion.

   Switching from diesel tractors to electric tractors powered by a combination of solar panels and wind turbines can significantly reduce carbon and harmful emissions. This transition leads to an annual reduction of approximately 93,980 kg of CO_2, 720 kg of NO_x, and 14.4 kg of PM, contributing to a cleaner and more sustainable environment.

13.15 ADDRESSING THE COMPLEXITY IN IMPLEMENTATION AND MAINTENANCE OF RENEWABLE-BASED ELECTRIC TRACTOR CHARGING SYSTEMS

The proposed architecture for renewable-based electric tractor charging presents a multifaceted system that requires careful planning, robust design, and meticulous maintenance. To fully address the potential challenges related to the complexity of such a system, a detailed technical approach is necessary. Here's a structured discussion on how to tackle these challenges:

Conducting Comprehensive Planning and Design: A detailed analysis of the site's solar insolation and wind patterns is crucial. This includes using Geographic Information System tools and meteorological data to map out optimal locations for solar panels and wind turbines. Soil tests and structural assessments ensure that the land can support the installation. Customizing the system involves calculating the exact energy requirements based on the operational profile of the electric tractors. This includes peak power demands, daily energy consumption, and seasonal variations. Sophisticated simulation software like HOMER or PVSyst can model various scenarios to optimize the design.

Investing in High-Quality Components and Technology: Opting for Tier-1 solar panels and certified wind turbines with proven performance records. Ensuring compliance with standards such as IEC 61215 for solar panels and IEC 61400 for wind turbines guarantees durability and efficiency. Implementing Supervisory Control and Data Acquisition systems for real-time monitoring. These systems provide data on energy production, weather conditions, and system health, enabling predictive maintenance and performance optimization.

Ensuring Robust Installation Processes: Hiring certified installers with experience in renewable energy projects is critical. Training programs provided by organizations like North American Board of Certified Energy Practitioners ensure that technicians are well-versed in best practices. Developing detailed Standard Operating Procedures for installation helps in maintaining consistency and quality. These procedures should cover every aspect from electrical connections to mechanical mounting and system commissioning.

Providing Comprehensive Training and Support: Designing a comprehensive training program that covers the operation of the renewable systems, troubleshooting common issues, and safety protocols. Interactive modules, hands-on training, and regular refresher courses can be beneficial. Establishing a local support network involving partnerships with renewable energy companies for technical support and spare parts supply. Having a dedicated helpline or online support portal can facilitate the quick resolution of issues.

Implementing Preventive Maintenance Strategies: Creating a detailed maintenance schedule that includes periodic inspections, cleaning of solar panels, lubrication of wind turbine components, and electrical system checks. Utilizing Computerized Maintenance Management System software to manage and track maintenance activities. Utilizing internet of things sensors and data analytics to predict potential failures.

Machine learning algorithms can analyze historical data to identify patterns that precede equipment failures, allowing for timely intervention.

Securing Adequate Financing and Incentives: Identifying and applying for government grants, subsidies, and incentives that support renewable energy projects. Working with financial advisors to explore green bonds, low-interest loans, and power purchase agreements. Allocating a portion of the budget to cover unexpected expenses, including potential upgrades or replacements. This includes setting aside funds for spare parts and emergency repairs.

Enhancing System Integration and Interoperability: Ensuring that solar panels, wind turbines, and battery storage systems are compatible and can be managed cohesively. Using hybrid inverters capable of handling multiple energy sources and integrating with the grid. Adopting industry standards such as Modbus or IEC 61850 for communication between different components. This facilitates seamless integration and future scalability.

Engaging in Continuous Improvement and Innovation: Keeping abreast of technological advancements in renewable energy. This includes newer, more efficient solar cells, advanced wind turbine designs, and improved battery storage technologies. Conducting regular performance evaluations using Key Performance Indicators such as energy yield, system efficiency, and downtime. Implementing feedback loops to continuously improve system performance.

Fostering Community and Stakeholder Involvement: Engaging with the local community through awareness programs about the benefits of renewable energy. Involving stakeholders in the planning process to ensure their needs and concerns are addressed. Partnering with academic and research institutions for technical expertise and innovative solutions. This collaboration can provide access to cutting-edge research and development resources.

Addressing the complexities associated with implementing and maintaining renewable-based electric tractor charging systems requires a multi-pronged approach that encompasses thorough planning, investment in quality components, robust installation processes, and comprehensive support systems. By systematically addressing these challenges, farmers can harness the full potential of renewable energy, ensuring long-term sustainability and substantial environmental benefits.

13.15.1 Integration of the Solution Architecture with Existing Agricultural Infrastructure

The solution architecture can be implemented with existing agricultural infrastructures with some measures and actions. At first, a center point should be chosen among a farming community comprising users of electric tractors. The centralized hub for charging should be built after considering factors like easy availability, shorter distance, and good connectivity to the existing grids. Then, the hub can be connected to both Conventional and Renewable grids. Normally, for agricultural purposes, the authorities in India are providing timely distribution of electric power at incentivized costs. So, the connection to the grids can be easily achieved for conventional grids. Also, some major areas in India, particularly the Chota-Nagpur plateau, Ganga Plateau, and Western Ghats areas, have renewable grids which easily aid in connecting to the renewable grids and the centralized charging hub. Additionally, the charging hub can also opt for a swapping facility, which is available in charging fleets in major cities. This can maximize the availability of charging. While going through the south-eastern part of the Theni district of Tamil Nadu for field surveys, we found that a large wind farm is spread among the farms. The majority of the farms were small to medium in size and are based on borewell-based irrigation. Some farms even had small- to medium-level solar-panel-based irrigation. The pump systems in the farms were well connected to a separate grid (as shown in Figure 13.42), which

FIGURE 13.42 Renewable and conventional grids at the farms with farmers at different farms in the same locality doing agricultural works.

provides electricity to farmers at particular duration (mostly morning time 9–11 AM) for irrigation purposes at an incentivized or null cost by the state government. This aids in the easier adaptation and implementation of the system. For seamless integration of the system at a large level, the government or the undertaking authority should collaborate with engineers, farmers, and all other associated stakeholders.

If the individual or a small group of farmers are willing, the same can be achieved. At a pilot scale, a centralized hub for 100–250 farmers can be implemented at a rate of 1–2 per district. This can be implemented as a feature on a large scale with reduced costs. In short, the whole integration process is analogous to charging EVs in cities for transportation (as Tesla has a large network in the USA). The optimization processes, algorithms, and methodologies used for the commute segment can be used with considerable modifications in the agriculture sector.

This proposed solution reduces the logistical difficulties experienced by farmers who would normally have to drive great distances to charge their electric tractors by bringing the charging point right to the farm. This on-farm charging paradigm offers a practical and effective way to electrify farming equipment while also meeting the specific needs of agricultural operations. The viability and efficacy of this Electric Tractor Charging Architecture have been guaranteed by addressing a number of important factors; as more people buy electric tractors, the infrastructure can be expanded because it is scalable. The charging stations' technological adaptability guarantees that they may be adjusted to suit different electric tractor models and changing industry norms. The project continues to place a high premium on regulatory compliance, keeping up with and abiding by all laws pertaining to the integration of renewable energy sources into agriculture and the charging of EVs.

The solution architecture is explained in detail to the Microsoft Dall – E Bing AI image creator, and the resultant images were found to be quite correlating with what the team thought as a solution architecture in their mind during the initial stage. Usage of such AI tools accelerates the solution framework development and enables in clear understanding with increased collaboration among the team (Figure 13.43).

13.16 CONCLUSION

At last, the combined charging topologies of Renewable and Conventional grid-based solution architecture are discussed in all aspects. Initially, the idea is modeled using MATLAB Simulink software as a process of model-based development and engineering. Later, a small PoC was designed and tested in conditions and environment mimicking the pilot level – large-scale infrastructure. The steps of the solution are derived based on all aspects (solar, wind, conventional grid, financial data) obtained from field surveys at different farms and institutions. The technical analysis of the same is discussed using analytical calculations by considering the available electric tractor products in the market. Financial and economic aspects such as initial investments, operational costs, and ROI over 10 years were also discussed. The environmental impact, with consideration of emissions, is also calculated. The potential challenges existing in implementing the solution architecture with the existing structure and the solutions to overcome were also analyzed. To familiarize farmers with the advantages of electric tractors and the ease provided by the envisaged charging

FIGURE 13.43 Realization of the solution architecture using Dall – E AI image creation.

infrastructure, an educational program has to be designed and implemented. The project is to hasten the acceptance and implementation of electric tractors in the agricultural sector by raising knowledge and comprehension. In conclusion, by tackling the crucial issue of the availability of EV charging stations, this project represents a revolutionary step toward the sustainable electrification of agriculture. With its optional CSR-built community charging/swapping hub and combined energy supply of renewable and conventional sources, the innovative Electric Tractor Charging Architecture is an example of how emissions can be reduced, energy efficiency can be increased, and the agricultural industry's reliance on fossil fuels can be reduced. The small-scale pricing structure that is put into place on farms not only solves real-world issues but also represents a paradigm change in the direction of a more ecologically friendly and sustainable agricultural future.

REFERENCES

Abdul Lateef Dar (2022). An introduction to electric tractor. doi:10.13140/RG.2.2.29165.26087

Bind, S. & Gupta, R. (2023). Integration of PV and wind energy with grid and to charge electric vehicles battery. In *IECON 2023–49th Annual Conference of the IEEE Industrial Electronics Society*, Singapore, Singapore, pp. 1–6. doi:10.1109/IECON51785.2023.10312282

Fathabadi, H. (2017). Novel grid-connected solar/wind powered electric vehicle charging station with vehicle-to-grid technology. *Energy*, 132. doi:10.1016/j.energy.2017.04.161

Jia, C., Qiao, W. & Qu, L. (2018). Modeling and control of hybrid electric vehicles: A case study for agricultural tractors. *2018 IEEE Vehicle Power and Propulsion Conference (VPPC)*, Chicago, IL, pp. 1–6. doi:10.1109/VPPC.2018.8604997

Li, H., Liu, H., Ji, A., Li, F. & Jia, Y. (2013). Design of a hybrid solar-wind powered charging station for electric vehicles. *2013 International Conference on Materials for Renewable Energy and Environment*, Chengdu, China, pp. 977–981. doi:10.1109/ICMREE.2013.6893835

Lo Franco, F., Ricco, M., Cirimele, V., Apicella, V., Carambia, B. & Grandi, G. (2023). Electric vehicle charging hub power forecasting: A statistical and machine learning based approach. *Energies*, 16, 2076. doi:10.3390/en16042076

Malik, A. & Kohli, S. (2020). Electric tractors: Survey of challenges and opportunities in India. *Materials Today: Proceedings*, 28. doi:10.1016/j.matpr.2020.04.585

Manojkumar, M., Porkumaran, K. & Kathirvel, C. (2014). Modeling of hybrid wind photovoltaic system with energy storage for rural application. In *2014 International Conference on Green Computing Communication and Electrical Engineering (ICGCCEE)*, Coimbatore, India, pp. 1–5. doi:10.1109/ICGCCEE.2014.6922427

Mastoi, M., Zhuang, S., Munir, H., Haris, M., Hassan, M., Muhammad, U., Bukhari, S. S. H. & Ro, J.-S. (2022). An in-depth analysis of electric vehicle charging station infrastructure, policy implications, and future trends. *Energy Reports*, 8, 11504–11529. doi:10.1016/j.egyr.2022.09.011

Melo, R., Antunes, F. L. M., Daher, S., Vogt, H., Albiero, D. & Tofoli, F. (2019). Conception of an electric propulsion system for a 9kW electric tractor suitable for family farming. In *IET Electric Power Applications*. doi:10.1049/iet-epa.2019.0353

Moghaddam, V., Ahmad, I., Habibi, D. & Phung, Q. V. (2017). Smart charging strategy for electric vehicle charging stations. *IEEE Transactions on Transportation Electrification*, 4, 76–88. doi:10.1109/TTE.2017.2753403

Rituraj, G., Mouli, G. R. C. & Bauer, P. (2022). A comprehensive review on off-grid and hybrid charging systems for electric vehicles. *IEEE Open Journal of the Industrial Electronics Society*, 3, 203–222. doi:10.1109/OJIES.2022.3167948

Scolaro, E., Beligoj, M., Estevez, M. P., Alberti, L., Renzi, M. and Mattetti, M. (2021). Electrification of agricultural machinery: A review. *IEEE Access,* 9, 164520–164541. doi:10.1109/ACCESS.2021.3135037

Solar Charged Electric Farming Tractors, Konstantinos Michail Akritidis, Master's Thesis, Department of Mechanical and Aerospace Engineering, University of Strathclyde – Engineering, 2015.

Yadav, R., Maurya, M. & Vishwanath, G. M. (2022). Implementation of a grid-connected solar-wind system with charging station for electric vehicles. In *2022 IEEE International Conference on Power Electronics, Drives and Energy Systems (PEDES),* Jaipur, India, pp. 1–6. doi:10.1109/PEDES56012.2022.10080206

Zhang, R., Horesh, N., Kontou, E. & Zhou, Y. (2023). Electric vehicle community charging hubs in multi-unit dwellings: Scheduling and techno-economic assessment. *Transportation Research Part D: Transport and Environment*, 120, 103776. doi:10.1016/j.trd.2023.103776

14 Assessment of Grid Impact Analysis for EV Charging

Preetha Krishnasamy, Titus Sigamani, Muralikrishnan Gopalakrishnan, and Karthikeyan Palaniyappan

14.1 INTRODUCTION

The COP 21, the Agreement of Paris (2016) is a wide-ranging and long-term goal agreement reached through governments to decrease the consequences of climate change. To tackle warming temperatures, India has committed to integrating 175 GW of green energy into the system by 2022. One of the ecological justifications for leveraging the widespread adoption of EVs as a way of mitigation is global warming. The transport industry accounts for approximately 23% of worldwide energy-related green house gases emissions. As a result, in the transportation sector, GHG emissions are considerably reduced; policymakers must expand the use of renewable energy while also replacing conventional vehicles with EVs. EV penetration in transport engineering is initiated in two ways, namely, Faster Adoption and Manufacturing of Electric Vehicles and Intended Nationally Determined Contributions.

Modern EV integration into the grid's distribution system will have no effect on the grid. However, EV penetration in the distribution grid from moderate to high levels would very certainly pose some operational and managerial issues. When seen as a simple unmanageable load, EV has an influence on the grid around peak load hours. Lower-voltage problems in largely radial networks, increased peak loads and energy losses, and, most likely, large voltage dips and load mismatches among phases in low voltage (LV) grids are some of the issues concerned. To avoid the above-mentioned issues due to the growth of EVs in distribution networks, two methods are preferred. The first step is to strengthen the current power infrastructure and model novel systems so as to fully accommodate EV integration, instead with a huge number of EVs. The secondary step is to develop and deploy improved techniques in distribution systems for EV charging management, as well as demand side management functions that fulfill the needs of the grid and their owners. The following sections will address the four major categories into which this section has been divided. Section 14.2 discusses about EV charging technology types; Section 14.3 inculcates the EVSE; Section 14.4 summarizes the different EV charging methods and systems; Section 14.5 instantiates the various EV charging models; Section 14.6 deals with the EV charging station requirements for grid infrastructure; Section 14.7 deals with the

DOI: 10.1201/9781003496946-14

issues, impacts and mitigations at the customer, Distribution Company (DISCOM) control center, and charging station; Section 14.8 discusses the impacts of the EV presence in systems; and finally, Section 14.9 concludes the overall assessment of grid impact analysis for EV charging.

14.2 EV CHARGING TECHNOLOGIES

The elements influencing the development of charging technologies are efficiency, EV range, cost, and charging type. It is crucial to realize that the technology used depends on the charging station location. For EV development, advanced battery system technologies are essential in terms of efficiency, EV range, prices, etc. Techniques and capacities for charging and discharging are crucial. To ensure EV use and hence raise their acceptability, they are accepted under various settings, thereby satisfying the needs of drivers (Figure 14.1).

On-board and off-board approaches are capable of power flow in both directions, therefore enabling both grid-based EV battery charging and power injection back into the grid.

Figure 14.2 outlines various EV charging modes, based on location and infrastructure, such as home or public charging.

Due to the restricted space on the vehicle and the placement of the charger component in the body, on-board charging has size and weight restrictions. Off-board charging is preferred for most level 1 and level 2 AC chargers involving movement of the off-board charger assembly and supplying the EV charging input with an alternating current–direct current (AC-DC) rectified output. This primarily pertains to quick charging.

The majority of first-generation EVs are capable of AC charging, which is commonly regarded as the preferred method of car recharging. EVs of the current generation are supported by AC and DC charging stations, specifically for faster charging, and preferably, DC charging.

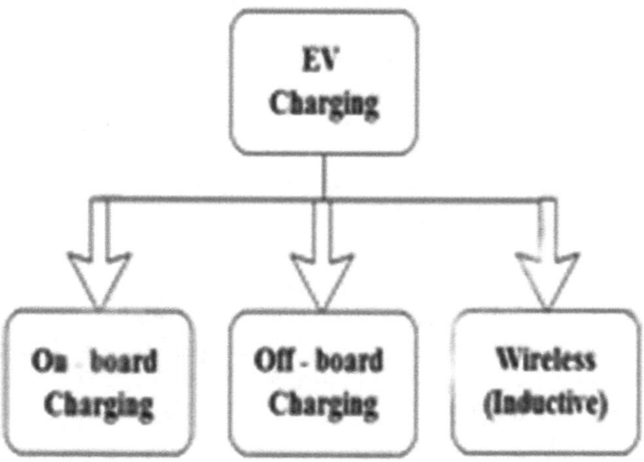

FIGURE 14.1 Types of EV charging.

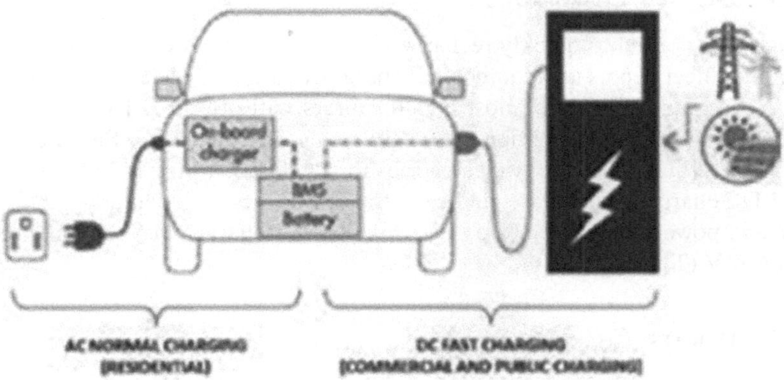

FIGURE 14.2 EV charging topology for the on-board and off-board.

14.2.1 Level 1 AC Charging

An EV is charged with about 15–20 A of current at Level 1 AC charging using a typical 120 V, single-phase, grounded, three-prong outlet (NEMA 5–15R standard plug) [1]. Ground fault interrupt devices and 15 A minimum branch protection circuits should be provided in Level 1 charging outlets. All EVs have different charging times based on the capacity of the on-board battery system and the operator's driving style. A car can be charged in a maximum of 8–10 hours (about 30–40 miles of electric driving); level 1 charging can be most effective. For instance, it takes up to 20–24 hours to fully recharge a Nissan Leaf at Level 1, compared to 8–10 hours for a Chevrolet Volt [2]. Assuming an outlet is accessible close to where the vehicle will be parked, Level 1 charging is free for EV owners because the Level 1 EV Supply Equipment (EVSE) is often included with the new vehicle.

The AC001 charger is the same as a Level 1 AC charger in India [3]. The term "AC001" refers to a charging point with a maximum output of 15 A and a maximum output power of 3.3 kW at a 230 V standard single-phase AC supply. The AC001 charger will be used with EVs that have on-board chargers.

14.2.2 Level 2 AC Charging

Both private and public applications use level 2 charging. Level 2 EVSEs require a single-phase, 208–240 V supply with a maximal continuous current of 80 A [1]. EVs are in the 60–80 mile range.

It takes roughly 3–7 hours to fully charge the range. A no-load make/break interlocking connection, additional grounding, personal defense system elements, and a safety breakout for the cable and connector are also necessary for level 2 charging services. At charging stations open to the public in India, Type 2 AC chargers with a minimum output of 22 kW are used [4]. However, since India isn't likely to install on-board chargers with larger capacities in the near future, the Standards Committee will approve the definition and construction of AC fast chargers beyond 3.3 kW in accordance with that.

14.2.3 DC Fast Charging

Fast charging is used to quickly recharge EV batteries and is likely to be available in corporate fleet depots and commercial charging stations. In addition to Level 1 or Level 2 charging connectors, most manufacturers will offer a fast-charge connection, providing owners the option of swiftly recharging their EVs. Based on power and voltage ratings, there are two categories for DC fast charging in India, including Level 1 DC chargers, which are public off-board chargers with output voltages of 48 V/72 V and power outputs of 10/15 kW, while Level 2 chargers have output voltages up to 1,000 V (Table 14.1).

14.2.4 Wireless Charging

A revolutionary technology called wireless charging enables EV owners to charge their cars while they are moving, using capacitive or inductive power transfer methods. The road systems that are wirelessly transmitted are coupled to the power generated by neighboring wind or solar resources. These cars might feature smaller batteries, which would lower their price and hasten their adoption. Efficient dynamical Wireless Power Transfer (WPT) technology is still in its infancy.

14.2.4.1 Inductive WPT

To charge the battery in an EV, inductive WPT uses coils that are enabled in both the road and the vehicle. However, ferrite cores are expensive and heavy and are necessary for magnetic flux steering and shielding. Additionally, the working frequency is restricted to 100 kHz to minimize losses in the ferrites, resulting in bigger coil sizes and lower power transfer densities. The drawback of this method is that extremely

TABLE 14.1
Types of Charging

Charging Level	Voltage	Charging Mode	Protection Type	Typical Power	Location
Level 1	120 V AC	—	None or breaker in cable	1.2–1.8 kW AC	Primarily residential in North America
Level 2	200–240 V AC	Mode 1	None	3.6–11 kW AC	Wall socket in Europe; primarily for 2 and 3 wheelers
		Mode 2	Pilot function and breaker in cable	3.6–22 kW AC	Home and workplace with cable or basic station
		Mode 3	Pilot function and breaker in a hardwired charging station	3.6–22 kW AC	Home, workplace, and public with the hardwired station
Fast charging	400–1,000 V DC	Mode 4	Monitoring and communication between the vehicle and EVSE	50 kW or more	Public, frequently, intercity

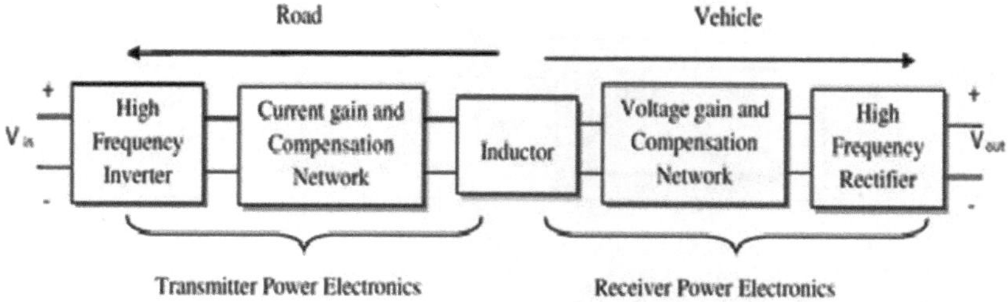

FIGURE 14.3 Inductive power transfer.

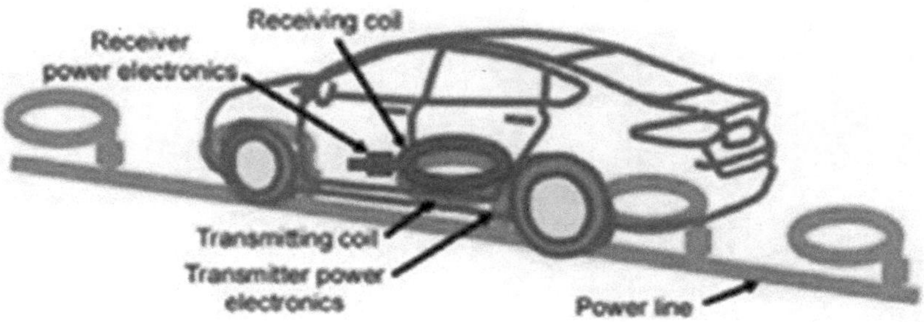

FIGURE 14.4 Schematic model for inductive wireless.

powerful coils are required to supply the necessary energy to the car in the brief intervals that the vehicle travels over a charging coil. Due to the aforementioned factors, inductive WPT has not yet achieved commercial viability (Figures 14.3 and 14.4).

14.2.4.2 Capacitive WPT

In order to recharge the battery in EVs, capacitive WPT employs coils that are enabled in roads and in EVs and are connected through electric fields. Due to the more focused form of the electric field, which reduces the need for electromagnetic field shielding, they could be preferable to inductive charging for certain applications. Due to the fact that this method of charging does not need ferrites, the charging systems may work at high frequencies, allowing for their compactness and lower cost. However, the power transmission will only be efficient at higher frequencies due to the minimal amount of capacitance across the road surface and the EV plates, making the design extremely difficult [5]. Even though EVs' capacitive charging satisfies Electro Magnetic (EM) requirements and tries to maintain optimal power transfer, it faces major limitations in attaining high efficacy at a higher power transfer density (Figures 14.5 and 14.6).

14.3 ELECTRIC VEHICLE SUPPLY EQUIPMENT (EVSE)

In EV infrastructure, providing electrical energy to recharge batteries requires an EVSE. The National Electric Code (NEC) defines EVSE as conductors, connectors, and power outlets. For wet or outdoor locations, Ground Fault Current Interruption

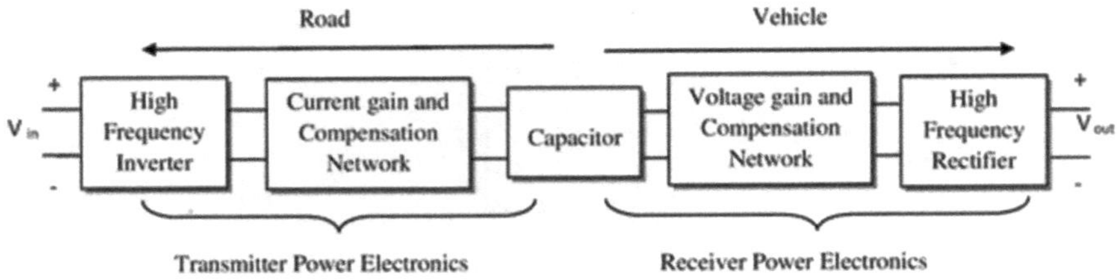

FIGURE 14.5 Capacitive power transfer.

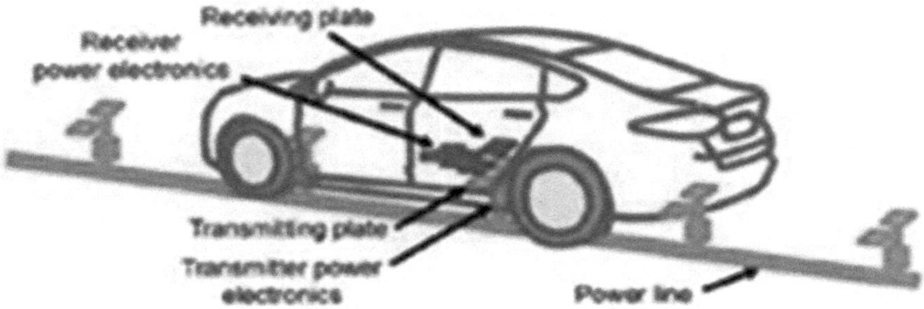

FIGURE 14.6 Capacitive WPT schematic model.

(GFCI) devices are required. GFCIs monitor current imbalances and de-energize circuits when current exceeds a predetermined value. Vehicle charge cords must carry integral ground fault protection devices, known as Charge Current Interruption Devices.

14.3.1 EV Charging Point Connectors

EVSE charging point connectors are divided into AC and DC charging types (Figure 14.7). Type-1 AC chargers can handle up to 7.4 kW, while Type 2 connectors can handle up to 44 kW. Combined charging system connectors can handle up to 43 kW AC and 400 kW DC. These connector technologies are suitable for India.

14.3.2 Cost Estimation of Charging Infrastructure

EV station installation cost in New Delhi includes fast charging hardware, location, installation, civil works, distribution network costs, Distribution System Operator (DSO) permissions, and operation and maintenance costs. Tariff rates for fast charging and residential vary, but a standard procedure is being started by DISCOMs in India. Southern California Edison has adopted a flat rate per kWh strategy for 5 years (Figure 14.8).

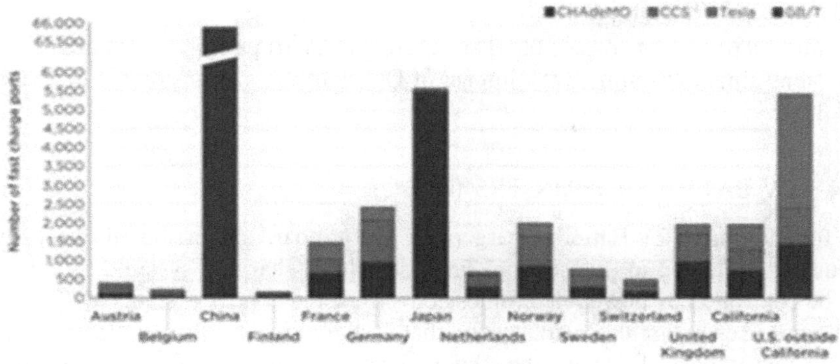

FIGURE 14.7 Plug type fast charging points in major EV market pools.

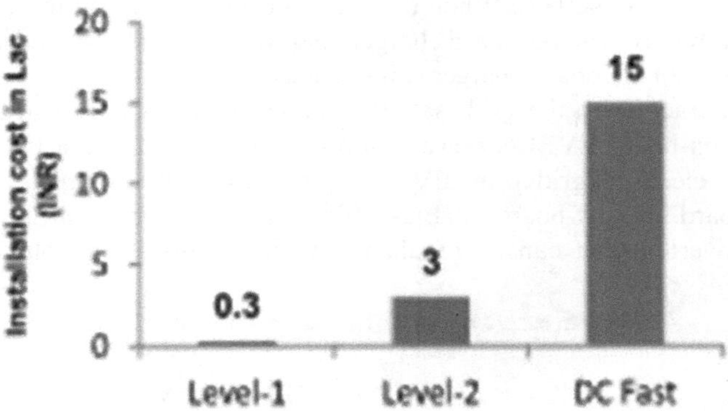

FIGURE 14.8 Estimated EV charging station Installation cost.

14.3.3 Tariff Structure for Fast Charging Stations

The real cost of electricity in India is challenging due to subsidies, central and state policy decisions, and lack of a market discovery mechanism. Adopting a tariff structure like San Diego Gas and Electric is necessary.

14.3.3.1 Residential Category

The proposed tariff system for EVs in India will be phased out by 2021 or 2022, with a focus on capacity and variable energy-based tariffs. A separate meter for EV charging is recommended, and a Time of Use tariff will be introduced to minimize peak feeder loading.

14.3.3.2 Commercial Category

The proposed tariff includes a capacity and energy-based component, with subsidized tariffs initially to encourage EV adoption, and a Time of Use tariff to reduce feeder loading during peak hours.

14.3.3.3 EV Charging Stations

The introduction of a two-part tariff structure and smart pricing tariff based on grid conditions aims to minimize the impact of DC charging stations on the grid during peak hours.

14.4 EV BATTERY CHARGERS (EVBC)

The functioning of EV Battery Chargers for the on-board and off-board systems will be further explained in this section. The EV battery charging system of the EVBC will be controlled by power electronics converters and control systems [6,7].

It comprises a digital control system that is shared by both power electronic converters, one at the grid side and the other at the battery side [8]. The power electronics devices are turned ON/OFF by gate pulses produced by the digital control system. Pulse Width Modulation (PWM) is typically used to generate the gate pulses using a closed loop control method. When these converters are located inside the EV, an EVBC is referred to as an on-board charger, and when they are put outside the EV, it is referred to as an off-board charger (Figure 14.9).

An EV interacts with the grid system using both an on-board and an off-board EVBC. The on-board EVBC's power converters will guarantee a two-way power flow from the electrical grid to the EV battery (Figure 14.10).

All on-board and off-board EVBCs utilize both grid side converters and battery-side converters that can be regulated by either current or voltage. Consider

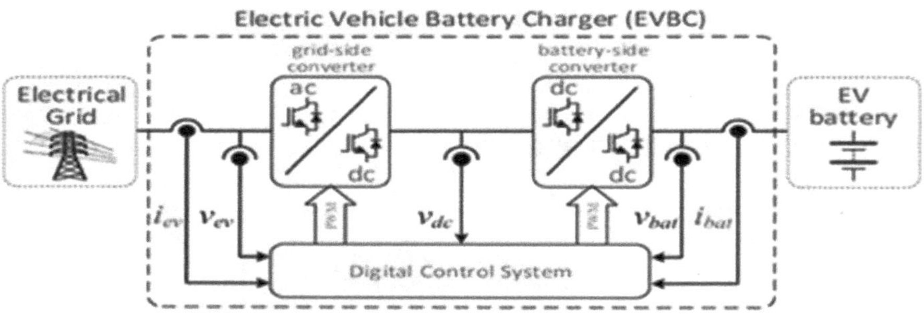

FIGURE 14.9 Block diagram of the EV battery charger.

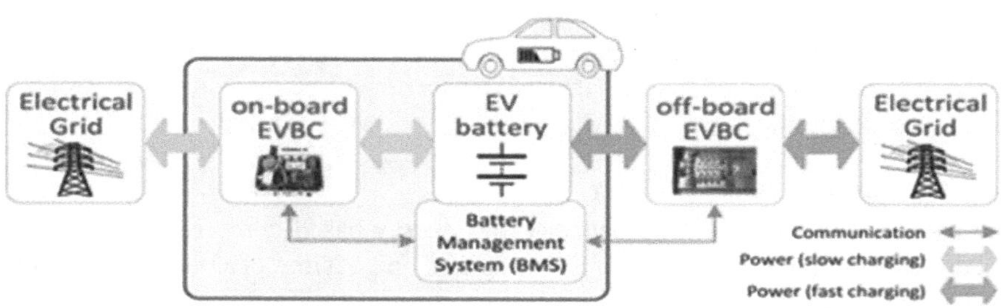

FIGURE 14.10 Operating mode of the charging system.

operating the battery in constant current mode until the State of Charge reaches 75%, and then in constant voltage mode until it is fully charged.

The Battery Management System, which sets the boundaries of current and voltage during the process of charging or discharging operations, defines the operating mode in constant contact with certain control algorithms. Additionally, on-board EVBCs are frequently intended for less power ratings than off-board EVBCs due to the volume and weight limits from an EV standpoint. However, converters are used to segregate the grid & battery for both on-board & off-board EVBCs. Indirectly, this will guarantee safe operation and lower operating voltage levels.

14.4.1 Power Electronics in EVs

Power device technology has advanced greatly during the last few decades. The Insulated-Gate Bipolar Transistor is almost entirely utilized for modern EVs among the existing power devices, including the thyristor, Gate Turn-Off Thyristor, Power Bipolar-Junction Transistor, etc.

In the first stages of an EV, both AC and DC motors are employed. Despite their greater price, the industry has recently placed more emphasis on AC motors due to their benefits, including their high-power rating, long range, and less risk of overheating during lengthy drives when compared to DC motors. An AC motor can be used as a generator to recharge its batteries. An AC motor can accelerate more quickly and can handle tougher terrain better.

The internal battery charger's power electronics are shown with an AC motor acting as a traction motor in Figure 14.11 [8]. It is created with a buck chopper and an AC/DC converter. The topology is changed from the traction operations to the charging operation with the aid of relays. The converter, which is connected to the

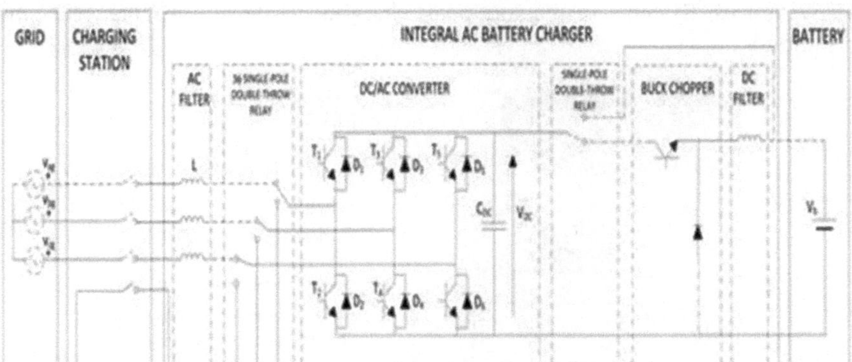

FIGURE 14.11 Power electronics of an internal battery charger [8].

battery during traction operation, is coupled to the traction motor via the three-phase relay. In this process, the converter functions as an inverter to convert the battery's DC power to AC power, which is then sent through the traction motor to power the EV. The converter serves as a rectifier during charging operations, connecting to the battery and AC charging port while excluding the motor that drives the traction through a relay. The Insulated-Gate Bipolar Transistor switch ON/OFF gating signal produced by the PWM technique will be used to regulate the converter operation in both the rectifier and the inverter. The dual function of the AC filter is to both decouple the grid from the switching voltage at the PWM rectifier's input and to lessen the high-frequency harmonics of electricity that are captured by the AC battery charger [9,10]. The goal of the AC filter is to compel the absorption of current harmonics to adhere to IEC 61000-3-2 and 61000-3-12 standards [11,12].

14.5 EV CHARGING MODELS (FIGURE 14.12)

14.5.1 Uncontrolled Mode of Charging

14.5.1.1 Public Charging Infrastructure

The EV Charging Point connects EVs to charging infrastructure, with a Final Customer Meter and utility meter installed by Utility DISCOM. Final Customer Meter measures energy consumption, peak consumption, and billing time, while Electronic Vehicle Management (EVM) measures energy consumption and billing periods. Power to EVs comes from grid, internal battery storage, or rooftop PV generation (Figure 14.13).

An energy management system termed as EV manager controller ensures reliable communication between the EV manager controller and on-board EV charge controllers. It measures battery state and offers charging options. On-board EV meters provide energy consumption information, and charging stations can have Battery Energy Storage Systems to manage peak charging while minimizing grid power consumption.

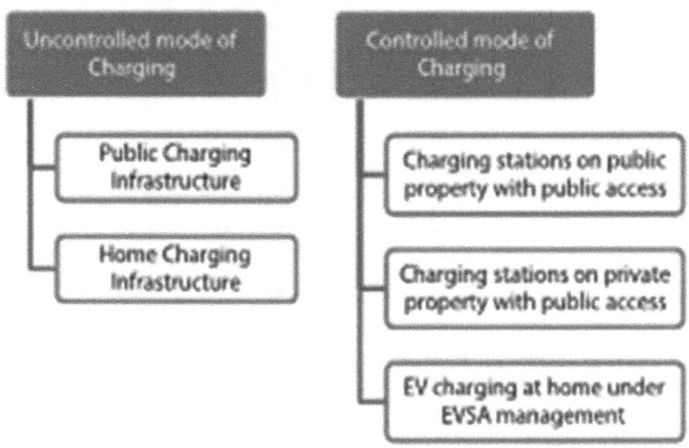

FIGURE 14.12 Different models of EV charging.

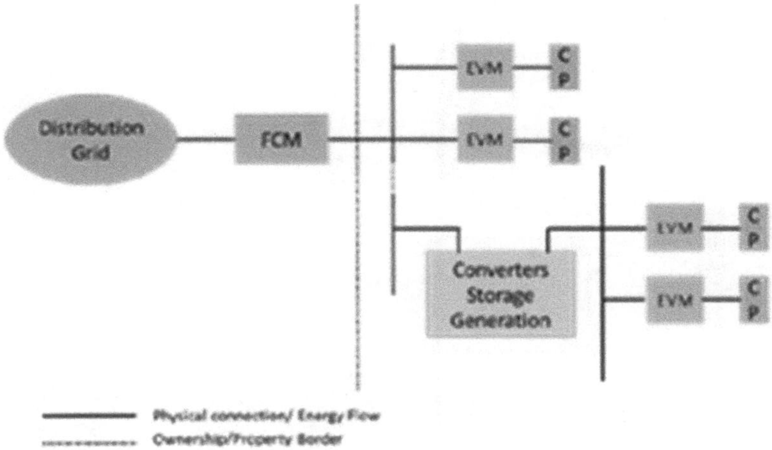

FIGURE 14.13 Public charging infrastructure.

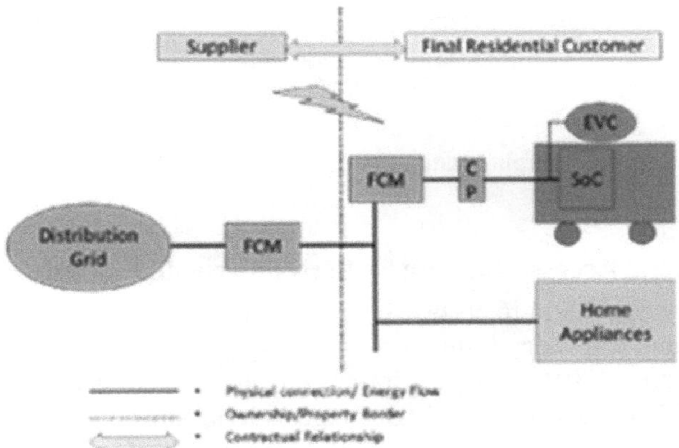

FIGURE 14.14 Home charging infrastructure.

14.5.1.2 Home Charging Infrastructure

From Figure 14.14, the residential customer will have a contract with DISCOM or Electric Vehicle Suitability Assessment (EVSA) which can control EV charging under two scenarios: peak and off-peak prices, or hourly time prices for integrated management.

14.5.2 Controlled Mode of Charging

14.5.2.1 Public Access Charging Stations on Public Property

As depicted in Figure 14.15, Local DISCOM plans to install EV charging stations on public property for low-cost and fast installation. Charge points will be accessible to all EVSAs without discrimination or monopoly practices. EV owners pay electricity bills and regulated network charges [13].

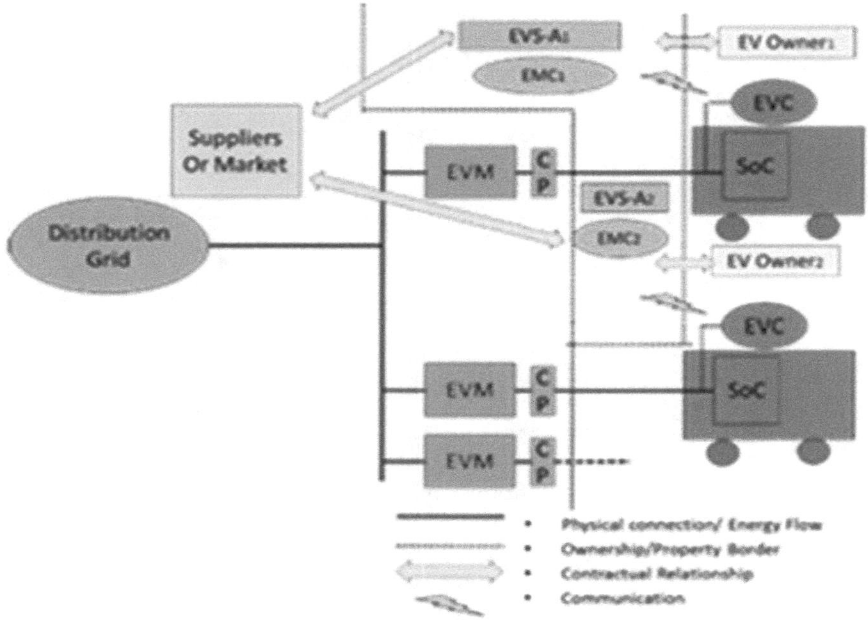

FIGURE 14.15 Public charging station.

14.5.2.2 Public Access Charging Stations on Private Property

As depicted in Figure 14.16, privately owned property with free vehicle parking requires a unique regulatory framework. A charging station owner, acting as Charge Point Manufacturer (CPM), installs infrastructure, buys electricity, and sells EV charging services. Additional equipment like storage batteries can optimize charging modes. Combining storage with local renewable energy sources can increase profits [14].

14.5.2.3 EV Charging under EVSA Management

EVSA as an intermediary agency buys and resells energy to EV owners under charging contracts. It could optimize energy by aggregating residential charging points, allowing separate pricing for transportation energy consumption and specific taxes (Figure 14.17).

14.6 EV BATTERY CHARGING STATION REQUIREMENTS FOR GRID INFRASTRUCTURE

A part of the EV infrastructure known as the EVSE provides electrical energy from a power source to recharge the EV batteries. The current NEC mandates the use of GFCI devices for damp and/or outdoor sites. The GFCI is a safety device that works by keeping an eye on the current imbalance between circuits with grounded and ungrounded conductors. When current to earth exceeds a preset value that

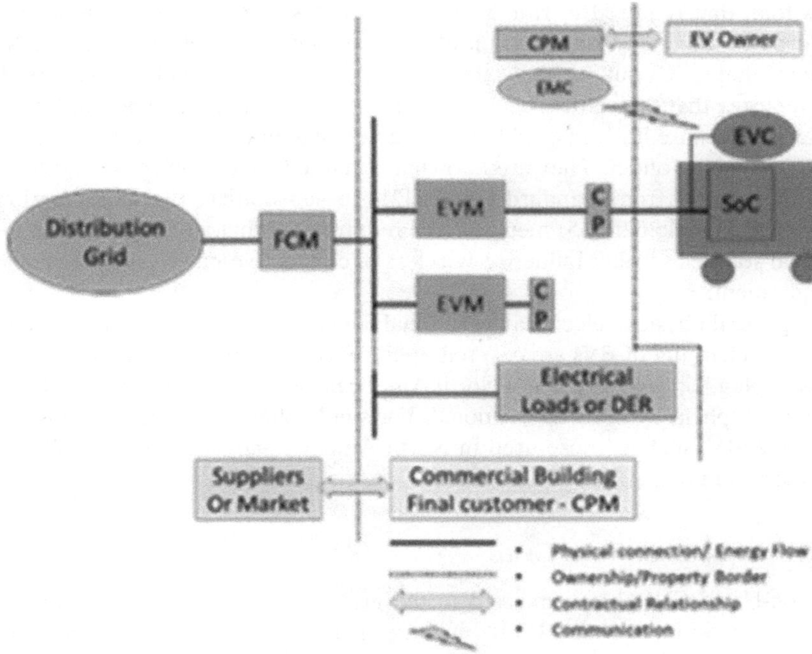

FIGURE 14.16 Privately owned charging station.

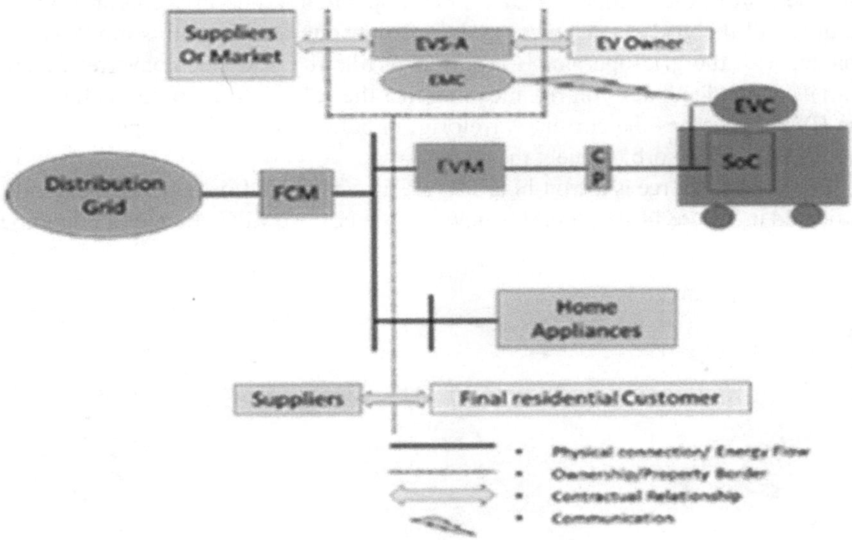

FIGURE 14.17 EV home charging under EVSA.

is less than that required to run the supply circuit's overcurrent safety device, it de-energizes the circuit or a portion of it within a set amount of time. The NEC mandates that GFCI outlets be located outside, in garages, and close to sinks. The NEC requires that the vehicle charging cord carry an inbuilt ground fault protection device because it is impossible to guarantee that an EV is always plugged into a GFCI-protected outlet. They are known as Charge Current Interruption Devices to set them apart from standard GFCIs [2]. To substantiate the aforementioned claim, NEC Article 625 [15] mentions the requirement that an EV be linked to an electrical source by either inductive-wireless or conductive-cable connected charging equipment.

The general physical, electrical, functional, and performance criteria that support conductive charging of EVs are covered by the SAE Recommended Practice J 1772 [16]. A "J plug" or SAE J 1772 is a North American electrical connector standard for EVs that is upheld by SAE International. For single-phase electrical systems using 120 V or 240 V, such as those used in North America and Japan, the J 1772–2009 connectors are intended.

14.6.1 Interaction with Grids

When on-board EVBC is connected to the grid via a smart meter with two-way communication, the on-board EVBC in G2V operating mode is the straightforward and typical method of operation as shown in Figure 14.18. EV battery charging is managed in accordance with the other connected loads when operating in controlled G2V mode [17].

For instance, the EV charging is managed as minimum if the overall load at home is at maximum, and vice versa. The stored energy in an EV battery will be fed back into the grid in the Vehicle-To-Grid (V2G) operation mode in accordance with the demands of the grid management system. To operate the EV in this mode, two-way contact with the grid operator is required. While the grid side converter functions with a current feedback control, meaning that the voltage is imposed by the grid and the EVBC specifies the current waveform, in both G2V and V2G operation modes, the EVBC will absorb or inject the current.

The voltage source is the on-board EVBC in Vehicle to Load operational mode, as indicated in Figure 14.19. To supply power to loads, the EVBC operates autonomously

FIGURE 14.18 On-board EVBC: G2V and V2G operation mode [8].

FIGURE 14.19 On-board EVBC: V2L operation mode [8].

FIGURE 14.20 On-board EVBC: V2H operation mode [8].

FIGURE 14.21 Off-board EVBC: G2V and V2G operation mode [8].

from the grid. In this mode, a voltage feedback system is used to operate the grid side converter [18].

The on-board EVBC functions as an uninterruptible power supply in Vehicle-2-Home (V2H) operation mode, as shown in Figure 14.20. To detect a power outage in this mode and even to some chosen priority loads, two-way communication between the smart house and EVBC is necessary.

With the exception of the power rating of the EVBC, as illustrated in Figure 14.21, off-board EVBC in G2V and V2G operation modes is comparable to on-board EVBC. For the V2G operation mode of the EVBC to function off-board, a bidirectional connection is required.

By regulating the drawal/injection of electrical power from/to the grid, off-board EVBC utilized for fleet charging and charge swapping batteries plays a crucial role in assisting with demand management of the grid.

It should be noted that while there is no charging activity, the off-board EVBC used in DC fast charging facilities might not be functioning. This opens up a new window of opportunity to correct issues with the distribution grid's non-linear loads, such as current harmonics, current imbalances, and power factor. Based on its rated capacity in both G2V and V2G operation modes, off-board EVBC may maintain the grid's power quality while the EV is being charged in a charging station. To address power quality challenges, Figure 14.22 illustrates the connectivity of an off-board EVBC close to industrial loads.

The off-board EVBC is enabled as a power quality compensator, which increases the system's robustness and dependability in its operation relative to the grid as illustrated in Figure 14.23.

Off-board EVBC and solar rooftop generation share the same grid side converter. Similar to this, Figure 14.24 shows a second battery storage system added on top of this one [19].

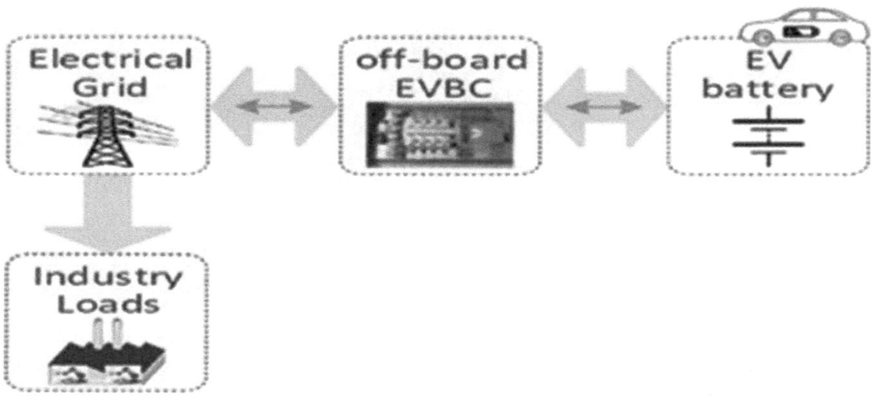

FIGURE 14.22 Off-board EVBC as power quality compensator [8].

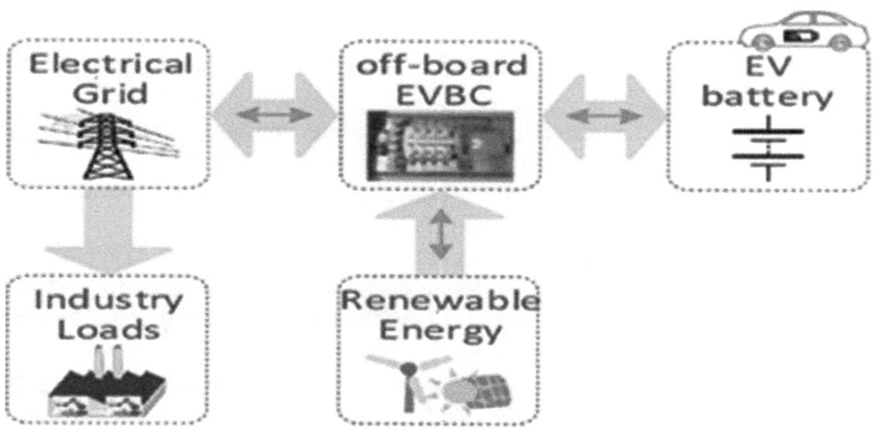

FIGURE 14.23 Off-board EVBC as power quality compensator with Renewable Energy System (RES) [8].

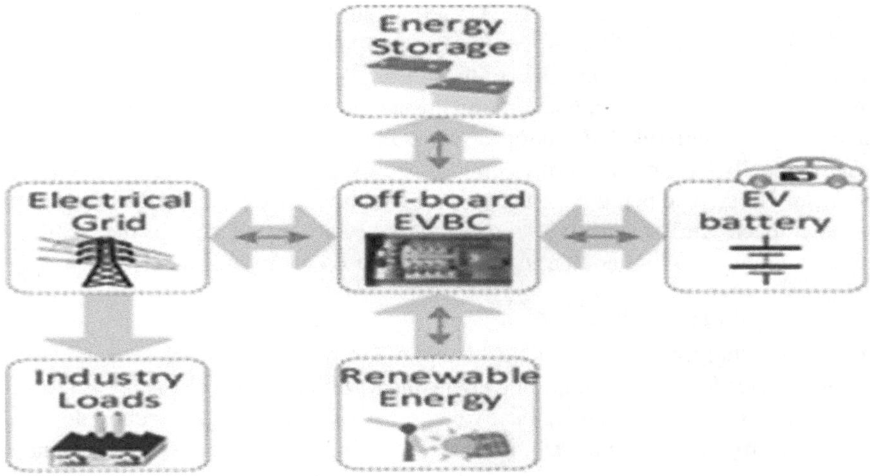

FIGURE 14.24 Off-board EVBC as power quality compensator with RES and storage [8].

14.7 ISSUES, IMPACTS, AND MITIGATIONS AT THE CUSTOMER SIDE, DISCOM CONTROL CENTER, AND THE CHARGING STATION

This sections deal with issues arising due to the integration of EVs with the grid and its impacts on the grid parameters and solutions to the impacts of EV charging at the customer side, DISCOM control center, and the charging station (grid side). The applicant must ensure a reliable protection system detects faults and isolates equipment, ensuring grid stability, and conduct an adequacy and stability study before connecting to the electricity system.

14.7.1 Impact on the Power Network

Unchecked EV charging can lead to a number of issues with the power network, such as voltage limit violations, component overloads, power system losses, phase imbalance, and power quality concerns. However, in addition to EV-specific factors like adoption rate and charging intensity, such as topology and current loading, the level of impact also depends on circuit-specific factors.

14.7.1.1 Component Overload

Expected load and temperature are important factors used, among others, to size the power network components. Because they do not benefit from spatial diversity, LV feeders and distribution transformers are most vulnerable to overloading from EV clusters. Transformers can be temporarily overloaded without risk, but if the operation is not balanced with sustained periods below the load rating, the transformer life will be shortened. Aging and the temperature of the winding hot spot are exponentially related. Loading and surrounding temperature are two factors that determine hot spot temperature.

Distribution transformers in the United States of America (USA), which frequently have a rating of 25 kVA, are the focus of most studies in this field. Based on an average life of 20.55 years, the study's findings indicate that uncontrolled charging can have a significant impact on transformer aging, with a projected transformer life of 6.7 years with two EVs charging [20]. On European distribution transformers, which typically have higher ratings up to 1,000 kVA, charging a single EV will have less of an effect. The percentage increase in loading over the load rating determines the EV impact. Similar sensitivity results might be anticipated, such as the significant effect on aging with higher charge rates and more EV charging.

14.7.1.2 Voltage Control

EV battery charging will raise the demand for power in distribution networks. A large voltage loss on distribution feeders due to a high EV uptake is projected [21]. When voltage control is needed, local voltage control is the simplest technique to use. The processing of voltage control actions varies from standard methods depending on the kind of grid. Reactive power regulation is insufficient in these networks to sustain effective system functioning, particularly in LV networks when the X/R index is low. It has been established that reactive power injection cannot effectively restore voltage sags. Reducing the load or adding active power is more effective in the current situation. The local control action is therefore based on a droop that regulates active power in accordance with voltage variances that may be caused by normal or abnormal grid operation.

A potential implementation of a voltage droop control in EV is shown in Figure 14.25. In stronger systems, this dead zone will be between 0.95 and 1.05 p.u., whereas it is anticipated to be between 0.9 and 1.1 p.u. DISCOM/DSO will implement coordinated voltage control using a process based on information gathered for voltage control logic.

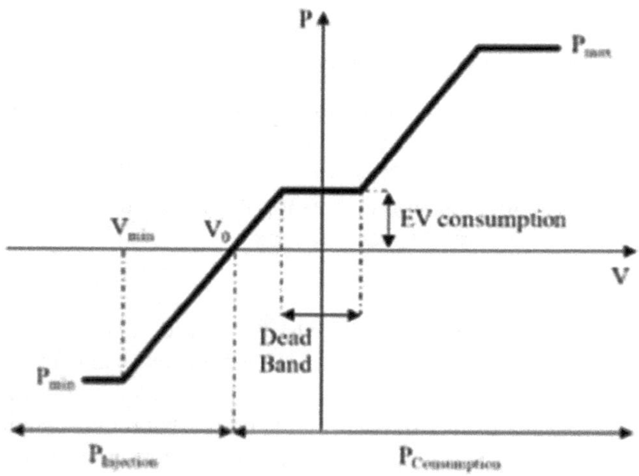

FIGURE 14.25 Voltage droop control in EV.

14.7.1.3 Voltage Limit Violation

For a brief period of time when an EV is first charging, a large inrush current will be drawn. Manufacturers will be accountable for any issues if this goes over the DSO guidelines. If this results in voltage level deviations, the DISCOM/DSO may request a soft starting to correct the voltage problems.

Due to power electronics equipment, the system will experience current harmonics, and if the grid impedance is high enough, the harmonics will have an impact on the grid voltage, which could result in harmonic violations in the grid.

The voltage drop on the medium voltage and LV network is proportionate to the load and is brought on by the increased power demand from EV charging. Due to the fact that transformers stepping down to LV typically do not use on-load tap changers, the effect of EVs on voltage drop is greater on the LV network. The most severe voltage limit violations will probably occur on rural networks, especially for users farthest from the transformer, as longer feeder lengths result in bigger voltage decreases. The location of the EV connection points is important, as well as the customer's location.

14.7.1.4 Reactive Power Issues

Since the battery's energy is in the form of DC, an AC-DC converter must convert AC to DC. To prevent reactive power flows that are wasteful, the usual guideline for converters is that the power factor shouldn't be less than 0.95. Higher reactive power flow when the power factor is lower. Reactive power flow in the line needs to be kept to a minimum since it strains the grid by increasing the system's overall current. The longer the assets are in use, the more heat is lost as a result of the increased current.

14.7.1.5 Power System Losses

Due to higher load requirements, the variable element of electrical system loss rises with EV charging. Given that variable losses are proportional to the square of current, the peak load period is important. Due to this non-linear relationship, power network operators may have a serious problem with the losses brought on by high market EV adoption rates. Studies in this field concentrate on the consequences of growing EV penetration.

According to one study, when EV penetration reached 50%, daily energy losses rose by 140%. Feeder length and phase imbalance also have an impact on power system losses, despite not being studied in the articles we evaluated. A PWM charger has THD values of 0.18% vs 28.16% for a charger based on thyristor rectification. Filters can be used to further lessen this distortion. Additionally, according to the manufacturers, the chargers have a sine-wave current draw and a unity power factor. Under normal operation, charging EVs with PWM chargers won't result in power quality issues, even with high charge rates and a large number of EVs.

14.7.1.6 Power Quality

EVBCs convert AC electricity to DC power using power electronics. Harmonic distortion in voltage and current can result from this conversion process. The THD of the current consumed by the EV will alter as a function of time when the EV charge controller transitions through various charging phases. When numerous EVs are connected in the same feeder, the distortion will rise. Transient loads arising due to overload charging in EV charging stations, voltage stability limits of EV charging, and HCRF due to custom power devices (eg: MSALC) installed at the EV charging stations needed to be considered [22–24].

14.7.1.7 Voltage Flicker

When an EV charger is attached or unplugged faster than the remaining generator in the grid can correct for the fluctuation, voltage flicker may occur. This causes voltage excursions, which customers may experience as light flashing or inconsistent motor operation. Additionally, the equipment that tries to maintain a constant voltage level may become worn out by these variations. Phase selectors or symmetrical three-phase charging of EVs can both reduce voltage imbalance.

14.7.1.8 System Imbalance

A system imbalance results from charging EVs from single-phase sources (households), which results in a differential in voltage and current from one phase to the next. Zero sequence components are produced by the system imbalance. This increases the neutral line in a system with a star connection or circulates in a system with a delta connection, which causes conductor heating.

14.7.2 Impacts of EV Charging on the Grid

The operation and planning of the power system will be impacted by the integration of EVs and the charging of large fleets connected to LV networks. The effects of these new loads, which were not anticipated during the planning stage of the electricity system, will be felt throughout the entire system and will be greatly influenced by regional variables such as the penetration of EVs, the energy mix, the grid topology, and the driving habits of EV owners, among others.

14.7.3 DISCOM Issues

The various DISCOM issues due to the penetration of EV in the grid are listed and explained.

14.7.3.1 Network Planning

Forecasting future load growth with EVs is more difficult. Grid reinforcements might be necessary, which might not be covered by the permitted earnings. If DSOs are in charge of building the infrastructure for charging, there is a considerable danger of loss unless cost recovery is guaranteed by law. In the absence of this, rival networks might emerge.

14.7.3.2 Grid Operation

The distribution system will use more power as EV penetration increases, necessitating the development of extra system infrastructure by the DISCOM. Inefficient EV charging, especially in long radial networks, may result in severe voltage drops on heavily loaded lines. EV inverters, which convert power from DC to AC, can lead to power quality issues, such as harmonics. Relying on EVs to deliver system services entails some non-compliance risks, which the DISCOM and EVS-A/CPM should manage and communicate.

14.7.3.3 Metering

EVs will require smart meters presently in use to adapt. A sizable roll-out of new meters would be necessary if dedicated meters are necessary to separate EV charging from other electricity consumption. The extent will rely on the adopted market models. Regulation should specify the function of DISCOMs and how to pay for the installation of new meters.

14.7.3.4 Tariff Setting

Energy tariffs may distort the incentives offered by distribution charges due to a lack of design freedom to create cost-reflective distribution charges. Some distribution areas may experience issues as a result of this.

14.7.4 DISCOM Opportunities

Increasing energy distribution could boost DISCOM revenues under certain regulatory frameworks. EV inverters can regulate local voltage and power, compensate for high Distributed Generation (DG) production during valley hours, and provide value to investments in demand side management, particularly in car parks and charging stations, providing another incentive for demand side management installation. EVSA may introduce new services relating to power flow control at the distribution level that could take the place of traditional operating procedures in order to reduce interruptions, control traffic, or schedule repair procedures. Large EVSA may interact with a DISCOM control center and offer services like demand forecasting, which makes it easier to operate the grid.

14.7.5 Mitigation of Impacts of EV Charging

The most likely effects of charging installations on the grid are transformer and distribution modifications, as was mentioned in the preceding section. These effects can be reduced using a variety of tactics, such as picking areas with minimal grid impact, smart charging, and fusing fast charging with energy storage. Below mentioned are some potential mitigation strategies.

14.7.5.1 Load Shifting Using Time of Use Tariff

Imposing a separate tariff during peak and off-peak time periods could be a key step in reducing the impact or even better management of the current peak demand for energy. This would allow for shifting demand from peak moments to lower demand periods.

14.7.5.2 Specific Network Mitigation Measures

By raising the load rating of components that are likely to be overloaded, network reinforcement can lower component overloads. Currently being evaluated as potential remedies for voltage limit violations are in-line voltage regulators and distribution transformer on-load tap changers. The location of the fast charging station will be crucial for network or transformer upgrades.

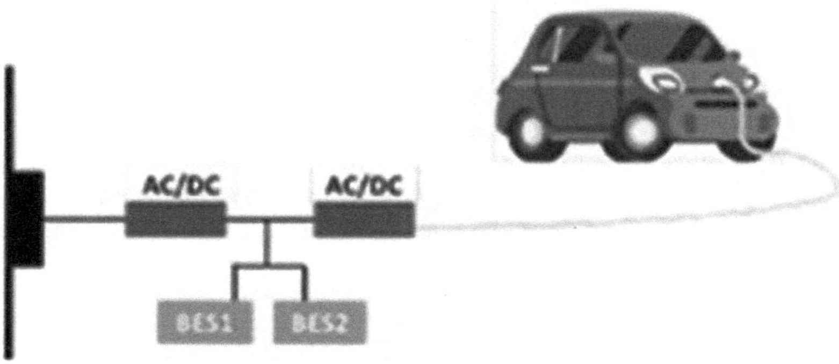

FIGURE 14.26 DC fast charging station with the Battery Energy Storage Systems.

14.7.5.3 Charging Control Strategies

The amount of power system losses, the number of overloaded components, and voltage limit breaches can all be changed by adjusting the EV charging time. The simplest charging management strategy is a dual tariff, but there are others as well. Dual tariff is a less sophisticated approach than smart charging, which involves the active oversight of EV chargers. The utility or a third-party aggregator may provide certain chargers instructions for when to start and stop charging or how much power they can draw. EV charging may be managed in real time while the network is being actively monitored, preventing voltage limit breaches and component overloads.

14.7.5.4 Usage of Battery Energy Storage Systems

In the design stage of DC fast charging stations for EVs, the mitigation of harmonics with Battery Energy Storage Systems is covered in reference [25,26]. Additionally, it suggests using an energy storage system within the DC fast charging station to reduce the detrimental effects on the grid, as shown in Figure 14.26. The proposed system lessens the burden on the LV grid and the duration of charging. In this, a DC fast charging station is created by joining two identical Battery Energy Storage Systems.

14.7.5.5 Voltage-Dependent Charging Strategies

Strategies for charging that depend on voltage will help to reduce the effects of a high EV penetration in the distribution system. Various controls utilized for voltage-dependent charging techniques are presented now a days. The charging time and efficiency will increase if an EV is attached to a LV feeder since it will be "penalized" with less reference current.

14.8 SYSTEM IMPACTS RESULTING FROM THE PRESENCE OF EVS

14.8.1 EV Energy Requirements and Load Profile Changes

The study identifies the maximum percentages of EVs that can be safely integrated into the urban network, based on charging strategies and weekly load diagram changes as mentioned in Figure 14.27. The Urban Network has 5,072, 7,186, and

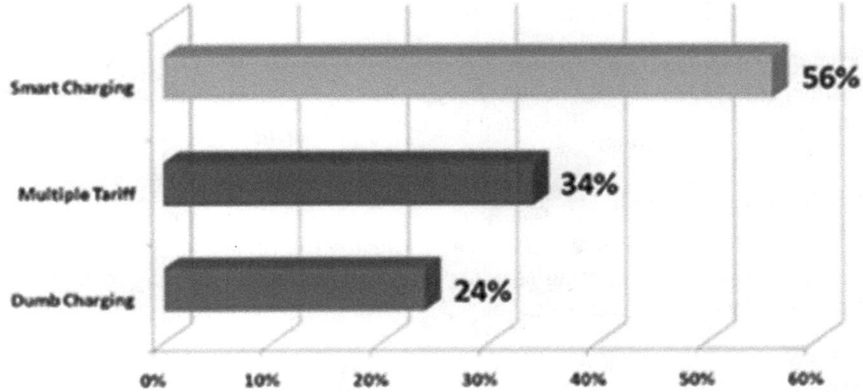

FIGURE 14.27 Urban network EV integration percentile (maximum).

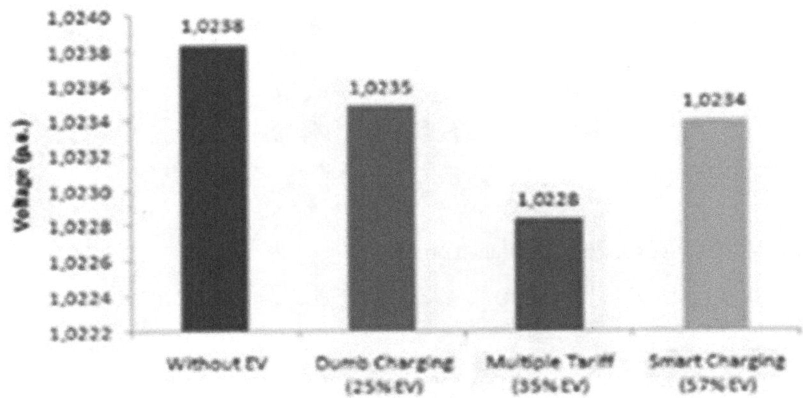

FIGURE 14.28 Urban network worst bus voltage profile.

11,836 EVs safe for dumb charging, multiple tariffs, and smart charging, respectively. Line/cable overloading limits EV integration, particularly in dumb charging and multiple tariff cases.

The network's peak load increases with dumb, multiple tariffs, and smart charging, with smart charging achieving a 57% integration rate for 12,047 EVs, resulting in a 3.6 MW increase.

14.8.2 Effects on Voltage Profiles

As illustrated in Figure 14.28, an urban network worst bus voltage profiles are obtained when maximum allowable EV integration is reached, varying from peak load hours. Additional EV demand reduces voltage values, but the high X/R ratio in medium voltage networks makes the impact of active power consumption less relevant. Urban areas are more susceptible to congestion issues.

14.8.3 Line/Cable Congestion Levels

The study shows that branches' load levels worsen with the growth of EVs in the grid, limiting further EV integration, and charging methods applied to the most congested branch at a specific hour as illustrated in Figure 14.29.

14.8.4 Energy Losses

As illustrated in Figure 14.30, the chart evaluates the impact of EV charging on network energy losses, plotting absolute losses for different charging methods and relative energy consumption values.

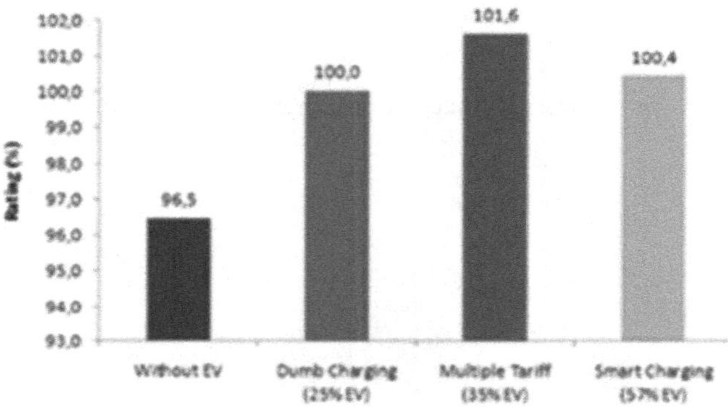

FIGURE 14.29 Network's worst branch rating.

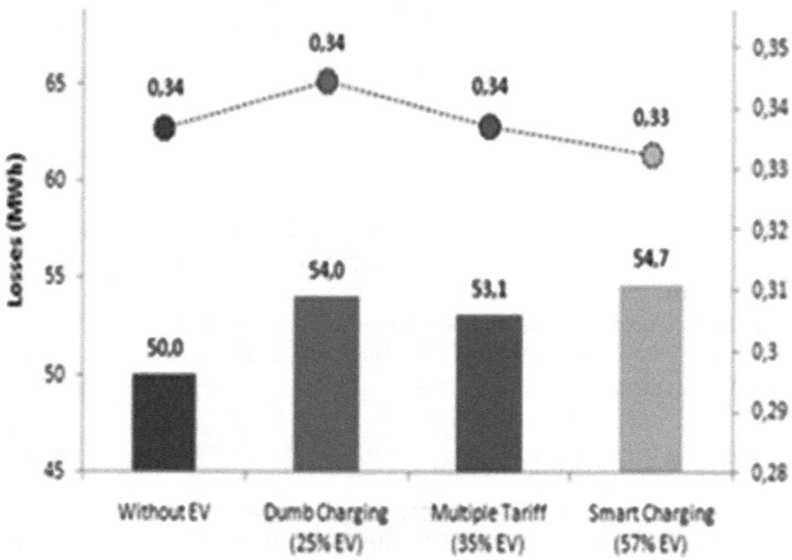

FIGURE 14.30 Energy losses in the network.

TABLE 14.2
Average Battery Sizes in India

Vehicle Type	Battery Rating (kWh)
Bus	120
4W – PV	15
3W – PV	4
4W – CV	15
3W – CV	7
2W	1.5

The system can handle EV penetration without affecting network infrastructures. The maximum number of EVs safely integrated depends on charging schemes. A smart charging strategy yielded better results, allowing higher integration levels without violating network technical restrictions. Urban networks face branch/transformer overloading problems faster than voltage drop issues, while rural networks have long radial lines causing voltage drops. EV integration changes network load profiles, and fast charging stations should be carefully analyzed for potential voltage violations or branch overloading.

The study calculates charging patterns for different vehicle categories using average battery size, efficiency, km run/day, distance per charge, charging cycles, full charging time, and charging rate at different locations, considering existing EV battery sizes in the Indian market (Table 14.2).

14.9 CONCLUSION

This dissertation will provide an overall review for the research community performing research on the grid impact of EV charging. Various EV charging technologies, factors affecting charging technologies, different methods of charging, EV charging station requirement for grid infrastructure, and the impacts of the grid due to charging of EV were considered in the review. Various power quality issues and their performance parameter such as THD and HCRF were also taken into consideration. Finally, the mitigation of impacts due to EV charging was considered in this review, thereby providing an overall assessment of the EV charging impact on the grid.

REFERENCES

1. Chittenden County Regional Planning Commission. (2014, June). Electric vehicle charging station guidebook: Planning for installation and operation. https://www.ccrpcvt.org/wp-content/uploads/2016/01/EV-Charging-Station-Guidebook.pdf
2. Electric Power Research Institute (EPRI). (2011). Transportation electrification: A technology overview. https://www.epri.com/research/products/1022649
3. Ministry of Heavy Industries & Public` Enterprises, Department of Heavy Industry. (2017, May 15). Standardization of protocol for charging infrastructure. Government of India.
4. Ministry of Power, Charging infrastructure for electric vehicles – Guidelines and standards, 14 December 2018. Available: https://www.mohua.gov.in/upload/whatsnew/5c6e472b20d0aGuidelines%20(EVCI).pdf

5. Richard, S. P. Titus, S., Efficient Energy management using Red Tailed Hawk optimized ANN for PV, Battery & Super Capacitor driven Electric Vehicle, *Journal of Energy Storage*, 2025, 117 (116126).
6. Ponraj, R. P., Sigamani, T., Ravindran, V., Analysis of multi input multilevel boost inverter circuit with optimal firing angles using dSPACE, *International Journal of Electrical Engineering & Technology*, 2023 18(06), pp. 1135–1145.
7. Ponraj, R. P., Sigamani, T., Subramanian, V., A developed H-bridge cascaded multilevel inverter with reduced switch count, *International Journal of Electrical Engineering & Technology*, May 2021, 16(03), pp. 1445–1455.
8. Monteiro, V., Afonso, J. A., Ferreira, J. C., Afonso, J. L., Vehicle electrification: New challenges and opportunities for smart grids, *Energies Journal*, 2018, 12, p. 118.
9. Muralikrishnan, G., Mohanty, N. K., Modified shunt active line conditioner using enhanced self-restoring technique with step size error elimination algorithm, *Advances in Intelligent Systems and Computing*, 2019, 846, pp. 237–259.
10. Muralikrishnan, M. G., Mohanty, N. K., Fundamental sequence component based hysteresis current controlled MSALC, *Energy Procedia*, 2017, 117, pp. 1093–1100.
11. Muralikrishnan, G., Mohanty, N. K., Performance analysis of modified shunt active line conditioner (MSALC) using notch adaptive filter, *International Journal of Innovative Technology and Exploring Engineering*, 2019, 8(7), pp. 1863–187.
12. Muralikrishnan, G., Mohanty, N. K., Performance analysis of modified shunt active line conditioner (MSALC) using kalman filter trained by neural network, *Journal of Advanced Research in Dynamical and Control Systems*, 2019, 11(2S), pp. 689–698.
13. Muralikrishnan, G. et al., Improved binary quantum based elk herd optimizer for optimal location and sizing of hybrid system in micro grid with electric vehicle charging station, *Journal of Renewable and Sustainable Energy*, 2024, 16(6), pp. 064101–064114.
14. Muralikrishnan, G. et al., Optimal allocation of renewable energies in distribution network considering flicker mitigation using improved approach, *Journal of Energy Storage*, 2025 116, e116034.
15. National Fire Protection Association, NFPA 70: National Electrical Code, Article 625 – Electric Vehicle Charging and Supply Equipment Systems, Quincy, MA: NFPA, 2023.
16. SAE, Electric vehicle and plug in hybrid electric vehicle conductive charge coupler J1772_201001, Line. Available: https://www.sae.org/standards/content/j1772_201001.
17. Nivedha, M. Titus, S., IoT-based monitoring of smart grid using high-gain converter with optimized maximum power point tracking, *International Journal of Electrical Engineering & Technology,* 2024, 106(3), pp.2297–2311. DOI:10.1007/s42835-022-01186-0.
18. Gabriel Santhosh Kumar, G. Titus, S., Hybrid Artificial Rabbit Optimization and Perturb & Observe MPPT for Grid-Connected PV System, *International Journal of Electric Power Components and Systems*, 2024, 52(11).
19. Vennila, C., Muralikrishnan, G., Malathi, G., Banoth, R., Raj, J. I. D., Designing and performance analysis of a concentrated solar power system in cold arid high DNI area, *International Journal of Intelligent Systems and Applications in Engineering*, 2024, 12(2s), pp. 475–486.
20. Gong, Q., Midlam-Mohler, S., Marano, V., Rizzoni, G., Study of PEV charging on residential distribution transformer life, *IEEE Transactions on Smart Grid*, 2011, 3(1), pp. 1–9.
21. Moreira, C. L., Rua, D., Extend concepts of MG by identifying several EV smart control approaches to be embedded in the smart grid concept to manage EV individually or in clusters, Mobile Energy Resources in Grids of Electricity, Deliverable D1.2, June 2010. Available: https://www.ev-merge.eu.
22. Muralikrishnan, G., Mohanty, N. K., Integration of CFA^2FB control scheme with modified shunt active line conditioner (MSALC) connected distributed system for mitigation of harmonics, *International Transactions on Electrical Energy Systems*, 2020, 30(12), p. e12652.

23. Muralikrishnan, G., Mohanty, N. K., Harmonic current reduction analysis of traditional PI and enhanced FoPID controller-based MSALC, *Journal of Electrical Engineering and Technology*, 2021, 16(4), pp. 1769–1777.
24. Muralikrishnan, G., Mohanty, N. K., Seven levels fuzzy rule-based controller for modified shunt active line conditioner, *International Journal of Electronics*, 2022, 109(4), pp. 617–651.
25. Gjelaj, M., Træholt, C., Hashemi, S., Andersen, P. B., Optimal design of DC fast-charging stations for EVs in low voltage grids, In *Proceedings of 2017 IEEE Transportation Electrification Conference*, Chicago, IL, 2017.
26. Arun Prasad, D., Muralikrishnan, G., Navaneethan, C., Meenatchi, S., A novel modified switched capacitor multilevel inverter using SARC-DQRLC controlling mechanisms for grid systems, *International Journal of Hydrogen Energy*, 2023, 77, pp. 40–53.

15 Renewable Energy-Integrated EV Charging with Advanced Power Conversion

Anitha Perumalsamy, Karthik Kumar Kalirajan, and Kamaraja Arunavathi Soundarapandian

15.1 INTRODUCTION

DC-DC converters play a pivotal role in electric vehicle (EV) battery applications, serving as vital components for voltage regulation, power management, and energy efficiency enhancement. These converters are instrumental in maintaining optimal output voltage levels, ensuring consistent performance, and protecting sensitive electronic systems. Their versatility in accommodating various battery chemistries and effectively distributing power among different subsystems contributes significantly to optimizing energy usage, extending driving range, and improving overall vehicle performance. Additionally, DC-DC converters facilitate the conversion of regenerated energy from regenerative braking into usable power, thereby enhancing efficiency and increasing range. Their compact and lightweight design not only maximizes space utilization but also minimizes weight, making them indispensable components in the design and operation of contemporary EVs.

Due to their adaptability, isolated-type DC-DC conversions and linked inductance conversions that make use of a high-frequency transformer are able to achieve substantial voltage gains [1]. A two-stage boosting conversion, a linked inductance, and a multiplying cell are suggested in [2] for use in an ultrahigh step-up DC-DC conversion. The multiplying cell unites the connected inductor's second side. A high-efficiency operation is also caused by the fact that the linked inductor's leaking power is properly transported to the outputs and reused. The converter's key benefits include a tiny inductance being required on the low voltage side, a minimal voltage load on the electric toggles, and a large voltage gain. An improved version of the traditional boosting conversion called a resonant modular multilevel DC-DC converter is suggested in [3]. Instead of the diodes and single-switch used in the traditional boost architecture, floated capacitance clamped submodules and switched-inductor cells serve as the inductance. A large voltage conversion ratio at moderate or lower duty cycles was achieved by combining the coupled inductance (CI) with other voltage enhancing methods, which allowed for the development of conversions.

 DOI: 10.1201/9781003496946-15

The reduced shifting energy losses due to the lower voltage load on the energy switching are an additional benefit of these conversions [4]. There have been several proposals for DC-DC conversions that use various boosted approaches to achieve higher voltage gains and increased conversion efficiencies [5]. DC-DC conversions that use switched-capacitor technology are compact, dependable, and capable of achieving high voltage gain. But these converters aren't very efficient since they work under severe shifting circumstances (with losses during shifting). Due to their high component count, multiple-stage DC-DC conversions are inefficient and bulky, and they also cost more to produce. In [6], large step-up DC-DC conversions were achieved by using the switched-capacitor approach. To get the most out of natural power resources, current-fed buildings are the way to go. Conversely, conversions with high ripples provide an incoming power that is neither uninterrupted nor constant, which disrupts the highest energy monitoring mechanism and is damaging to sensitive inputs [7]. High step-up conversions have been newly produced by integrating certain voltage-boosting methods with Z-source and quasi-Z-source systems. For conventional voltage-fed and current-fed inverters, the issues of shoot-through and low output voltages, together with the need to attain a buck-boost voltage gain, led to the first proposal of Z-source and quasi-Z-source designs [8]. The design of the suggested conversion in [9] also makes use of the two cells with switched-capacitors and the power-storing cell. Lower duty cycle values are produced by them, leading to significant voltage increases. Also, the voltage load on the energy components is negligible, being less than 50% of the supply voltages. Subsets of the short circuit currents include voltage multiplier (VM) blocks, which are affordable and effective circuitry with basic architectures that produce a large DC voltage value. VM blocks consist of a mix of diodes and capacitance. Producing higher inrush electricity charges or discharging the capacitor is a major issue with these conversions. A workaround for this problem is to combine the VM components with connected induction devices [10]. Since they are more space and money efficient than separated structures, multi-input nonisolated DC-DC conversions are a favorite in the field of hybrid power integration. This article introduces a tiny dual-input hybrid step-up DC-DC converter. The switching inductance approach is the basis of dual-input hybrid step-up DC-DC converter's operation [11]. It follows that reversible DC-DC conversions based on a large voltage converter ratio are required to link the lower super capacitor voltages to the higher DC-link voltages. The traditional reversible DC-DC conversions run with a very high duty cycle—<0.1 when the converters are in the step-down phase and >0.9 when they are in step-up mode—to achieve a higher voltage conversion ratio. In addition, the converters' performance and transient functioning are compromised during severe duty cycle operations [12]. An offshore wind farm-specific high-power step-up DC/DC conversion using transformers, half-bridge submodules, thyristors, and diodes is suggested in [13]. Fewer devices, lower prices, fewer losses, and less electricity storing demand are comparable to other promising DC/DC architectures.

VMs, switching capacitors, switching inductors, and transmitted connectors are some of the boosted strategies used by step-up DC-DC conversions [14] to achieve large voltage gains. Nevertheless, the efficiency and energy densities of these circuits are diminished due to the numerous devices used in their operation under harsh shifting situations. A novel DC-DC conversion with a shared connected switching quasi-Z-source is suggested in [15]. The suggested conversion differs from the standard pseudo Z-source converter in that it employs an extra switch and a diode. A connected inductance and switching capacitance unit generates various advantages in the suggested conversion. These characteristics allow for a constant incoming current wave, little voltage load on semiconductors (switches and diodes), and the capacity to raise voltages within a narrow duty cycle region. One of the requirements for DC-DC converters in a green energy generating system is a constant incoming electricity, in addition to a high voltage gain. The input power source may be protected to some extent by maintaining a constant input current. Furthermore, it helps make the maximum power point tracking (MPPT) of photovoltaic panels a reality [16]. Because of its constant input current and straightforward design, the classic boosted conversion is all the rage.

For low-power applications requiring high-gain voltages, including high-intensity mobility discharging lights and battery-powered light emitting diode light structures, a nonisolated DC-DC high step-up converter using a passive switched-inductor-capacitor system is suggested in [17]. Activated switch experiences a lower voltage load and transmission energy losses because of the suggested circuit's ability to provide a larger gain voltage with a shorter duty cycle. A reversible electrical car charging station uses a DC-DC reversible step-up overlapped converter connected by a center capacitance in a transmitted architecture to charge and discharge an electrical vehicle's batteries. A single (coupled) framework is suggested in this paper that incorporates all switch periods into one model. This will enable more precise evaluation and improved regulator development [18]. One way to greatly enhance the voltage gains of a conventional DC-DC conversion is to include a CL cell and choose an appropriate CL turns ratios n [19–21]. Right now, the CL's turns ratios n controls the output voltages. Keep in mind that the converter's poor efficiency is due to its high leakage inductance energy losses and big turns ratios n [22,23].

The existing literature on power converters for EV charging and renewable energy integration may fall short in several key areas. First, many converters lack the capability to achieve the high voltage conversion ratios necessary for efficient integration of renewable energy sources in EV charging systems. Second, conventional designs may not adequately address the issue of voltage stress on semiconductor components, potentially leading to reliability issues and reduced lifespan. Third, while some converters attempt to reduce MOSFET switching losses, they may not do so effectively, resulting in lower overall efficiency. Finally, the potential for energy recycling from leakage inductance in converters may not have been fully explored in prior research, representing a missed opportunity for improving efficiency and reducing waste heat generation.

Despite advancements in power electronics and converter design, there remains a critical gap in the development of a converter system that can efficiently integrate renewable energy sources into EV charging infrastructure while simultaneously minimizing voltage stress on semiconductor components and reducing switching losses. Existing converters may not fully meet the demands of modern EV charging systems, which require high efficiency, reliability, and compatibility with renewable energy sources. The research community has yet to develop a converter system that can achieve an exceptionally high voltage conversion ratio in a quasi-quadratic form, while also addressing key challenges such as voltage stress, switching losses, and energy recycling. This research aims to fill this gap by proposing a novel QQHBCIC design that offers significant advancements over existing converters. By thoroughly exploring the operational principles, design considerations, and efficiency analysis of the proposed converter, this research seeks to establish a new benchmark for efficiency and performance in power converters for EV charging and renewable energy integration. The key contributions of this research are as follows:

1. Developed a novel QQHBCIC, offering advanced functionality compared to traditional designs.
2. Achieved an exceptionally high voltage conversion ratio in a quasi-quadratic form, surpassing existing converter capabilities.
3. Implemented innovative quasi-resonant operation and utilized CL leakage inductance to effectively reduce MOSFET switching losses.
4. Utilized energy stored in the leakage inductance for output capacitor charging, minimizing waste heat generation, and improving overall efficiency.
5. Conducted comprehensive operational principle exploration and comparative analysis with existing converters in various conduction modes.
6. Demonstrated rapid dynamic response, robustness to load impedance variations, and efficient voltage regulation, ensuring suitability for industrial and EV charging applications.

The article is structured as follows: Section 15.2 delves into assessing converters through propositional and modal analysis, exploring innovative converter architecture, the operational overview of the proposed QQHBCIC system, steady-state performance evaluation, high voltage conversion efficiency in continuous conduction mode (CCM), semiconductor voltage stress analysis in CCM, and insights into current stresses in CCM. Section 15.3 evaluates the dynamic performance of power converters in discontinuous conduction mode (DCM), detailing the conduction boundary between continuous and discontinuous modes and providing operational insights into DCM. Section 15.4 focuses on design insights to enhance the efficiency of the QQHBCIC for quasi-resonant performance, including strategic design factors and theoretical efficiency analysis. Section 15.5 introduces the novel PICR control strategy, analyzing its stability and implementation. Section 15.6 discusses the implementation of the proposed circuit configuration using a MATLAB-based approach, detailing method-specific settings and providing simulation results for the innovative converter design. Finally, Section 15.7 concludes the article.

15.2 ASSESSING CONVERTERS: INSIGHTS FROM PROPOSITIONAL AND MODAL ANALYSIS

15.2.1 Innovative Converter Architecture

Figure 15.1 shows the structure of the proposed designs. This converter could offer a boost/buck conversion ratio owing to its constant current source and minimal voltage strain on the primary power MOSFET as seen in Equation (15.1):

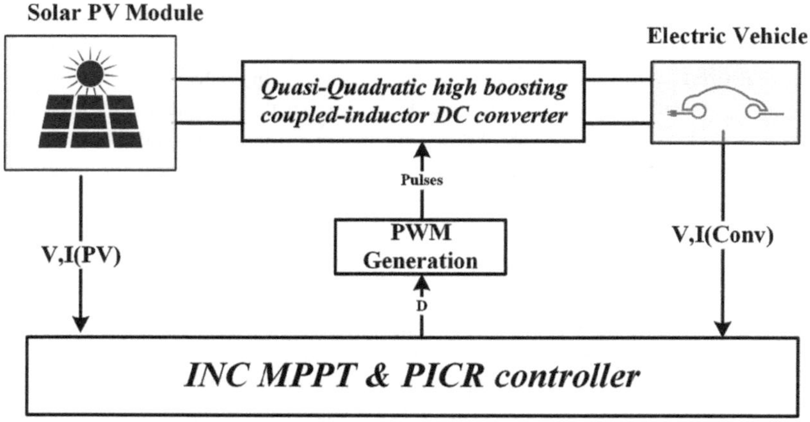

FIGURE 15.1 Schematic illustration of the proposed configuration.

$$m_{ccm(Ideal)} = \frac{D_{st}}{(1 - D_{st})^2} \tag{15.1}$$

When it comes to triggering switches, D stands for duty cycles. This converter lacks an excessive voltage conversion ratio, while having a quadratic formula for the increase in voltage. The optimal voltage enhancement of this conversion, $m=5.2$, is achieved, for instance, in conditions of a high duty ratio $D_{st}=0.23$. The renewable energy source's lower voltage stage, nevertheless, makes ultra-voltage boost step-up converters frequently required. Using CI is an appropriate, easy, and highly efficient way to increase the converter's duty ratio while maintaining high voltage conversion proportions, as previously stated. The proposed equivalent circuitry is shown in Figure 15.2. To get an even higher voltage gain proportion, the suggested converter stacks on the second side of CI toward the end of the converter. In the suggested QQHBCIC, two triggering switches may be used at the same time, a supply inductance called L_s, a current injector, four semiconductor diodes labeled D_A, D_B, D_C, and D_D, and four capacitances labeled C_A, C_B, C_C, and C_D. To improve the voltage ratio of gain and decrease the stress of voltage on the primary power switches, the conversion device utilizes the use of a regenerating, passively clamping capacitance C_B. Furthermore, the output diodes D_D, MOSFETs S_A and S_B, and the quasi-resonating operations (QRO) among the leakage inductive properties of the CI, C_A, and C_B all contribute to an electrical current with a sinusoidal pattern. This gets rid of the diode's inverse recovering loss and lowers the triggering turn-off inefficiencies.

15.2.2 Operational Overview of the Proposed QQHBCIC System

A number of conditions are introduced in order to streamline the steady-state evaluation for the QQHBCIC.

(1) All of the semiconductors used in the conversion are perfect. (2) Because of their size, all capacitances are considered to have relatively stable potentials over a

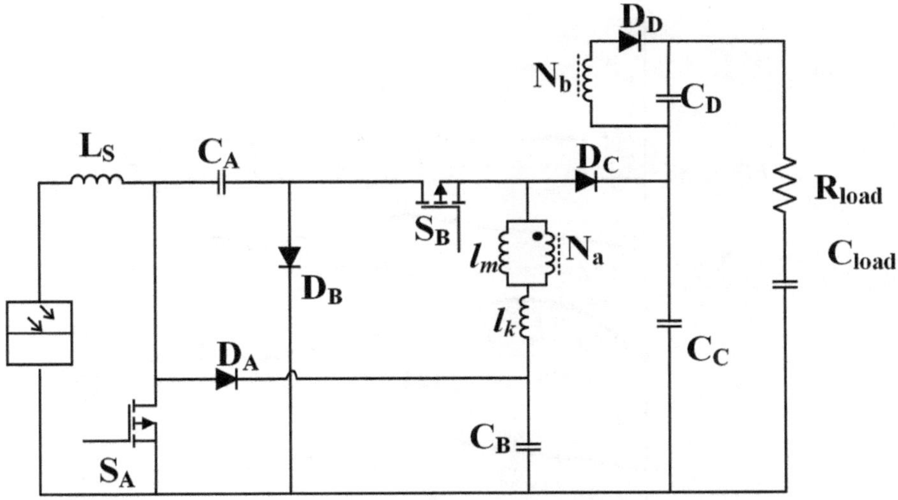

FIGURE 15.2 Schematic depicting the configuration of the proposed circuit layout.

single transitioning cycle, and any potential waves are disregarded. The coupling coefficients $K = l_m / (l_m + l_k)$ and turns ratios $N = N_2 / N_1$ are used to represent the CI. On the main side, there is a magnetic inductance (l_m) (3) and a merged leakage inductance (l_k). (4) The current fluctuations of the inputs and magnetically attracting inductors are disregarded since they are deemed sufficiently significant. During one transitioning phase, Figure 15.3 shows the idealized key patterns of the component currents and voltages of the suggested converter operating in CCM. In addition, as can be seen in Figure 15.4a–c, the suggested QQHBCIC operates through three distinct phases during every shifting cycle.

Phase I [$t_0 < t < t_1$]: While diode D_D is in its turn-ON states, triggering switches S_A and S_B start conducting identically at time $t = t_0$. Concurrently, V_{CA}, V_{CB}, and $V_{CA} - V_{CC}$ alternately bias diodes D_A, D_B, and D_C, correspondingly. In this phase, Figure 15.4a shows the present flow route of the QQHBCIC. After receiving energy from the supply source of voltage V_S via S_A, the current flowing via the input inductance L_S expands. The leakage inductance on the main side of the CI (l_k), together with the balanced (C_A) and clamped (C_B) capacitances, allows the conversion unit to function in QRO during this phase. The currents of both triggering switches S_A and S_B, as well as diode D_D, grow in a symmetrical fashion with the aid of these resonating tanks. This lessens the diode D_D's reversal recovery losses and drastically cuts down on triggering switches' losses when turned off. When the magnetic inductance receives voltages from capacitances C_A and $C_B(V_{CA} + V_{CB})$, the current through it grows linearly. Conversely, via CI and D_D, energy was transferred to the resultant capacitance C_D from the leaking of the inductive element, balanced capacitance C_A, and clamped capacitance C_B. Also, the resultant load is supplied in serial by the capacitor that outputs C_C.

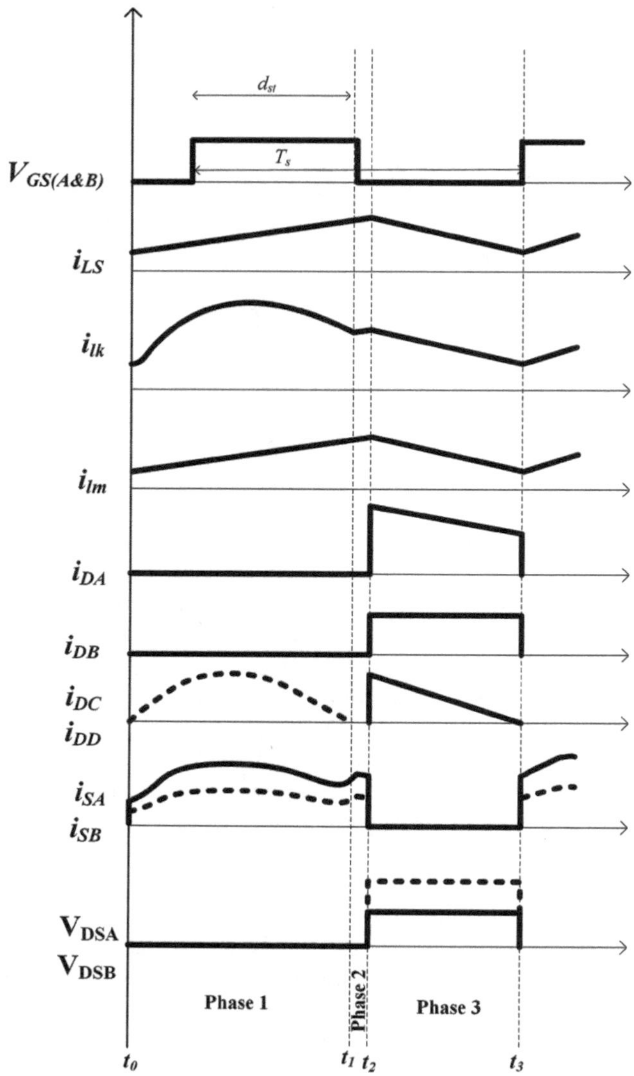

FIGURE 15.3 Essential steady-state waveforms depicting the operational phases of the innovative QQHBCIC across various time intervals.

Using these equations, we can get the frequency that is resonant as seen in Equations (15.2)–(15.6):

$$V_{LS} = V_S \tag{15.2}$$

$$V_{lm} = K\left(V_{CA} + V_{CB}\right) \tag{15.3}$$

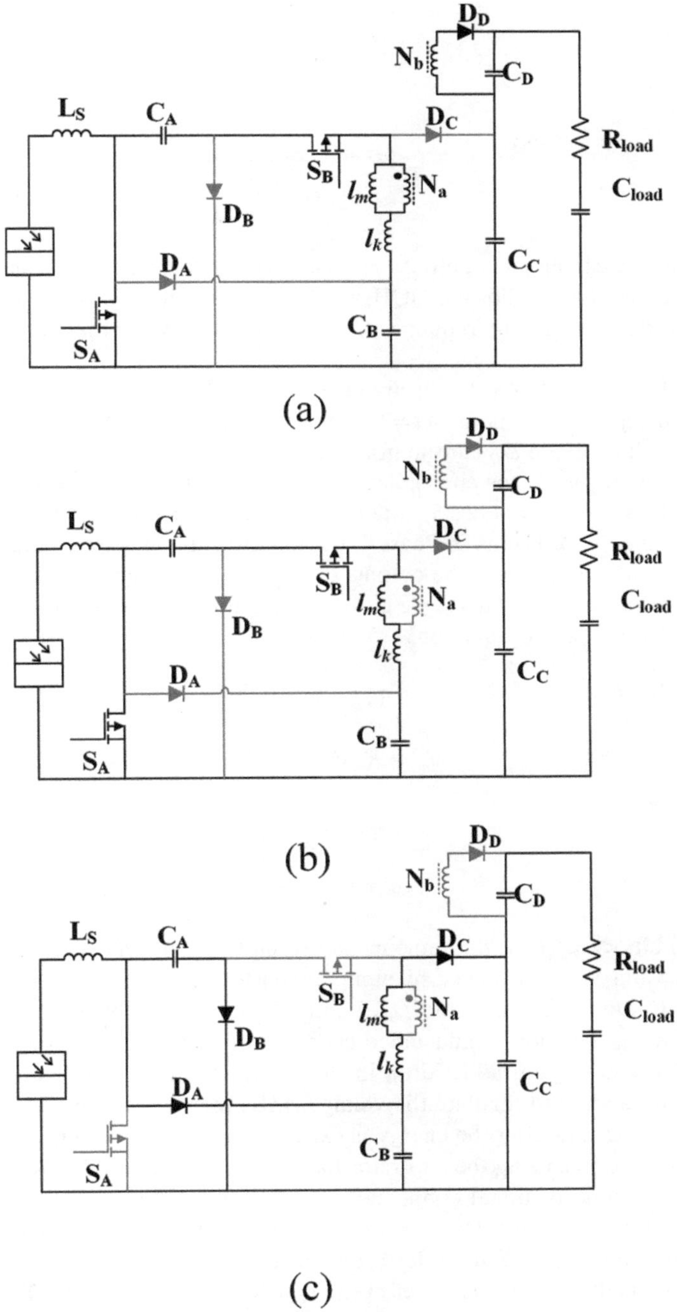

FIGURE 15.4 Path of the current flow in the proposed QQHBCIC across a single switching cycle in CCM: (a) Phase I [$t_0 - t_1$], (b) Phase II [$t_1 - t_2$], and (c) Phase III [$t_2 - t_3$].

$$i_{SB} = i_{lk} \tag{15.4}$$

$$i_{SA} = i_{lk} + i_{LS} \tag{15.5}$$

$$F_r = \frac{1}{T_r} = \frac{1}{2\pi\sqrt{l_k \dfrac{C_A C_B}{C_A + C_B}}} \tag{15.6}$$

The leakage inductor coupling coefficients (K) and resonating frequency (F_r) are defined herein. For the QQHBCIC to function in resonating mode, the converter's resonating frequency ($F_r > F_s$) must be greater than its switching frequencies.

Phase II [$t_1 < t < t_2$]: As demonstrated in Figure 15.4b, the transitional time period begins at the moment $t = t_1$, which is whenever QRO is completed and the resultant diode D_D current gradually approaches zero, lacking reversing recovery losses. At the current level, there is no difference between the currents flowing via switches S_A, the leakage inductance (i_{lk}), and the magnetic inductance (i_{lm}). Just as in Phase I, the inputs of the magnetic inductances' currents are amplified. The outputs of capacitance C_C and C_D also provide current to the loads. This stage allows for the formulation of the subsequent formulas as seen in Equations (15.7)–(15.10):

$$V_{LS} = V_S \tag{15.7}$$

$$V_{lm} = K\left(V_{CA} + V_{VB}\right) \tag{15.8}$$

$$i_{SB} = i_{lm} \tag{15.9}$$

$$i_{SA} = i_{lm} + i_{LS} \tag{15.10}$$

Phase III [$t_2 < t < t_3$]: At the same instant when $t = t_2$, the ON states of switching S_A and S_B are achieved by simultaneously providing a bias that is forward to diodes D_A, D_B, and D_C. Figure 15.4c shows that throughout this time span, the supply inductance energy moves through capacitances C_A and C_B, causing a gradual drop in their current. Capacitances C_A and C_B work in tandem to regulate the voltage stress throughout the main electrical switches S_A. After the energy is extracted from the magnetic inductors, it is used for charging the capacitor that outputs electricity C_C. As a result, there is a unidirectional connection between the currents in the magnetic and leaking inductive devices. It should be noted that the resultant diodes D_C will be turned off at the lowest reversal recovery losses since the magnetic inductive current is lower at the conclusion of this phase. The phase expresses the subsequent equations as seen in Equations (15.11)–(15.13):

$$V_{LS} = V_S - V_{CA} = V_S - V_{CB} \tag{15.11}$$

$$V_{lm} = K(V_{CA} - V_{CC}) \tag{15.12}$$

$$I_{DC} = I_{lk} \tag{15.13}$$

The suggested converters can experience resonant functioning in two distinct states: under resonances ($T_r / 2 < D_{st}T_s$) and over resonances ($T_r / 2 > D_{st}T_s$). The critical phase working ($T_r / 2 \approx D_{st}T_s$) of the conversion is where the triggering and diode reversal recovery deficits are at their lowest, and this will be explored in the following section. The conversion device goes through two distinct operational phases, I and III, during essential and non-resonant operations.

15.2.3 Steady-State Performance Evaluation of the QQHBCIC

15.2.3.1 High Voltage Conversion Efficiency in CCM

Phase II's time change in CCM functioning is so quick that it's easy to overlook. All four parameters from the second part are taken into account in order to streamline the constant-state evaluation. The supply inductance (L_S) voltage is equalized to the voltages of the center capacitance (C_A and C_B) by using the inductive volt-second balancing technique as seen in Equation (15.14):

$$V_{CA} = V_{CB} = \frac{V_S}{1 - D_{st}} \tag{15.14}$$

According to which D_{st} is the duty ratio of triggering switches S_A and S_B. Instead, the following is the expression for the resultant quasi-stage voltage gains in CCM derived via the volt-second balances for the magnetic inductive devices (l_m) as seen in Equations (15.15) and (15.16):

$$V_{CC} = \frac{(1 + D_{st})V_S}{(1 - D_{st})^2} \tag{15.15}$$

$$V_{CD} = NK(V_{CA} + V_{CB}) = \frac{2NKV_S}{(1 - D_{st})} \tag{15.16}$$

Accordingly, the suggested converter's stable potential conversion rate is obtained through combining (15.15) and (15.16) in the following way:

$$m_{ccm(Ideal)} = \frac{1 + D_{st} + 2NK(1 - D_{st})}{(1 - D_{st})^2} \tag{15.17}$$

As a case study, the optimal voltage increase in the suggested converter is $m = 18.76$ when $D_{st} = 0.23$ and $N = 1$, which is 1.5 times more than the conversion rate in traditional conversions (where $m = 5.2$ in $D_{st} = 0.23$). And it's obvious that the leakage inductance modest coupling coefficient ($K = 0.8$ and $K = 0.7$) may reduce the voltage gain ratios. Nevertheless, as this parameter had no substantial impact on the conversion rate at low CI power levels, it may be fixed to one ($K \approx 1$) in Equation (15.17). A perfect voltage conversion rate for the suggested QQHBCIC is therefore given as seen in Equation (15.18):

$$m_{ccm(Ideal)} = \frac{1 + D_{st} + 2N(1 - D_{st})}{(1 - D_{st})^2} \tag{15.18}$$

On top of that, the QQHBCIC's voltage conversion rate grows both continuously with respect to the duty period and proportionately with respect to the turns value for CI. A high voltage conversion rate may therefore be achieved by the suggested converters.

15.2.3.2 Semiconductor Voltage Stress Analysis in CCM

Among the key indications of conversions to pick the right elements is voltage stress. Capacitance C_B and C_A work in tandem to limit the highest voltage strain on the primary triggering switches S_A in the suggested QQHBCIC. The potential stresses between switching S_A and S_B and diodes D_A, D_C, and D_D are determined using the information from the preceding computations as seen in Equations (15.19)–(15.22):

$$V_{DSA} = \frac{V_S}{1 - D_{st}} = \frac{V_{load}(1 - D_{st})}{1 + D_{st} + 2N(1 - D_{st})} \tag{15.19}$$

$$V_{DSB} = V_{CC} = \frac{V_{load}(1 + D_{st})}{1 + D_{st} + 2N(1 - D_{st})} \tag{15.20}$$

$$V_{DC} = \frac{2V_S}{(1 - D_{st})^2} = \frac{2V_{load}}{1 + D_{st} + 2N(1 - D_{st})} \tag{15.21}$$

$$V_{DD} = \frac{2NV_S}{(1 - D_{st})^2} = \frac{2NV_{load}}{1 + D_{st} + 2N(1 - D_{st})} \tag{15.22}$$

Voltage strains across semiconducting devices are clearly oppositely related to each turns ratio of CI (N), as may be shown from (15.19) to (15.22). Consequently, the potential strains are substantially decreased at a bigger turns ratio.

15.2.3.3 Tackling Semiconductor Strain: Insights into Current Stresses in CCM

The mean values of the input voltages and magnetic inductance currents are determined taking into account a lossless performance and ignoring the current rippling of magnetic elements in CCM operations as seen in Equations (15.23) and (15.24):

$$I_{S(Avg)} = mI_{load} \tag{15.23}$$

$$I_{lm(Avg)} = \frac{I_{load}}{1 - D_{st}} \tag{15.24}$$

wherein I_{load} represents the resultant current under load and mI_{load} optimum represents the steady voltage increase. There is no relationship between the turns ratios of CI

(N) and the typical current traveling via the magnetic inductors, as stated in (15.24). The CI component size is going to be drastically reduced because of this. The current strains that flow via the diodes may also be identified using the ampere-second rule on the capacitance C_C and C_D in the following way as seen in Equations (15.25)–(15.28):

$$I_{DC(Max)} = \frac{I_{load}}{1 - D_{st}} \tag{15.25}$$

$$I_{DD(Max)} = \frac{\pi I_{load}}{2D_{st}} \tag{15.26}$$

$$I_{DA(Max)} = \frac{(1 + ND_{st})I_{load}}{(1 - D_{st})^2} \tag{15.27}$$

$$I_{DB(Max)} = \frac{(D_{st} + 2N - 3ND_{st})I_{load}}{(1 - D_{st})^2} \tag{15.28}$$

For several resonating states, namely without QRO, under resonances, and critical phase ($T_r / 2 \approx D_{st}T_s$), Figure 15.5 shows the current form of the converter triggering switches and output diode D_D. The sine pattern of output diode D_D's current pattern is clearly caused by the QRO effects. Figure 15.2 shows that the QQHBCIC's critical function reduces the switches' turn-OFF losses through $2NI_{load} / D_{st}$ and removes the resultant diode's reversal recuperation loss. The effectiveness may be greatly enhanced as a result. The highest and minimum root-mean-square (rms) levels of the switch current are determined by taking critical phase operations into account by applying Equations (15.4), (15.5), (15.23), and (15.24) and are given by (15.29)–(15.34):

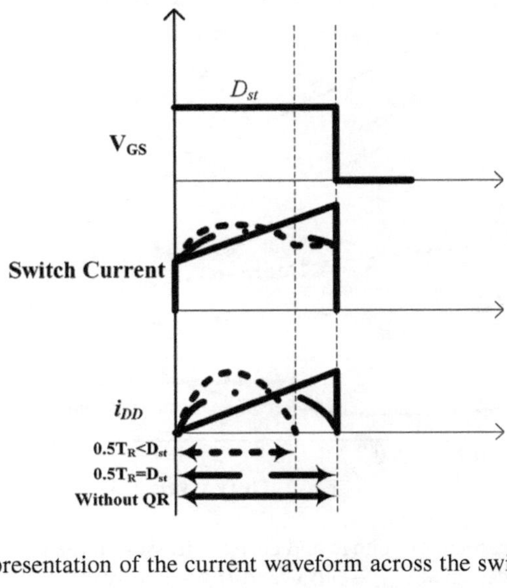

FIGURE 15.5 Representation of the current waveform across the switch and output diode D_D during QRO.

$$I_{SA}(t) = mI_{load} + \frac{I_{load}}{1-D_{st}} + \frac{\pi N I_{load}}{2D_{st}} \sin(\omega_r t) \tag{15.29}$$

$$I_{SA(Max)} = \frac{I_{load}}{1-D_{st}} + \frac{\pi N I_{load}}{2D_{st}} + mI_{load} \tag{15.30}$$

$$I_{SA(rms)} = I_{load} \sqrt{D_{st} \left(\frac{2(1+N-ND_{st})}{(1-D_{st})^2} \right)^2 + \frac{(N\pi)^2}{2D_{st}} + \frac{4N(1+N-ND_{st})}{(1-D_{st})^2}} \tag{15.31}$$

$$i_{SB}(t) = \frac{I_{load}}{1-D_{st}} + \frac{\pi N I_{load}}{2D_{st}} \sin(\omega_r t) \tag{15.32}$$

$$I_{SB(Max)} = \frac{I_{load}}{1-D_{st}} + \frac{\pi N I_{load}}{2D_{st}} \tag{15.33}$$

$$I_{SB(rms)} = I_{load} \sqrt{\frac{D_{st}}{(1-D_{st})^2} + \frac{2N}{1-D_{st}} + \frac{(N\pi)^2}{8D_{st}}} \tag{15.34}$$

Figure 15.6 also shows the primary triggering switches of the suggested QQHBCIC under various voltages and current strains, with $V_{load} = 160$ V and $P_{load} = 800$ W.

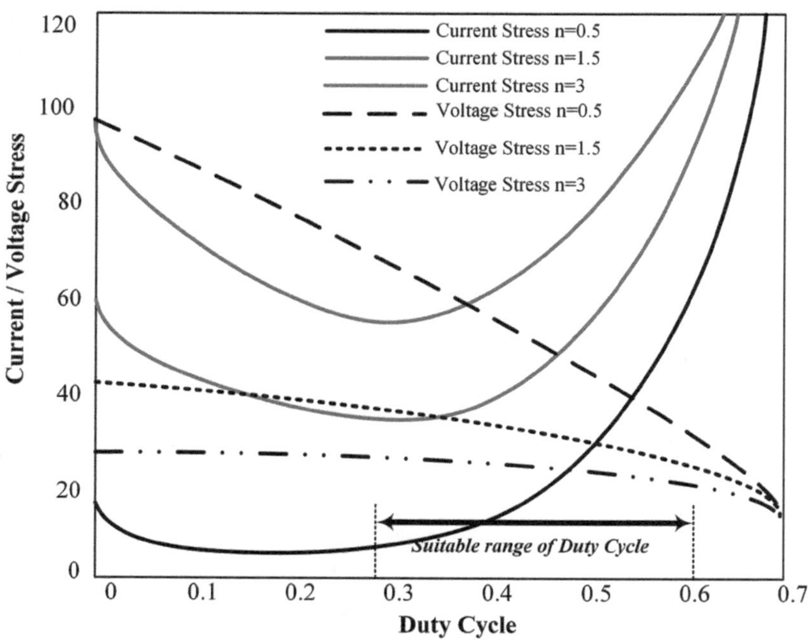

FIGURE 15.6 Comparison of voltage and current stresses on the primary power switch in the novel QQHBCIC design at $V_{load} = 160$ V and $P_{load} = 800$ W.

This graphic shows that increasing the CI's turns ratios reduces the voltage strain on the primary control switch despite increasing current stress. In addition, the ideal duty ratio band for reducing the primary switch's voltages and current strains is from 0.3 to 0.6.

15.3 DYNAMIC PERFORMANCE EVALUATION OF DCM IN POWER CONVERTERS

15.3.1 Deciphering the Conduction Boundary: Bridging Continuous and Discontinuous Modes in Power Conversion

The suggested converters would always enter the state of discontinuous conduction (DCM) during certain circumstances, including very low output loads, a decrease in the frequency of switching, and tiny supply inductance. This state transitions through the CCM to the DCM boundary situation when the supply inductance L_A current drops down to 0 before the subsequent switching phase. In Figure 15.7, we can see the current as well as voltages of L_A during boundary operations.

Both (a) and (b) of Figure 15.8 show the standard deviations of the load side resistances and capacitances and output current as a function of D_{st} and N under the boundary conditions. Figure 15.8c shows that the supply current ripple curve has standardized values. This graph shows that when the CI's turns ratio decreased, the

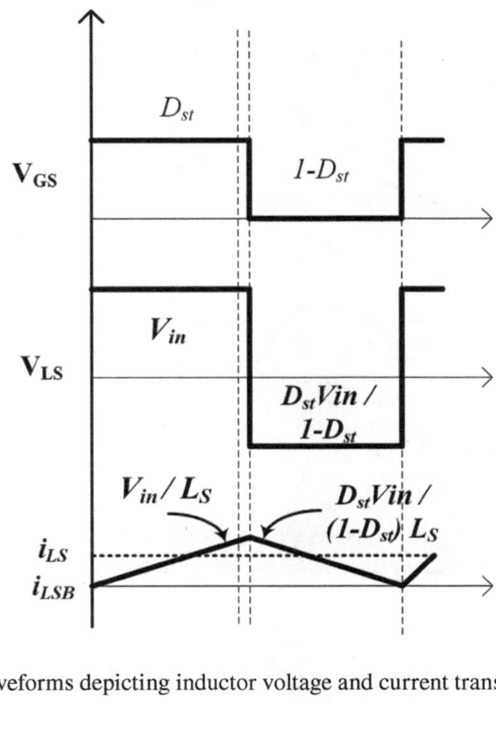

FIGURE 15.7 Waveforms depicting inductor voltage and current transitions between CCM and DCM.

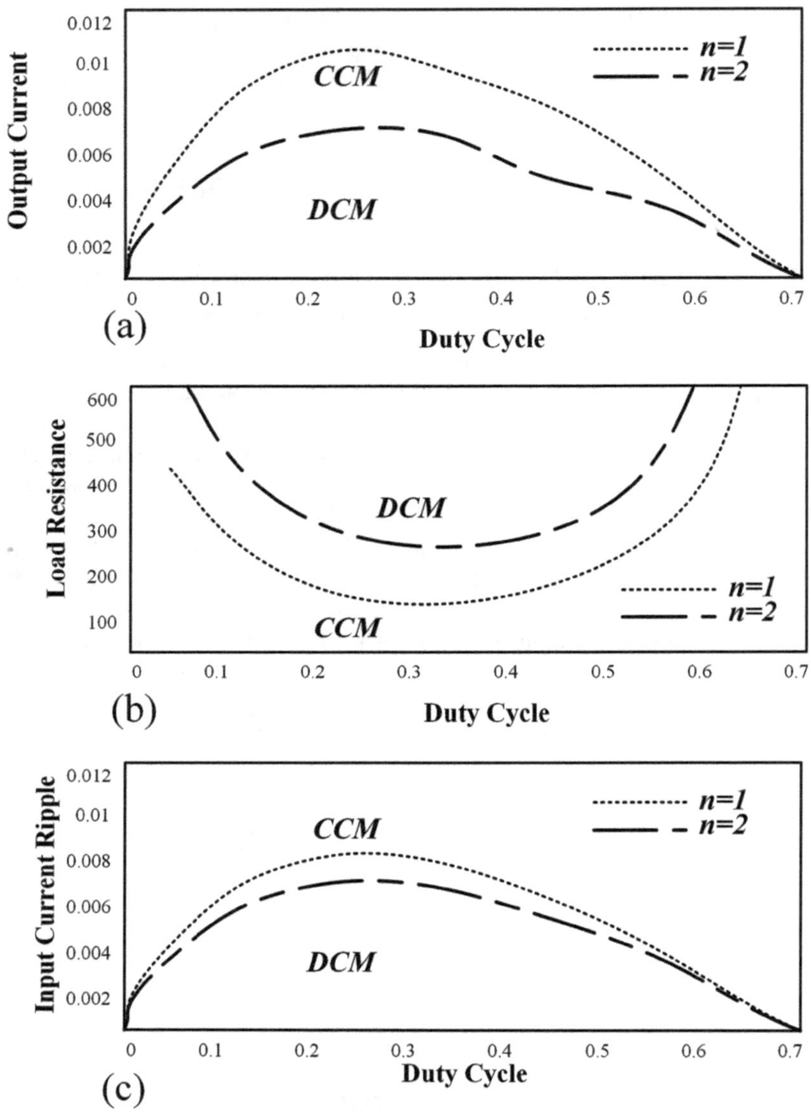

FIGURE 15.8 Standardized (a) output load current, (b) output load resistance, and (c) input current fluctuation with respect to D_{st} and N.

source of current ripples decreased as well. It is also possible to express the standardized current source ripples as seen in Equation (15.35):

$$\frac{\delta i_S}{\dfrac{V_{load}}{L_S F_s}} = \frac{D_{st}\left(1-D_{st}\right)^2}{1+D_{st}+2N\left(1-D_{st}\right)} \tag{15.35}$$

15.3.2 Unveiling the Dynamics: DCM Operational Insights

Figure 15.9 displays the voltages and current patterns of L1 throughout the DCM performances of the suggested converters.

We may obtain the time length of Phase III (15.1) utilizing the volt-second balancing for the supply inductance as seen in Equations (15.36) and (15.37):

$$D_{st}V_S + \delta_A (V_S - V_{CA}) = 0 \tag{15.36}$$

$$\delta_A = \frac{D_{st}V_S}{V_{CA} - V_S} \tag{15.37}$$

Using the volt-second balancing on the magnetic inductance (l_m) yields the following potential for capacitors C1 as well as seen in Equation (15.38):

$$V_{CA} = \frac{1 - D_{st}}{1 + D_{st}} V_C \tag{15.38}$$

By switching out (15.38) for (15.37) and rearrangement, we get Equations (15.39) and (15.40):

$$\delta_A = \frac{D_{st}(1 + D_{st})}{(1 - D_{st})m_C - 1 - D_{st}} \tag{15.39}$$

$$m_C = \frac{V_C}{V_S} \tag{15.40}$$

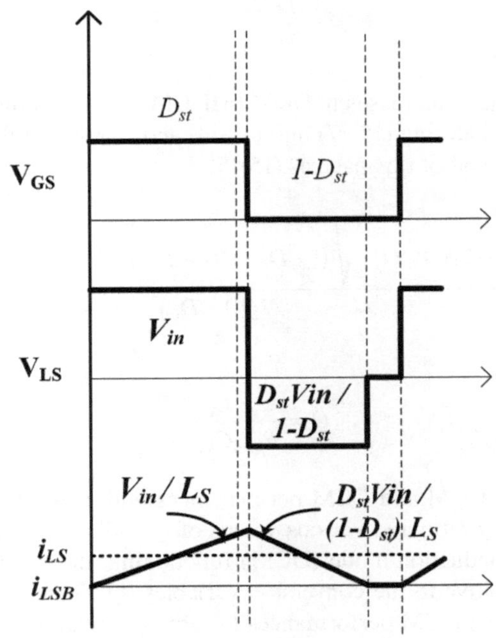

FIGURE 15.9 Waveforms depicting inductor voltage and current in DCM.

In DCM, the resultant semistage voltage gain is given by taking into account Phase I while employing (15.38):

$$m_D = \frac{V_{CD}}{V_S} = 2N\left(\frac{1-D_{st}}{1+D_{st}}\right)m_C \tag{15.41}$$

By combining Equations (15.40) and (15.41), we can get the constant voltage gain proportion of the suggested converters in DCM (MDCM) as seen in Equation (15.42):

$$m_{dcm} = m_C + m_D = m_C\left(1+2N\left(\frac{1-D_{st}}{1+D_{st}}\right)\right) \tag{15.42}$$

This is the result of the semistage voltage increase in relation to the overall voltage gains in DCM, as shown in Equation (15.43):

$$m_C = \frac{m_{dcm}}{1+2N\left(\frac{1-D_{st}}{1+D_{st}}\right)} \tag{15.43}$$

With reference to (15.2) and Figure 15.9, the mean value of the source current, which is the same as the supply inductance current, is expressed as in (15.44):

$$I_S = \frac{V_S D_{st} T_S}{2L_S}\left(D_{st}+\delta_A\right) \tag{15.44}$$

The converter's voltage increases in DCM (MDCM) as a consequence of D_{st}, N, and Q may be found by entering (15.37) into (15.44) and reordering the variables, which are displayed at the end of this page as (15.45):

$$m_{dcm} = \frac{\left((1+D_{st}+2N-2ND_{st})Q\right)+\sqrt{((1+D_{st}+2N-2ND_{st})Q)^2+4Q(1-D_{st})\frac{D_{st}^{\ 2}}{2}(1-D_{st})}}{2Q(1-D_{st})} \tag{15.45}$$

$$Q = \frac{F_s L_S}{R_{load} C_{load}} \tag{15.46}$$

The comparison of DCM and CCM performances with respect to voltage increase as a function of duty ratio and various values of Q and n is shown in Figure 15.10. In discontinuous conduction mode (DCM) functioning, the voltage rate of gains is obviously very sensitive to the conversion variables (Q). That is why it is common practice to discourage DCM performances as seen in Equation (15.46).

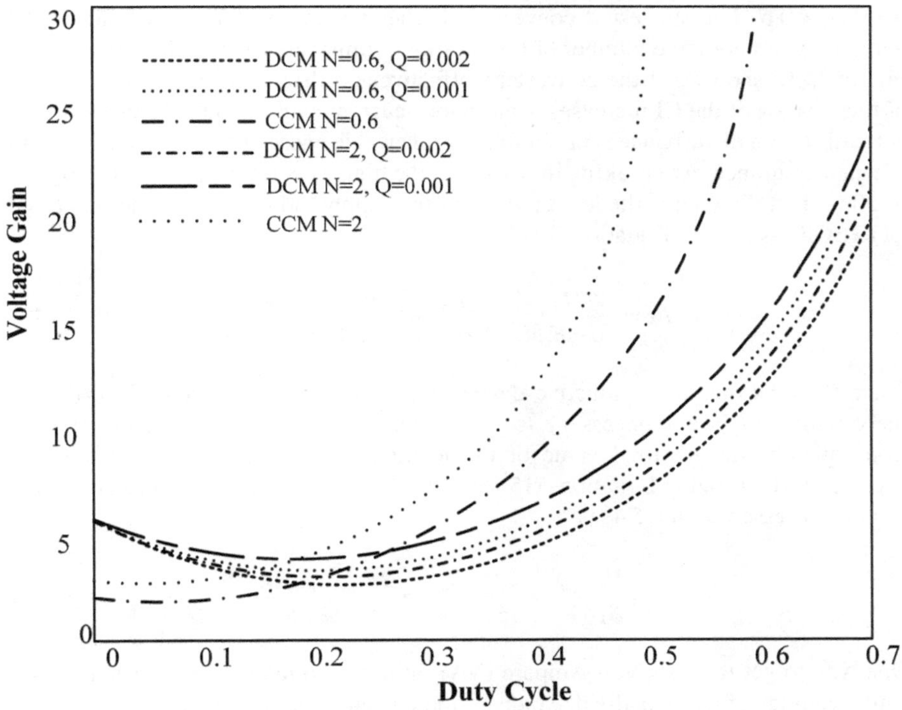

FIGURE 15.10 Comparative analysis of voltage gain in different operating phases across varying Q and N parameters.

15.4 EVALUATING EFFICIENCY: DESIGN INSIGHTS INTO QQHBCIC FOR QUASI-RESONANT PERFORMANCE ENHANCEMENT

15.4.1 Strategic Design Factors Enhancing Performance in the QQHBCIC: A Quest for QRO Excellence

The following is the formula for calculating the turns ratios of the CI (N) utilizing the constant voltage increase as seen in Equation (15.47):

$$N = \frac{m(1-D_{st})^2 - 1 - D_{st}}{2(1-D_{st})} \tag{15.47}$$

Careful consideration of D_{st} and N is required when choosing metrics for performances, particularly those pertaining to the QQHBCIC's components' voltages and current stressors. The ideal duty ratio values for minimizing voltages as well as current strain over the primary triggering switches, as shown in Figure 15.7, are

$0.3 < D_{st} < 0.6$. The suggested converter's higher voltage gain ratios mean that CI with an extremely large number of turns ratios is unnecessary, which is great news. Figure 15.12 shows that the converter's effectiveness drops when the total number of turns ratios of the CI increases. Continuous current extraction using such sources with minimum disturbance is preferable for renewable energy source's efficiency and lifespan assurance. After taking into account the highest permissible current rippling of 20%, the following is the lowest value of the supply inductance for the suggested QQHBCIC as seen in Equation (15.48):

$$L_S = \frac{V_S D_{st}}{\delta I_{LS} F_s} > \frac{V_S D_{st}}{20\% i_{LS} F_s} > \frac{R_{load} C_{load} D_{st}}{20\% m^2 F_s} \tag{15.48}$$

Since the CI's magnetic inductive device (l_m) is situated in the central portion of the circuitry, it is not necessary to develop it to produce lower current ripples. Consequently, the nominal value of the leaking inductance is also substantially decreased. By utilizing Equations (15.14) and (15.24), the lowest possible values of l_m can be expressed as in (15.49):

$$l_m = \frac{V_{lm} D_{st}}{\delta I_{lm} F_s} > \frac{2 V_{CA} D_{st}}{75\% i_{lm} F_s} > \frac{2 R_{load} C_{load} D_{st}}{75\% m F_s} \tag{15.49}$$

One way to get the CI's Volt-Ampere (VA) ratings is to take the root-mean-square (rms) readings of its second-side voltages and currents and use them as in (15.50):

$$VA_{CI} = \frac{N \pi P_{load}}{1 + D_{st} + 2N(1 - D_{st})} \sqrt{\frac{(1 - D_{st})}{2}} \tag{15.50}$$

Component A_p, which can be described as a combination of the window's winding area and the area of the cross-sections, is frequently utilized to determine the appropriate core magnetic properties as expressed in Equation (15.51):

$$A_p = \left\{ \frac{k_I L \widehat{I^2} \sqrt{1+\beta}}{b_{Max} k_t \sqrt{k_u \delta T}} \right\}^{\frac{8}{7}} 10^8 \tag{15.51}$$

In this context, k_u denotes the window's consumption factors, k_t is a constant with a range of 48–200, β denotes the proportion of iron losses to copper losses, L stands for the values of the inductor, $\hat{I}$ denotes the maximum current, k_i denotes the current pattern factors, and δT denotes the permissible increase in temperature. For ferrous Mn–Zn, $b_{Max} = 0.3$. Two output capacitors, C_C and C_D, are built according to the power outputs and the permissible voltage ripples, which is typically 1% of V_{load} as expressed in Equations (15.52) and (15.53):

$$C_C = \frac{P_{load}}{V_{load} \delta V_{CC} F_s} > \frac{P_{load}}{V_{load} 1\% V_{CC} F_s} \tag{15.52}$$

$$C_D = \frac{P_{load}}{V_{load}\delta V_{CD}F_s} > \frac{P_{load}}{V_{load}1\%V_{CD}F_s} \tag{15.53}$$

Electrostatic capacitors with a greater capacitive ratio are preferable since they often have a lower resistance. In accordance with the greatest possible current and critical phase operations ($T_r / 2 = D_{st}T_s$), the suggested converter's intermediary capacitances C_A and C_B are designed as in Equations (15.54) and (15.55):

$$C_A = C_B = \frac{2I_{load}}{\delta V_{CA}(1-D_{st})F_s} > \frac{2D_{st}I_{load}}{(10-15\%)V_{CA}(1-D_{st})F_s} \tag{15.54}$$

$$C_A = C_B = \left(\frac{D_{st}}{F_s\pi}\right)^2 \frac{2}{l_k} \tag{15.55}$$

Also, the capacitances C_A and C_B have VA ratings of (15.56) and (15.57), respectively, which can be seen below:

$$VA_{CA} = \frac{(1-D_{st})P_{load}}{1+D_{st}+2N(1-D_{st})}\sqrt{\frac{D_{st}}{(1-D_{st})^2}+\frac{(N\pi)^2}{8D_{st}}+\frac{N}{1-D_{st}}+(1-D_{st})\left(\frac{D_{st}+2N-3ND_{st}}{(1-D_{st})^2}\right)^2} \tag{15.56}$$

$$VA_{CB} = \frac{(1-D_{st})P_{load}}{1+D_{st}+2N(1-D_{st})}\sqrt{\frac{D_{st}}{(1-D_{st})^2}+\frac{(N\pi)^2}{8D_{st}}+\frac{N}{1-D_{st}}+(1-D_{st})\left(\frac{D_{st}(1+N)}{(1-D_{st})^2}\right)^2} \tag{15.57}$$

15.4.2 Theoretical Efficiency Analysis of the Proposed Converter Design

Most power losses occur in converter elements due to parasitic resistances, which may be found in magnetized equipment, MOSFETs, capacitances, and diodes. There are three primary components to the triggering losses of switches: turning-ON, turning-OFF, and conducting losses. The conducting losses of the switches in the suggested QQHBCIC are determined by (15.58):

$$P^{loss}_{S(Condu)} = r_{DSA(on)}I^2_{SA(rms)} + r_{DSB(on)}I^2_{SB(rms)} \tag{15.58}$$

As an example, the constant drain-to-sources ON-resistance is denoted as $r_{DS(on)}$. As a function of the tensions between the sources and drain potentials, the triggering switches S_A and S_B's turning-OFF and turning-ON states lose as much energy as (15.59) and (15.60):

$$P^{loss}_{S(on)} = \frac{1}{2T_S}\left[I_{DSA(ON)}V_{DSA}t_{ONA} + I_{DSB(ON)}V_{DSB}t_{ONB}\right] \tag{15.59}$$

$$P_{S(off)}^{loss} = \frac{1}{2T_S}\left[I_{DSA(OFF)}V_{DSA}t_{OFFA} + I_{DSB(OFF)}V_{DSB}t_{OFFB}\right] \tag{15.60}$$

wherein the MOSFETs' switch trigger ON and OFF changeover times are represented by toff and ton, correspondingly. And during the ON and OFF moments, accordingly, the states of the switching current are I_{DSA} (on or off) and I_{DSB} (on or off). Figure 15.5 shows that the currents at the triggering MOSFETs' turning-OFF instances are drastically lowered as a result of the suggested converter's QRO capability. One way to describe the copper losses is by taking into account the intrinsic resistance of the magnetic elements of the converters (L_A and CI) as expressed in Equation (15.61):

$$P_{Magn.\ Cu}^{loss} = R_{LA}I_{LS(rms)}^2 + R_{lm}I_{lm(rms)}^2 \tag{15.61}$$

Furthermore, the ON state power loss of a semiconductor diode is given by a relationship with the highest forward drop potential (V_f) and is denoted as (15.62):

$$P_{Diode}^{loss} = V_f V_{d(avg)} \tag{15.62}$$

Significantly, with greater current levels, the potential fall of diodes rises. It is possible to represent the losses in capacitance in the following manner due to the corresponding series resistances (R_{esr}) (15.63):

$$P_{Capacitor}^{loss} = R_{esr}I_{C(rms)}^2 \tag{15.63}$$

Therefore, the QQHBCIC is determined to have an estimated efficiency of (15.64):

$$\eta = \frac{P_{load}}{P_{load} + P_{loss}} \tag{15.64}$$

Figure 15.11 shows how the suggested QQHBCIC's QRO capability affects effectiveness enhancement. Increasing its conversion effectiveness is clearly possible with the help of the QQHBCIC's resonating capability. Figure 15.12 also shows the effectiveness and optimum voltage increase profiles computed for CI with respect to duty ratio and various turns ratios. Decreases in the converter efficiency are seen when CI duty ratios and turns ratios increase. The highest power-handling capability is further limited by the abrupt loss in the converter performance that occurs at high rates of duty as a result of the higher current and high voltages. Particularly in the traditional flyback conversion, this sharp decline in efficiencies happens in all of the higher step-up conversions. The suggested converter's higher power-handling capability is the result of its reduced power losses and greater voltage gain ratio.

Prior sections examined the benefits and drawbacks of using simulation tools vs doing research in real-life situations. Figure 15.13 shows the results of the evaluation of the component losses utilizing Computations (15.58) to (15.64).

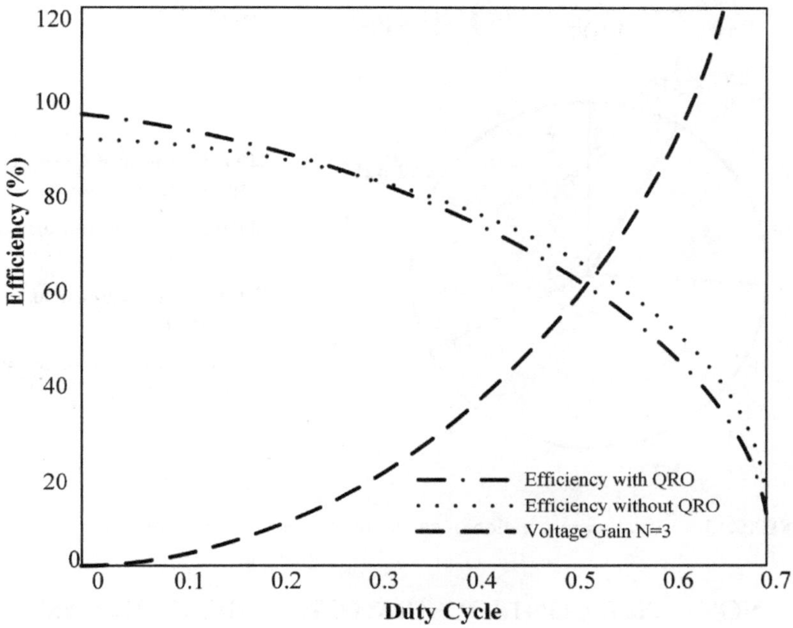

FIGURE 15.11 Impact of QRO on the efficiency enhancement of the proposed QQHBCIC.

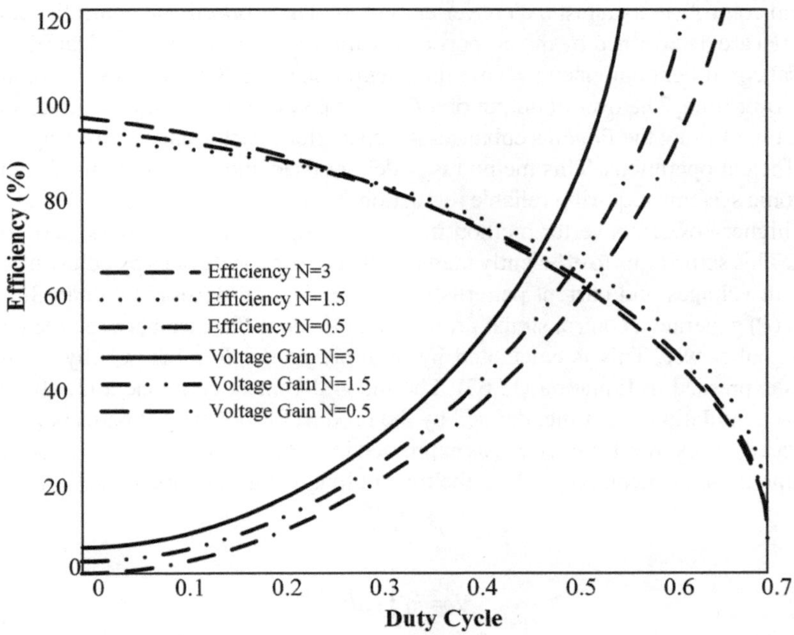

FIGURE 15.12 Efficiency and optimal voltage enhancement compared across duty cycles and varying CL turns ratios.

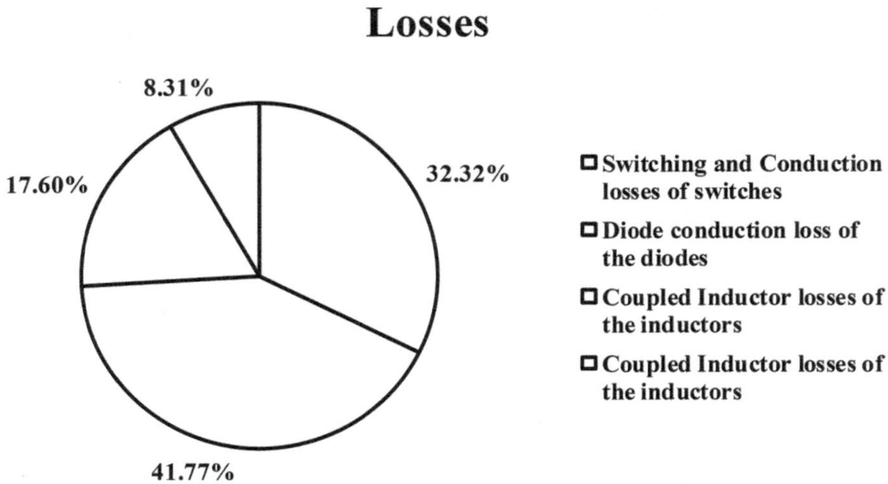

FIGURE 15.13 Visual representation of calculated losses for each component.

15.5 NOVEL PICR CONTROL STRATEGY: STABILITY ANALYSIS AND IMPLEMENTATION

Electrical power converters use PICR-based controller techniques, using proportional–integral controllers to adjust the converter outputs. The proportional controller components (P) are determined by the proportional gains (k_p) and errors (e_r). Meanwhile, the integral controller components (I) use the integration gains (k_i) to address accumulated errors over time. The overall output of a *PI* control system is the sum of these components. Fine-tuning the *PI* gains enhances the converter's performance, ensuring precise and efficient operations. This method is widely praised and commonly used in power electronic systems requiring reliable regulation.

A higher-power converter method in modern QQHBCIC converters also utilizes PICR. This setup aims to efficiently manage DC conversion devices by adjusting based on input voltages and current patterns over time. The proportionally controlled elements (P) generate a control signal proportional to the difference between the desired and actual power. This is calculated by multiplying the P gains (k_p) by the errors (e_r) as expressed in Equation (15.65). The integral control component (I) eliminates error accumulation over time, defined by the product of the integral gains (k_i) and the integrated errors over time ($\int e_r dt$) as expressed in Equation (15.66). The PI controllers combine these elements to produce the final output using Equation (15.67):

$$P = k_p e_r \tag{15.65}$$

$$I = k_i \int e_r dt \tag{15.66}$$

$$picr_{out} = P + I \tag{15.67}$$

Optimizing the *PI* values (k_p, k_i) in the PICR method enhances controller configuration for DC responses. This optimization reduces reactive energy consumption and improves DC conversion efficiency. Integrating PICR technologies into the control setup of upgraded DC converters allows for precise and dynamic control, resulting in better power usage and performance.

The Bode graph in Figure 15.14 demonstrates the system's stability characteristics. At a cutoff frequency of 4.12e+03 rad/s, the system shows a gain margin of −67.4 dB, indicating good stability. Additionally, a phase margin of 90° at 2e+07 rad/s further supports reliable operation. These margins highlight the system's resilience to uncertainties and disruptions. In summary, the Bode plot analysis confirms that the chosen system exhibits desirable stability properties. The transfer function derived from Equations (15.68) and (15.69) can be expressed using the criteria in Table 15.1:

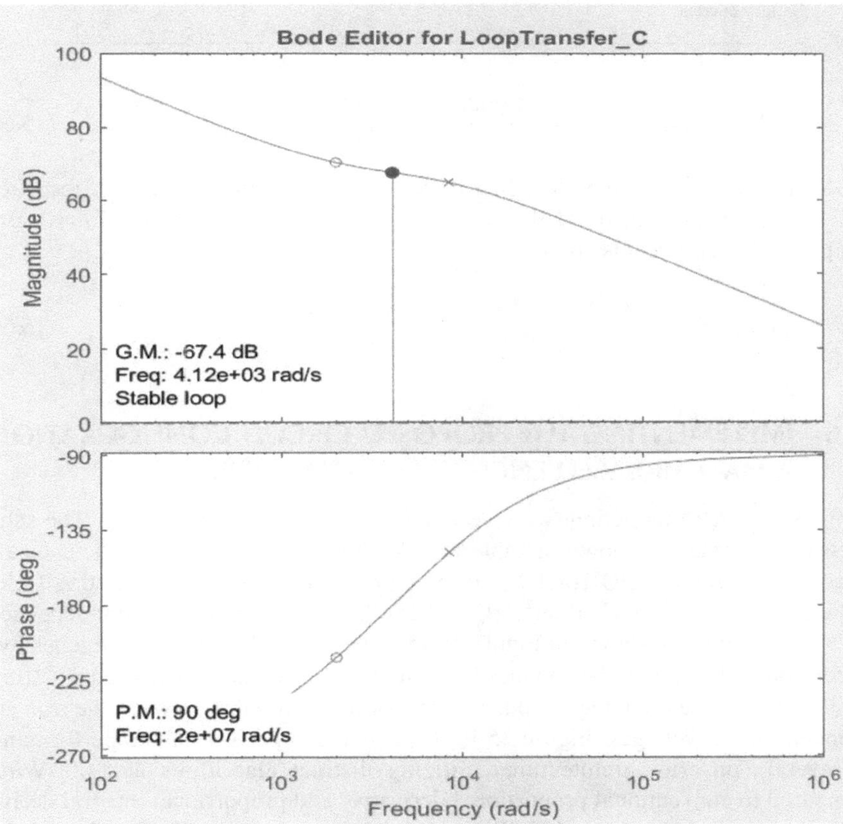

FIGURE 15.14 Analysis of stability attributes: Bode plot representation for the configured converter configuration.

TABLE 15.1
Enhancing Performance through Design Parameter Optimization

	Software Settings
Parameters	**Values**
Power rating	0.8 kW
Input voltage (V_S)	40 V DC
Load voltage (V_{load})	160 V DC
Load current (I_{load})	4.93 A DC
Inductors (L_s, L_m, L_k)	188 μH; 56 μH; 13 μH
Coupled inductance (CI)	$r_A = r_B = r_C = 1.12\ \Omega$, $n_A = 75\ \mu H$, $n_B = 25\ \mu H$
Capacitor C_A, C_B, C_C, and C_D	$C_{A\&D} = 320\ \mu F$; $C_B = 560\ \mu F$; $C_C = 600\ \mu F$
Duty ratio (d_{st})	0.23
K_p, K_i, and K_d	1.6, 88 and 0
Battery nominal voltages	48 V (Package of 12); $R_{BATY} = 8.29\ \Omega$; $C_{BATY} = 1\ \mu F$
Battery current	12 Ah

$$G_{OL\_QQHBCIC}(S) = \frac{1.998e^{7}S^{8} + 3.398e^{15}S^{7} + 6.796e^{18}S^{6}}{S^{9} + 1.701e^{8}S^{8} - 1.446e^{12}S^{7}} \tag{15.68}$$

Equation (15.34) gives the closed-loop transferring functions with unit feedback, which may be utilized to get the constants of $K_p = 1.6$, $K_i = 88$ and $K_d = 0$ after using the Ziegler–Nichols approach:

$$G_{CL\_QQHBCIC}(S) = \frac{3.197e^{07}S^{09} + 5.437e^{15}S^{8} + 1.117e^{19}S^{7} + 5.98e^{20}S^{6}}{S^{10} + 2.02e^{08}S^{09} + 5.435e^{15}S^{8} + 1.117e^{19}S^{7} + 5.98e^{20}S^{6}} \tag{15.69}$$

15.6 IMPLEMENTING THE PROPOSED CIRCUIT CONFIGURATION: A MATLAB-BASED ENCLOSED SYSTEM APPROACH

Using MATLAB® modeling, we investigated the suggested conversion. The schematic of the system is shown in Table 15.1. With a threefold boost in efficiency, the solar array raises the QQHBCIC's power supplies from 40 V DC and 20 A DC to 160 V DC and 4.93 A DC, all at 25°C and 1,000 W/m^2 irradiance. Two independent PICR techniques are shown in Figure 15.15 as being used to activate switches by a PI controller. To find the best values for k_p and k_i, we tweak the PICR gain settings until we minimize the integral square errors of the difference between the true and intended output voltages. Figure 15.16 displays a comparison of the performance of several converter architectures utilizing distinct algorithms at 1,000 W/m^2. Compared to conventional proportional derivative and proportional integral derivative controllers, the suggested PICR is more efficient, reaching 98.6%. Figure 15.17 displays the effect of dependent upon time irradiation on a steady temperature of 25°C. To improve the distribution of electricity and load management, MPPT makes use of the improved incremental conductance approach, as illustrated in Figure 15.18. Once MPPT is implemented, the features of PV and IV are shown in Figure 15.19a. The QQHBCIC receives 800 W of electricity at 40 V DC and 20 A DC from the solar panel array.

15.6.1 Method-Specific Settings: Establishing Parameters in Advance

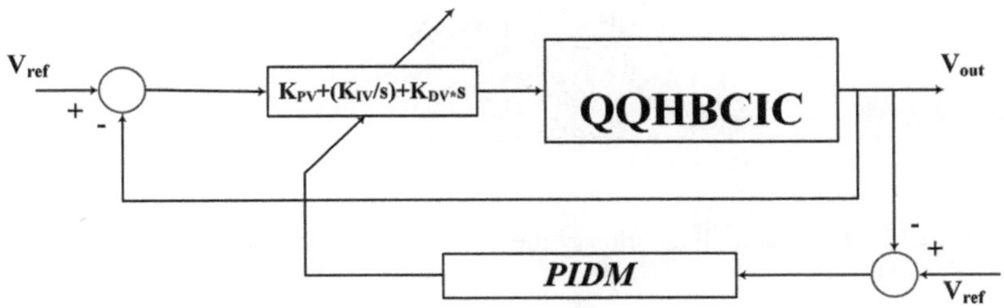

FIGURE 15.15 Operational diagram of the PICR configuration.

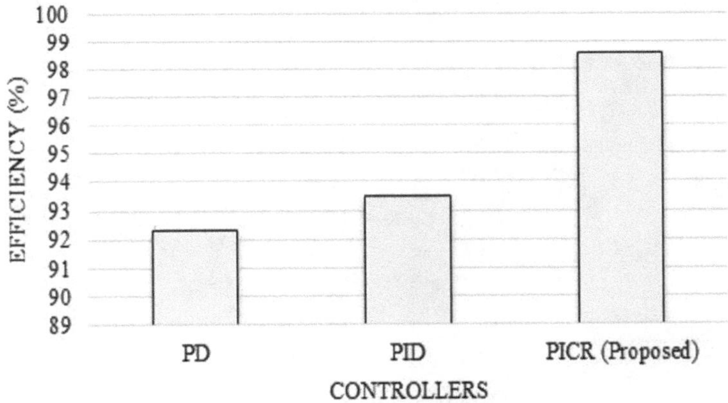

FIGURE 15.16 Efficiency comparison among different control mechanisms.

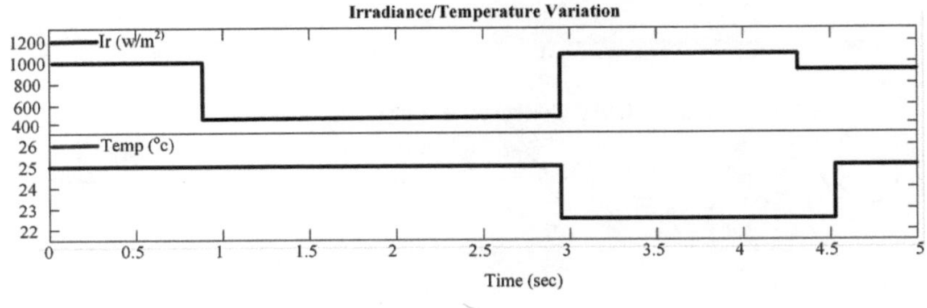

FIGURE 15.17 Graph showing solar irradiance and thermal fluctuations.

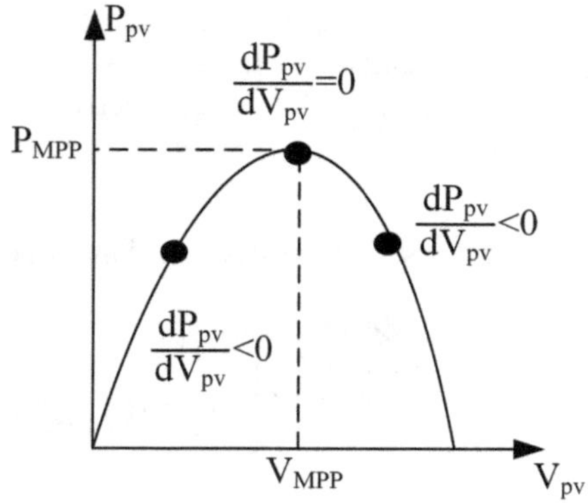

FIGURE 15.18 Schematic illustration of the incremental conductance technique.

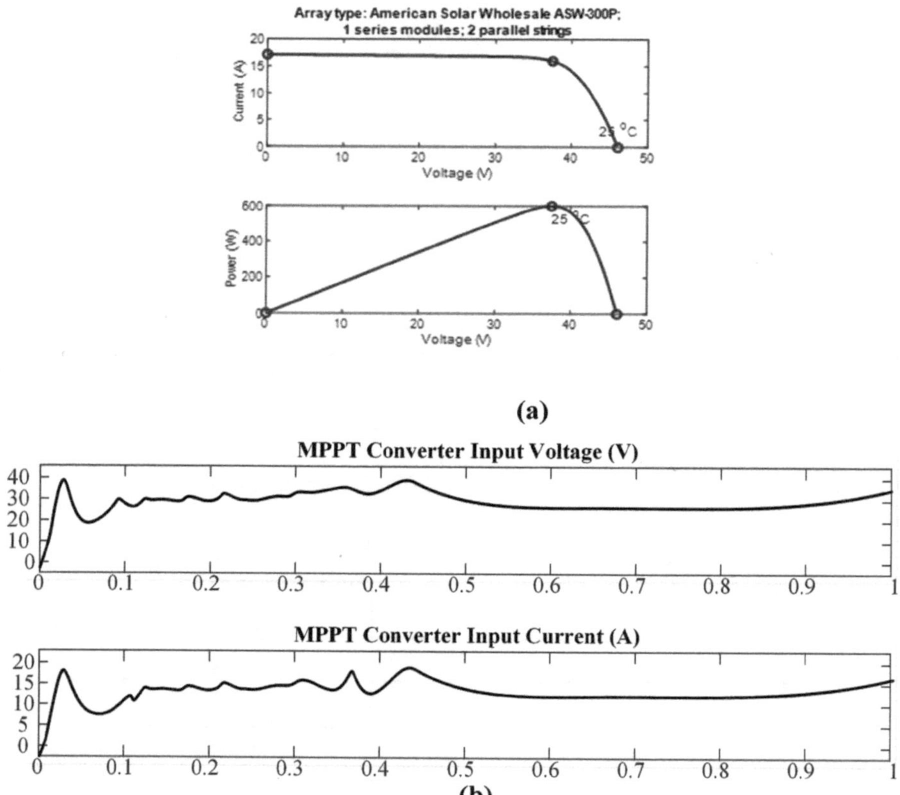

FIGURE 15.19 (a) Characteristics of current–voltage and power–voltage and (b) input voltage and current profiles for the MPPT converter.

15.6.2 Detailed Examination and Simulation Results of the Innovative Converter Design in MATLAB

This study aims to enhance the adaptability of the Smart Independent Solar Battery Charger (QQHBCIC) by integrating a feedback system using a PI-built MPPT approach and adjusting with PICR. The modified incremental conductance method is used for MPPT. Source voltage and current to the MPPT converters are illustrated in Figure 15.19b. Figures 15.20a, b, 15.19a, and b show the source voltage modification from 40 to 160 V using MPPT and conventional *PI* controls. Maximum overshoot percentages and prolonged settling periods are demonstrated in Figure 15.20a and b, with load voltages of 160 V DC and a current of 4.93 A DC due to tuning with conventional *PI*. Figure 15.21 depicts the modeled fluctuations in output voltages with standard *PI*. Using PICR, the system achieves a quick rise time of 0.03 seconds, a settling time of 0.09 seconds, and zero steady-state error, as shown in Figure 15.22a and b. The charging current and steady-state voltage output profiles of the QQHBCIC with the proposed MPPT and PICR are also shown in Figure 15.22a and b. Comparing the QQHBCIC's output signal response with MPPT, PICR, and controls across tuning parameters, the PICR method performs better with less damping. Figures 15.23 and 15.24 indicate that PICR significantly reduces ripple, producing oscillations between 160.035 and 159.99 V DC, much lower than other methods, with a ripple of just 0.045 V. Figure 15.24 details the source voltage adjustment from 40 to 60 V at $t=0.41$ seconds and its restoration at $t=0.46$ seconds to evaluate the dynamic behavior of a solar-photovoltaic-interconnected QQHBCIC with a 160 V load voltage. Additionally, the load is adjusted from 7.29 to 8.29 Ω. The modeling

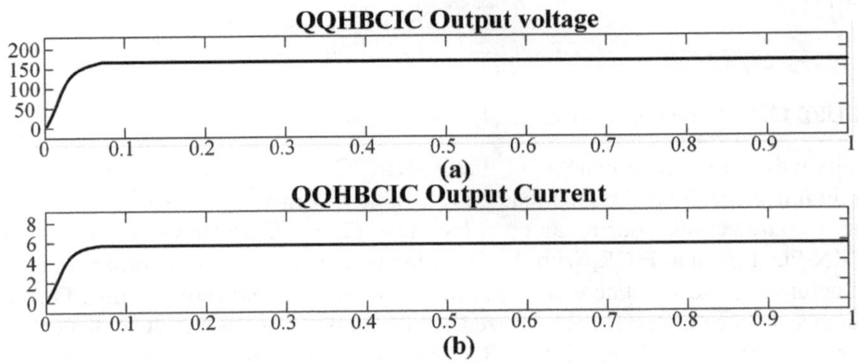

FIGURE 15.20 Conventional PI controller outcomes showing (a) voltage levels and (b) load current.

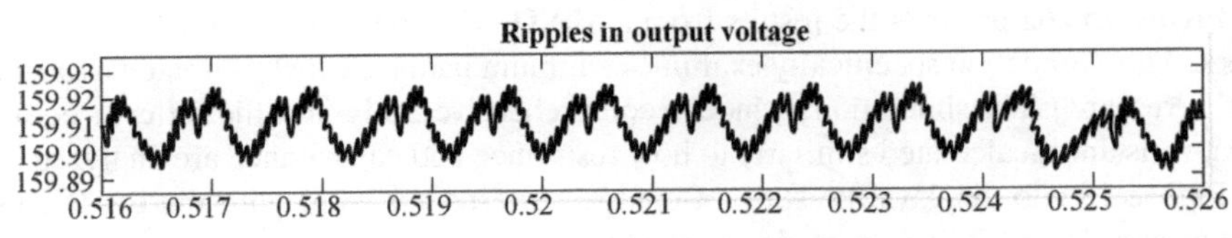

FIGURE 15.21 Simulated variations in the output voltages of a conventional PID controller.

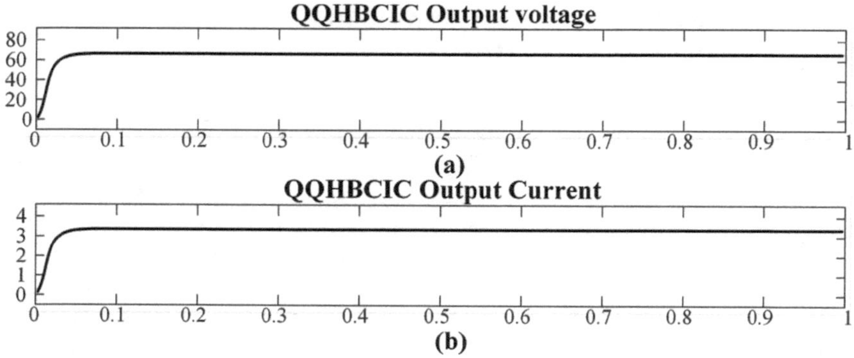

FIGURE 15.22 Results of PICR: (a) voltage readings and (b) current measurements.

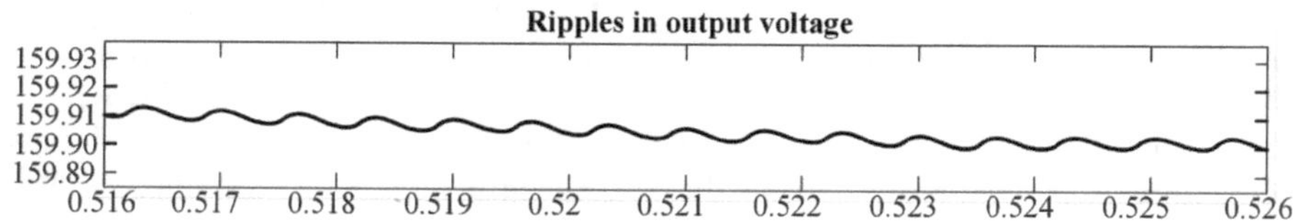

FIGURE 15.23 Simulated fluctuations in output battery voltages using PICR control.

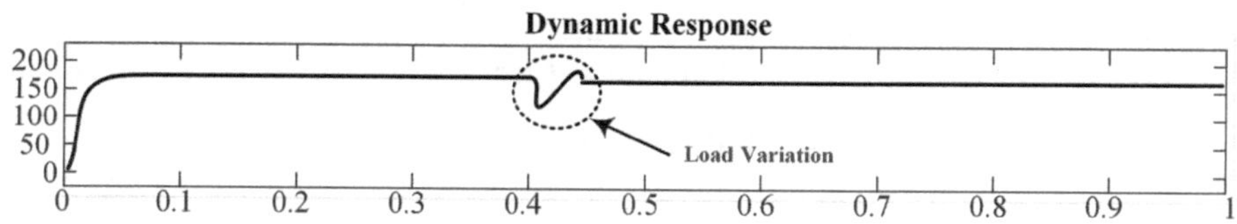

FIGURE 15.24 Transient response analysis of the innovative converter.

results indicate multiple benefits of the QQHBCIC, including lower source voltage maximum overshoot and higher load power. Table 15.2 investigates variations in irradiation characteristics during an open-loop test. Figure 15.25 shows response times for ZN-PD, PID, and PICR, with the PICR technique outperforming others in terms of fluctuations, steady-state error, maximum overshoots, and settling time. Further, reference voltage changes are analyzed for optimal efficiency in PICR demonstrations using the QQHBCIC. Table 15.3 presents a study of dynamic load changes for the proposed converter. Table 15.4 compares performance using traditional and innovative algorithms. Table 15.5 highlights the proposed converter's unique qualities and benefits compared to other topologies.

Figure 15.26a presents the results from a MATLAB simulation of a battery framework. This simulation specifically examines a lithium battery, which is treated as a load in this research. The simulation includes a comprehensive analysis of the battery's equivalent resistance, calculated similarly to how resistance and capacitance are analyzed.

Figure 15.26b displays the state-of-charge patterns achieved with the PICR control system. The PICR system demonstrates superior and more consistent charging patterns compared to other standard controllers. This indicates that the PICR control

TABLE 15.2
Variations in Solar Irradiance during Open-Loop Assessment

Irradiations (W/m^2)	Duty Cycle	V_s (V)	I_s (A)	V_{load} (V)	I_{load} (A)	η (%)
700	0.38	40	18.5	158	4.11	87.8
900	0.4	40	19.2	159	4.3	89
1,000	0.23	40	20	160	4.5	90
1,200	0.3	40	20.1	162	4.51	90.8

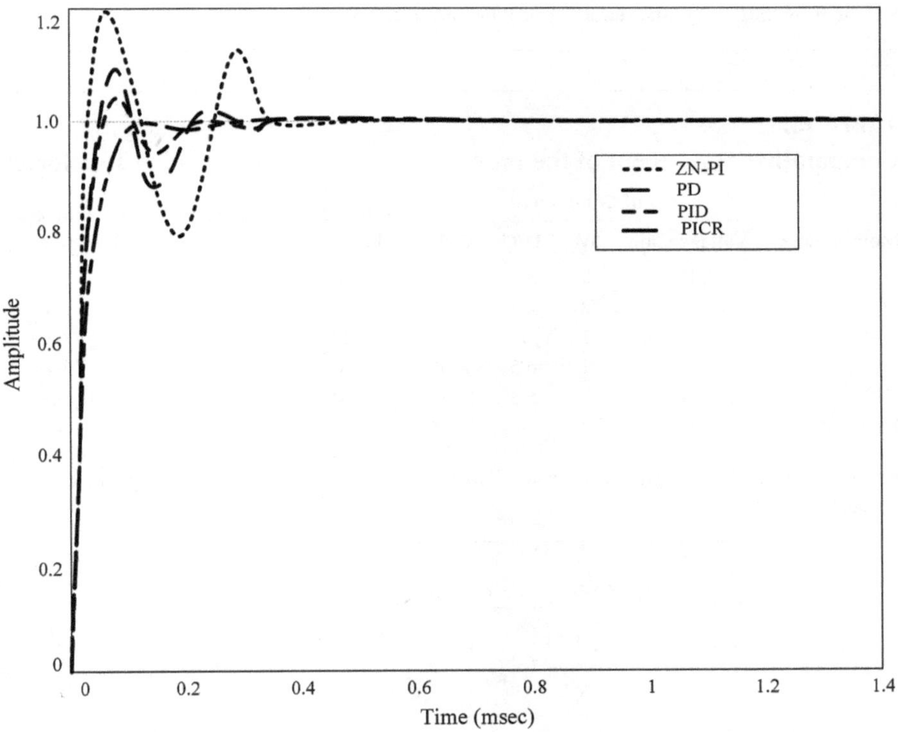

FIGURE 15.25 Comparative step response analysis of PD, PID, and PICR controllers.

TABLE 15.3
Variable Load Dynamics in the Innovative Converter

Loads (%)	V_s (V)	I_s (A)	V_{load} (V)	I_{load} (A)	Efficiency (%)
30	40	6	160	1.4	91.8
60	40	12	160	2.82	94
90	40	18	160	4.35	96.8
100	40	20	160	4.93	98.6

TABLE 15.4
Performance Comparison of Topologies Using Traditional and Innovative Algorithms

Technique Types	k_p	k_i	k_d	RT (Seconds)	SSE (V)	ST (Seconds)	PO (%)	Efficiency @ 1,000 W/m² (%)
PD	1.9	0	0.23	0.04	0.8	0.7	1.7	92.3
PID	3.69	103	0.45	0.034	0.09	0.5	1.6	93.5
PICR (proposed)	1.6	88	0	0.03	0.009	0.09	1	98.6

RT, rise time; SSE, steady-state error; ST, settling time; PO, peak overshoot.

TABLE 15.5
Comparative Assessment of the Proposed Converter with Existing Topologies

Methodology	Voltage Gain	No of Components			Voltage Stress in Switch	Current Stress in Switch	Efficiency (%)
		SW	D/C	CI/L			
[1]	H	2	3/3	–/2	M	M	93
[7]	H	2	4/3	$2^{2W}/3$	M	M	94
[14]	H	1	4/5	$1^{3W}/3$	M	L	95.4
[17]	H	1	6/5	–/4	M	M	94.8
[19]	H	2	5/5	$1^{3W}/1$	M	M	96.2
Proposed	H	2	4/4	$1^{2W}/1$	L	L	98.6

SW, switch; D, diode; C, capacitor; CI, coupled inductor; L, inductor; H, high; M, medium; L, low; NA, not applicable.

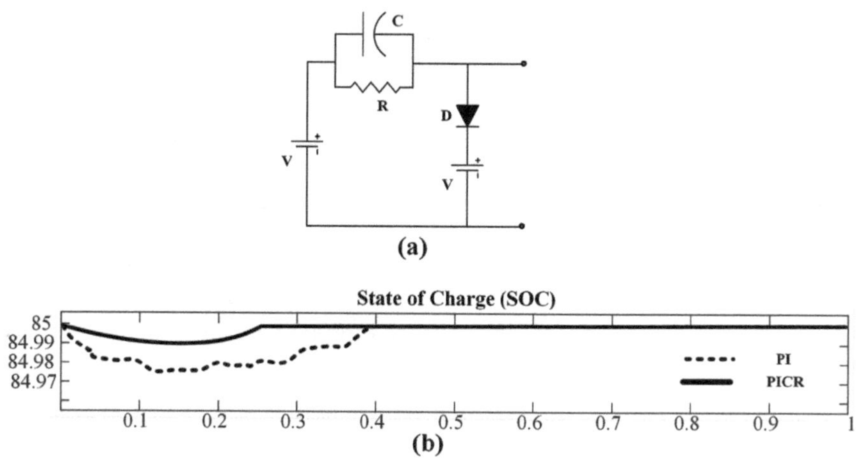

FIGURE 15.26 (a) Resistance capacitance model of the battery and (b) battery state of charge.

TABLE 15.6
Efficiency Comparison between Open-Loop and Closed-Loop Simulation

Software Implementations (800 W)					
Open-Loop Structure			**Closed-Loop Structure**		
P_s (W)	P_{load} (W)	η (%)	P_s (W)	P_{load} (W)	η (%)
800	730.6	90.8	800	788.8	98.6

system provides a more effective and reliable method for managing the charging process of lithium batteries.

Table 15.6 provides a comparative analysis of efficiency between open-loop and closed-loop simulations. This table highlights the differences in performance and efficiency when using different control strategies, showcasing the benefits of the closed-loop system in achieving higher efficiency in battery charging processes.

15.7 CONCLUSION

In conclusion, this study introduces a novel QQHBCIC that significantly enhances EV charging efficiency by integrating renewable energy sources. The converter's innovative design not only achieves a high voltage conversion ratio but also minimizes voltage stress and switching losses, leading to exceptional efficiency and performance. To advance this research, future work will focus on several key areas. First, long-term performance studies will be conducted to assess the durability and reliability of the QQHBCIC in real-world scenarios. Second, the integration of the converter with various renewable energy sources, such as wind and hydro power, will be explored to evaluate its adaptability and effectiveness across different energy inputs. Finally, the converter will be tested under diverse environmental conditions and load profiles to understand its robustness and dynamic response further. These future research directions will provide a comprehensive framework for optimizing the QQHBCIC and expanding its applicability, thereby contributing to a more sustainable and efficient EV charging infrastructure.

REFERENCES

1. A. Rajabi, A. Rajaei, V.M. Tehrani, P. Dehghanian, J.M. Guerrero and B. Khan, A non-isolated high step-up DC-DC converter using voltage lift technique: Analysis, design, and implementation, *IEEE Access* 10 (2022), pp. 6338–6347.
2. M. Rezaie and V. Abbasi, Ultrahigh step-up DC-DC converter composed of two stages boost converter, coupled inductor, and multiplier cell, *IEEE Transactions on Industrial Electronics* 69 (2022), pp. 5867–5878.
3. X. Zhu, L. Jiang, B. Zhang and K. Jin, The resonant modular multilevel DC-DC converter adopting switched-inductor cells for high step-up ratio, *IEEE Journal of Emerging and Selected Topics in Industrial Electronics* 10 (2022), pp. 6634–6647.

4. S. Habibi, R. Rahimi, M. Ferdowsi and P. Shamsi, An impedance-source-based soft-switched high step-up DC-DC converter with an active clamp, *IEEE Transactions on Power Electronics* 39 (2024), pp. 3712–3723.
5. M.R.S. De Carvalho, E.A.O. Barbosa, F. Bradaschia, L.R. Limongi and M.C. Cavalcanti, Soft-switching high step-up DC-DC converter based on switched-capacitor and auto-transformer voltage multiplier cell for PV systems, *IEEE Transactions on Industrial Electronics* 69 (2022), pp. 12886–12897.
6. R. Rahimi, S. Habibi, P. Shamsi and M. Ferdowsi, A three-winding coupled-inductor-based dual-switch high step-up DC-DC converter for photovoltaic systems, *IEEE Journal of Emerging and Selected Topics in Industrial Electronics* 3 (2022), pp. 1106–1117.
7. H.S. Gohari, S. Abbasian, N.A. Mardakheh, K. Abbaszadeh and F. Blaabjerg, Coupled inductor-based current-fed ultra-high step-up DC-DC converter featuring low input current ripple, *IEEE Transactions on Circuits and Systems II: Express Briefs* 71 (2024), pp. 887–891.
8. R. Rahimi, S. Habibi, M. Ferdowsi and P. Shamsi, Z-Source-based high step-up DC-DC converters for photovoltaic applications, *IEEE Journal of Emerging and Selected Topics in Industrial Electronics* 10 (2022), pp. 4783–4796.
9. T. Jalilzadeh, N. Rostami, E. Babaei and M. Maalandish, Nonisolated topology for high step-up DC-DC converters, *IEEE Journal of Emerging and Selected Topics in Industrial Electronics* 11 (2023), pp. 1154–1168.
10. A. Nafari and R. Beiranvand, An extendable interleaved quasi Z-source high step-up DC-DC converter, *IEEE Transactions on Power Electronics* 38 (2023), pp. 5065–5076.
11. S. Athikkal, B. Chokkalingam, S.I. Ganesan, B. Lehman and T. Brunelli Lazzarin, Performance evaluation of a dual-input hybrid step-up DC-DC converter, *IEEE Transactions on Industry Applications* 58 (2022), pp. 3769–3782.
12. A.K. Singh, A.K. Mishra, K.K. Gupta and Y.P. Siwakoti, High voltage gain bidirectional DC-DC converters for supercapacitor assisted electric vehicles: A review, *CPSS Transactions on Power Electronics and Applications* 7 (2022), pp. 386–398.
13. X. Zhao, B. Li, B. Zhang and D. Xu, A high-power step-up DC/DC converter dedicated to DC offshore wind farms, *IEEE Transactions on Power Electronics* 37 (2022), pp. 65–69.
14. S. Hasanpour, Y.P. Siwakoti and F. Blaabjerg, A new high efficiency high step-up DC/DC converter for renewable energy applications, *IEEE Transactions on Industrial Electronics* 70 (2023), pp. 1489–1500.
15. A. Samadian, M.G. Marangalu, H. Tarzamni, S.H. Hosseini, M. Sabahi and A. Mehrizi-Sani, High step-up common grounded switched quasi Z-source DC-DC converter using coupled inductor with small signal analysis, *IEEE Access* 11 (2023), pp. 120516–120529.
16. T. Jin, X. Yan, H. Li, J. Lin, Y. Weng and Y. Zhang, A new three-winding coupled inductor high step-up DC-DC converter integrating with switched-capacitor technique, *IEEE Transactions on Power Electronics* 38 (2023), pp. 14236–14248.
17. X. Zhu, K. Ye, K. Liu and B. Zhang, Nonisolated high step-up DC-DC converter with passive switched-inductor-capacitor network, *IEEE Journal of Emerging and Selected Topics in Industrial Electronics* 10 (2022), pp. 6444–6456.
18. W.C. Leal, M.O. Godinho, R.F. Bastos, C.R. De Aguiar, G.H.F. Fuzato and R.Q. MacHado, Cascaded interleaved DC-DC converter for a bidirectional electric vehicle charging station, *IEEE Transactions on Industrial Electronics* 71 (2024), pp. 3708–3717.
19. T. Jin, J. Lin, H. Li, X. Yan, Y. Weng and X. Mao, A novel three-winding coupled inductor-based high step-up DC-DC converter for renewable energy application, *IEEE Journal of Emerging and Selected Topics in Industrial Electronics* 11 (2023), pp. 4477–4490.

20. M. Zhang, Z. Wei, M. Zhou, F. Wang, Y. Cao and L. Quan, A high step-up DC–DC converter with switched-capacitor and coupled-inductor techniques, *IEEE Journal of Emerging and Selected Topics in Industrial Electronics* 3 (2022), pp. 1067–1076.
21. A. Asghari and Z.J. Yegane, A high step-up DC-DC converter with high voltage gain and zero-voltage transition, *IEEE Transactions on Industrial Electronics* 71 (2023), pp. 6946–6954.
22. A. Shoaei, K. Abbaszadeh and H. Allahyari, A single-inductor multi-input multilevel high step-up DC-DC converter based on switched-diode-capacitor cells for PV applications, *IEEE Journal of Emerging and Selected Topics in Industrial Electronics* 4 (2022), pp. 18–27.
23. R. Rahimi, S. Habibi, M. Ferdowsi and P. Shamsi, An interleaved high step-up DC-DC converter based on integration of coupled inductor and built-in-transformer with switched-capacitor cells for renewable energy applications, *IEEE Access* 10 (2022), pp. 34–45.

16 Blockchain and AI Solutions for Smart EV Charging

S. Muthukaruppasamy, G. Arun Sampaul Thomas, J. Nandha Gopal, and E. Parimalasundar

16.1 INTRODUCTION

Modern civilization relies on electric cars to solve environmental, energy, health, economic, and technical issues. Their adoption is crucial for sustainability, energy independence, and a clean, healthier, and more resilient future for generations. Due to their potential to solve problems and promote sustainability, EVs are important in modern society. In a civilized society, EVs are leading in the following areas:

Environmental Sustainability: EVs have zero tailpipe emissions, lowering air pollution and transportation's health and environmental impacts. As climate change concerns grow, EVs are vital to lowering greenhouse gas emissions and attaining carbon neutrality [1].

Energy Independence: EVs use solar, wind, and hydroelectric power to decrease fossil fuel needs. This boosts energy independence, resilience, and domestic, renewable energy consumption by reducing oil imports [2].

Public Health: In metropolitan areas with high automobile air pollution, EVs benefit public health and air quality. EVs reduce respiratory ailments and other air pollution-related health issues by removing tailpipe emissions of nitrogen oxides and unstable biological composites. Battery technologies, electric drivetrains, and car connections evolve as EVs become mainstream. This advances energy storage, efficiency, and electric propulsion systems, enabling more sustainable and efficient transportation. All these variables accelerate into cleaner and more reliable energy systems [3].

Economic Opportunities: The growth of the EV industry creates economic opportunities through job creation, innovation, and investment. From manufacturing and infrastructure development to research and development, the EV ecosystem stimulates economic activity and drives growth in related industries, contributing to job creation and economic prosperity [4].

Grid Integration and Energy Flexibility: EVs can serve as grid-connected storage devices, offering opportunities for demand-side management, load balancing, and grid stabilization. Through vehicle-to-grid (V2G) technologies and intelligent charging solutions, EVs can provide flexibility to the

DOI: 10.1201/9781003496946-16

electricity grid, support renewable energy integration, and optimize energy use, enhancing grid resilience and efficiency [5].

Consumer Benefits: EVs offer several benefits to consumers, including lower operating costs, reduced maintenance requirements, and a smooth driving experience. With advancements in charging infrastructure and battery technology, EVs are increasingly affordable, commodious, and comprehensible to a broader range of consumers, driving greater adoption and market penetration [6].

16.1.1 Emerging Technologies in EV's Infrastructure-AI and Blockchain

As the world evolves to more sustainable transportation, the rise of EVs is driving a critical need for robust charging infrastructure. However, traditional charging networks face several challenges, including managing variable energy demands, integrating renewable sources efficiently, and providing a seamless user experience. This is where the emerging technologies of Artificial Intelligence (AI) and blockchain play a transformative role.

Role of AI in EVs' Infra: In the fast-changing world of sustainable mobility, AI and smart EV charging infrastructure are beacons of innovation and efficiency. As electric mobility gains momentum, the integration of AI is guaranteed to revamp the way we manage, optimize, and utilize EV charging networks, ushering in a new era of intelligent mobility solutions [7]. Traditional EV charging systems often grapple with challenges such as grid congestion, peak demand management, and user convenience. AI is a transformative technology that empowers charging infrastructure with predictive analytics, real-time optimization, and adaptive control capabilities, enabling smarter, more responsive charging experiences. AI systems examine weather forecasts, immense volumes of data, including energy convention, grid dynamics, and customer predilections, using machine learning (ML) algorithms [8]. By harnessing insights from this data, AI optimizes charging schedules, allocates resources efficiently, and adapts to changing conditions in real time, maximizing the performance and reliability of EV charging infrastructure. AI can enhance route planning for EV drivers, considering factors like traffic conditions, climate patterns, and real-time charging station availability, ultimately increasing the practicality and convenience of EVs. By leveraging ML algorithms, AI systems can accurately forecast energy demands, enabling intelligent load balancing and efficient use of charging resources.

Role of BC in EVs' Infra: As electric mobility gains traction worldwide, the integration of blockchain holds the promise of addressing key challenges and unlocking new opportunities for efficiency, security, and innovation [9]. Traditional EV charging systems face various hurdles, including concerns over transaction security, billing transparency, and interoperability between charging networks. Blockchain, a decentralized, immutable ledger system, can solve these problems and transform EV charging. Blockchain provides a decentralized platform with a safe stand for the ecosystem of EV transactions and data [10]. Its distributed ledger architecture enables energy trading, safe payment processing, and green energy credit monitoring. By banishing the requirement for central point authorities, Blockchain fosters a more democratized and resilient EV charging network. Blockchain technology fundamentally alters the paradigm of EV charging infrastructure by providing a secure, transparent, and interoperable platform for managing charging transactions, energy trading, and data exchange. By leveraging cryptographic principles and decentralized consensus mechanisms, Blockchain ensures trust, integrity, and resilience in the charging ecosystem, flagging the tactic for an efficient, smarter future.

Fusion of AI and BC in EVs' Infra: In an era marked by the imperative requisite for sustainable solutions, the convergence of cutting-edge technologies is reshaping the landscape of transportation like never before. At the forefront of this revolution are AI and blockchain technologies, poised to redefine how we envision and implement EV infrastructure [11].

The implementation of EVs represents a vital step toward lowering carbon pollution, fighting climate change, and ushering in a cleaner, greener future. However, the widespread amalgamation of EVs into daily life hinges not only on advancements in battery technology and vehicle design but also on the development of robust infrastructure capable of supporting their proliferation. AI and blockchain transform EV infrastructure, from charging networks to grid control. AI's ability to analyze large data and create real-time decisions is changing charging schedules, maintenance demands, and grid energy flows. Blockchain technology enables decentralized, secure transactions, peer-to-peer energy exchange, and vehicle charging station communication [12].

The union of AI and blockchain technologies holds immense potential for creating an intelligent and decentralized EV charging ecosystem [12]. By leveraging the strengths of both technologies, a truly smart EV infrastructure can be developed, addressing the challenges of efficient energy management, renewable energy integration, and seamless user experiences.

i. **AI-Driven Intelligent Charging Management:**

The core of a smart EV infrastructure dwells in AI-powered intelligent charging management systems. These systems can employ ML algorithms to examine historical data on energy consumption patterns, weather conditions, traffic patterns, and user behavior. By uncovering intricate correlations and trends, AI models can accurately forecast energy demands across different charging stations and time periods. Based on these predictions, the AI system can dynamically optimize the allocation of charging resources, ensuring efficient load balancing and minimizing energy waste. Additionally, AI can enable advanced route planning for EV drivers, suggesting optimal charging stops and providing real-time updates on station availability and wait times. This stage of intelligent planning can significantly improve the convenience and range confidence of EV users.

ii. **Blockchain-Enabled Decentralized Energy Trading:**
Complementing AI's predictive capabilities, blockchain technology introduces a secure and transparent framework for decentralized energy trading within the EV infrastructure. Through smart contracts on the blockchain, EV owners can participate in peer-to-peer energy transactions, buying and selling energy directly from renewable sources or other EV owners. This decentralized approach removes the necessity for centralized interventions, decreasing transaction price and fostering a more democratized energy market. Additionally, blockchain's immutable ledger allows for the tracking and certification of green energy credits, incentivizing the acquisition of sustainable energy resources and enabling carbon credit trading models.

iii. **Autonomous Valet Charging and V2G Integration:**
Advanced use cases, like autonomous valet charging and V2G integration, are enabled by AI and blockchain. Self-driving EVs can navigate to charging stations with AI, while blockchain allows safe and transparent payments. Blockchain-based smart contracts can also enable V2G models, where EVs inject extra energy into the grid during peak demand. AI systems can optimize bidirectional energy flows to meet grid demands and EV owners' desires.

iv. **Data Integrity and Privacy Considerations:**
As AI models rely heavily on data for training and decision-making, blockchain's inherent data integrity and transparency features become invaluable. By storing AI training data and model updates on the Blockchain, the integrity and provenance of the data can be verified, ensuring the reliability of AI-driven decisions within the EV infrastructure. However, it is significant to discuss privacy concerns surrounding the use of personal data, such as driver locations and charging patterns. Blockchain's cryptographic features and decentralized architecture can potentially enable privacy-preserving solutions, allowing AI models to leverage data while maintaining individual privacy.

Through the harmonious integration of AI and blockchain technologies, a smart EV infrastructure can be realized, optimizing energy consumption, enabling efficient renewable energy integration, and providing a seamless and sustainable transportation experience for EV users.

In this article, we venture into a journey to explore the synergies between AI, blockchain, and EV infrastructure. We delve to address challenges in EV adoption and optimize EV charging infrastructure using AI and blockchain technologies, promising initiatives, and prospects of this dynamic intersection, uncovering the transformative potential that lies at the nexus of both technology and smart EV transportation.

The remaining part of this article is organized as follows: Section 16.2 delivers an outline of the global landscape of EV stock and sales. Section 16.3 outlines methodologies of EV charging. Section 16.4 discusses the main challenges that EV charging networks face today. Section 16.5 describes the emerging digital technologies in EVs. Section 16.6 discusses harnessing the power of AI and Blockchain in the EV sector. Section 16.7 involves the research perspective of future trends and outlook, which exposes Federated learning in the context of blockchain networks and how to adapt protocols like Open Charge Point Protocol (OCPP) to use both technologies for security and energy management. Finally, the conclusion concludes the integration of Blockchain and AI technologies which presents exploration of smart EV charging infrastructure, yields several key benefits, including enhanced security, interoperability, optimizing charging schedules, reducing congestion, adapting user preferences, improving user experience and grid management, transparency and trust in transactions, addressing concerns regarding data integrity and cyber security in EV charging networks.

16.2 AN OVERVIEW OF THE GLOBAL LANDSCAPE OF EV STOCK AND SALES

EV sales have grown rapidly worldwide due to government incentives, environmental restrictions, technological advances, and changing customer tastes. The EV business attracts investors. Here is a glimpse of the global landscape.

1. **Rapid Growth:**
   EV sales are rising worldwide, year-over-year, in several locations. EV sales have grown faster in recent years as manufacturers debut new EVs and expand their EV offerings. Government subsidies, supporting regulations, and rising consumer demand have propelled EV sales in emerging economies like China and Europe. In 2023, the market predicts 14 million global EV sales, a 60% increase from 2022. Due to this, Tesla, BYD, and Volkswagen stock prices are rising [13].
2. **Regional Trends:**
   By 2022, 16.5 million electric automobiles will make up 0.8% of the worldwide vehicle stock. China has almost 9 million EVs, followed by Europe with 4.5 million and the US with 2.6 million. Norway has the highest number of EVs, over 20%, which leads to significant EV incentives. The global EV fleet might reach 116 million by 2030 under Bloomberg New Energy Outlook's Economic Transition Scenario.
   i. **China**: China sells the most electric automobiles worldwide. Chinese EV usage has grown rapidly due to government subsidies, incentives, and restrictions.
   ii. **Europe**: Europe is another major market for EVs, with countries like Norway, the Netherlands, and Germany leading in EV adoption rates. Stringent emissions standards, along with government incentives and charging infrastructure investments, have driven EV sales in Europe.
   iii. **North America**: While the United States has been slower to adopt EVs compared to other regions, EV sales are gradually increasing, fueled by consumer interest, regulatory mandates, and automakers' commitments to electrification.

iv. **Other Regions**: EV sales are also growing in other regions such as Japan, South Korea, and select countries in Latin America and the Middle East. Government policies, infrastructure development, and market dynamics vary across these regions, influencing EV adoption rates.

3. **Vehicle Segments:**

Passenger Cars: Passenger cars constitute the majority of EV sales globally, with a wide range of models available from various automakers. The increasing affordability, range, and performance of electric passenger cars have contributed to their growing popularity among consumers.

Commercial Vehicles: The adoption of electric commercial vehicles, including buses, trucks, and vans, is also on the rise, driven by factors such as lower operating costs, environmental benefits, and fleet electrification initiatives in urban areas.

Two-Wheelers: Electric bicycles, scooters, and motorcycles are popular modes of transportation in many regions, especially in densely populated urban areas. The affordability and convenience of electric two-wheelers make them attractive options for commuting and short-distance travel.

4. **Future Outlook:**

The anticipated advanced battery technology, expanded charging infrastructure, and supportive government regulations will keep the global EV industry booming.

Market Competition: Established and emerging EV automakers are vying for market share. This competition is boosting EV innovation, lowering prices, and increasing customer choice.

Policy Support: Government policies and regulations will continue to shape the EV industry. Incentives, subsidies, emissions standards, and infrastructure expenditures will affect global EV adoption and market dynamics.

As the world moves toward cleaner, more sustainable transportation, EV sales are changing globally, offering opportunities for development and innovation.

16.2.1 Factors Propelling the Global Rise of EVs

In recent years, EV inventories have risen as worldwide demand for sustainable transportation options has grown. Government incentives, environmental concerns, technological advancements, and changing customer tastes drive EV adoption. Here are the key points about EV stocks and their global use.

Market Trends: Many regions have seen a steady rise in EV sales in recent years. Growth has increased the valuation of EV makers and linked firms, raising their stock prices.

Investor Interest: EV stocks are popular among investors looking for exposure to the burgeoning electric car market. Stock values have increased as investors bet on electric mobility, EV production, battery technology, charging infrastructure, and associated companies.

Volatility: EV stock values fluctuate sharply due to market sentiment, regulatory changes, and macroeconomic considerations. Investors should be cautious while investing in EV stocks due to large price fluctuations.

Competition: The EV industry is getting more competitive due to conventional manufacturers and the participation of newcomers. Established automakers are increasing EV production, while start-ups and tech businesses are entering the market with revolutionary EV designs and technology.

Regulatory Environment: Government rules shape the EV industry. Tax credits, rebates, and subsidies for EV consumers can boost sales, but pollution requirements and fuel economy rules might affect manufacturers' investment decisions and product offers.

Charging Infrastructure: EV utilization is dependent on charging infrastructure availability and accessibility. Countries with extensive charging networks and fast-charging stations have greater EV adoption rates because drivers trust the ease and dependability of charging.

Urban vs. Rural Usage: Cities with shorter travel distances and better charging facilities use EVs more frequently. Due to range anxiety and infrastructure concerns, rural EV adoption may be gradual, as there are fewer charging alternatives and greater driving distances.

Environmental Advantages: EVs diminish greenhouse gas, air pollution, and reliance on fossil fuels, which encourages their adoption. As more people become conscious of climate change, they are turning to EVs as a cleaner, greener option to autos.

Technological Advances: Battery, electric drivetrain, and car connectivity improvements make EVs more feasible, economical, and appealing. We predict a rise in EV usage as technology advances and costs decrease, propelling global market development and acceptance.

16.2.2 Impact of EV Adoption and Usage Worldwide

In 2022, electric cars avoided around 0.8% of global CO_2 emissions from road transport. Countries with higher EV uptake have seen greater emissions reductions from transportation. For instance, Norway reduced sector emissions by 18% between 2019 and 2021 due to EVs. Total electricity consumption from EVs is still relatively small globally at around 80 TWh in 2022, or 0.3% of total electricity demand. However, clusters of high EV adoption in certain cities/regions can create localized grid impacts requiring smart charging management. Miles traveled by EVs grew over 40% annually from 2019 to 2021 as more EVs entered the roads globally.

16.2.2.1 Factors Driving Further Adoption

i. Battery cost declines and increased production capacities, making EVs more affordable.
ii. Expanding public charging infrastructure and increasing EV model availability.
iii. Tightening CO_2/pollution regulations incentivize automakers to electrify vehicle fleets.

iv. Government policies like fuel economy standards, EV subsidies, and eventual phase-out of internal combustion engines.

Electric automobiles are an excellent financial opportunity and a practical solution to the environmental problems caused by conventional transportation. The global outlook for EV inventories and usage is positive in the long run due to the increasing adoption of electric transportation by governments, manufacturers, and consumers. AI and blockchain may handle rising power consumption, renewable charging, and the experiences of EV drivers and owners as the stock and usage of EVs grow.

16.3 METHODOLOGIES OF EV CHARGING

Methodologies of charging covers (i) type of charging, (ii) level of charging, and (iii) mode of charging.

16.3.1 Type of Charging

i. **Conductive Charging:**
   An EV charger charges the battery with a power wire or cable. The charger or cable connects to the built-in charger to power the battery, as shown in Figure 16.1.
ii. **Non-conductive Charging:**
   It charges EV batteries without contacting the vehicle. Chargers, also known as induction pads, send kHz radio frequency signals to EV receiving pads for wireless charging. Subterranean frequency power wires below the surface or charging stations provide inductive charging or magnetic induction for EV batteries. Road-powered or wireless EVs charge wirelessly. Safe cordless charging is its key feature. The slow AC charging, no DC charging, and high cost are all demerits of this charging. Figure 16.2 depicts typical non-conductive wireless charging [14].
iii. **Solar-Powered Charging:**
   EVs may be charged by sunlight. The on-board charger uses solar photo voltaic modules directly linked by cable or wire, but it also uses conductive charging. Unidirectional power flow (G2V) solar charging systems as

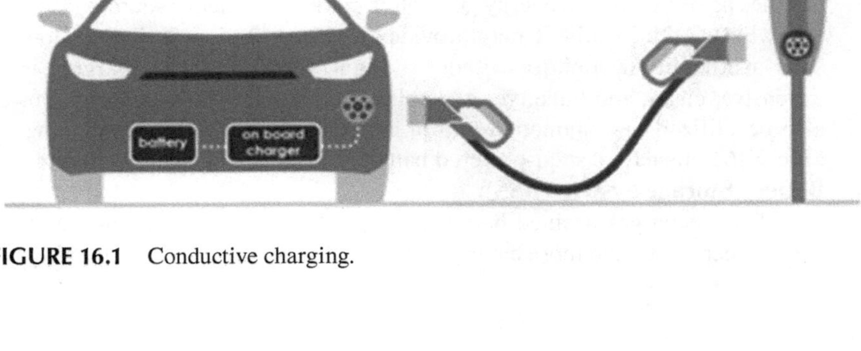

FIGURE 16.1 Conductive charging.

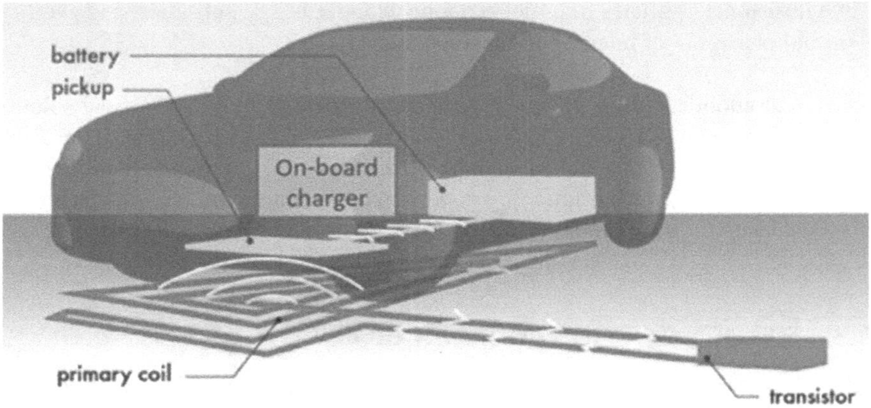

FIGURE 16.2 Non-conductive charging.

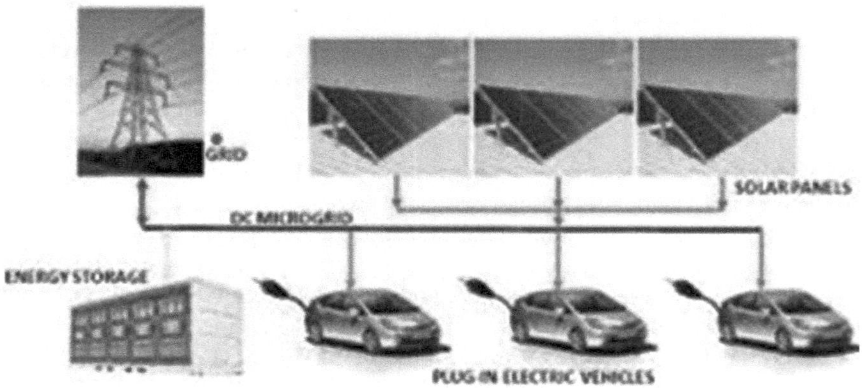

FIGURE 16.3 Solar-powered charging.

shown in Figure 16.3 charge DC devices exclusively during the day but use free electricity.

iv. **Battery Energy Storage System (BESS):**

BESS stores electrical energy in a battery or set of batteries. The BESS has the capability to save energy generated from green energy sources (solar and wind). Additionally, it may provide electricity during periods of low solar availability at night or without sunlight. BESS-powered charging is expensive, clean, and huge, yet it supplies electricity 24/7. The BESS can also be utilized in conjunction with a converter during V2G power flow. Figure 16.3 displays a solar-powered battery energy EV charging diagram.

v. **Battery Storage System (BSS):**

A BSS recharges drained batteries. This BSS has fixed charging time, simple operations, and more battery life, and it has optimized slow charging, but it also has higher battery prices and expensive changing systems. We use the BSS for 2-W and 3-W EVs because of their lighter batteries, contrary to 4-W batteries.

vi. **Pantograph Solution:**

It charges heavy-duty EV batteries, such as e-buses, lorries, trams, and trains, conductively. Only top-mounted shaving batteries charge. Pantograph charging offers multiple power levels, faster charging, automated operation, and rugged construction. Its drawbacks include its high cost, restricted DC charging compatibility, and heavy-duty electric car use only.

16.3.2 Level of Charging

People know that India and most other countries use 230 V/phase, i.e., 400V line to line, 50 Hz. The US, Canada, South Korea, Japan, etc. utilize 60 Hz and 120V per phase. Electric vehicle charging system manufacturers can make them for their grids. EVCS has Level 1, Level 2, DC Charging, and Superchargers as per Figure 16.4. Charging levels are briefed as below.

i. **On-board Charger (Level 1):**

This type of charger is utilized in countries with a power grid operating at 120V and a frequency of 60 hertz. The maximum current per phase is 12 A or 15 A, and the power is 1.44–1.90 kW. It will not work in India because the grid's nominal voltage is root mean square of 220V–240V at 50 Hz about a 20-hour charge. The charging process is slow, the initial cost is low, and the peak demand is quiet.

ii. **The Level 2 (On-board Charger):**

It requires 19.2–22 kW, 240V per phase, and the ampere is 80 A. In India, it is applicable at home which is slow and ordinary charging. It takes about 8 hours to charge.

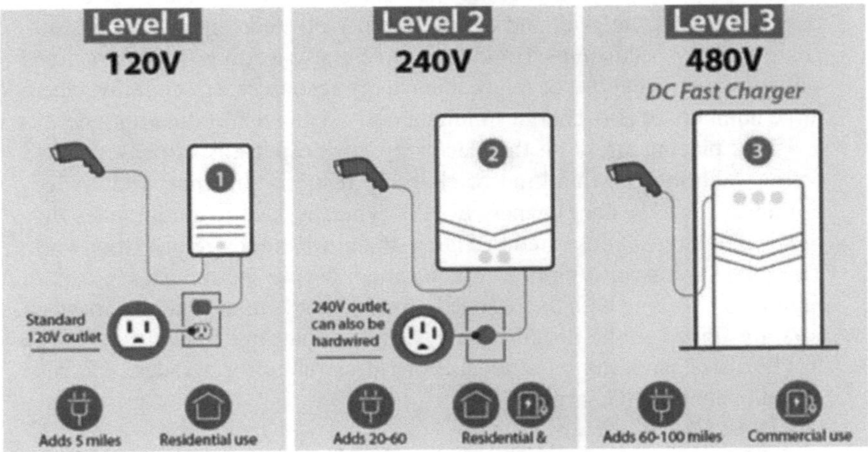

FIGURE 16.4 Level 1, 2, and 3 charging.

iii. **DC Charging:**

Charging is from 50 to 1,500V DC, and 80 to 400 A. The power level will be 48–400 kW+. It is good for public charging stations. It takes 0.5 hours to charge. It relies on battery capacity and charge-power state of charging (SOC). Merits are greater charging power and faster charging. Demerits are higher prices; regulatory complexity; and safety and security requirements.

iv. **Off-Board Superchargers:**

Superchargers only charge TESLA EVs. It supports 120 V, 240 V, and 400 V or greater charging. Power is 120–250 kW. Superchargers are Direct Current Fast Charging. Charging takes 0.30 hours. This is determined by the SOC and strength of the battery at the charging power. Advantages are extremely fast charging. Drawbacks are eccentric design and applicable in TESLA EVs only. DC and superchargers charge EV batteries faster than AC levels 1 and 2. Tesla Superchargers are Level 4, and DC chargers are Level 3.

16.3.3 Mode of Charging

The two primary types of charging modes are unidirectional and bidirectional charging. Only energy from the grid is transferred to EVs during unidirectional charging (also known as coordinated, off-peak, and uncoordinated charging). Bidirectional charging allows electricity to move from an EV to the grid or loads like homes, as well as back again. The collection of various methods is shown in Figure 16.5. This figure illustrates how varying EV charging or discharging techniques can alter the distribution system's overall power demand profile [15].

i. **Uncoordinated Charging:**

Uncoordinated charging often involves fixed charging rates determined by the capabilities of the charging station or outlet. These rates may vary depending on the voltage, current, and power output of the charging equipment. Uncoordinated charging may not take into account factors like energy rate, demand of the grid, and the availability of green energy. As a result, charging may occur indiscriminately, potentially leading to peak demand spikes or inefficient use of renewable energy resources. In scenarios where large numbers of EVs charge simultaneously without coordination, there is a risk of placing stress on the electricity grid, especially at peak demand hours. Additionally, this kind of charging restricts the appropriate adoption level for EVs since owners of EVs typically charge automobiles during peak hours, as illustrated in Figure 16.6a, when they return from work. The total power consumption of consumers during the day is depicted in Figure 16.6a, along with the change in total power demand that occurs when EVs are linked to the distribution system for charging (shown in Brown). It will also hasten the requirement for infrastructural upgrades. It is also called uncontrolled charging [16].

ii. **Off-Peak Charging:**

Off-peak charging, also known as scheduled charging or timed charging, refers to a feature available in many EVs and charging stations that

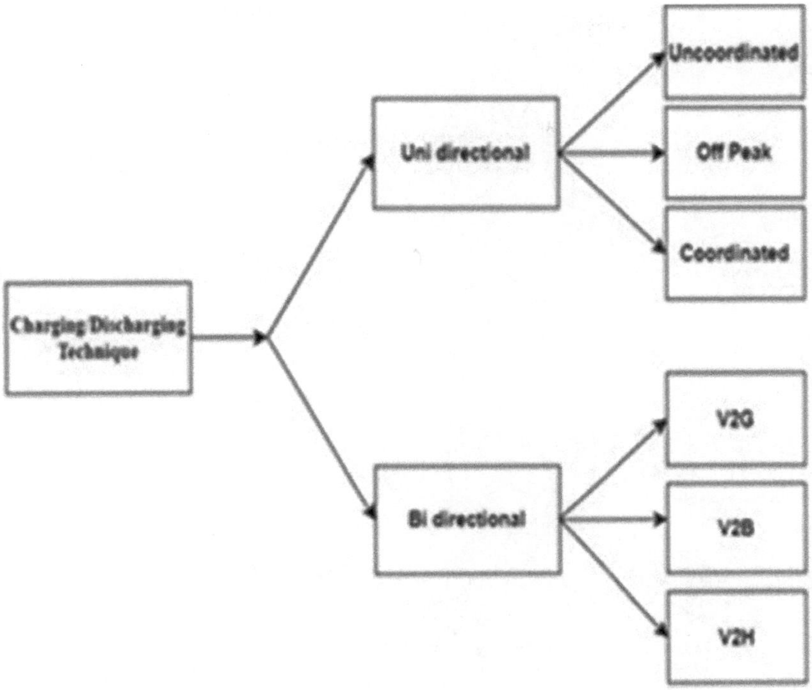

FIGURE 16.5 Collection of charging and discharging methods.

allows users to set a specific time for their vehicle to begin charging. Instead of immediately initiating the charging process upon plugging in the vehicle, users can schedule the charging to start at a later time, such as during the off-peak period or while electricity rates decrease. EV owners should charge vehicles during lower price periods, as seen in Figure 16.6b, to maximize the benefits to both EV owners and utilities. EV owners can access the settings or interface of their vehicle or charging station to specify the desired start time for charging. This can often be done through a mobile app, the vehicle's on-board computer, or the charging station's control panel. Users select the time at which they want charging to commence. This could be during the early hours of the morning, overnight, or at any other time when electricity demand is typically lower, and rates may be more favorable. Once the scheduled time is reached, the charging process begins automatically without the need for manual intervention. The vehicle or charging station initiates the flow of electricity to the EV's battery according to the predetermined schedule. It is also called as indirect controlled charging. This charging method is also called as controlled charging.

iii. **Coordinated Charging:**

The intelligent control of EV charging procedures is known as coordinated charging to ameliorate energy consumption, grid stability, and

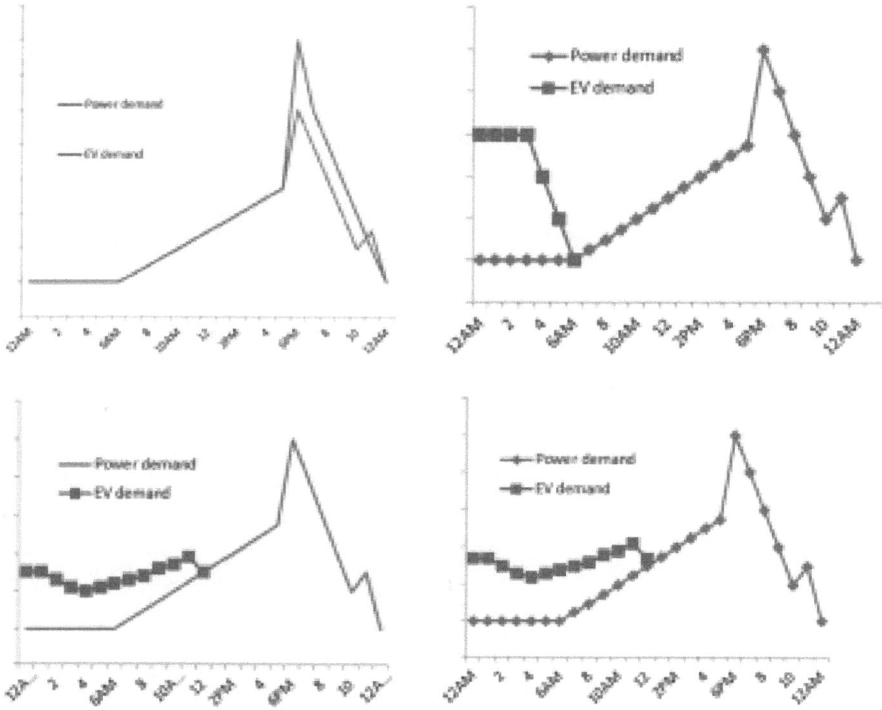

FIGURE 16.6 Overall power demand graph using charging as well as discharging techniques: (a) uncoordinated charging; (b) off-peak charging; (c) coordinated charging; and (d) V2G, V2B, and V2H.

cost-effectiveness. Unlike uncontrolled charging, where EV owners simply plug in their vehicles and charge without active management, coordinated charging systems employ various strategies to coordinate and regulate charging activities based on factors such as grid demand, electricity prices, and renewable energy availability. Coordinated charging systems can adjust charging rates and schedules in real time based on grid conditions. This includes prioritizing charging during periods of low demand or high renewable energy production to maximize the use of clean energy resources. Coordinated charging enables participation in demand response programs, where EV charging can be temporarily reduced or shifted during times of peak demand to support grid stability. In exchange, EV owners may receive incentives or credits from utilities or grid operators. EV owners profit from lower charging costs, while utilities gain from less distribution system stress and losses, improved quality of power, and EV shifts of load to off-peak hours, which causes the valley filling as illustrated in Figure 16.6c. Coordinated charging relies on advanced charging infrastructure equipped with communication capabilities, such as smart meters, communication protocols (e.g., OCPP), and data analytics platforms. These technologies enable seamless coordination between EVs, charging stations, and the grid [17,18]. Coordinated charging systems often allow EV owners to specify their charging preferences and priorities. This may include setting preferred charging times, specifying minimum charge levels needed by a certain time, or overriding default settings for immediate charging when necessary. By enabling smarter, more flexible charging practices, coordinated charging supports grid modernization efforts, enhances energy efficiency, and accelerates the transition to a sustainable transportation system.

iv. **Vehicle to Grid (V2G):**

V2G technology enables EVs to serve as mobile energy storage units that can both receive and supply electricity to the grid. This bidirectional flow of energy allows EV batteries not only to charge from the grid but also to discharge energy back into the grid when needed. When plugged into a charging station, an EV can draw electricity from the grid to charge its battery, as in traditional charging scenarios. With V2G capability, the EV battery can also discharge electricity back into the grid when conditions warrant. This occurs when there is excess energy stored in the vehicle's battery, or when the grid requires additional power to balance supply and demand. EVs can discharge stored energy during periods of peak demand, reducing the need for additional electricity generation or grid infrastructure. By distributing stored energy during times of high demand or grid congestion, V2G helps stabilize the grid and improve overall reliability. EVs can respond to frequency deviations in the grid by injecting or absorbing power as needed, helping to maintain grid stability. V2G technology can aid in the integration of renewable energy sources by storing excess renewable energy when it surpasses demand and returning it to the grid when required. As seen in Figure 16.6d, V2G can offer a variety of grid functions, including frequency management, spinning reserve, facilitating the integration of additional renewable energy system peak demand shaving, and valley filling [19]. V2G requires advanced charging infrastructure capable of bidirectional power flow and communication with the grid. This includes V2G-enabled charging stations equipped with the necessary hardware and software to manage energy flows between EVs and the grid. However, challenges such as battery degradation, regulatory barriers, and interoperability issues still need to be addressed for widespread adoption of V2G technology. Despite these challenges, V2G holds significant promise as a key component of future smart grids and sustainable transportation systems.

v. **Vehicle to Building (V2B):**

V2B technology allows EVs to supply electricity to buildings, acting as mobile energy storage units that can power homes, businesses, or other facilities. V2B leverages the bidirectional charging capabilities of EV batteries to enable them to discharge stored energy back into the building's electrical system when needed. EVs charge their batteries either from the grid or from renewable energy sources such as solar panels, wind turbines, or other on-site generation systems. When the building requires additional electricity or during peak demand periods, V2B-enabled EVs can supply power to the building by discharging energy from their batteries. V2B technology requires integration between the building's electrical system and the EV charging infrastructure [20]. This may involve the installation of bidirectional chargers capable of managing energy flows between EVs and the building. In the event of a power outage or emergency, V2B-enabled EVs can provide backup power to critical systems in the building, such as lighting, refrigeration, or communication equipment. Building owners or operators may incentivize EV owners to participate in V2B programs by offering financial incentives, reduced energy costs, or other benefits.

vi. **Vehicle to Home (V2H):**

V2H technology allows EVs to supply electricity to residential homes, acting as a backup power source or supporting the homes energy needs during peak demand periods. V2H systems leverage the bidirectional charging capabilities of EV batteries to enable them to discharge stored energy back into the home's electrical system when needed. EVs charge their batteries either from the grid or from renewable energy sources such as solar panels installed on the home. When the home requires additional electricity or during power outages, V2H-enabled EVs can supply power to the home by discharging energy from their batteries. V2H technology requires integration between the home's electrical system and the EV charging infrastructure. This may involve the installation of bidirectional chargers capable of managing energy flows between EVs and the home.

16.4 THE MAIN CHALLENGES THAT EV CHARGING NETWORKS FACE TODAY

The rise of EVs is creating a demand for a robust and user-friendly charging network. However, several challenges are hindering the widespread adoption of EVs, all stemming from limitations in the current charging infrastructure. Here are some of the main obstacles faced by EV charging networks today:

i. **Poor Charging Infrastructure:**

Absolutely, poor infrastructure is indeed one of the primary challenges facing EV charging networks today. In many regions, especially rural or less densely populated areas, there's a scarcity of charging stations. This lack of coverage can discourage potential EV owners from making the switch, as they may be concerned about finding a charging point when needed, particularly during longer journeys. Even in areas where charging infrastructure exists, accessibility can be an issue. Charging stations may not be conveniently located, making it inconvenient for EV owners to access them during their daily routines or travel plans. The availability of different charging speeds at various stations can be confusing and inconvenient for EV drivers. While some stations offer fast charging, others may only provide slower options, leading to uncertainty and longer waiting times for charging [21]. Reliability is crucial for EV drivers, who rely on the charging infrastructure to keep their vehicles powered.

Instances of out-of-service or malfunctioning charging stations can disrupt travel plans and erode trust in the reliability of EVs. Addressing the challenge of poor infrastructure requires significant investment in expanding and upgrading charging networks. Governments, utilities, private companies, and other stakeholders need to work together to develop comprehensive strategies to build out charging infrastructure, particularly in underserved areas.

ii. **V2G Interoperability:**

V2G interoperability refers to the ability of EVs and associated charging infrastructure to communicate and exchange energy with the electrical grid in a standardized and seamless manner. Interoperability requires the establishment of standardized communication protocols that enable EVs and charging infrastructure from different manufacturers to communicate effectively with each other and with the grid. V2G interoperability also entails compatibility between the hardware and software components of EVs and the charging infrastructure. This includes ensuring that EVs are equipped with bidirectional charging capabilities and that charging stations support bidirectional power flow. V2G systems must be compatible with the electrical grid's infrastructure and operations. This involves adhering to grid standards and regulations, as well as implementing grid-friendly charging and discharging strategies to ensure grid stability and reliability. Interoperable V2G systems must prioritize security and privacy considerations to protect against cyber threats and unauthorized access to EVs and the grid [22]. This includes implementing robust authentication, encryption, and data privacy measures. Achieving V2G interoperability requires collaboration among automakers, charging infrastructure providers, utilities, standards organizations, and regulatory agencies. Also, establishing common standards, protocols, and certification processes is essential to ensure seamless interoperability across the V2G ecosystem.

iii. **Station Performance and Downtime:**

The performance and uptime of EV charging stations are critical factors in ensuring the reliability and effectiveness of the charging infrastructure. The rate at which an EV charging station can replenish an EV's battery determines its performance. Faster charging speeds, such as those offered by DC fast chargers, are desirable for reducing charging time and increasing convenience for EV owners [23]. The ability of a charging station to consistently operate without issues or failures is crucial for maintaining user satisfaction and confidence. Reliable charging stations minimize downtime and ensure EV owners can access charging when needed. Downtime can inconvenience EV owners who rely on charging stations for their daily transportation needs. Unexpected outages may disrupt travel plans and lead to frustration for charging station operators, resulting in downtime and lost revenue opportunities. Each hour of downtime represents potential revenue that cannot be generated from charging sessions. Persistent downtime or unreliable performance can damage the reputation of charging station networks and deter EV owners from using them in the future. Positive user experiences are crucial for building trust and loyalty among EV drivers. Overall, ensuring optimal performance and minimizing downtime are essential for providing reliable and convenient charging experiences for EV owners, supporting the growth of EV adoption, and building confidence in charging infrastructure.

iv. **Payment Method Integration:**

Integration of payment methods is crucial for EV charging stations to facilitate convenient and seamless transactions for EV owners. Accepting credit and debit cards is one of the most common payment methods at EV charging stations. It offers convenience for users who may not carry cash and allows for quick and easy transactions. Radio-frequency identification cards or key fobs can be used for contactless payments at charging stations. Users can tap their radio-frequency identification cards on the reader to initiate and authorize charging sessions. Payment method integration requires robust security measures to protect users' financial information and prevent fraud. Compliance with industry standards and regulations such as PCI DSS is essential to safeguard sensitive data. Encryption and tokenization should be implemented to secure payment transactions and protect users' financial data. By integrating a diverse range of payment methods and addressing integration challenges effectively, EV charging stations can provide convenient, secure, and user-friendly payment experiences for EV owners, ultimately supporting the widespread adoption of EVs [24].

v. **Power Issues While Charging:**

Power issues with charging EVs can arise from various sources, impacting the efficiency, safety, and reliability of the charging process. Charging an EV during periods of high demand on the electrical grid can result in power issues. Grid congestion may lead to voltage drops, frequency fluctuations, or limited availability of charging power, particularly in densely populated areas or during peak electricity demand hours. Malfunctions in the charging station's hardware or software can result in power-related issues during charging. This includes problems such as faulty connectors, damaged cables, software glitches, or communication errors between the EV and the charging station. To mitigate power issues during EV charging, it's essential to ensure proper electrical infrastructure, use compatible charging equipment, maintain charging stations regularly, and consider factors such as voltage stability and grid demand [25].

vi. **Charging Station to Network:**

The connection between EV charging stations and a broader network is crucial for ensuring their functionality, accessibility, and management. EV charging stations are equipped with communication capabilities that allow them to connect to external networks. These networks can include central management systems operated by charging network providers, utilities, or third-party service providers. Charging stations use various communication technologies to connect to networks, including Ethernet, Wi-Fi, cellular, or dedicated communication protocols such as OCPP. These connections facilitate real-time monitoring, control, and data exchange between charging stations and network operators. Charging network operators utilize central management systems to monitor and control charging stations remotely. A central management system allows operators to track charging session data, manage billing and payments, perform diagnostics, and update software or firmware across the charging network [26]. Network-connected charging stations generate valuable data on charging behavior, energy consumption, and station performance. Network operators utilize data analytics tools to analyze trends, identify usage patterns, and optimize network operations. Reporting capabilities enable operators to generate insights for stakeholders, track key performance indicators, and comply with regulatory requirements.

Significant research endeavors have addressed the aforementioned issues by incorporating cutting-edge digital technologies such as AI and blockchain to revamp the architecture of a safe EV charging system.

16.5 EMERGING DIGITAL TECHNOLOGIES IN EVS

The rapid rise of EVs presents both opportunities and challenges for building a robust and efficient charging station network. Several emerging digital technologies are poised to address these challenges and optimize EV charging infrastructure. As these technologies evolve and converge, we can expect a future where charging EVs is as seamless and convenient as refueling gasoline-powered cars today. This section discusses new technologies such as AI and blockchain technology, as well as charging interoperability in both private and public spaces.

16.5.1 AI: Toward Intelligent EV Charging

AI might be the single most important technical innovation of the present decade. We have utilized AI to transform almost every sector. EV charging is not exceptional; AI will play a significant role in the quick expansion of EV charging networks. As depicted in Figure 16.7, AI will transform the EV charging industry in the following ways.

16.5.1.1 Predict User Charging Behavior

AI and ML techniques can play a significant role in predicting user charging behavior for EVs, which is crucial for optimizing the utilization of charging infrastructure and improving the overall user experience. ML algorithms can be trained on historical data, including charging patterns, location data, user preferences, and other relevant factors, to build predictive models. These models can forecast charging demand, identify peak hours, and anticipate the likelihood of a user needing to charge at a particular location and time. AI can analyze user data, such as driving habits, commute patterns, and charging history, to create personalized user profiles. These profiles can be used to predict individual charging needs and preferences, enabling targeted recommendations and personalized services.

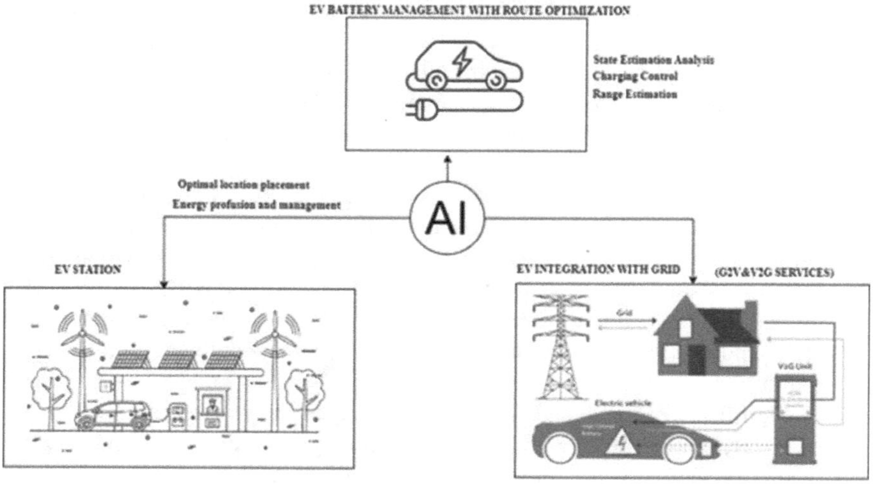

FIGURE 16.7 Role of AI: intelligent EV charging.

16.5.1.2 Optimize Charging Schedules and Provide Dynamic Pricing

AI can play a pivotal role in optimizing charging schedules and providing dynamic pricing for EV charging stations. AI algorithms can analyze historical data, user behavior patterns, and contextual factors (such as weather, events, and traffic patterns) to forecast charging demand at different locations and times. This demand forecasting can help anticipate peak periods and allocate resources accordingly. By combining demand forecasts with real-time data from charging stations and the electricity grid, AI systems can optimize charging schedules to balance the load across multiple stations. This can involve prioritizing charging for certain vehicles, staggering charging times, or directing users to less congested stations, thereby reducing strain on the grid and maximizing the utilization of available resources. AI can enable dynamic pricing strategies that adjust charging rates based on supply and demand. During periods of high demand or limited electricity supply, prices can be increased to incentivize users to charge at off-peak times or alternative locations. Conversely, prices can be lowered during periods of low demand or high renewable energy generation, encouraging EV adoption and promoting the integration of sustainable energy sources.

16.5.1.3 Battery Development and Testing

AI can accelerate the research and development process, optimize battery performance, and improve quality control measures. AI can analyze data from the battery manufacturing process, including factors such as temperature, humidity, and material properties, to identify correlations and patterns that influence battery performance and quality. This information can be used to optimize manufacturing parameters, reduce defects, and improve consistency in battery production. AI can contribute to the development of efficient recycling and reuse processes for batteries.

ML algorithms can analyze material compositions, identify valuable components, and optimize recycling processes to recover valuable materials and reduce environmental impact. AI-powered image recognition and pattern recognition algorithms can be used for automated visual inspection of battery components and assemblies, enabling early detection of defects or manufacturing irregularities, and ensuring consistent quality control.

16.5.1.4 Choosing a Location and Installing Charging Stations

AI can help identify the most strategic locations for charger deployment, ensuring efficient utilization of resources and meeting the evolving charging needs of EV users. AI techniques like multi-criteria decision analysis and constraint programming can evaluate potential charging station locations based on a range of criteria, including accessibility, proximity to high-demand areas, grid capacity, land availability, and zoning regulations. Optimization algorithms can identify the optimal combination of sites that maximize coverage, minimize infrastructure costs, and ensure equitable distribution of charging facilities.

16.5.1.5 Optimize V2G Services

AI algorithms can analyze real-time grid data, including electricity demand, renewable energy production, and weather forecasts. Based on these factors, they can predict peak demand periods and optimize V2G power flow. This helps to balance the grid by strategically discharging power from EVs during peak hours and charging them when demand is lower. AI coordinates V2G services with the availability of renewable energy sources, such as solar or wind power, to facilitate the efficient integration of these intermittent sources into the grid. By leveraging EVs as energy storage devices, AI can help balance the intermittency of renewable energy and improve grid stability, promoting sustainable energy practices. By leveraging AI for optimizing V2G services, grid operators can effectively manage the complexities of bidirectional energy flow, balance energy supply and demand, and facilitate the seamless integration of EVs into the power grid.

16.5.2 Blockchain Technology: EV Charging Management and Decentralized Energy Trading

A decentralized encrypted network is used to record any transaction involving digital assets as delineated in Figure 16.8a. The technology behind the blockchain is a dividend and trustworthy distributed ledger. It was first created to record monetary transactions. It verifies transactions instantly and guarantees payments using safe encryption methods. Consequently, the blockchain has sparked significant interest in several industries, such as the trading of energy, where all participants in the network are considered peers. The blockchain technology facilitates a decentralized network that eliminates the need for intermediaries, resulting in reduced operational costs. This, in turn, provides a faster, more secure, and cost-effective approach to transactions in the sector of energy. A way to trade energy using blockchain technology was suggested as a way to create a competitive market for energy trading. However, the

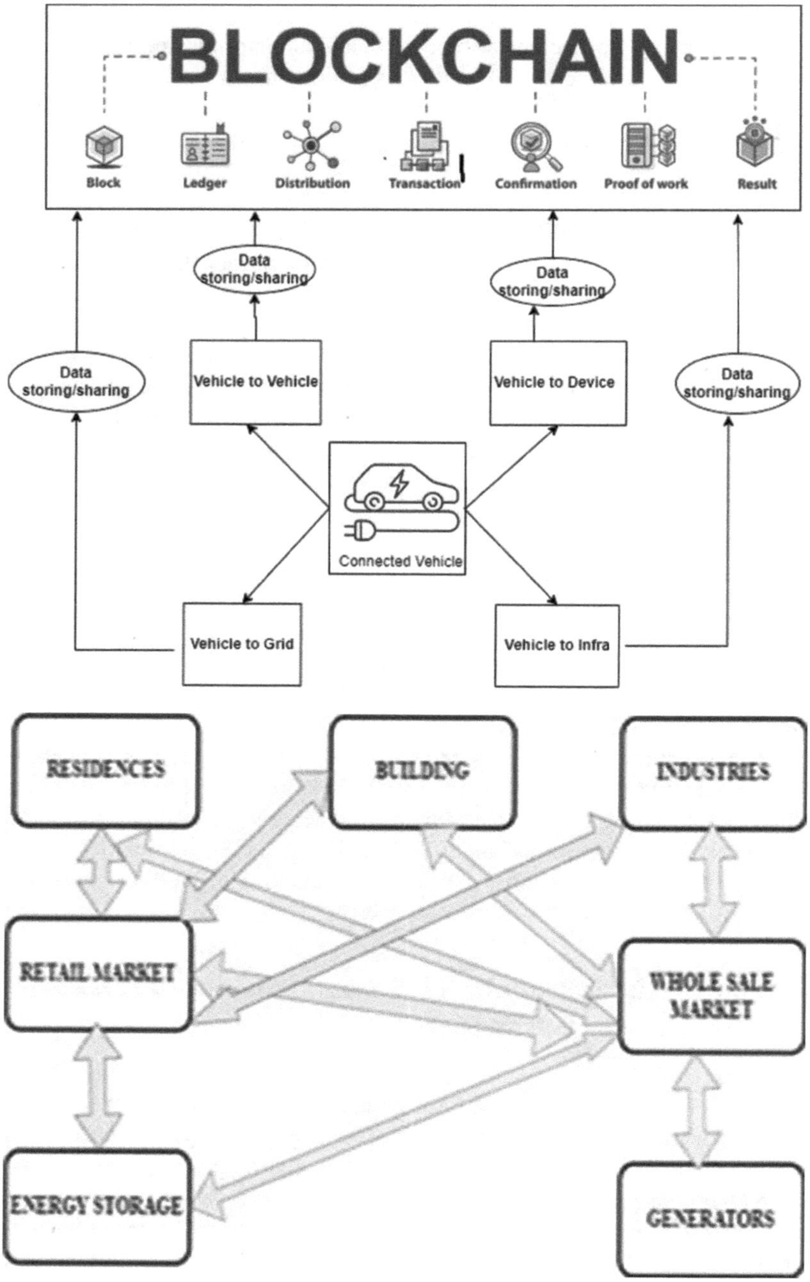

FIGURE 16.8 (a) Illustration of blockchain-enabled EV. (b) Transaction of the energy system market.

blockchain is only utilized to store information and keep track of deals [27]. The market economic assessment for local energy trading is done. Open-source deals govern blockchain smart contracts. We can also control business energy supply and demand. Smart contracts are designed to automate complex and dispersed workflows in the energy industry, allowing for the self-execution of operations with multiple stages. In conclusion, uncontrolled EV charging/discharging may disrupt grid operation. On a blockchain-based trading platform, efficient distributed grid activities require excellent scheduling.

16.5.2.1 Energy Trading System Model

EVs and tiny businesses energy generators can power the transition to a decentralized transactional energy market as depicted in Figure 16.8b, where users can buy, sell, and trade extra power just like any other commodity. The Blockchain-Based Energy Trading Model facilitates energy transactions and agreements without the need for a hierarchical system structure or information sharing [28]. Therefore, all the various electricity consumers, including residential, commercial, and industrial users, are connected to both the retail market for individual customers and the wholesale market to big-scale electricity producers. A decentralized energy trading concept utilizing blockchain technology enables prosumers to engage with electricity trading inside the grid. This approach allows prosumers to have autonomy in a blockchain power exchange system. It allows users to input and extract energy in order to engage in the trading platform of the smart grid public blockchain.

In the blockchain-based energy trading paradigm, the term "components" refers to all the generators and electrical load things that are linked to the retail markets. Every individual part has the ability to publish and send the command to charge or discharge to the trading platform of the smart grid blockchain. EVs may be charged and discharged using a programmed charging system. This is to facilitate the immediate activation or deactivation of power transmission, as directed by EVs, presuming the use of advanced switch technology. The energy distributors in the blockchain power trading system consist of power plants, small renewable producers, storage facilities, and EVs. In addition, various power loads such as residential areas, hospitals, and EVs are all linked with a common blockchain power trading network. Data transmission on the blockchain network occurs at intervals of thirty minutes. The components have the ability to determine the price of the energy they generate in order to encourage consumers to equalize supply and demand. Traditional power producers are linked to the wholesale market. It facilitates trading with significant energy demand offerings, as seen in Figure 16.1. In addition to facilitating wholesale market transactions in the traditional grid system, transactional energy proposes a strategy for organizing retail participants through numerous frequent transactions that are automatically executed by a blockchain-enabled platform. This approach aims to reduce the centralized aspects of the future grid system [28]. Interchange of data remains consistent across many entities, including distributed energy resources, a sizable generator, non-conventional energy generators, EV, power sellers, agents, and network operators. The transactions may be conducted across retail and wholesale marketplaces, so ensuring equal opportunities for all participants. In addition, the transactions must also consider the distribution and transmission restrictions as well as other limitations on the grid.

Assume the EV status matrix Z as follows:

$Z_{i,t} = 1$, if EV is charged at time t
$Z_{i,t} = 0$, if not
Z-EV status matrix

The power consumption of an EV is defined by the initial state of charge (SOCini) of its battery and the expected state of charge (SOCexp) after charging. Thus, the formulation may be stated as follows:

$$E_{ev}(t) = \sum_{i=1}^{I}\left(z_{i(t)}\left(SOC_{\text{expec}}(i) \pm SOC_{\text{inti}}(i)\right)\right)$$

Next, we establish the entire domestic load by combining the demand for EV (charging, discharging) and the range of load excluding EVs. This will allow us to construct the problem of EV charging:

$$E_{sum}(t) = E_{load}(t) + E_{ev}(\text{t})$$

P_{load} represents the power load generated or consumed by the loads in the grid network. By using the comprehensive utility function for the local region, we may employ optimization approaches to accomplish the desired outcome, including load flattening, peak shaving, and privacy preservation.

16.6 HARNESSING THE POWER OF AI AND BLOCKCHAIN IN THE EV SECTOR

From the standpoint of the energy market, the blockchain offers the technological underpinnings and trading platform for decentralized markets in energy that are available to private, open prosumers and have improved automated processes, safety, and privacy [29]. From a managerial point of view, AI helps control systems to make tactical decisions that optimize system performance and accomplish certain objectives, such as increasing profit generation, reducing carbon emissions, and anticipating system uncertainties. Figure 16.9 shows the framework for integrating blockchain technology and AI with prosumers' and consumers' assimilation in EVs. The EV energy sector is rapidly evolving. Advancements in technology like AI and blockchain are enabling new innovations that have the potential to transform how we produce, distribute, and consume energy. Let us explore the key technologies that can be harnessed in the energy industry.

16.6.1 Optimizing Energy Management with AI

AI can analyze vast amounts of data from sensors and internet-connected devices to optimize energy management. ML algorithms can detect patterns and predict energy usage more accurately, allowing for better efficient operations, asset upkeep, and

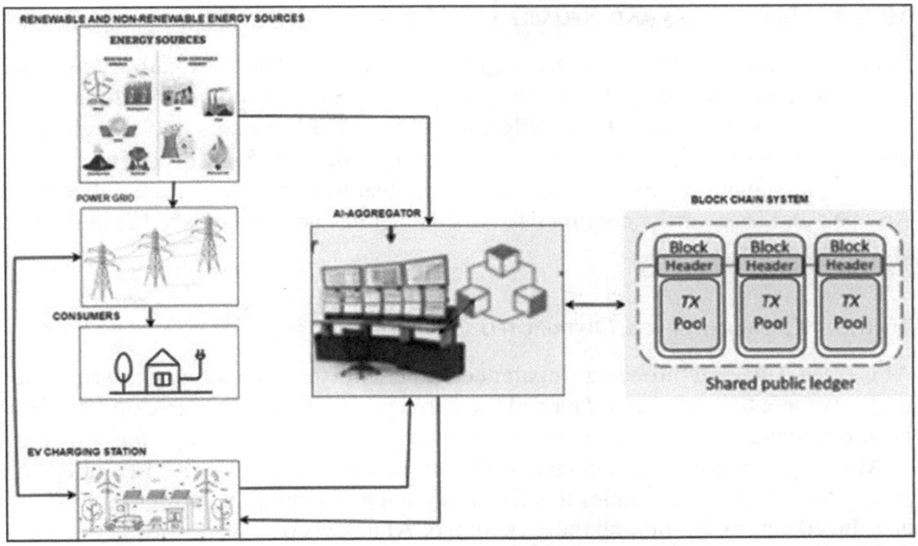

FIGURE 16.9 AI and Blockchain in the EV sector.

demand prediction. For example, AI can help to adjust energy supply and demand in real time across smart power grids. It can also identify anomalies and issues before disruptions or outages occur.

16.6.2 Decentralizing Grids with Blockchain

Blockchain, the technology behind cryptocurrencies like Bitcoin offers a decentralized and immutable ledger for recording transactions. This could disrupt traditional power grid models by enabling decentralized energy networks. Blockchain can engage energy trading between suppliers and customers in microgrids. Smart contracts automate processes, while cryptography enhances security. Blockchain also enables tracking of renewable energy certificates. Overall, it increases transparency and efficiency in energy distribution.

16.6.3 Enhancing Renewable Energy Integration

As non-conventional energy sources like sun and wind gain share, effectively integrating them into legacy grid systems becomes more complex because the supply fluctuates based on weather conditions. AI can analyze massive troves of data from weather forecasts, market dynamics, and production facilities to optimally manage and distribute renewable energy. Blockchain enables decentralized marketplaces where households with solar panels can sell excess power to their neighbors. These innovations will be key to overcoming intermittency challenges with renewables.

16.6.4 Transparent and Reliable Data Sources

In order to enhance the secure sharing of data among numerous organizations, it is crucial to prioritize the openness and dependability of data sources. With the synchronization of the nodes' whole ledger, competent blockchain technology guarantees data transparency on the chain. It also assures data tracing through the use of transaction signatures, time stamps, and other features after certificate validation. An open and trustworthy communication channel has been established among several participants.

16.6.5 Capabilities for Distributed Computing

AI is often delivered through a singular computer entity or computing platform. Due to the exponential growth of data and the corresponding rise in complexity of computation, conventional computing platforms struggle to meet the computing demands of AI on their own. It results in increased hardware and maintenance costs for enterprises [30]. Blockchain enables the decentralization of computing power through its distributed nature, facilitating the execution of AI models on a worldwide network of decentralized nodes and enabling decentralized computing.

16.6.6 Revolutionizing EVs and Charging Infrastructure

The rise of EVs will put pressure on power grids. AI and blockchain can help to coordinate EV charging patterns and integrate charging stations with grid infrastructure. AI can optimize EV charging schedules and routes based on real-time energy prices. Blockchain can authenticate transactions at charging stations while distributing demand across the grid to avoid power surges. This will enable large-scale adoption of EVs.

In summary, leveraging technologies like AI and blockchain will lead to major disruptions across the energy sector. They enable optimized energy management, decentralized power networks, better integration of renewables, and next-generation charging infrastructure to accelerate the transition to cleaner and more resilient energy systems. Companies that embrace these innovations will thrive in the energy landscape of the future.

16.7 REAL-WORLD USE CASE/ONGOING PROJECT

One notable example is the partnership between Elia Group, BMW Group, and the blockchain startup Energy Web Foundation. They collaborated on a blockchain-based project called "Share&Charge", aiming to create a decentralized network of EV charging stations. The platform utilizes blockchain for transparent, secure transactions and integrates AI for optimizing charging station utilization and predicting maintenance needs. By combining AI for predictive analytics and blockchain for secure, transparent transactions, this integrated approach enhances the efficiency, reliability, and user experience of EV charging stations, paving the way for smarter, more sustainable transportation infrastructure.

16.7.1 Detailed Exploration of a Real-World Use Case or Project: Elia Successfully Demonstrates Transmission Service Operator (TSO)-Visible EV Charging in Collaboration with Energy Web and BMW Group

As depicted in Figure 16.10, TSO-visible EV charging session has been successfully demonstrated by Elia Group, the Belgian and German TSO, in collaboration with Energy Web, a non-profit organization that specializes in building operating systems for energy grids, and BMW Group. This session has laid the groundwork for improved EV charging, including charging that is solely powered by renewable energy and rapid switching between suppliers [31].

The successful demonstration builds on Elia's ongoing work with Energy Web on the E-Mobility Dashboard, an open platform made to make it easier to find EVs and charging stations. This platform lets everyone get reliable, safe data using the Open Charging Network, energy Web's e-roaming hub. Because of this, it is a lot easier for EV drivers to share data with people in the energy industry, like TSOs, suppliers of energy, and adaptable providers of services. While an EV is powered up, the local grid worker usually gets information about charging as a whole, not details about each car. The same names can be used to make payments to other market players easily. This cuts down on waste and unnecessary exchange of personal information.

Energy Web's operating system facilitates the allocation of digital identities to EVs and stores them in the cars' digital wallets. This allows the car to safely communicate its identity to the TSO without disclosing the driver's identity. The demonstration highlights the ability to enhance visibility and security among participants in the charging process. Increasing the visibility of various assets in the system is crucial for enhancing EV charging and facilitating future solutions, such as charging with 100% renewable energy. "Through the utilisation of Energy Web's blockchain-powered operating system, TSOs can now gain visibility at the individual device level, a capability that was

FIGURE 16.10 Public fast-charging point in Berlin.

previously unavailable", stated Jesse Morris, the CEO of Energy Web. EV owners will also benefit from a future system that automatically recognizes their car when they charge it, regardless of the charging operator. They will also have the assurance that the energy used to charge their vehicle comes only from sources that are sustainable.

Kai Schmied from Elia stated that BMW Group is participating in this groundbreaking experiment, which is the first of its kind for TSO-visible electric car charging. "We are excited to continue collaborating with Energy Web on these innovative charging solutions". Currently, there is a lack of individual monitoring of EV charging operations, which hinders the ability to actively control these activities. In order to facilitate flexibility, use cases, or provide new services like supplier switching at the charging station, grid operators require access to individual charging operations and a secure identification of the parties involved.

Energy Web is an international, member-led non-profit organization that is expediting the shift to carbon-free, customer-focused energy systems by harnessing the power of free and open-source digital technology. The Energy Web Decentralized Operating System allows any energy asset, regardless of ownership, to actively engage in any energy market. The Energy Web Chain is the first public AI-blockchain designed specifically for the energy industry. It serves as the foundation for the Energy Web Decentralized Operating System technology stack. The Energy Web ecosystem consists of prominent services, grid administrators, renewable energy providers, commercial energy purchasers, IoT/telecom administrators, and other stakeholders.

The above use case demonstrates how AI and blockchain technologies can synergize to drive innovation in the EV industry, addressing challenges such as infrastructure scalability, operational efficiency, and data security.

16.8 CHALLENGES & SOLUTION, ECONOMIC FEASIBILITY & COST-BENEFIT ANALYSIS, AND FUTURE DIRECTION

In order to integrate AI technology into smart and sustainable EV charging infrastructure, it is imperative to tackle the associated technical and regulatory challenges. This section addresses the primary roadblocks to AI integration in EVs. It offers topics for future study to aid in both technological and economic feasibility and cost-benefit study, as well as to promote the rise of AI-powered EVs.

16.8.1 Challenges and Solution

16.8.1.1 Privacy Concerns and Data Management-AI Empowers Blockchain-Privacy Concerns and Data Management in EV Charging

Many criteria are involved in the creation and execution of blockchain, and security, efficiency, decentralization, and other factors must be traded off. These choices can be made more easily using AI technology, which can also optimize the blockchain for improved performance and control. Furthermore, AI may lessen mistakes brought on by human error and increase the intelligence of blockchain applications. As we all know, blockchain is nearly impossible to attack unless an attacker has the bulk of mining privileges. However, decentralized apps constructed on the blockchain technology have less secure programs and functionalities. For instance, hackers regularly withdrew money from the Decentralized Autonomous Organization event [32] by exploiting vulnerabilities in smart contracts, which lost $50 million.

AI technology opens up intriguing options for the brainy blockchain system security defense, as well as providing security and technical assistance throughout the blockchain transaction lifecycle. AI technology is employed to build smart contracts by itself, reducing their susceptibility to the greatest extent feasible. In addition, data mining along with additional technologies are utilized to analyze the smart contract's shortcomings, and big data analysis is utilized to detect dangerous flaws to minimize economic damages incurred by hackers. By incorporating AI into the blockchain, smart contracts could become more capable and intelligent. They can also be able to develop a more comprehensive code through constant practice and learning, which can change the way blockchain smartcontracts function. The effectiveness of blockchain-based applications and the rapidity of data queries have both greatly increased through ongoing learning and practice.

16.8.1.2 Dealing with Uncertainty and Risk

Precise estimate and prediction of energy use in e-transportation systems is hampered by uncertainty related to elements such as user behavior, charging facilities, and fatigue of the battery pack [33]. Probabilistic models, uncertainty quantification methods, and risk analysis systems help us to address these uncertainties and related hazards. AI stands out as a viable answer among these methods, especially reinforcement learning (RL). By using real-time feedback, RL algorithms enable adaptive decision-making; so, e-mobility systems are empowered to maximize their energy consumption management. To reduce battery wear, RL can model user behavior to determine ideal charging schedules based on preferences and prior patterns, optimize charging and discharging profiles, and intelligently allocate resources to optimize charging infrastructure utilization. E-mobility systems can properly manage uncertainty, increase effectiveness in operation, and guarantee dependable performance by including RL models. Future difficulties in e-mobility energy management consist, in general, in data availability and quality, model complexity and scalability, real-time prediction, renewable energy integration, and uncertainty management. Dealing with these difficulties calls for multidisciplinary cooperation, sophisticated ML methods, strong data infrastructure, and regulatory support to help dependable and effective e-mobility systems grow.

16.8.1.3 Regulatory Challenges and Standardization Challenges

Standardizing EV charging infrastructure has regulatory and standardization challenges. A thorough analysis of challenges would be beneficial for stakeholders and policymakers.

16.8.1.3.1 Regulatory Challenges

1. **Data Privacy and Security**: AI and blockchain involve handling large amounts of sensitive data. Regulations such as general data protection regulation in Europe and various data protection laws globally impose strict requirements on how data is collected, stored, and used.

2. **Liability and Accountability**: Determining liability in case of accidents or malfunctions involving AI-controlled or blockchain-managed systems is complex. Clear regulations are needed to allocate responsibility among stakeholders.
3. **Ethical Use of AI**: Regulations may need to address ethical concerns related to AI decision-making, bias in algorithms, and transparency in AI-driven processes, especially in critical infrastructure like EV charging.
4. **Interoperability**: Ensuring that AI systems and blockchain platforms can interoperate with existing infrastructure and across different jurisdictions is crucial for seamless operation and scalability.
5. **Impact on Labor and Employment**: AI deployment and automation could disrupt traditional job markets. Regulations may be required to manage workforce transitions and ensure fair employment practices.

16.8.1.3.2 Standardization Challenges

1. **Technical Standards**: Establishing technical standards for AI algorithms, blockchain protocols, and interoperability with EV charging infrastructure is essential for compatibility, security, and reliability.
2. **Blockchain Scalability and Performance**: Scaling blockchain networks to handle large-scale EV charging transactions efficiently while maintaining performance and security standards is a significant technical challenge.
3. **Consensus Mechanisms**: Determining consensus mechanisms in blockchain networks that are energy-efficient and secure, especially considering the energy consumption concerns associated with some blockchain implementations.
4. **Integration with Grid Management Systems**: Standards for integrating AI and blockchain systems with existing energy grid management systems to optimize energy distribution and ensure grid stability.
5. **User Experience and Accessibility**: Developing user-friendly interfaces and standard protocols for EV drivers to access and pay for charging services via blockchain platforms, ensuring a seamless experience.

16.8.1.3.3 Addressing These Challenges

1. **Collaboration**: Governments, industry stakeholders, and standards organizations must collaborate to develop cohesive regulatory frameworks and technical standards.
2. **Pilot Projects and Testing**: Conducting pilot projects to test AI and blockchain applications in real-world settings can help identify regulatory gaps and technical challenges early on.
3. **Education and Awareness**: Raising awareness among policymakers, regulators, and the public about the benefits and risks of AI and blockchain technologies is crucial for informed decision-making.
4. **Adaptive Regulations**: Establishing adaptive regulatory frameworks that can evolve with technological advancements and market dynamics to foster innovation while ensuring safety and consumer protection.

By addressing these regulatory and standardization challenges, stakeholders can foster an environment conducive to the safe and effective deployment of AI and blockchain technologies in EV charging stations, contributing to sustainable mobility and energy management solutions.

16.8.2 Economic Feasibility and Cost-Benefit Study

The economic feasibility and cost-benefit study of implementing AI and blockchain in EV charging infrastructure are crucial considerations for stakeholders, including governments, energy providers, and technology developers. Below is a summary of essential factors and aspects.

16.8.2.1 Economic Feasibility

1. **Initial Investment**:
   - **AI**: Costs include the development of AI algorithms for demand prediction, scheduling optimization, and predictive maintenance.
   - **Blockchain**: Initial investment covers setting up blockchain infrastructure for secure transactions, smart contracts, and data management.
2. **Operational Costs**:
   - **AI**: Ongoing costs involve maintenance of AI models, data storage, and computing resources for real-time processing.
   - **Blockchain**: Costs include transaction fees, network maintenance, and energy consumption (depending on the consensus mechanism).
3. **Training and Integration**:
   - Costs associated with training staff on AI technologies and integrating AI systems with existing EV charging infrastructure.
   - Integration costs for blockchain platforms with existing payment systems and regulatory frameworks.
4. **Scalability and Upgradation**:
   - Costs to scale AI models and blockchain networks as EV adoption and charging station networks grow.
   - Upgradation costs to keep pace with technological advancements and regulatory changes.

16.8.2.2 Cost-Benefit Study

1. **Benefits**:
   - **Operational Efficiency**: AI can optimize charging schedules, reduce wait times, and improve energy management, potentially lowering operational costs.
   - **Transaction Efficiency**: Blockchain can streamline payment processes, reduce transaction fees, and enhance transparency and security.
   - **Customer Experience**: Improved user interfaces and seamless transactions can enhance user satisfaction and loyalty.
2. **Energy Savings**:
   - AI-enabled predictive analytics can optimize energy usage, reducing peak demand and optimizing grid integration.

- Blockchain can facilitate peer-to-peer energy trading and incentivize renewable energy usage, contributing to sustainability goals.

3. **Revenue Generation**:
   - Potential revenue streams from offering premium AI-driven services (e.g., predictive maintenance alerts, personalized charging plans).
   - Additional revenue from blockchain-enabled services such as secure data sharing and fee management.
4. **Risk Mitigation**:
   - Enhanced security and transparency provided by blockchain can mitigate risks associated with fraud, cyberattacks, and data breaches.
   - AI-driven analytics can improve risk management through real-time monitoring and predictive maintenance.

The following factors are also considered to determine economic feasibility and cost-benefit analysis:

1. **Regulatory Compliance**:
   - Costs associated with ensuring compliance with data privacy laws (e.g., GDPR) and regulatory standards for energy and transportation sectors.
2. **Market Adoption and Acceptance**:
   - Potential challenges in user acceptance and adoption of new AI and blockchain technologies, requiring education and demonstration of benefits.
3. **Environmental Impact**:
   - Assessing the environmental footprint of AI algorithms and blockchain networks, including energy consumption and carbon emissions.

Conducting a thorough economic feasibility and cost-benefit analysis is essential before implementing AI and blockchain in EV charging infrastructure. It involves assessing initial investments, ongoing operational costs, potential benefits in efficiency and revenue generation, as well as addressing regulatory and environmental considerations. Stakeholders should collaborate closely to ensure that the deployment of these technologies aligns with strategic objectives for sustainable and efficient mobility solutions.

16.8.3 Future Direction

16.8.3.1 Scalability of Blockchain

The main issue lies in the inherent trade-off between **decentralization**, **security**, and **scalability**, often referred to as the "blockchain trilemma". Decentralization, a key strength of blockchain, requires every node to maintain a copy of the entire blockchain, making it inherently slower than centralized systems. Securing this consensus among all nodes further adds to the processing load. As a result, increasing transaction throughput often comes at the expense of either decentralization or security. Several promising approaches are being explored to overcome the scalability challenge.

1. **On-Chain Scaling**:
   - This straightforward solution simply enlarges the blocks within the blockchain, allowing more transactions per block. However, it increases storage requirements and processing power needed for each node, potentially compromising decentralization. Blockchain is divided into smaller partitions (shards), enabling parallel transaction processing. This improves scalability but raises concerns about inter-shard communication and security guarantees. Block space usage is optimized by separating transaction signatures from transaction data, allowing more transactions within the same block size.
2. **Off-Chain Scaling**:
   - Lightning Network creates a second-layer network built on top of the main blockchain to facilitate micropayments and faster transactions. Security relies on the underlying blockchain but introduces potential risks associated with the second-layer protocol. Plasma is similar to lightning but uses smart contracts on the main chain to anchor child chains with their own rules and faster processing. It adds adaptability but makes design and security more complicated. With the use of two-way bridges, sidechains can function as separate blockchains that are linked to the main chain. Allows for more rapid transactions and the testing of various consensus algorithms, but raises concerns about the safety of bridge operations.

   Quest for Optimal Blockchain Scalability Solutions: Combination of on-chain and off-chain solutions can leverage the strengths of each while mitigating their individual limitations. The quest for optimal blockchain scalability solutions is ongoing. Each approach has its advantages and trade-offs, and the ideal solution varies based on use case and desired properties. Active research and development are continuously exploring new frontiers, such as Directed Acyclic Graphs and Byzantine Fault Tolerance consensus mechanisms, which hold promise for further advancements. While blockchain scalability remains a critical challenge, the emergence of diverse solutions demonstrates the ongoing efforts to overcome it. As these solutions mature and gain wider adoption, we can expect blockchain technology to unlock its full potential and revolutionize various industries by enabling secure, transparent, and scalable transactions.

16.8.3.2 Advancements in AI Algorithms

i. **ML Algorithms:**

The utilization of supervised learning algorithms highlights the potential of AI and ML approaches to enhance the security of data in EVs by enhancing the accuracy of pretext detection and preventing their occurrence. Further, application of AI approaches is discussed, such as deep learning networks and standard ML algorithms, for diagnosing and detecting faults in electric car motors [34]. Unsupervised learning techniques, such as clustering and anomaly detection, can be employed to uncover hidden patterns and abnormalities in datasets related to EVs. Clustering algorithms can be used to group automobile settings based on driving behavior or to divide the customer pool for more precisely targeted marketing. Anomaly detection algorithms have the ability to discover unusual patterns of data and promptly notify operators of potential cybersecurity concerns or system failures. These technologies have demonstrated significant efficacy in several fields, including spacecraft systems, aircraft engines, and marine vessel engines.

ii. **Neural Networks and Deep Learning:**

The advancement of AI, mostly propelled by advanced neural network structures such as deep learning methods, has greatly enhanced its capacity to effectively address intricate difficulties. Significant advancements have been made in the domains of parallel processing of data and machine vision, which are crucial for many activities related to EVs. Within these domains, we have observed the superior performance of neural networks, namely those that are based on the transformer architecture, compared to earlier models. A significant advancement has been observed in the field of computer vision, particularly in the effectiveness of Convolutional Neural Networks for tasks such as object identification, face detection, picture categorization, and localization. In the field of automated driving, the digital vision revolution has made it possible to precisely recognize objects and lanes. This is achieved by using Convolutional Neural Networks to replace traditional object recognition methods with a faster and more accurate approach. Convolutional Neural Network-based system vision resulted in significant improvements in target detection and semantic segmentation. Both autonomous driving systems and applications for EVs can reap the benefits of predictive maintenance and time-series forecasting. These tasks can be effectively handled by Recurrent Neural Networks and their subset, Long Short-Term Memory networks. These networks can analyze previous sensor data to forecast future performance indicators, such as component wear or battery decay over time. Long Short-Term Memory networks are particularly good at capturing long-term dependencies and modeling sequential data. For example, they have been successfully used to predict the remaining useful life of lithium-ion batteries. Recurrent Neural Networks have also been widely used for short-term load forecasting in electricity power networks.

Road Ahead of AI Algorithms: The emergence of advanced AI algorithms has led to significant advancements in the field of EVs. These developments have the ability to revolutionize the automobile sector and greatly impact the lives of drivers worldwide. This is achieved through the utilization of deep learning techniques and insights based on data.

16.8.3.3 Recommended Future Work

i. **Federated Learning in the Context of Blockchain Networks Is an Intriguing Area of Research:**

Federated learning on the blockchain would enable a decentralized network with dispersed AI nodes. Federated learning within a blockchain network offers a promising approach to create a more efficient, secure, and privacy-preserving EV charging ecosystem. Federated learning allows charging stations to train AI models on their local data (charging habits, usage patterns) without directly sharing it. This protects user privacy while enabling stations to collaboratively improve AI models for tasks like predicting demand and optimizing price. Predicting Demand accurately predicts charging demand at specific stations, allowing for better resource allocation and infrastructure planning. Optimizing pricing involves developing dynamic pricing models based on real-time demand and grid conditions, encouraging off-peak charging and efficient energy usage. A blockchain can establish a secure marketplace where anonymized and aggregated data from Federated learning models can be securely shared. This allows stakeholders like utility companies and charging station providers. Utility Companies gain insights into overall charging patterns to improve grid management and integrate renewable energy sources. Charging station providers benchmark their performance against anonymized industry data and identify areas for improvement. Finally, Federated learning allows stations to collaboratively train models without compromising privacy. This accelerates the development of AI-powered solutions for the EV charging ecosystem. Blockchain ensures secure and verifiable data exchange, fostering trust among participants in the network.

ii. **How to Adapt Protocols like OCPP to Use Both Technologies for Security and Energy Management Is a Further Captivating Study:**

Robust authentication and authorization mechanisms within OCPP prevent unauthorized access and misuse of the charging infrastructure. Upgradation of communication protocols used in OCPP ensures secure and encrypted communication between charging stations, backend systems, and EVs. Implementation of Transport Layer Security or Datagram Transport Layer Security protects data integrity and confidentiality during data transmission. Extension of OCPP is to support the edge computing capabilities for local processing and decision-making at charging stations. Edge computing platforms can perform real-time analysis of sensor data, detect anomalies, and respond to critical events without relying on centralized backend systems, improving responsiveness and reliability. Extended OCPP implementations with emerging standards and protocols should ensure interoperability and compatibility in the EV ecosystem. Industry stakeholders and standardization bodies should align OCPP extensions with emerging technologies and promote seamless integration with other systems and devices.

16.9 CONCLUSION

In conclusion, the integration of Blockchain and AI technologies presents a promising avenue for standardizing the EV charging ecosystem. Through our exploration of smart EV charging infrastructure, we have identified several key benefits, including enhanced security, interoperability, and efficiency. Blockchain's decentralized ledger offers transparency and trust in transactions, addressing concerns regarding data integrity and cybersecurity in EV charging networks. Moreover, AI algorithms optimize charging schedules, reduce congestion, and adapt to user preferences, ultimately improving user experience and grid management. The standardization enabled by these technologies not only fosters seamless interoperability among charging stations but also accelerates the adoption of EVs by assuaging range anxiety and streamlining the charging process. This not only benefits individual consumers but also contributes to the broader goal of sustainable transportation and decarbonization efforts. Looking ahead, further research and development are warranted to address remaining challenges such as scalability, privacy, and regulatory frameworks. Additionally, collaboration among stakeholders, including governments, industry players, and academia, is crucial to realize the full potential of smart EV charging infrastructure. In essence, the convergence of Blockchain and AI technologies represents a paradigm shift in the EV charging landscape, heralding a future of cleaner, smarter, and more accessible transportation.

REFERENCES

1. Chan, C. C. (2002). The state of the art of electric and hybrid vehicles. *Proceedings of the IEEE, 90*(2), 247–275.
2. Alrubaie, A. J., Salem, M., Yahya, K., Mohamed, M., & Kamarol, M. (2023). A comprehensive review of electric vehicle charging stations with solar photovoltaic system considering market, technical requirements, network implications, and future challenges. *Sustainability, 15*(10), 8122.
3. Hsu, Y. M., Ji, D. Y., Miller, M., Jia, X., & Lee, J. (2023). Intelligent maintenance of electric vehicle battery charging systems and networks: Challenges and opportunities. *International Journal of Prognostics and Health Management, 14*(3). https://doi.org/10.36001/ijphm.2023.v14i3.3130
4. Hossain, M. S., Kumar, L., El Haj Assad, M., & Alayi, R. (2022). Advancements and future prospects of electric vehicle technologies: A comprehensive review. *Complexity*, 2022, 1–21.
5. Patil, P. (2019). Electric vehicle charging infrastructure: Current status, challenges, and future developments. *International Journal of Intelligent Automation and Computing*, *2*(1), 1–12.
6. Aruna, P., & Vasan Prabhu, V. (2021). Evolution and recent advancements in electric vehicle (EV) technology. In *Emerging Solutions for e-Mobility and Smart Grids: Select Proceedings of ICRES 2020* (pp. 91–109). Springer Singapore.
7. Chen, Q., & Folly, K. A. (2022). Application of artificial intelligence for EV charging and discharging scheduling and dynamic pricing: A review. *Energies*, 16(1), 146.
8. Sheeba, T. B., Sharanya, C., Nayanatara, C., Indumathi, S. K., Kalins, K., & Rajathi, G. I. (2022, October). Deep learning enabled smart charging technology for electric vehicles. In *AIP Conference Proceedings* (Vol. 2527, No. 1). AIP Publishing.
9. Okwuibe, G. C., Li, Z., Brenner, T., & Langniss, O. (2020). A blockchain based electric vehicle smart charging system with flexibility. *IFAC-PapersOnLine, 53*(2), 13557–13561.
10. Knirsch, F., Unterweger, A., & Engel, D. (2018). Privacy-preserving blockchain-based electric vehicle charging with dynamic tariff decisions. *Computer Science-Research and Development, 33*, 71–79.
11. Rane, N., Choudhary, S., & Rane, J. (2023). Blockchain and artificial intelligence (AI) integration for revolutionizing security and transparency in finance. Available at SSRN 4644253. http://doi.org/10.2139/ssrn.4644253

12. Yaqub, R., Ahmad, S., Ali, H., & Asar, A. U. (2020). AI and blockchain integrated billing architecture for charging the roaming electric vehicles. *IoT, 1*(2), 382–397.
13. Trovão, J. P. (2020). Automotive electronics market evolution. *IEEE Vehicular Technology Magazine, 15*, 107–118.
14. Mahesh, A., Chokkalingam, B., & Mihet-Popa, L. (2021). Inductive wireless power transfer charging for electric vehicles – A review. *IEEE Access, 9*, 137667–137713.
15. Falvo, M. C., Sbordone, D., Bayram, I. S., & Devetsikiotis, M. (2014, June). EV charging stations and modes: International standards. In *2014 International Symposium on Power Electronics, Electrical Drives, Automation and Motion* (pp. 1134–1139). IEEE, Ischia, Italy.
16. Muharam, A., Pratama, M., Ismail, K., Kaleg, S., Kurnia, M. R., & Hapid, A. (2016, October). A development of smart metering infrastructure for electric vehicle charging point. In *2016 International Conference on Sustainable Energy Engineering and Application (ICSEEA)* (pp. 27–33). IEEE, Jakarta, Indonesia.
17. Mayfield, D., & Ohio, C. F. (2012). *Siting Electric Vehicle Charging Stations*. Editor Carlotta Collette, Virginia, USA.
18. Peter, V. D. B., Tom, T., Noshin, O., & Joeri, V. M. (2016). Developments and challenges for EV charging infrastructure standardization. *World Electric Vehicle Journal, 8*(2), 557–563.
19. Blake, C., Buttner, W., & Rivkin, C. (2010). Vehicle codes and standards: Overview and gap analysis. https://www.nrel.gov/docs/fy10osti/47336.pdf
20. Chynoweth, J., Chung, C. Y., Qiu, C., Chu, P., & Gadh, R. (2014, February). Smart electric vehicle charging infrastructure overview. In *ISGT 2014* (pp. 1–5). IEEE, Washington, DC, USA.
21. Tu, H., Feng, H., Srdic, S., & Lukic, S. (2019). Extreme fast charging of electric vehicles: A technology overview. *IEEE Transactions on Transportation Electrification, 5*(4), 861–878.
22. Ferreira, J. C., Monteiro, V., & Afonso, J. L. (2013, November). Dynamic range prediction for an electric vehicle. In *2013 World Electric Vehicle Symposium and Exhibition (EVS27)* (pp. 1–11). IEEE, Barcelona, Spain.
23. Falvo, M. C., Sbordone, D., Bayram, I. S., & Devetsikiotis, M. (2014, June). EV charging stations and modes: International standards. In *2014 International Symposium on Power Electronics, Electrical Drives, Automation and Motion* (pp. 1134–1139). IEEE, Ischia, Italy.
24. Etezadi-Amoli, M., Choma, K., & Stefani, J. (2010). Rapid-charge electric-vehicle stations. *IEEE Transactions on Power Delivery, 25*(3), 1883–1887.
25. Bai, S., Yu, D., & Lukic, S. (2010, September). Optimum design of an EV/PHEV charging station with DC bus and storage system. In *2010 IEEE Energy Conversion Congress and Exposition* (pp. 1178–1184). IEEE, Atlanta, GA, USA.
26. Qureshi, K. N., Alhudhaif, A., & Jeon, G. (2021). Electric-vehicle energy management and charging scheduling system in sustainable cities and society. *Sustainable Cities and Society, 71*, 102990.
27. Isa, A. V., & Kinaci, B. F. (2022, November). Challenges and opportunities of blockchain in electric vehicles. In *International Conference on Engineering Sciences* (Vol. 11, p. 12). Bakü.
28. Liu, H., Lv, Z., Song, Z., Zhou, S., Liu, W., & Fang, M. (2022). Application of blockchain technology in electric vehicle charging piles based on electricity Internet of Things. *Wireless Communications and Mobile Computing*. https://doi.org/10.1155/2022/8533219
29. Shibl, M., Ismail, L., & Massoud, A. (2021). Electric vehicles charging management using machine learning considering fast charging and vehicle-to-grid operation. *Energies, 14*, 6199.

30. Hua, W., Chen, Y., Qadrdan, M., Jiang, J., Sun, H., & Wu, J. (2022). Applications of blockchain and artificial intelligence technologies for enabling prosumers in smart grids: A review. *Renewable and Sustainable Energy Reviews*, *161*, 112308.
31. Ledger Insights (2022). BMW demos recharging cars using energy web blockchain. https://www.ledgerinsights.com/bmw-demos-recharging-cars-using-energyweb-blockchain/
32. Mehar, M. I., Shier, C. L., Giambattista, A., Gong, E., Fletcher, G., Sanayhie, R., & Laskowski, M. (2019). Understanding a revolutionary and flawed grand experiment in blockchain: The DAO attack. *Journal of Cases on Information Technology (JCIT)*, *21*(1), 19–32.
33. Lipu, M. H., Hannan, M. A., Hussain, A., Ayob, A., Saad, M. H., Karim, T. F., & How, D. N. (2020). Data-driven state of charge estimation of lithium-ion batteries: Algorithms, implementation factors, limitations and future trends. *Journal of Cleaner Production*, *277*, 124110.
34. Da Silva, F. L., Nishida, C. E., Roijers, D. M., & Costa, A. H. R. (2019). Coordination of electric vehicle charging through multiagent reinforcement learning. *IEEE Transactions on Smart Grid*, *11*(3), 2347–2356.

17 Thermal and Efficiency Comparison of Wide-Bandgap Semiconductor Devices in EV Charger

Lanka Sridutt, Suyash Prakash, Saundarya Priyadarshini, and Kanimozhi Gunasekaran

17.1 INTRODUCTION

In recent years, there has been a shift toward electric vehicles (EVs) over conventional fuel vehicles [1]. This is mostly due to the need for a shift to renewable energy from conventional energy sources [2]. As the requirement for energy is increasing rapidly, conventional energy sources fail to meet the requirements. In the transportation sector, EVs are playing a crucial role in promoting the shift from conventional energy sources to renewable energy sources. Power converter circuits are implemented to charge the batteries from the energy supplied through the grid. It is at these circuits that maximum power loss occurs, and the need to prioritize efficiency rises. Power converter circuits use semiconductor devices to do their intended operation. With time, the technologies owing to the construction of semiconductor devices have been changing, and new technologies and compounds have improved the efficiency and longevity of these circuits. As these power circuits deal with high switching speeds and power, it is necessary for the devices in the circuit to be able to handle the thermal stress being placed on them during operation. In recent years, silicon carbide (SiC) and gallium nitride (GaN) [3] devices have proven to be good alternatives for conventional silicon (Si) devices.

SiC and GaN devices fall under the category of Wide-BandGap (WBG) semiconductor devices. To understand how WBG semiconductor devices are different from conventional devices, we need to know what a bandgap means. In chemical elements, electrons reside at certain fixed energy levels; these are called energy states or energy bands. All electrons in one band have the same energy level. Electrons can be pushed to higher energy bands by providing them with energy. The bandgap is the energy difference between the conduction band and the dielectric band.

DOI: 10.1201/9781003496946-17

WBG semiconductor devices work as better alternatives for conventional Si devices due to the following reasons: WBG semiconductor devices have a wider bandgap than conventional Si semiconductor devices, which allows them to operate at higher voltages, frequencies, and temperatures (up to 300°C). WBG devices handle higher power densities. (Power density implies handling more power in a smaller area on the board.) These devices allow faster electron mobility, thus allowing for operation at a higher switching frequency. Although these reasons are very exciting to hear and pique our interest in these new devices and tempt us to completely switch to WBG semiconductor devices, they too have their share of concerns, which will be addressed in Section 17.3 of this chapter. Being a fairly new technology, WBG devices are still a work in progress. This chapter will deal with the detailed analysis and comparison of SiC MOSFETs vs. GaN MOSFETs, which includes:

1. Thermal handling and thermal stress management
2. Power loss comparison for the designated EV Charger
3. Transient and steady-state thermal load analysis of the devices

This chapter is organized in the following manner: Section 17.1 deals with the two-wheeler onboard charger. Section 17.2 covers information about the workings and a basic understanding and comparison between WBG semiconductor devices. Parts 17.3 and 17.4 focus on the thermal modeling of the Buck Boost (BB) converter circuit and the equations used to calculate the thermal losses. Sections 17.5 and 17.6 deal with the simulation findings and the comparison between the semiconductor devices, followed by the conclusion in Section 17.7.

17.2 TWO-WHEELER ONBOARD CHARGER

The proposed EV Charger [2] in this chapter is for charging two-wheeler batteries. The proposed EV Charger consists of three major parts, which can be summarized using the block diagram referred to in Figure 17.1 and following the block diagram, in Figure 17.2 is the block diagram of the whole circuit of the onbroad chargers (OBC):

The resonant tank in purpose does Zero Voltage Switching operation to reduce switching losses and thus improve the efficiency of the charger. A brief overview of the different blocks of the referred OBC is beneficial to understand the function each block provides in the charger.

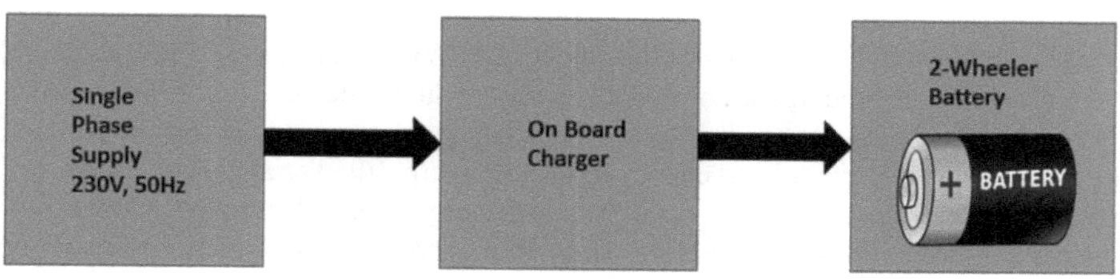

FIGURE 17.1 Connections between the supply and battery.

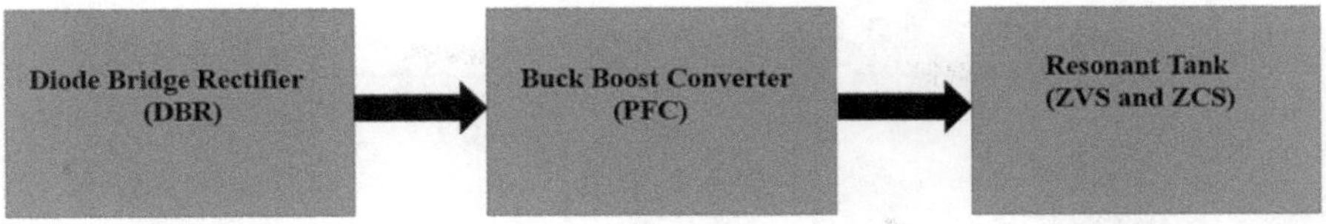

FIGURE 17.2 Block diagram of the onboard charger.

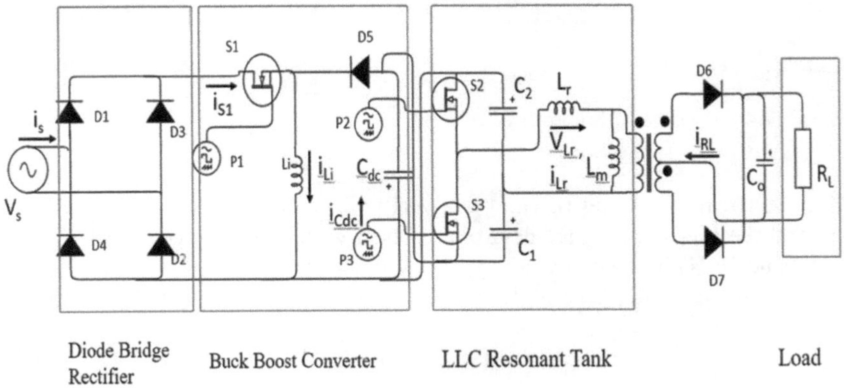

FIGURE 17.3 Circuit diagram of the OBC.

1. **The Diode Bridge Rectifier:** The diode bridge rectifier takes an AC voltage input from the supply and converts it into a variable DC output voltage.
2. **The BB Converter (Operating as a Buck Converter):** The BB converter operates in Buck operation to provide the DC link voltage to the following block of the charger (HB LLC Resonant Converter). The primary operation of the BB converter is to provide for the power factor correction [4] of the input voltage and current [5].
3. **Half Bridge LLC Resonant Converter** [6]**:** The HBLLC Resonant Tank is responsible for isolating the charger and the battery as the battery could discharge after reaching a certain state of charge. This could damage the charger as it doesn't have reverse current protection. Hence, the resonant converter provides isolation [7] between the charger and the battery. To employ the isolation, a transformer equipped with a ferrite core to minimize hysteresis and eddy current losses with a winding ratio of 14:5 is employed. The HBLLC converts the DC link voltage from the BB converter to high-frequency AC, which is rid of harmonics by the LLC Resonant Tank [8]. The output is a sine wave of a resonant frequency. The voltage is then fed to the transformer where it is stepped down and passed to the center tapped full wave rectification to feed the load.

Figure 17.3 presents the entire circuit of the OBC, and Figure 17.4 showcases the BB converter, which will be the main discussion of this chapter. P1, P2, and P3 are the gate pulses for the MOSFETs S2, S2, and S3, respectively.

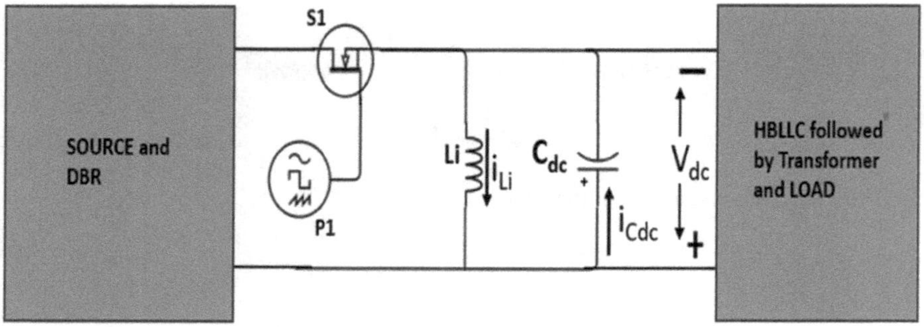

FIGURE 17.4 Proposed BB converter.

The electrical stress and thermal analysis will be centered around the BB converter and the thermal management comparison with different MOSFET Switches (Si, SiC, and GaN) in the converter (S1) [9].

The following are the general design equations of the conventional BB converter.

$$V_a = -(V_s * k)/(1-k)\ V \tag{17.1}$$

Where the −ve sign indicates that the voltage at the output is of reverse polarity of the input source.

The average current through the inductor is given as

$$I_l = I_a/(1-k)\ A \tag{17.2}$$

The inductor value and the capacitance critical values are obtained from the expressions:

$$L = (V_s * k)/(f * \Delta I)\ H \tag{17.3}$$

$$C = (I_a * k)/(f * \Delta V_c)\ F \tag{17.4}$$

where V_s = DC source voltage, V_a = average output DC output voltage, k = duty ratio, I_a = average output current, f = switching frequency, ΔI = ripple in inductor current, and ΔV_c = ripple in capacitor voltage.

17.3 WBG DEVICES

Wideband devices are the new alternative for traditional silicon devices [10]. Silicon devices are still used in many of the power electronics devices, but silicon is reaching its threshold limit of output and application necessity performance, and hence, WBG devices appear to be the next viable option. WBG devices, as their name suggests, have a WBG, in the order of 2–4 eV. The bandgap of an element is the energy difference between the conduction band and the valence band.

In the current market, silicon carbide (SiC) [11] and the recent gallium nitride (GaN) semiconductors are being preferred for applications. Although they both are good technologies, their use is application specific as each of them has its own tradeoffs for benefits. Comparison of the semiconductor devices will be based on the following material properties:

1. **High Temperature Range of Operation**: WBG devices are capable of working in higher temperatures and hence are a better alternative to Si devices, considering the high-power necessities of modern technology. WBG devices have control of their conduction operations even at high temperatures; hence, they are preferred over basic Si devices.
2. **Bandgap**: Bandgap is temperature dependent; i.e, when the temperature of the semiconductor rises, the band gap decreases; hence, having a wider bandgap gives the system more reliability and control on the switching operation of the power device as the device won't cascade into conduction mode on higher operating temperatures. This is very important when the device is being used in high-power applications.
3. **Critical Field**: It is a measure of the voltage blocking capability of the semiconductor. A higher critical field means the semiconductor can withstand higher current through it before conducting heavily and damaging itself.
4. **Electron Mobility**: It is a measure of the movement speed of an electron in the lattice of the semiconductor crystal [12]. A higher electron mobility enables the semiconductor to perform better in higher-switching-frequency applications.
5. **Thermal Conductivity**: It is a measure of how efficiently the semiconductor can conduct heat from junction to case. The higher the thermal conductivity, the better it is for the device, as it will cool down faster and perform better.

Comparing some of the important characteristics required in semiconductors, the following chart for Si, SiC, and GaN is given in Figure 17.5. On comparing conventional Si devices with the WBG devices, it becomes clear why silicon is being replaced by new devices.

The high bandgap of the WBG devices inevitably makes them more suitable for high temperature and power applications. Moreover, WBG devices can bear more critical field stress and hence work as better switches in applications than conventional silicon, once again when the device has to handle high voltage stress across itself. Electron mobility indicates the switching ability of the device (its ability to enter into conduction and stop conduction). The electron mobility is one aspect where Si outperforms SiC but again is overperformed by GaN [13,14]. Thermal conductivity is again outperformed by WBG devices compared to conventional silicon. Although Si has a better thermal conductivity than GaN (only minorly), the other characteristics outweigh the inclination toward Si compared to WBG devices. With this comparison analysis, it is apparent that WBG devices outperform conventional silicon in the majority of the characteristics.

Silicon carbide (SiC) and gallium nitride (GaN) devices offer significant advantages over traditional silicon (Si) devices. They can reduce power losses by up to 90% during AC–DC and DC–AC power conversions, enhancing overall efficiency. These devices operate at performance levels up to ten times higher than Si-based devices, which boosts high-power performance. Additionally, SiC and GaN can function at much higher temperatures, improving system reliability. They also enable the creation of smaller, lighter systems with reduced life cycle energy use, opening up opportunities for new applications. Furthermore, their ability to operate at higher frequencies than Si-based devices allows for more compact and cost-effective product designs.

On comparing SiC and GaN, the prediction of the application-specific device choice becomes easier. A thorough conclusion can be drawn from the radar chart given in Figure 17.5 and the characteristic Table 17.1. However, comparison of the properties is justified for reasonable backing and purpose statements of the choice of WBG for the required purpose:

1. SiC and GaN almost have the same bandgap.
2. Critical field is one aspect where GaN dominates SiC. Having a higher critical field indicates the ability of GaN to handle more voltage stress across it compared to SiC.

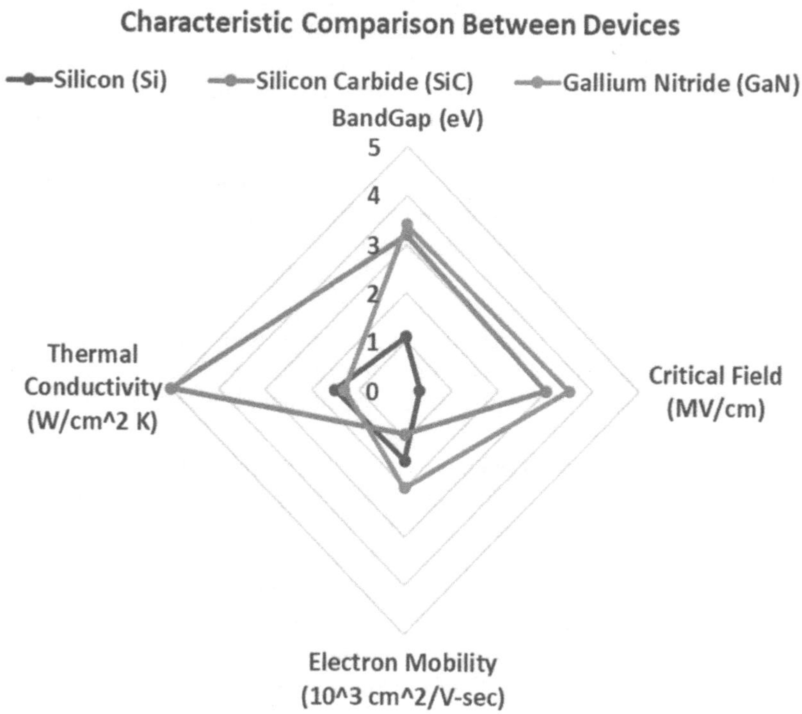

FIGURE 17.5 Radar chart comparison of Si (blue), SiC (red), and GaN (gray) for different material properties (thermal conductivity, electron mobility, critical field, and bandgap).

TABLE 17.1
Characteristics Table of Semiconductor Devices [16]

Material Property	Si	SiC	GaN
Bandgap (eV)	1.1	3.2	3.4
Critical field (MV/cm)	0.3	3	3.5
Electron mobility ($10^3 cm^2/V$-sec)	1.45	0.9	2
Thermal conductivity (W/cm^2 K)	1.5	5	1.3

3. Electron mobility is dominated by GaN across the semiconductor board. This makes GaN the best choice for applications that require high switching frequency.
4. Thermal conductivity is when SiC triumphs over GaN. For high-power applications, SiC has a better thermal conductivity, and hence, its junction-case thermal conductivity helps the device to work better by heat dissipation. This implies that the power density of SiC is better than GaN. GaN devices lack thermal conductivity in comparison to SiC and hence don't perform well in high-power applications.

A basic conclusion drawn from the comparisons reveals that GaN, although being predominantly better in many aspects than SiC [15], still has a lot of time to mature to the level of SiC and must be improved in thermal conduction to be able to replace SiC in applications. One more comparison point between SiC and GaN is the production process. Semiconductor devices are prepared from wafers. These wafers need to be thin (in micrometers) and large in diameter (in millimeters 25.4–300) to accommodate the production of more chips from each wafer, thus making a financial benefit.

Semiconductors are formed by stacking a number of layers of these wafers on top of each other, analogous to a building. The process of fabrication of these wafers involves growing an epitaxial layer over a substrate. If the epitaxial layer and the substrate are of the same material, then the fabrication process is called homoepitaxy; otherwise, it is referred to as heteroepitaxy. Production of homoepitaxial SiC wafers [16] is quite sophisticated, and the end product yields a good crystallographic match between the epitaxial layer and the substrate, thus resulting in a sound electrically and thermally conductive path from the top to the bottom of the wafer. In contrast, though there are homoepitaxial GaN wafers, they are in early stages of production and still need to be worked upon to produce wafers that have a smaller number of crystal defects. As such, the current trend is to employ a heteroepitaxial approach for GaN [17], where the epitaxial layer is of GaN but the substrate is a non-native SiC or Si substrate [18]. The non-native SiC or Si substrate thus requires higher manufacturing costs to account for the different crystal lattice structure between the epitaxy and the substrate.

A set of fundamental problems were, and still are, related to a lack of single-crystalline high-quality native GaN substrates.[19] Homoepitaxial GaN devices are still early and are very hard to grow; this is due to the lack of high-quality native GaN substrates. The main barrier for obtaining large, high-quality bulk crystals of GaN is related to the thermodynamic properties of GaN and, in particular, its extreme melting conditions and very high decomposition pressure of nitrogen at elevated temperatures necessary for crystallization. Thus, it is preferred to grow GaN on SiC or Si substrates, albeit there are crystal lattice differences, which need to be accounted for, which add additional materials and processing costs.

Though the crystallographic defects can be tackled by introducing a buffer layer, it only adds to the complexity of the production process and the costs. It is for this reason that GaN devices are still preferred with a heteroepitaxy wafer type [18,19]. Additional considerations involve the different thermal constants, thermal expansions, stress management of the various materials, etc.; with more parameters, the production becomes strenuous.

These complications indicate the inefficiency of GaN to replace its counterpart SiC in applications. With this, the conclusion arrives that each semiconductor technology has its place in the current market, and it is too early for one of them to triumph over the other. Application-specific devices will be employed, keeping efficiencies and cost parameters in regard to the industries. A trivial conclusion from the above discussion would be that for low-power/voltage and high-frequency applications, GaN would be a preferred choice, and for high voltage, high-power switching applications, SiC would be the preferred choice [20]. The Figure 17.6 area chart would summarize the application-specific dominance of each semiconductor device in consideration in this article (Si, SiC, and GaN).

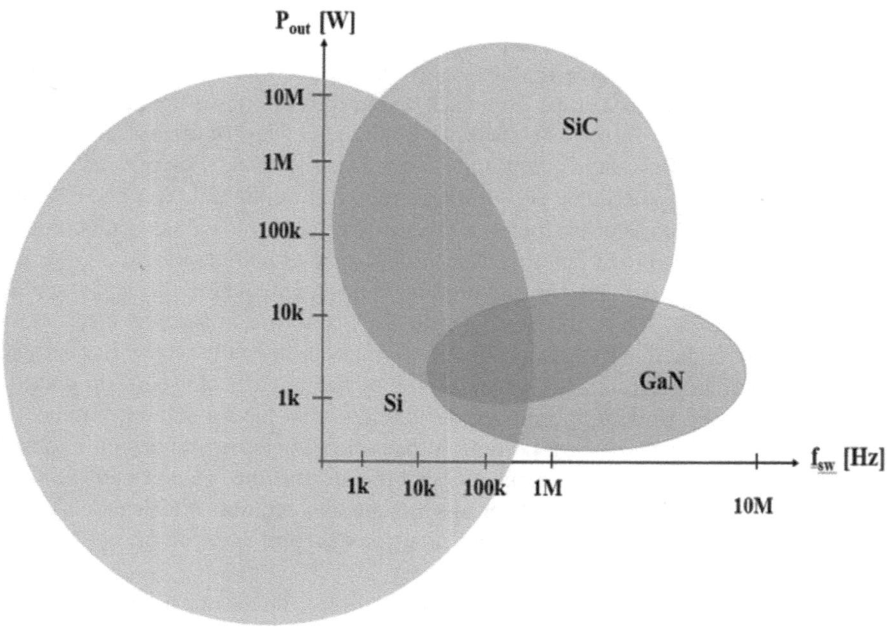

FIGURE 17.6 Area chart of predominant semiconductor devices based on application power and switching frequency necessity: Si (green), SiC (blue), and GaN (peach).

17.4 THERMAL MODELING AND ANALYSIS

The thermal modeling and analysis of the concerned EV Charger [21] should be able to confirm the conclusions of the semiconductor devices and conclude which semiconductor device is better for the respective switches in the charger. The power factor correction [21] of the charger is achieved by the BB converter. Therefore, focus will be put on the losses encountered by replacing the switch of the converter with the different semiconductor devices available. The predefined PLECS models of the semiconductor devices have been made available by the manufacturer, thus easing the simulation process.

It is important to analyze the losses incurred by dc/dc choppers in different topologies and choose the topology that meets the highest efficiency with minimum component count. As choppers are used in OBEVC for power factor correction purposes [22], it is necessary to analyze their reliability and robustness for long-term purposes. Thermal modeling in the current chapter is used to compare the performance characteristics of different semiconductor devices and choose a more viable option as the switch for the chopper. The thermal losses incurred in a power switch are the conduction losses and the switching losses.

17.4.1 Conduction Losses

These are proportional to the on-state resistance of semiconductors and the DC resistance of magnetic components. They might remain relatively steady or change with temperature and load:

$$\text{Conduction loss: } P\text{cond} = \left(I^2 S_{rms} * R_s\right) \tag{17.5}$$

where R_s = on-state resistance and IS_{rms} is the rms drain current through the switch. F_{sw} is the switching frequency, t_r and t_f are the rise and fall times of the power switches, respectively, V_{ds} is the drain-source voltage across the switch, and I_{savg} is the average current through the switch. Switching losses arise when power semiconductors switch between on and off states. They are impacted by the switching frequency, gate drive circuit design, and semiconductor properties. Predefined PLECS models for the switches according to the required ratings are available from manufacturers and have been directly used in the simulation, thus making the output more accurate.

17.4.2 Switching Loss

$$P_{sw} = \left(V_{ds} * I_{savg} * \left(t_r + t_f\right) * \frac{f_{sw}}{2}\right) \tag{17.6}$$

The switching losses and conduction losses flow out of the MOSFET via the switch as heat, analogous to the circuit shown in Figure 17.7. The resistances from junction to case, case to heatsink, and heatsink to ambience, along with the thermal capacitance

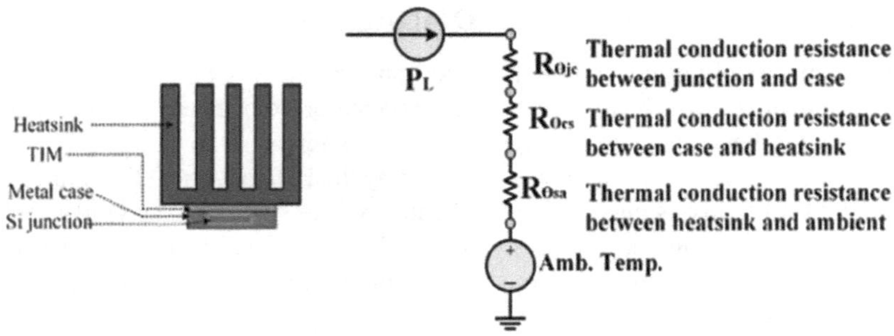

FIGURE 17.7 Thermal representation of the switch [23].

values (given in datasheets of the switch), account for what is called the thermal impedance of the switch. Though PLECS provides the option of filling in these values manually from the datasheet, PLECS predefined modules from the manufacturer reduce the hassle of dealing with these constants.

Various load and duty ratio fluctuations would allow for a good picture of the thermal performance of the switch in the chopper and thus a good thermal analysis of the chopper. The power efficiency, which is the ratio of the output power to the input power, will depend on the switching losses, which in turn depend on the switching frequency of the converter. The switching frequency f_{sw} should be high to reduce the values and sizes of capacitances and inductances. The designer must compromise on these conflicting requirements. In general, f_{sw} is higher than the audio frequency of 18 kHz [20,23].

The efficiency calculation, which includes the conduction loss of the converter, does not consider the switching loss due to turn-on and turn-off of practical converters. The efficiency of a practical converter varies between 92% and 99% [23]. A BB regulator provides output voltage polarity reversal without a transformer. It has high efficiency. Under a fault condition of the transistor, the di/dt of the fault current is limited by the inductor L and will be V_s/L. Output short-circuit protection would be easy to implement. However, the input current is discontinuous and a high peak current flows through MOSFET S1 [20,23].

17.5 SIMULATION RESULTS

PLECS standalone is used for thermal analysis. The component library of PLECS provides not only electrical but also thermal, magnetic, and mechanical properties of the power conversion system, improving the accuracy of real-time simulation. PLECS also provides the option to account for power losses, which thus promotes accurate simulations. The software considers both switching and conduction losses during simulation, maintaining ideal switching behavior without compromising simulation speed. These tables make the simulations more accurate as the tables themselves are plotted with actual device operation. Manufacturers provide ready-made PLECS models for their products, thus making the simulations easier and more accurate. The PLECS circuit model is depicted in Figure 17.8, and the specifications of the BB converter on the OBEVC are shown in Table 17.2.

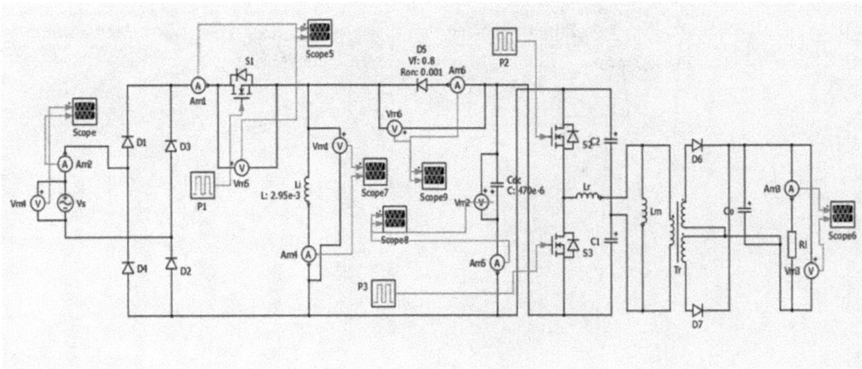

FIGURE 17.8 Circuit design in PLECS.

TABLE 17.2
Specifications of the BB Converter

Parameter	Value	Parameter	Value
Input voltage source	230V, 50 Hz	Output power (P_{out})	520 W
Inductor (L_i)	2.7 mH	Output voltage (V_o)	52V –53V
Capacitor (C_{dc})	19.1 μF	Switching frequency of S1	50 kHz
Resonant inductor (L_r)	29.6 μH	Silicon switch	APT8014L2FLL (Microsemi)
Magnetizing inductance (L_m)	178 μH	Silicon carbide	G3R40MT12D (GeneSiC)
Output capacitor (C_o)	470 μF	Gallium nitride	GS-065-060 (Infineon)

In the steady-state analysis, for the full-load condition (refer to Figure 17.9a–e), it can be observed that for the ideal simulation, the average current through the MOSFET switch is 1.042 A and the average voltage across the switch is 200.6V. The average current through the diode is 1.0685 A, and the average voltage across the diode is –278.07V. Finally at the output capacitor C_{dc}, the average current through it is –0.16226 Amps and the average voltage is –284.872V. This voltage is fed to the HBLLC converter of the charger. The average current through inductor L_i is 2.143 A and the average voltage across L_i is 0.1767V as presented in Figure 17.10a and b.

17.5.1 Thermal Modeling Results

Analyzing the thermal models of the three switches, the average and peak values of the thermal characteristics can be obtained. The same graphs are referred to in Figure 17.11a–d. For a conventional silicon switch, the average heat sink temperature is 25.1099°C, the peak value is 25.1348°C, the average junction temperature is 31.5813°C, and the peak value is 32.6786°C. The average conduction loss for the switch is 3.923 W, and the peak value is 6.985 J. The average switching loss for the switch is 10.7 μW, and the peak value is 0.372 mJ. For the SiC switch (refer

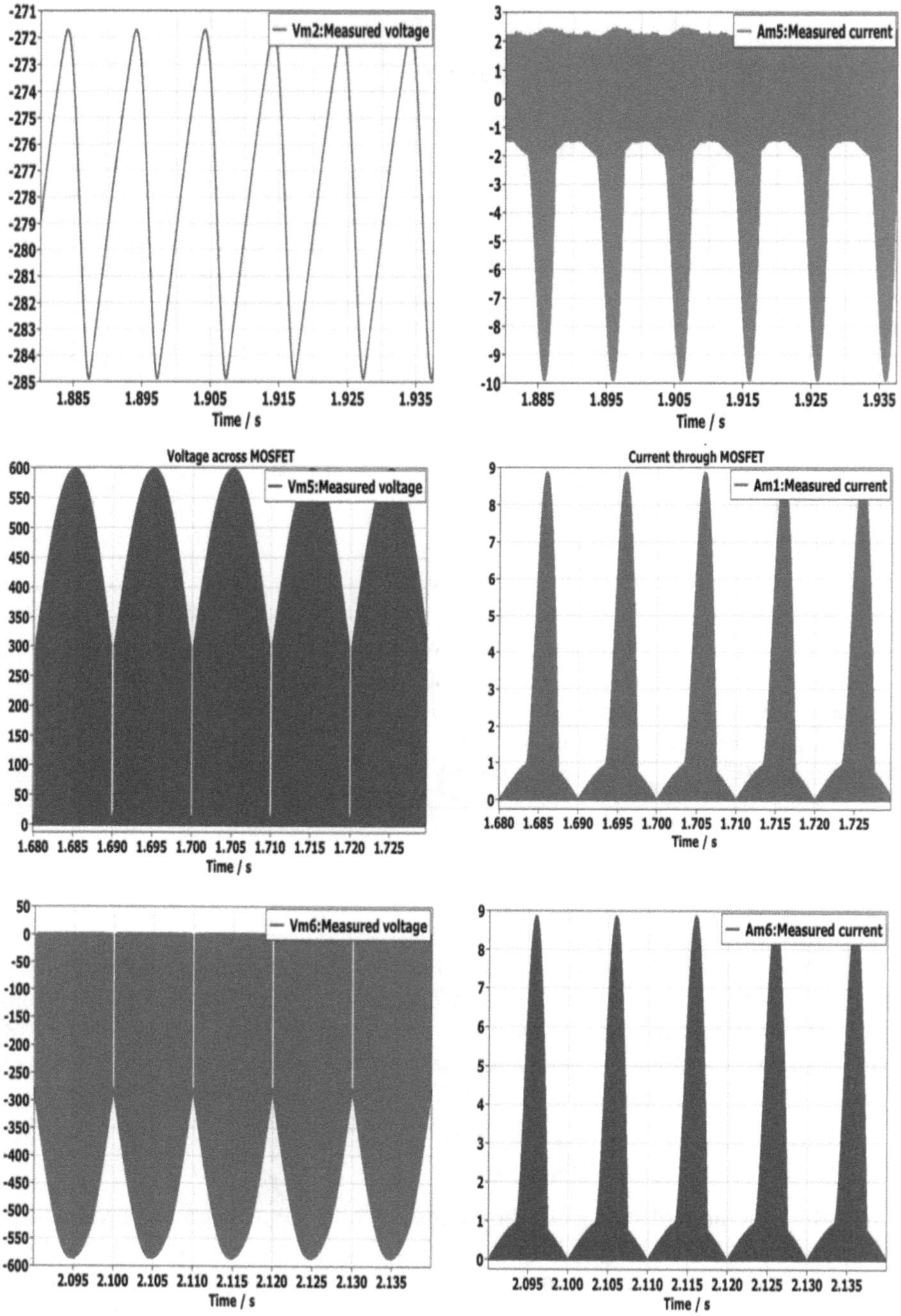

FIGURE 17.9 (a) Voltage across output capacitor C_{dc}. (b) Current through output capacitor C_{dc}. (c) Voltage across MOSFET S1. (d) Current through MOSFET S1. (e) Voltage across diode D5. (f) Current through diode D5.

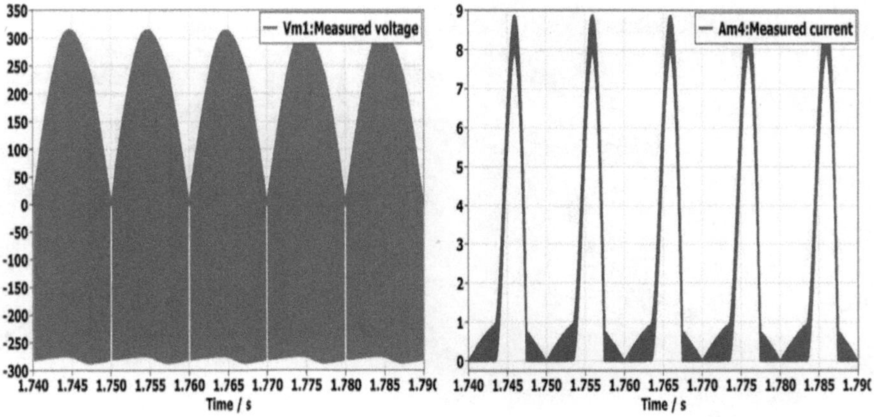

FIGURE 17.10 (a) Voltage across inductor Li. (b) Current through inductor Li.

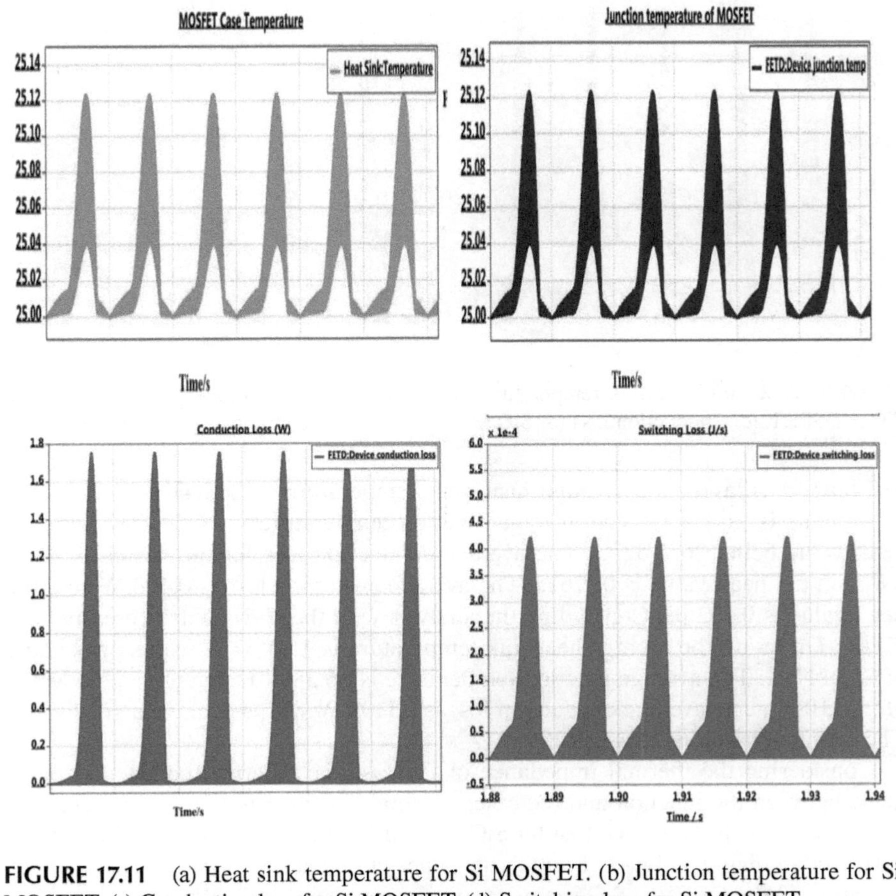

FIGURE 17.11 (a) Heat sink temperature for Si MOSFET. (b) Junction temperature for Si MOSFET. (c) Conduction loss for Si MOSFET. (d) Switching loss for Si MOSFET.

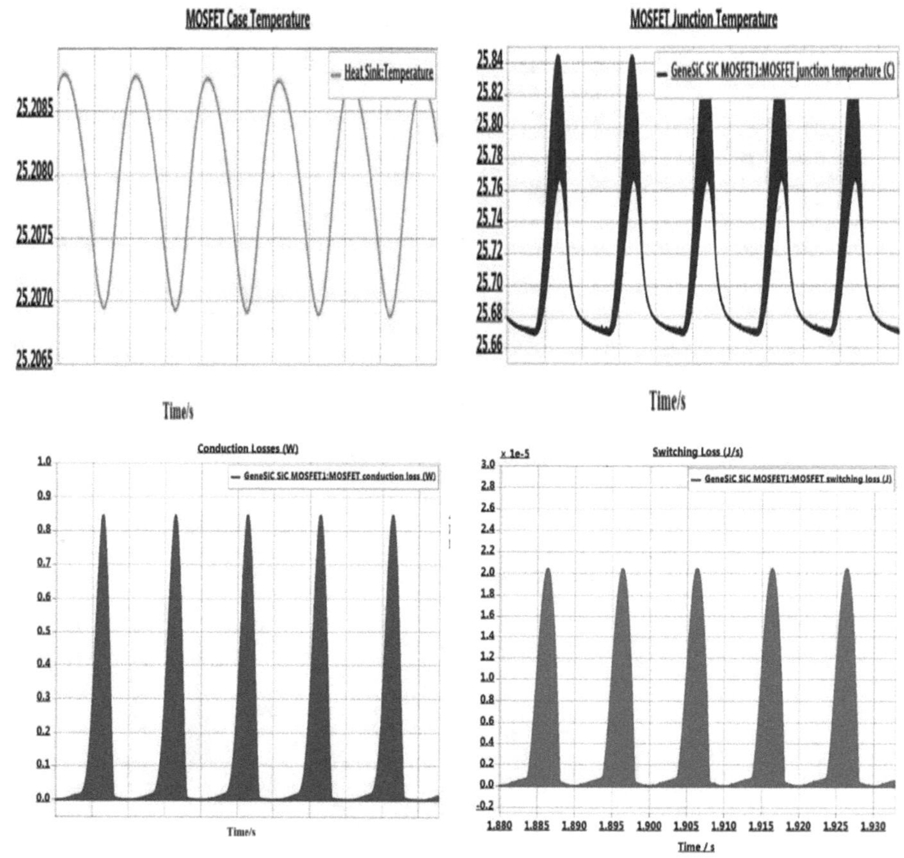

FIGURE 17.12 (a) Heat sink temperature for SiC MOSFET. (b) Junction temperature for SiC MOSFET. (c) Conduction loss for SiC MOSFET. (d) Switching loss for SiC MOSFET.

to Figure 17.12a–d), the average heat sink temperature is 26.0434°C, while the peak value is 26.114°C. The average junction temperature is 26.6732°C with the peak value being 26.7462°C. The average conduction loss for the switch is 0.144 μW, and the peak value is 0.1761 J. The average switching loss is 0.403 W, and the peak value is 0.572 μJ. Concluding the analysis with the GaN switch (presented in Figure 17.13a–d), the average heat sink temperature is 25.0157°C with a peak value of 25.0217°C. The average junction temperature is 26.3898°C, having a peak value of 27.5318°C. The average conduction loss is 0.1103 W with a peak value of 1.744 J. The average switching loss is 0.124 μW.

Considering the thermal impedance of the case, i.e., thermal conduction resistance between the junction and the case, the total loss incurred in the conventional Si power device is 21.944 W, that for SiC is 1.061 W, and that for GaN is 3.078 W as portrayed in Figure 17.14a–c. From these values, it is clear that for high-power applications as in the OBEVC, SiC is a better option compared to GaN and Si.

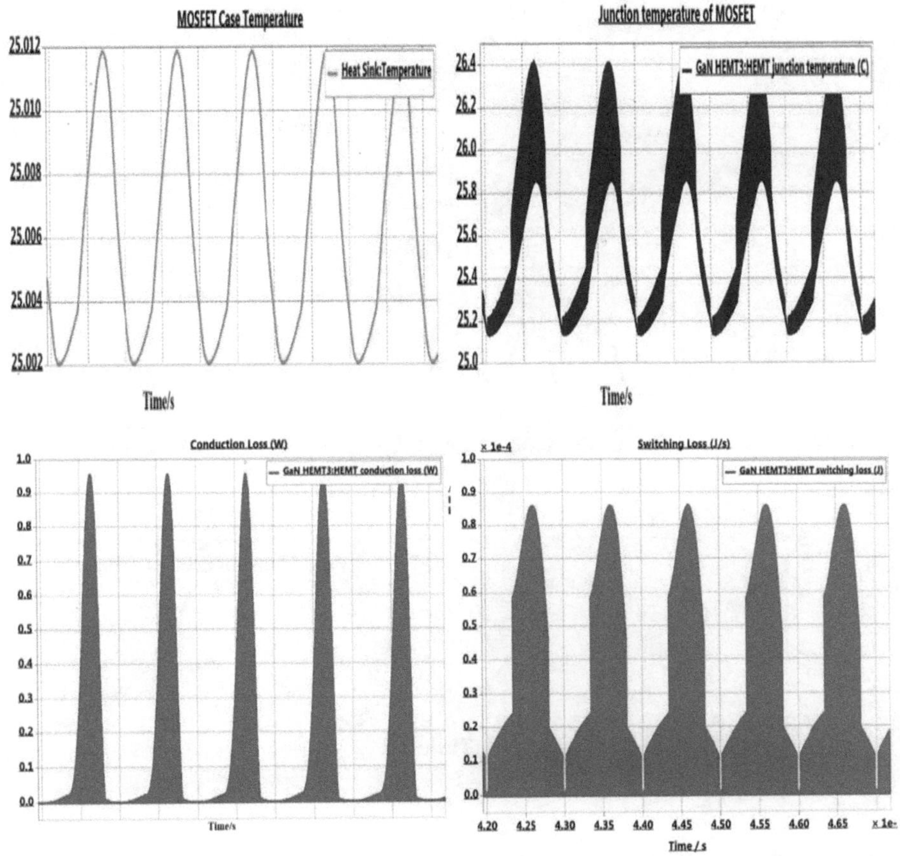

FIGURE 17.13 (a) Heat sink temperature for the GaN switch. (b) Junction temperature for the GaN switch. (c) Conduction loss for the GaN switch. (d) Switching loss for the GaN switch.

17.5.2 Comparison

Table 17.3 presents the thermal characteristics comparison of various devices in the converter operated under full-load conditions. Figures 17.15 and 17.16 show the trend comparison for various conditions of switch when the converter is operated under different load conditions.

From the trends given in Figure 17.15a–d, it can be observed that, in Figure 17.15a, for varying loads, the junction temperature of the silicon switch reaches the highest temperatures in comparison to GaN and SiC (SiC being the least), in Figure 17.15b, and the switching loss is again the highest for Si followed by GaN with SiC having the least switching loss among the three. Figure 17.15c depicts the conduction loss where Si has the maximum conduction loss, and SiC and GaN almost share the same values. Figure 17.15d concludes the comparison of the switches on the

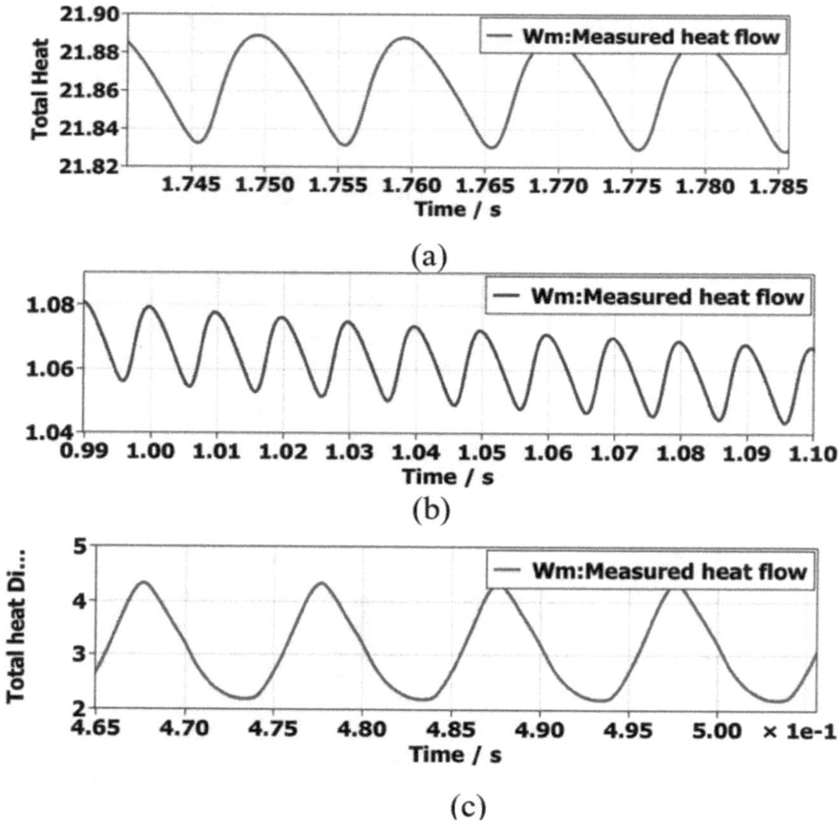

FIGURE 17.14 (a) Total heat dissipation from the SiC switch. (b) Total heat dissipation from the Si switch. (c) Total heat dissipation from the GaN switch.

TABLE 17.3
Comparison of Thermal Characteristics of Various Semiconductor Devices

Parameters/Configuration	Si MOSFET	SiC MOSFET	GaN MOSFET
Heat sink temperature (C)	25.1099	26.0434	25.0157
MOSFET junction temperature (C)	31.5813	26.6732	26.3898
MOSFET conduction loss (W)	3.923	0.144	0.1103
MOSFET switching loss (W)	10.7 μ	0.403 μ	0.124 μ

fixed duty cycle (0.49) by comparing and displaying the trend in the total loss of the switches, where Si tops the losses, followed by GaN, and the least losses are accounted for by SiC.

Figure 17.16a–d depicts the same characteristics for the switches but with a variable duty cycle/ratio. Figure 17.16a displays the junction temperature of the various

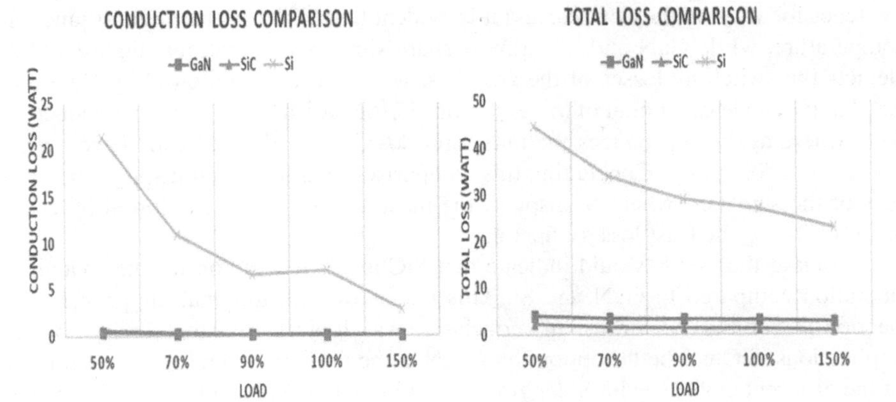

FIGURE 17.15 (a) Junction temperature comparison for various loads. (b) Switching loss comparison for various loads. (c) Conduction loss comparison for various loads. (d) Total loss comparison for various loads.

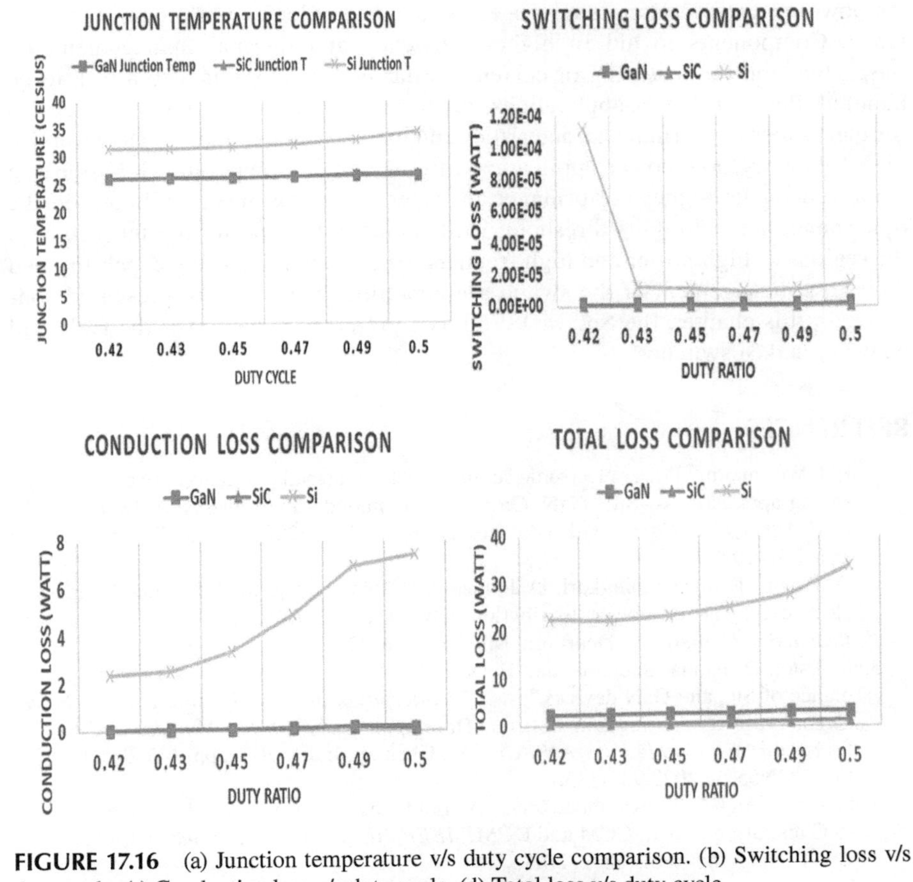

FIGURE 17.16 (a) Junction temperature v/s duty cycle comparison. (b) Switching loss v/s duty cycle. (c) Conduction loss v/s duty cycle. (d) Total loss v/s duty cycle.

switches for varying duty cycles, and it is evident that Si attains the highest junction temperature, while GaN and SiC almost share similar temperatures. Figure 17.16b depicts the switching losses of the switches, where Si leads, followed by GaN and SiC having the least amount of losses. Figure 17.16c indicates the trend in conduction loss, where again Si produces the maximum loss, while SiC and GaN have a very comparable loss value. Concluding this comparison, Figure 17.16d displays the total loss of the switches, where Si displays the most amount of loss, followed by GaN, with SiC being the least loss dissipator.

A look at the trends would indicate that SiC is a better option for the switching operation compared to GaN and Si. This resonates with the analysis provided in Section 17.5 of this chapter, where a conclusion was drawn stating that for high-power applications, SiC is a better option than GaN at the present technology and maturity of the different devices, with Si lagging far behind the two in terms of efficiency and thermal performance.

17.6 CONCLUSION

As power electronic circuits advance, various technologies will come into existence. Components to aid in higher efficiency and thermal management are inevitable, and the maturing of current available technology is also a task to be handled. For high-power applications, the current trend is to employ SiC devices for their superior thermal characteristics and maturity in the market. However, for high-frequency, low-power applications, the neophyte competitor GaN is being employed for its higher performance in faster switching means. Silicon, on the other hand, is reaching its threshold; while being able to perform almost in both the regions of high power and high frequency, its thermal handling capability and the stress management of the switch are appalling. Thus, for the presented converter in this chapter, the SiC MOSFET is a preferred choice over the GaN and conventional Si switches.

REFERENCES

1. P. J. Wellmann, "Power electronic semiconductor materials for automotive and energy saving applications—SiC, GaN, Ga_2O_3, and diamond," *Zeitschrift Für Anorganische Und Allgemeine Chem.*, vol. 643, no. 21, pp. 1312–1322, Nov. 2017, doi: 10.1002/zaac.201700270.
2. G. Putrus, P. Suwanapingkarl, D. Johnston, E. Bentley, and M. Narayana, "Impact of electric vehicles on power distribution networks," in *2009 IEEE Vehicle Power and Propulsion Conference*. Dearborn, MI, 2009, pp. 827–831.
3. R. Patel, B. Panda, Snehalikaand P. Dash, "A comprehensive analysis on the performance of SiC and GaN devices," in *2022 International Virtual Conference on Power Engineering Computing and Control: Developments in Electric Vehicles and Energy Sector for Sustainable Future (PECCON)*. Chennai, India, 2022, pp. 1–5, doi: 10.1109/PECCON55017.2022.9851186.
4. S. F. Lim and A. M. Khambadkone, "A simple digital DCM control scheme for boost PFC operating in both CCM and DCM," *IEEE Transactions on Industry Applications*, vol. 47, no. 4, pp. 1802–1812, 2011.

5. S. Bhagyashri and G. Kanimozhi, "Soft switching bridgeless rectifier for NEV battery charger," *International Journal of Renewable Energy Research*, vol. 10, no. 2, pp. 988–997, 2020.
6. R. Pandey and B. Singh, "Resonant LLC converter for E-bike charger," in *2019 International Conference on Electrical, Electronics and Computer Engineering (UPCON)*, Aligarh, India, 2019, pp. 1–6.
7. B. Singh, S. Singh, A. Chandra and K. Al-Haddad, "Comprehensive study of single-phase AC-DC power factor corrected converters with high-frequency isolation," *IEEE Transactions on Industrial Informatics*, vol. 7, no. 4, pp. 540–556, Nov 2011.
8. R. Beiranvand, B. Rashidian, M. R. Zolghadri and S. M. H. Alavi, "A design procedure for optimizing the LLC resonant converter as a wide output range voltage source," *IEEE Transactions on Power Electronics*, vol. 27, no. 8, pp. 3749–3763, 2012.
9. M. P. Waffler and J. W. Kolar, "Multi-objective optimization and comparative evaluation of Si soft-switched and SiC hard-switched automotive DC–DC converters," in *Proceedings of the 35th Annual Conference of the IEEE Industrial Electronics Society*, Nov. 2009, pp. 3814–3821, doi: 10.1109/IECON.2009.5415123.
10. Y. Tsao et al., "Ultrawide-bandgap semiconductors: Research opportunities and challenges," *Advanced Electronic Materials*, vol. 4, no. 1, p. 1600501, Jan. 2018, doi: 10.1002/aelm.201600501.
11. Y. Kishor and R. N. Patel, "Thermal modelling and reliability analysis of recently introduced high gain converters for PV application," *Cleaner Energy Systems*, vol. 3, p. 100016.
12. A. Khan, J. M. Van Hove, J. N. Kuznia, and D. T. Olson, "High electron mobility GaN/AlxGa1−xN heterostructures grown by low-pressure metalorganic chemical vapor deposition," *Applied Physics Letters*, vol. 58, no. 21, pp. 2408–2410, May 1991, doi: 10.1063/1.104886.
13. T. Hirose et al., "Dynamic performances of GaN-HEMT on Si in cascade configuration," in *Proceedings of the IEEE Applied Power Electronics Conference and Exposition (APEC)*, Mar. 2014, pp. 174–181, doi: 10.1109/APEC.2014.6803306.
14. M. Meneghini et al., "GaN-based power devices: Physics, reliability, and perspectives," *Journal of Applied Physics*, vol. 130, no. 18, p. 181101, Nov. 2021, doi: 10.1063/5.0061354.
15. S. Khoshzaman and I. Hahn, "A performance comparison of GaN FET and silicon MOSFET," in *2021 22nd IEEE International Conference on Industrial Technology (ICIT)*, Valencia, Spain, 2021, pp. 127–133, doi: 10.1109/ICIT46573.2021.9453693.
16. P. P. G. Microsemi, "Gallium Nitride (GaN) versus Silicon Carbide (SiC) in High Frequency (RF) and Power Switching Applications," Digi-Key Electronics, 2014. [Online]. Available: https://shop.richardsonrfpd.com/docs/rfpd/Microsemi-A-Comparison-of-Gallium-Nitride-Versus-Silicon-Carbide.pdf
17. Y. Zhong et al., "A review on the GaN-on-Si power electronic devices," *Fundamental Research*, vol. 2, no. 3, pp. 462–475, May 2022, doi: 10.1016/j.fmre.2021.11.028.
18. G. Gupta et al., "1200 V GaN switches on sapphire substrate," in *Proceedings of the IEEE 34th International Symposium on Power Semiconductor Devices and ICs (ISPSD)*, May 2022, pp. 349–352, doi: 10.1109/ISPSD49238.2022.9813640.
19. I. Grzegory et al., "The homoepitaxial challenge: GaN crystals grown at high pressure for laser diodes and laser diode arrays," in B. Gil (Ed.), *III-Nitride Semiconductors and Their Modern Devices*, Series on Semiconductor Science and Technology. Oxford: Oxford Academic, 23 Jan. 2014, doi: 10.1093/acprof:oso/9780199681723.003.0002
20. Z. Li, X. Yang, Y. Li, N. Wang, S. Zhang, H. Chen and J. Ma, "A high efficiency DC/DC converter with SiC devices and LLC topology for charging electric vehicles," in *2018 Asian Conference on Energy, Power and Transportation Electrification (ACEPT)*. Singapore, Oct 2018, pp. 1–7.

21. R. Pandey and B. Singh, "A cuk converter and resonant LLC converter based e-bike charger for wide output voltage variations," *IEEE Transactions on Industry Applications*, vol. 57, no. 3, pp. 2682–2691, May–June 2021, doi: 10.1109/TIA.2021.3066089
22. P. Shrisrinivas, S. Patra, P. S. Sundar and G. Kanimozhi, "Frontend AC-DC Bidirectional Converter for EV," in *2023 IEEE 20th India Council International Conference (INDICON)*. Hyderabad, India, 2023, pp. 439–446, doi: 10.1109/INDICON59947.2023.10440682.
23. A. Singh, G. Maurya, D. K. Jha and G. Kanimozhi, "Implementation of an E-bike charger using cuk converter and resonant LLC converter," in *2023 International Conference on Computer, Electronics & Electrical Engineering & Their Applications (IC2E3)*. Srinagar Garhwal, India, 2023, pp. 1–6, doi: 10.1109/IC2E357697.2023.10262756.

18 Solar-Powered Inductive Charging System for Marine Applications

S. Kiruthika and P. Sakthi

18.1 INTRODUCTION

The exploration of UWPT represents a frontier in engineering, poised to revolutionize various underwater applications. However, this domain encounters significant obstacles, primarily concerning charging time and power transfer efficiency, particularly when reliant on solar energy sources. The underwater environment presents unique challenges, including increased attenuation and limited power transmission capabilities, making the optimization of WPT systems an intricate task. Addressing these challenges is crucial, prompting the development of innovative solutions. This study focuses on integrating the transformer induction concept and adaptive robotic technology to mitigate the limitations encountered in underwater WPT. By leveraging these concepts, the objective is to create a WPT system capable of not only enhancing charging efficiency but also adapting to the dynamic underwater conditions, thereby improving overall performance. The proposed system's design emphasizes achieving high efficiency and unity power factor, crucial for effective power transfer. A primary goal is to establish a prototype that can successfully transmit an output voltage of approximately 15V over an 8 cm air gap with a maximum sliding distance of 10 cm. Additionally, the utilization of both solar and grid energy sources adds a versatile dimension to the system, allowing for optimized power generation, essential for sustained operations in challenging underwater environments. This introduction sets the stage for exploring the novel approach of integrating transformer induction and adaptive robotics in addressing the critical challenges faced in underwater WPT. The subsequent sections delve deeper into the technical aspects and experimental validations of this innovative system, aiming to demonstrate its feasibility and potential for revolutionizing power transfer in underwater settings.

18.2 WIRELESS POWER TRANSFER

The transmission of electrical energy from a power source to an electrical load, such as an electrical power grid or a consuming device, without using distinct human-made conductors is known as WPT, wireless energy transmission, or electromagnetic power transfer. Time-varying electric, magnetic, or electromagnetic fields are used in various power transmission systems, collectively called "wireless power". When

DOI: 10.1201/9781003496946-18

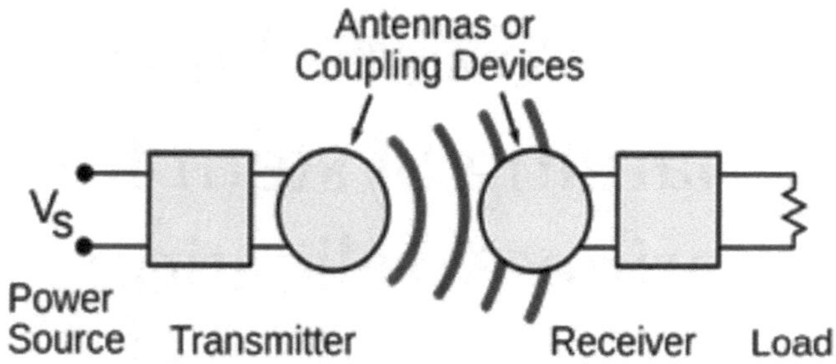

FIGURE 18.1 Block diagram of a wireless power system.

using WPT, field energy is sent from a wireless transmitter that is linked to a power source across a gap to one or more receivers, where it is transformed back into an electrical current and consumed. When using cables to connect electrical devices is impractical, dangerous, or not feasible, wireless transmission can be helpful. The two primary types of wireless power techniques are radiative and non-radiative. Power is transferred by magnetic fields using inductive coupling between wire coils or by electric fields using capacitive coupling between metal electrodes in near-field or non-radiative approaches. The most popular wireless technique is called inductive coupling, and its uses include smartcards, RFID tags, electric toothbrush chargers, implantable cardiac pacemaker chargers, and inductive powering or charging of electric buses and trains. The development of wireless charging systems in Figure 18.1 for portable computers, digital music players, and cell phones – as well as other mobile and handheld computing devices – without the need for wall plugs is a current area of focus. Power beaming, also known as far-field or radiative techniques, uses electromagnetic radiation beams, such as microwave or laser beams, to transport power. Longer distance energy delivery is possible with these methods, but they need to be aimed directly at the recipient. Wireless-powered drone aircraft and solar-powered satellites are two suggested uses for this kind of device.

The term “wireless power transfer” refers to a group of various technologies that use electromagnetic fields to transport energy. The technologies shown in the table below vary in terms of the electromagnetic energy they use, such as time-varying electric fields, magnetic fields, radio waves, microwaves, infrared or visible light waves, and whether the transmitter must be aimed (directed) at the receiver. They also differ in terms of the distance over which they can transfer power efficiently. The components of a wireless power system typically include a “transmitter” that is connected to a power source, such as a mains power line, and that transforms the power into a time-varying electromagnetic field; additionally, the system may include one or more “receiver” devices that receive the power and transform it into direct current (DC) or alternating current (AC) electric current that is used. An “antenna” of some kind transforms the incoming power into an oscillating electromagnetic field at the transmitter. This context uses the term “antenna” rather loosely; it could refer to a laser that produces light, a coil of wire that produces a magnetic field, or a metal plate that produces an electric field. The oscillating fields are converted to an electric current at the receiver by a comparable antenna or coupling device. The oscillations’

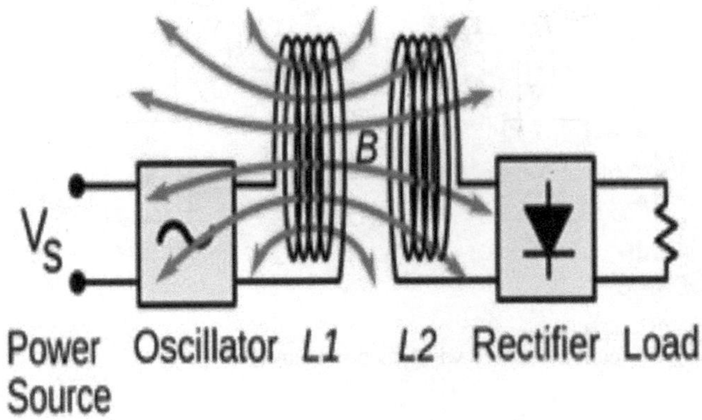

FIGURE 18.2 Inductive wireless power system.

frequency, expressed in hertz (f), is a crucial factor in identifying the kinds of waves. The waves that transfer energy across the gap have a wavelength of $\lambda = c/f$, which is determined by frequency, with c representing the speed of light.

18.3 INDUCTIVE WPT SYSTEM

A magnetic field is used in inductive coupling, also known as electromagnetic induction or inductive power transfer (IPT), to transfer electricity between wire coils. The transformer shown in Figure 18.2 is created by joining the coils of the transmitter and receiver. By using Ampere's law, an AC through the transmitter coil ($L1$) produces an oscillating magnetic field (B). As the magnetic field passes through the receiving coil ($L2$), Faraday's law of induction causes an alternating electromagnetic force (voltage) to be induced, which causes an AC current to flow through the receiver. Either the induced AC drives the load directly or a rectifier in the receiver converts it to DC, which drives the load. Because transmission efficiency increases with frequency, most systems use an electronic oscillator to generate a higher frequency AC current that powers the coil. A few systems, like electric toothbrush charging stands, operate at 50/60 Hz, so the transmitter coil receives direct AC mains current.

18.4 RESONANT INDUCTIVE WIRELESS POWER SYSTEM

Resonant inductive coupling, also known as electrodynamic coupling, strongly coupled magnetic resonance, is a type of inductive coupling in which magnetic fields (B, green) transfer power between two resonant circuits (tuned circuits), one in the transmitter and one in the receiver (see the diagram, right). In Figure 18.3, the resonant circuit is made up of a self-resonant coil, another resonator with an internal capacitance, or a coil of wire coupled to a capacitor. Both are adjusted to have the same resonance frequency. Similar to how a vibrating tuning fork can cause sympathetic vibration in a distant fork tuned to the same pitch, the resonance between the coils can significantly boost coupling and power transfer.

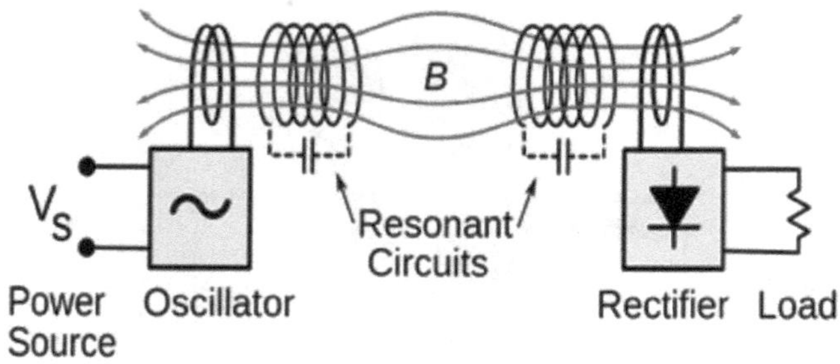

FIGURE 18.3 Resonant inductive wireless power system.

18.5 SURVEY OF THE WPT SYSTEM

There is an increasing need to offer contactless or wireless power for a variety of applications, from high-power battery charging devices to low-power biomedical implants. Power can be transferred from one system to another without any physical contact, thanks to the well-known method known as IPT. The unique bidirectional IPT system presented in this study is especially well-suited for applications where two-way power transfer is advantageous, including vehicle-to-grid systems and plug-in electric cars (EVs). Through loose magnetic coupling and no physical connections, the proposed IPT system enables the simultaneous and controlled charging or draining of many EVs. The amount and direction of power transfer between EVs or between many systems can be adjusted by modulating the phase or amplitude of the voltages generated by each system's converters, as demonstrated by a mathematical model that is presented (Song et al. 2019). A large air-gap bidirectional WPT charger for electric vehicles is suggested in this chapter. In order to provide continuous frequency functioning under wide air gaps without the need for an additional current chopper, it is regulated by pulse width modulation with a self-resonant frequency created by a self-inductance and resonant capacitor. A 6.6-kW prototype with an air gap of 12–20 cm was used to confirm the viability of the suggested approach. An alternate method for a large air-gap bidirectional WPT EV charger is proposed in this chapter. With a frequency created by the self-inductance and resonant capacitor, it is managed by self-resonant pulse width modulated. No additional current chopper is needed with the suggested approach, which offers constant frequency operation independent of air gaps. Power transfer is facilitated under broad air gaps by its increasingly high gain as the air gap increases (Lee and Han 2015).

One emerging trend in mobile devices is wireless battery charging. Large power applications, however, are the subject of comparatively fewer research projects. Traditional electric vehicle battery charging, which depends on the usage of a plug-in charging mechanism, has this significant issue. An LC tank can be used to cooperatively transfer data, such as emergency messages, vehicle identifying numbers, and battery status, between two isolated units in this paper's proposed wireless power and bidirectional data transmission strategy. To provide power flow and AC to the load through mutual inductance, the primary unit uses an inverter. Zero-voltage switching is the method used by the primary unit to receive data, while the secondary unit transfers data by varying the load's current. By cutting the current curve, the primary unit transmits commands to the secondary unit, which then uses the carrier wave's period to decode the commands. Remarkably, the design approach allows it to respond to an emergency by relaying the command to the main unit to perform an emergent halt (Huang and Lin 2018). An efficient 6.78 MHz active rectifier and DC-DC converter for magnetic resonant A4WP applications is shown in this study, along with a full-CMOS Wireless Power Receiving Unit. Resonant wireless charging applications are the target market for the highly efficient active rectifier with a Delay Locked Loop circuit that is being suggested. The resonant frequency of this circuit is 6.78 MHz. The AC input voltage controls how each MOSFET in the suggested rectifier is turned on and off. The power efficiency of the system will be weakened by reverse leakage current resulting from internal blocks such as the comparator, gate driver, level shifter, and voltage limiter delaying their processing time. To eliminate reverse leakage current and maximize power efficiency, the suggested active rectifier uses a Delay Locked Loop to make up for the delay brought on by internal blocks. Additionally, Negative Impedance Circuit was used to reduce switching loss and increase power efficiency.

For WPT systems, an adaptive on/off delay-compensation technique is suggested to enhance the efficiency of CMOS active rectifiers. Power conversion efficiency, voltage conversion ratio, and output voltage ripple of active rectifiers with either a parallel-resonant or a series-resonant circuit at the secondary coil are explored in relation to the impacts of the on/off delays. Both on and off-delays are accurately corrected against PVT changes and mismatches by adding two feedback loops to the active diodes to create the switched-offset currents for the comparators adaptively. A parallel-resonant secondary in a fully integrated active rectifier for biological applications was created as a design example using a standard 0.35 μm CMOS technology. For a 500 Ω load resistor, the measured power conversion efficiency is greater than 89.1% and the measured voltage conversion ratio is greater than 90% when the AC input is between 1.8V and 3.6V. Specifically, while utilizing the suggested approach, the power conversion efficiency is raised by 9% in comparison to the active rectifier (Lin Cheng et al. 2016). In order to improve IPT converter performance, compensation is essential. Simplified control circuitry, lower component ratings, and increased power conversion efficiency are the outcomes of an IPT converter's ability to simultaneously achieve load-independent constant output voltage current, near-zero reactive power, and soft power switch switching with appropriate compensation at certain frequencies. Optimization of the converter design is challenging since consistent output voltage or current is contingent upon the transformer's characteristics, which are sometimes limited by space. This study presents a family of higher-order compensation circuits for IPT converters that accomplish any required constant-voltage or constant-current output with almost low reactive power and soft switching, freeing the design from the limitations imposed by the transformer characteristics. For the desired transfer function, unrestricted by transformer settings, a detailed derivation of the compensating approach is provided. To validate the analysis, three distinct output specifications are built into prototypes of constant-current IPT setups based on a single transformer (Xiaohui Qu et al. 2017).

18.6 EXISTING WPT SYSTEM

This research was undertaken in the backdrop of an era where power consumption is surging due to a rapid increase in electronic gadgets; power transmission, or charging as it is more commonly known, is an urgent necessity. WPT has seen tremendous advancement and is being used in many different fields. All technology, though, may be made even more practically successful by reaching maximum efficiency. With this research, we hope to improve the efficiency of underwater WPT by creating a prototype model. In order to accomplish this, we make an effort to reduce the transfer's losses. The project's objectives are to create a working prototype of an underwater WPT system from scratch, comprehend the basic ideas behind it, and enhance its efficiency. The class E amplifier inverter circuit, gated inverter circuit, and DC power supply make up the transmitter design. The bridge rectifier and regulator, which are connected to the battery, make up the receiver design. The system receives power from a 12V, 20mA DC power source (battery). This is then sent to the class E amplifier after going via the gating (inverter) circuit. The high-frequency inverter is the key component of the suggested contactless power transfer system.

The circuit that converts a DC voltage input into an AC voltage output is known as an inverter. An high frequency inverter is used to generate high-frequency AC signals; in this case, it converts our 12V DC power output into a 12V 50Hz AC power output. For contactless underwater power transfer applications, we are using class E amplifiers in this area. Over an air gap, mutual inductance magnetic coupling is used in IPT to transfer power from a fixed main source to one or more movable pickup units. Recent years have seen a wide range of industrial and domestic applications adopt the wireless charging method, resulting in a variety of uses including contactless charging of electric vehicle batteries, induction cooking, bidirectional IPT, vehicle-to-grid power systems, medical implant systems for cells, phone charging, avionic applications, dynamic charging systems, power and data transfer systems, and special applications. By contrasting power level, gap distance, operational frequency, and other factors, an empirical study of wireless power transmission across various applications and technologies was presented. Nevertheless, the majority of the research focuses only on power transmission; it's suggested that the IPT method has been used to charge car batteries and exhibits excellent operating efficiency.

The dual-side control method in the system design uses an additional wireless communication device to send data from the secondary side to the primary side. It is handled by outer and inner control loops in the architecture. The IPT system uses contactless technology to transfer power and transmit information in a drill machine; however, due to the improved power transfer safety of two separated units, it is important to continuously monitor battery condition feedback to the primary unit. For power and data transmission, respectively, it has two sets of coils. The technique has two working frequencies, though, which might make the devices quite expensive. The authors have put forth a technique for identifying load parameters in the IPT system, which is crucial for setting up a reliable and effective wireless power supply for kitchen appliances. Using the energy injection mode and free resonant mode to examine the feedback signals for load information is one transient load detection technique that may be used to identify load parameters and loading circumstances. This is an easy and dependable procedure. It does not, however, cover the data transfer function. In order to improve power transfer efficiency and lower electromagnetic inference noise, the work described here suggests an entirely new contactless power and bidirectional data transmission approach employing a fly back converter that makes use of the zero-voltage switching technology. Data is received under zero-voltage switching by the primary unit, and it is modulated by the secondary unit by instantly altering the load current. By using energy injection and free resonant modes, the authors provide a transient load detection approach to detect load situations alone. The primary unit trims current to send commands, and the secondary unit receives the commands and decodes the data. Though the structure and purpose of our method are entirely different, they have a similar structure. Here, bidirectional digital data transfer is taken into account. The main purpose of this work's motivation is to provide an illustration of the construction of an efficient power and bidirectional data transmission scheme that may be used in the vehicle.

Recently, there has been a lot of interest in WPT, which uses near-field magnetic coupling and was invented by Tesla. Applications for it are numerous and include electric vehicles, portable electronic device battery chargers, and biomedical implants. IPT and resonant wireless power transfer are two subtypes of WPT. An IPT system functions as a tightly connected air-core transformer because the primary and secondary coils are positioned closely and precisely aligned with one another. Magnetic resonance compensates for leakage inductance in a resonant WPT system, allowing for effective power transmission even in cases when the coils are loosely linked. A4WP, or the Alliance for Wireless Power, and the Digital Object Identifier have been created. A4WP creates resonant WPT standards that attempt to provide spatial independence and multi-device simultaneous charging, but the specifications released by WPC and PMA are typically considered as IPT solutions, such as the "Qi" standard. As the resonant frequency fs of the LC tank, ISM band frequencies like 6.78 or 13.56MHz are typically chosen. A secondary coil $L2$ can be adjusted in series or parallel with the capacitor $C2$, and the power amplifier drives the primary coil $L1$ to create magnetic fluxes that induce an AC voltage at $L2$. Coil geometry, coil alignment, and coil distance all affect the coupling coefficient k. In order to power up the loading circuits, the rectifier cascaded with the tuned circuit transforms the AC voltage into a DC voltage. Higher energy loss can heat up a charger or an implant. The rectifier's design influences the efficiency of the system. For this reason, active diodes made using CMOS transistors and comparators take the role of passive diodes with forward voltage dips of 0.7V. The voltage conversion ratio and power conversion efficiency are increased as a result of the lower turn-on resistances.

18.7 PROPOSED WPT SYSTEM

The goal of the planned UWPT system is to make it easier for a ship to provide energy to an underwater vehicle in Figure 18.4. Power is generated and wirelessly transmitted via water on the ship using a system that includes a solar panel, battery, inverter, and transmitter coil. This energy is captured by a receiver coil on the car, which then transforms it into electricity for components like motor controllers and LED lights using a rectifier and voltage regulator. Voltage sensors keep an eye on

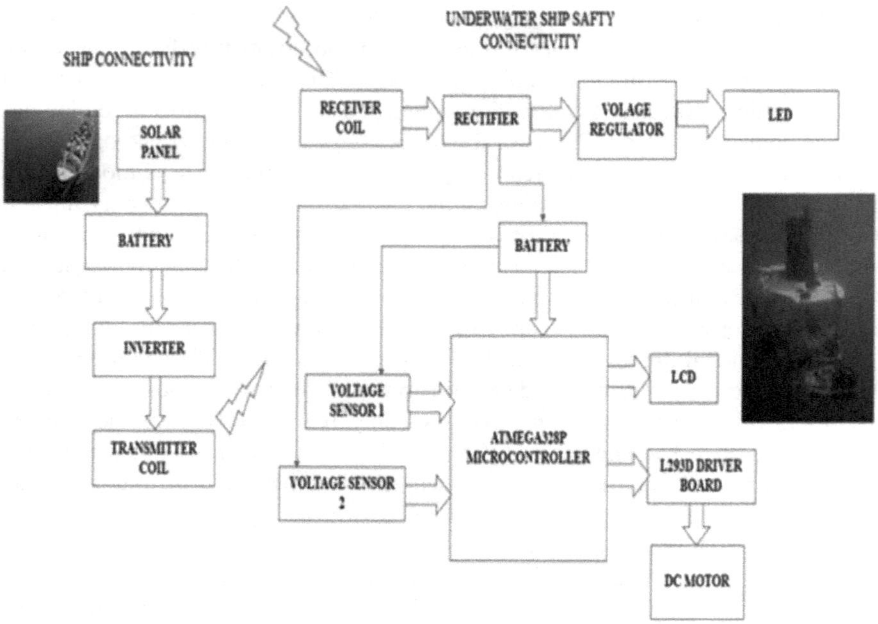

FIGURE 18.4 Block diagram of the WPT system.

battery levels and provide information to an Arduino that is linked to an LCD for a real-time display. In order to enable vehicle driving based on received power and battery status, the Arduino also connects with a motor control board. Underwater vehicle operation is made possible by an all-inclusive system that combines power management, control mechanisms, and UWPT technology.

18.7.1 Transmitter Section

A solar panel installed on the ship's surface captures sunlight and converts it into electrical energy. This energy can be used to power various systems on the ship or stored in a battery for later use. The solar panel can be connected to a battery system. The battery serves as an energy storage device, storing excess energy generated by the solar panel. It ensures a continuous power supply, especially when there's insufficient sunlight or during nighttime. An inverter is necessary to convert the DC electricity generated by the solar panel and stored in the battery into AC electricity. AC is commonly used for power transmission. The transmitter coil is a crucial part of the UWPT system. It's connected to the inverter and generates an alternating magnetic field. This field carries the electrical energy, which is then transmitted wirelessly through the water to a receiver coil. Sunlight is captured via a solar panel mounted on the exterior of the ship, which then produces electricity. The energy can either be stored in a battery for later use or used to power other systems on the ship. A battery system may be linked to the solar panel. Extra energy produced by the solar panel is stored in the battery, which acts as an energy storage device. It guarantees a constant power source, particularly at night or in situations with little sunlight. To convert the DC electricity produced by the solar panel and stored in the battery into AC electricity, an inverter is required. Power transfer using AC is ubiquitous. The UWPT system's transmitter coil is an essential component. With the inverter connected, it produces an alternating magnetic field. This field transports the electrical energy, which is subsequently wirelessly transferred to a receiver coil through the water.

18.7.2 Receiver Section

The receiver coil is positioned on the underwater vehicle and is designed to capture the wirelessly transmitted energy from the transmitter coil on the ship. It receives the alternating magnetic field and converts it back into electrical energy. The AC received by the receiver coil needs to be converted into DC for use by the vehicle's systems. A rectifier circuit is used for this purpose, converting the AC signal into a pulsating DC signal. The pulsating DC signal from the rectifier needs to be regulated to ensure a stable voltage output. A voltage regulator circuit is employed to maintain a constant voltage level required for the vehicle's components. LEDs can serve as indicators to show the status of the received power, providing visual feedback on whether the system is receiving and converting power effectively. Voltage sensors can monitor the voltage level of the battery connected to the receiver system. These sensors provide information about the battery's charge level, ensuring it stays within safe operating limits. The undersea vehicle's receiver coil is mounted there with the purpose of receiving the energy that the ship's transmitter coil wirelessly transmits. It absorbs the alternating magnetic field and transforms it into electrical energy once more. For use by the vehicle's systems, the receiver coil must transform the AC it received into DC. For this, an AC signal is rectified into a pulsing DC signal using a rectifier circuit. For the rectifier's pulsing DC signal to produce a stable voltage output, regulation is required. The circuit of a voltage regulator is used to keep the voltage level necessary for the vehicle's components constant. When used as indicators, LEDs can provide a visual representation of the power received, allowing the user to see if the system is successfully receiving and converting power. Voltage sensors have the ability to track the voltage of the battery that is attached to the receiver system. The battery is kept within safe operating limits, thanks to the information these sensors provide regarding its charge state. The voltage level of the battery can be shown on an LCD panel by using an Arduino microcontroller to interpret data from the voltage sensors. It provides real-time battery status information by serving as an interface between the sensors and the display unit. The undersea vehicle's motors and other equipment can be operated and controlled via this board. It takes signals from the Arduino to supply the power and control required to operate the car's systems.

18.8 HARDWARE IMPLEMENTATION

A solar panel installed on the ship's surface captures sunlight and converts it into electrical energy. This energy can be used to power various systems on the ship or stored in a battery for later use. The solar panel can be connected to a battery system. The battery serves as an energy storage device, storing excess energy generated by

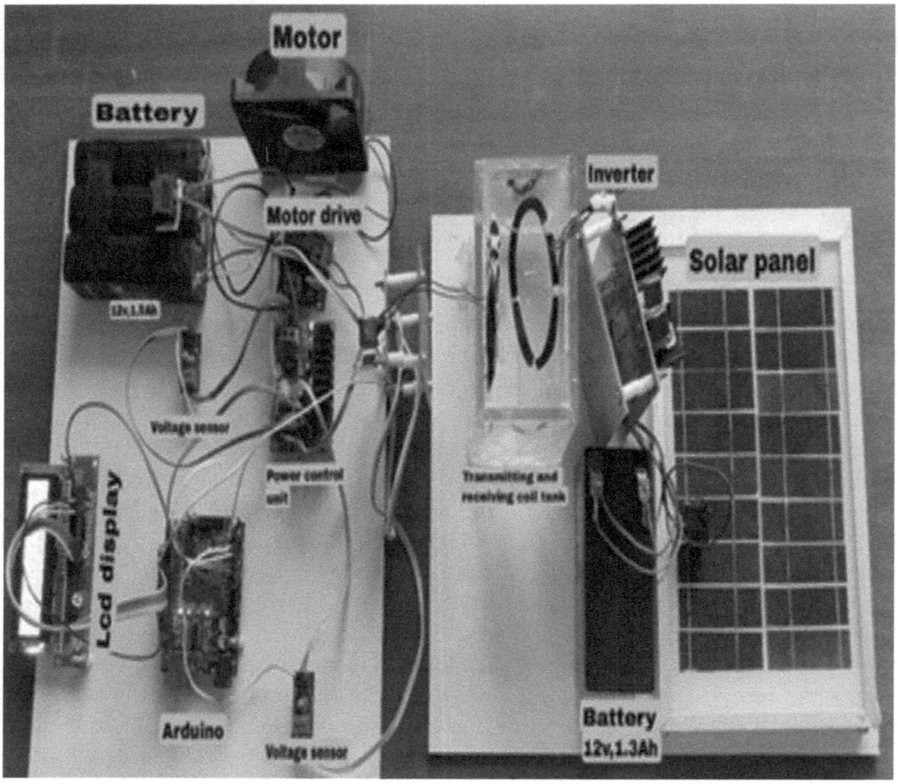

FIGURE 18.5 Hardware working model prototype.

the solar panel. It ensures a continuous power supply, especially when there's insufficient sunlight or during nighttime. An inverter is necessary to convert the DC electricity generated by the solar panel and stored in the battery into AC electricity. AC is commonly used for power transmission. The transmitter coil is a crucial part of the UWPT system in Figure 18.5. It's connected to the inverter and generates an alternating magnetic field. This field carries the electrical energy, which is then transmitted wirelessly through the water to a receiver coil. The receiver coil is positioned on the underwater vehicle and is designed to capture the wirelessly transmitted energy from the transmitter coil on the ship. It receives the alternating magnetic field and converts it back into electrical energy. The AC received by the receiver coil needs to be converted into DC for use by the vehicle's systems. A rectifier circuit is used for this purpose, converting the AC signal into a pulsating DC signal. The pulsating DC signal from the rectifier needs to be regulated to ensure a stable voltage output. A voltage regulator circuit is employed to maintain a constant voltage level required for the vehicle's components. LEDs can serve as indicators to show the status of the received power, providing visual feedback on whether the system is receiving and converting power effectively. Voltage sensors can monitor the voltage level of the battery connected to the receiver system. These sensors provide information about the battery's charge level, ensuring it stays within safe operating limits. An Arduino microcontroller can be used to process data from the voltage sensors and display the battery's voltage level on an LCD screen. It acts as an interface between the sensors and the display unit, providing real-time information about the battery's status. This board can be used to control and drive motors or other devices in the underwater vehicle. It receives signals from the Arduino and provides the necessary power and control to run the vehicle's systems.

18.8.1 Solar Panel

An apparatus that transforms sunlight into electrical energy is a solar panel, sometimes referred to as a photovoltaic panel. It is an essential part of solar power systems and is crucial to the process of using the sun to produce sustainable energy. A vital component of renewable energy systems, solar panels offer a sustainable and environmentally friendly substitute for conventional fossil fuels. These have extensive applications in both grid-tied and off-grid systems, where electricity is stored in batteries for future use and sent back into the utility grid. As time has gone on, solar panels have become more and more popular as a cost-effective and technologically advanced option for homes, businesses, and governments wishing to switch to greener energy sources. A solar panel's solar cell, which is usually composed of semiconductor materials like silicon, is its fundamental component. Photons, or light particles from sunlight, dislodge electrons from the semiconductor material on the surface of the solar cell, resulting in the creation of an electric current. We call this mechanism the photovoltaic effect. From small panels used for household applications to enormous arrays used in utility-scale solar power plants, solar panels are available in a variety of sizes and capacities. The quantity of sunshine a solar panel receives, its size, and the efficiency of its solar cells all affect how much electricity it can generate.

18.8.2 Battery

In order to operate electronic gadgets without being dependent on an electrical source all the time, batteries are essential for delivering portable power. Electrons are transferred from the anode to the cathode of the battery through a chemical reaction that occurs during charging. This process stores energy in the battery. When an electrical current is supplied to a device through the battery, a chemical reaction known as the reverse reaction takes place. This allows electrons to go from the cathode to the anode and powers the device. Different battery types, such as alkaline, lithium-ion, lead-acid, and nickel-metal hydride, have distinct chemistries that are best suited for particular uses. To suit the requirements of various devices, ranging from little home appliances like remote controls to more substantial ones like cars and even power grids for storing renewable energy, they are made in a variety of sizes, capacities, and voltages. They are indispensable to our everyday existence, powering everything from automobiles and emergency backup systems to cell phones and laptops. Two electrodes, a cathode (negative) and an anode (positive), as well as an electrolyte that facilitates ion movement between them, are found inside a battery

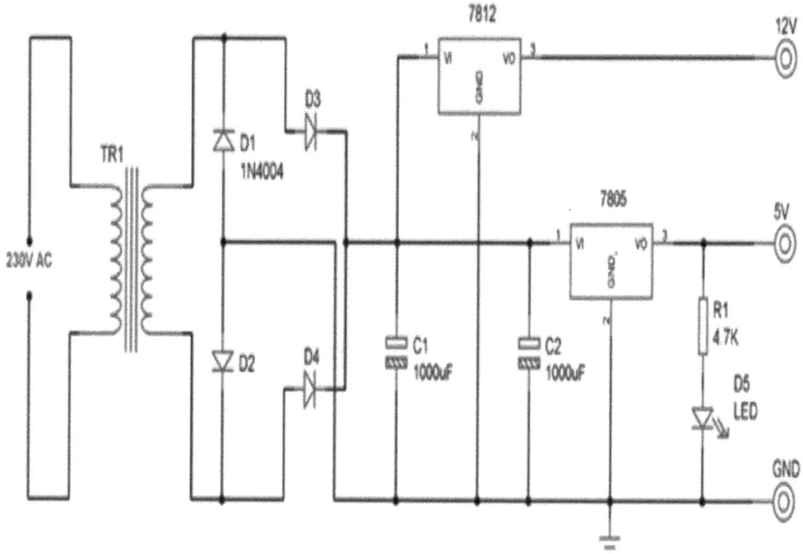

FIGURE 18.6 Power supply unit.

18.8.3 Power Supply Unit

A power supply unit is a crucial part of electronic systems and devices because it transforms electrical energy from a source – like a battery or wall outlet – into a form that can power the different parts of the device. It supplies the required voltage, current, and frequency to guarantee dependable and steady functioning in Figure 18.6. This kind transforms the DC required by electrical equipment from AC in wall outlets. To accomplish this conversion, it usually has a regulator, rectifier, and transformer. The maximum output capacity of the power supply is expressed in watts (W), and this determines its rating. To ensure that the linked components run smoothly and efficiently, it's imperative to choose a power supply unit with enough wattage to satisfy their demands.

In order to protect the connected equipment and the power supply itself from potential harm, power supplies frequently have a variety of precautions installed, including short-circuit, overvoltage, and overcurrent protection. Many electronic gadgets, including computers, laptops, televisions, mobile phones, and industrial machinery, require power supply units as essential parts. They are essential to providing steady, dependable electricity that these devices need to operate properly.

18.8.4 Arduino UNO

The Arduino Uno's ATmega328P microcontroller is its brains. It contains 2 KB of SRAM, 1 KB of EEPROM, and 32 KB of flash memory for program storage. There are 14 digital input/output pins on the Arduino Uno in total. Six of them are suitable for pulse width modulation output. Six analog input pins on it can be converted to digital inputs. Because these pins can read analog signals, they can be used with sensors and other analog devices. The ATmega328P has a frequency of operation of 16MHz, which is sufficient for a variety of applications. The Arduino Uno can be powered through the VIN pin using a regulated power source, an external power supply, or a USB connection. It may be connected to a computer for programming and serial communication, thanks to its USB interface. The board gets electricity from this interface as well. The board has a voltage regulator built in to make sure the microcontroller and other attached parts have a steady 5V supply. Restarting the software from scratch is possible with the reset button. Many sensors and modules can be used with the Arduino Uno because it runs on 5V. Its compatibility with an extensive array of sensors, shields, and modules renders it adaptable and appropriate for an extensive array of applications. The Arduino Integrated Development Environment allows for the programming of the Arduino Uno through the use of a streamlined version of C/C++. Open-source hardware and software form the foundation of the Arduino Uno. This indicates that you can change and distribute the design files and source code without restriction. The user and development community for the Arduino Uno is sizable and vibrant. For help, a wealth of internet forums, tutorials, and resources are at your disposal.

18.8.5 Voltage Senor

An electrical gadget called a voltage sensor gauges the voltage present in an electrical circuit. Electrical system monitoring and control are aided by the information it offers regarding the potential difference between two points in the circuit. Voltage sensors are crucial parts of many systems, from electronic gadgets and automotive systems to power distribution and renewable energy systems.

18.8.6 Inverter

A power inverter, often known as an inverter, is a type of electrical equipment or circuitry that converts DC to AC. A device's or circuitry's design determines its input voltage, output voltage and frequency, and total power handling. There is no power produced by the inverter; the DC source provides power. A power inverter can consist of all electrical components, or it can combine electronic circuitry with mechanical components like a rotating equipment. There are no moving parts involved in the conversion process with static inverters.

18.8.7 Hardware Output

Through the use of panels placed on the ship's exterior, solar energy is collected and stored in battery systems, providing a reliable power source even in isolated or dangerous underwater places. By utilizing the alternating magnetic field produced by the ship's power source, the transmitter and receiver coils work together to enable WPT. Through rectification and voltage control, the undersea vehicle's receiving coil takes this energy and transforms it back into electrical power, ensuring system compatibility. In order to enable operators to appropriately monitor power conversion and battery state, the system also incorporates safety features, including voltage sensors and LED alarms. All things considered, this integrated UWPT system guarantees consistent performance, effective power transmission, and elevated safety during maritime operations and exploration. The discussion makes it clear that fundamental issues with underwater vehicle powering are handled by combining solar panels, battery storage, and WPT technology. The system's dependability in isolated and hazardous places makes it useful for a variety of underwater applications, including operations and exploration. In addition, the inclusion of safety features such as LED alerts and voltage sensors demonstrates a commitment to ensuring both human safety and system durability. Because it provides a long-lasting and efficient power source, this underwater power delivery system is a viable answer to the power supply needs of underwater vehicles and has the potential to totally alter maritime research and operations. Increased research and development to improve efficiency and scalability may widen the scope.

18.8.8 Input Voltage

A reasonably stable DC power source that can provide enough current to meet the systems' projected power demands is necessary for a conventional power inverter device or circuit. The inverter's purpose and design determine the input voltage. A rechargeable 12V lead-acid battery and an automotive electrical outlet are common sources of power for smaller consumer and business inverters, such as 12 VDC. 24, 36, and 48 VDC are typical household energy system standards. When photovoltaic solar panels are the source of power, 200–400 VDC is used. In car-to-grid systems, power from electric vehicle battery packs is delivered between 300 and 450 VDC. High-voltage DC power transmission system, with the inverter operating at hundreds of thousands of volts.

18.8.9 Output Waveform

A circuit design of an inverter determines whether it produces a square wave, modified sine wave, pulsed sine wave, pulse width modulated wave, or sine wave. Modified sine wave and sine wave are the two most common commercialized waveform kinds of inverters as of 2007. To generate home plug-in voltage from a lower-voltage DC source, there are two fundamental methods. The first one generates a higher-voltage DC using a switching boost converter before converting to AC. Using a line-frequency converter to provide the output voltage, the second approach transforms DC to AC at the battery level.

18.8.10 Output Frequency

Typically, a power inverter device's AC output frequency is 50 or 60 Hz, which is the same as the typical power line frequency. For optimal transformer efficiency, the frequency can be significantly higher if the devices or circuit's output needs to be further conditioned (stepped up, for example).

18.8.11 Output Voltage

A power inverter's AC output voltage is frequently controlled to match the grid line voltage, which is normally 120 or 240VAC at the distribution level, even in the event that the load the inverter is driving varies. Because of this, the inverter can power a lot of gadgets that are meant to run on regular line power. Additionally, some inverters support constantly varying or adjustable output voltages.

18.8.12 Output Power

Usually measured in watts or kilowatts, a power inverter's total power rating is provided. Indirectly, this indicates the amount of power required from the DC source and the amount of power that the inverter will be able to supply to the device. From 150 to 3,000 W is the usual range for smaller, more widely used consumer and commercial devices that simulate line power. Applications for inverters don't always revolve around power delivery; occasionally, a subsequent circuit or device uses the frequency and/or waveform characteristics (Table 18.1).

For a conventional power inverter device or circuit, a DC power source that can deliver enough current to meet the system's anticipated power requirements and is reasonably stable is required. The input voltage is determined by the inverter's intended usage and design. One example is the 12 VDC utilized by smaller inverters for consumers and businesses, which often get their power from a 12V lead-acid battery that can be recharged or from an automotive electrical outlet. When solar energy is employed, the conventional standards for home energy systems are 24, 36, and 48 VDC between 200 and 400 VDC. The voltage drop in vehicle-to-grid systems using electric car battery packs is 300–450 VDC.

In systems where the inverter is part of a high-voltage DC power transmission system, the voltage drop can reach hundreds of thousands of volts in Figure 18.7. The AC output frequency of a power inverter device is usually 50 or 60 Hz, which is the same frequency as the traditional power line. For the best transformer efficiency, a much higher frequency is needed.

TABLE 18.1
Transmitting and Receiving Voltage Level Output

Charging Percentage Level (%)	Battery's Transmitting Voltage Level in (V)	Receiving Voltage Level in (V)
100	12.88	5.84
98	12.90	6.06
94	12.93	7.04
90	13.07	8.94

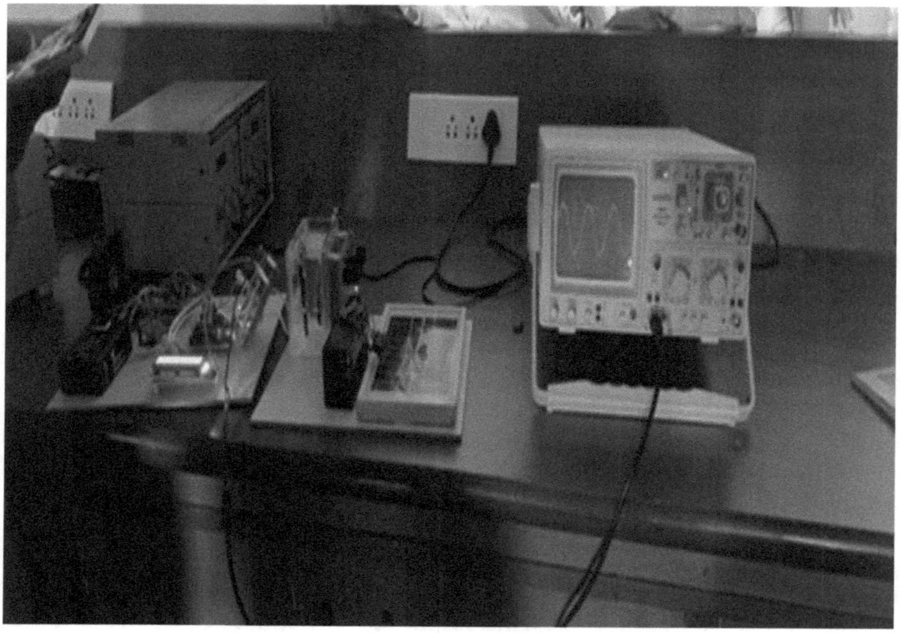

FIGURE 18.7 Transfer coil output waveform in CRO.

18.9 CONCLUSION

In conclusion, the development of an efficient UWPT system, integrating transformer induction and adaptive robotic technology, represents a significant stride in overcoming critical challenges associated with power transmission in aquatic environments. The culmination of this research underscores the potential for transformative advancements in various fields reliant on sustained power sources beneath the water's surface. This achievement not only addresses the hurdles of charging time and power transfer efficiency but also establishes a foundation for enhanced operations in underwater settings. Furthermore, the unity power factor attainment and integration of diverse energy sources, encompassing solar and grid energy, showcase the adaptability and reliability of the system. This adaptability is paramount for ensuring consistent power availability and optimized energy generation, crucial for sustained applications across various underwater fields. The amalgamation of cutting-edge technologies in this research marks a pivotal step forward in underwater power transmission, offering unprecedented opportunities for oceanography, marine research, infrastructure maintenance, aquaculture, communication, search and rescue operations, environmental monitoring, and more. The adaptability and efficiency of this WPT system pave the way for advancements in exploration, research, and industry, promising greater efficiency, autonomy, and sustainability in underwater operations. As this technology continues to evolve, its potential impact on underwater applications is substantial. Through ongoing innovation and refinement, the prospects for reliable, efficient, and autonomous power transmission beneath the waves hold immense promise, unlocking new frontiers for exploration, research, and sustainable utilization of our underwater resource.

REFERENCES

L. Cheng, W.-H. Ki, Y. Lu, and T.-S. Yim, "Adaptive on/off delay-compensated active rectifiers for wireless power transfer systems," *IEEE Journal of Solid-State Circuits*, vol. 51, no. 3, pp. 712–723, 2016.

C.-C. Huang and C.-L. Lin, "Wireless power and bidirectional data transfer scheme for battery charger," *IEEE Transactions on Power Electronics*, vol. 33, no. 6, pp. 4679–4689, 2018.

J.-Y. Lee and B.-M. Han, "A bidirectional wireless power transfer EV charger using self-resonant PWM," *IEEE Transactions on Power Electronics*, vol. 30, no. 4, pp. 1784–1787, 2015.

X. Qu, Y. Jing, H. Han, S.-C. Wong, and C. K. Tse, "Higher order compensation for inductive-power-transfer converters with constant-voltage or constant-current output combating transformer parameter constraints," *IEEE Transactions on Power Electronics*, vol. 32, no. 1, pp. 394–405, 2017.

Y. Song, U. K. Madawala, D. J. Thrimawithana, and M. Vilathgamuwa, "Three-phase bidirectional wireless EV charging system with high tolerance to pad misalignment," *Special Issue: Advanced Technologies Utilised in Wireless Power Transfer Systems, IET Power Electronics*, vol. 12, no. 10, pp. 2697–2705, 2019.

19 Bidirectional Power Converters in Renewable Energy

A Comprehensive Review

Krishnaveni Subramani

19.1 INTRODUCTION

As energy demands grow, it's crucial to develop and expand alternative energy sources. RES like solar, wind, hydro, and geothermal are key to addressing both the need for sustainable energy and the challenge of reducing environmental impact. They offer a way to produce electricity with minimal pollution and a lower carbon footprint compared to fossil fuels. Integrating RES into our energy systems can also contribute to energy security and economic growth by creating jobs and reducing dependence on imported fuels [1]. Indeed, the intermittent nature of RES like PV systems and wind turbines can make it difficult to ensure a stable and reliable electricity supply [2–7]. Addressing these challenges typically involves several strategies:

- Advanced Power Electronics: Devices such as inverters and converters are crucial for managing the variable outputs from RES. They help in converting the generated electricity to a form suitable for the grid and ensuring that the voltage and frequency remain stable.
- Control Strategies: Effective control systems are necessary to optimize the performance of RES. This includes predictive algorithms, real-time monitoring, and adjustments to respond to changing weather conditions and energy demands.
- Energy Storage Systems: Storage solutions, such as batteries or pumped hydro storage, are vital for balancing supply and demand. They store excess energy generated during periods of high production and release it during periods of low production, thereby smoothing out the variability.
- Grid Integration: Upgrading and smartening the grid infrastructure can help accommodate the decentralized and variable nature of RES. Smart grids and microgrids can enhance flexibility and reliability.
- Hybrid Systems: Combining different types of RES (e.g., solar and wind) can mitigate the issues of intermittency, as different sources may complement each other's production patterns.

DOI: 10.1201/9781003496946-19

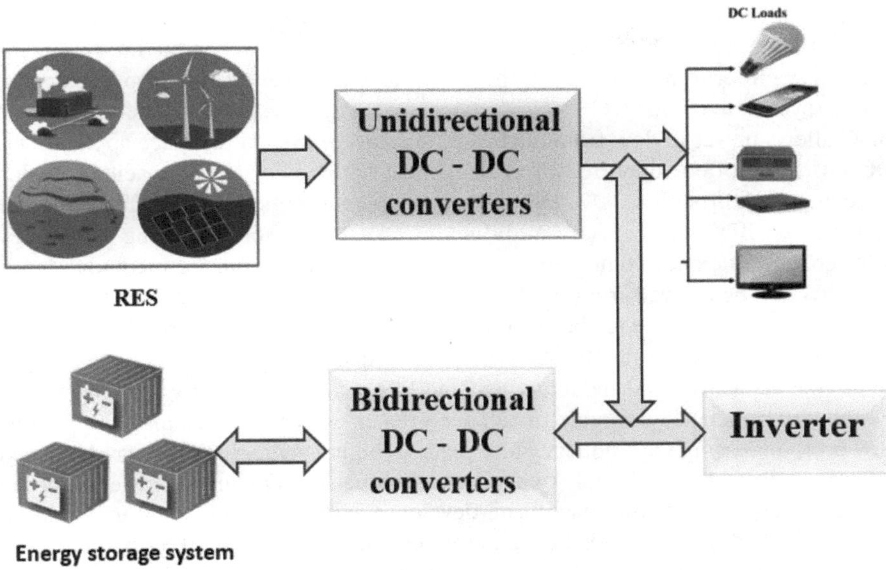

FIGURE 19.1 RES system.

- Demand Response: Implementing demand response strategies can help align energy consumption with periods of high renewable energy generation, further balancing the load.

Developing and integrating these technologies effectively can help overcome the limitations of RES and support a more reliable and sustainable energy future, and the RES system is illustrated in Figure 19.1.

As a result of RES, the demand for microgrids and nanogrids is rising in preference over traditional power transmission systems. This shift is due to their lower maintenance costs, reduced environmental impact, decreased power loss in transmission and distribution, and improved voltage regulation. If we can minimize the costs associated with power generation and storage, conventional power systems may become obsolete. Consequently, microgrids and nanogrids are likely to become dominant in the power sector. These grids are characterized by their integration of local generation sources, which enhances reliability and allows them to operate both independently (islanded mode) and in conjunction with the main grid. Local generation sources may include solar energy, wind energy, and diesel generators, with the exception of diesel generators, all of which are environmentally friendly. Moreover, the current transportation model is unsustainable due to its heavy reliance on fossil fuels. To address this issue, electric vehicles have been introduced as a solution which also requires bidirectional DC–DC converters to manage power flow from the energy sources to the motors, ancillaries, and electronic control units.

A bidirectional converter (BDC) is a device that facilitates the transfer of power between a DC bus system and an energy storage system in both directions, depending on the application. Researchers are continually working to enhance the efficiency of BDCs. BDCs operate in two modes based on load demand. When the DC bus needs additional power to meet the load, the BDC enables power to flow from the energy storage system to the load. Conversely, when there is excess power in the DC bus, the BDC allows power to flow from the bus to the storage system. In space applications, DC buses are often supported by rechargeable batteries or supercapacitors, which are connected through a BDC. Depending on the requirements, either an isolated or non-isolated BDC can be used. Additionally, BDCs with soft switching technology offer greater efficiency compared to those using standard switching methods.

BDCs can be derived from the basic converters like buck, boost, and buck-boost DC–DC converters, where the load voltage varies according to the duty cycle of the main power switch [8–10]. As switching frequencies decrease, passive components like capacitors and inductors become larger and more costly, making higher switching frequencies desirable to achieve greater power densities and lower costs. However, increasing the frequency also leads to higher switching losses and reduced efficiency. To address this, soft switching techniques such as zero-voltage switching and zero current switching have been developed, although they add complexity to the circuit. Additionally, wide bandgap semiconductor materials like silicon carbide (SiC) and gallium nitride (GaN) offer an alternative to traditional silicon (Si) power switches. SiC and GaN enable higher power outputs and operating frequencies, which can reduce switching losses and the size of passive components compared to conventional insulated gate bipolar juction transistor (IGBT) and metal oxide semiconductor field effect transistor (MOSFET) devices. While silicon-based power electronics are well-established and allow for the integration of power and control circuitry into a single chip, wide bandgap materials are still in the process of development, making them more expensive [11,12].

Various reviews have classified BDCs by power switch type, power rating, topology, frequency of operation, and applications [13–17]. This paper reviews bidirectional DC–DC converters and their control techniques, focusing on their use in renewable energy applications. Bidirectional DC–DC converters can be classified into two primary groups based on their configurations: isolated and non-isolated topologies, and the demand for new converters in renewable energy applications has driven the development of newly derived converters.

Electrical energy is transferred via non-isolated topologies without magnetic isolation. Although the non-isolated topologies lack the benefits of galvanic isolation and do not require a transformer isolation, like a high voltage gain ratio step-up, they are advantageous due to less complicated configuration and higher efficiency and don't experience the depression benefits of magnetic interference from galvanic isolation or a large weight. These qualities make them appropriate when weight and size are especially significant considerations in applications. On the other hand, isolated topologies essentially transform a DC voltage into an AC voltage waveform which passes through a high-frequency converter and is converted back to a DC waveform. In isolated topologies, the voltage gain is generally higher, but still the process of transformer design and reducing leakage inductance effect are important since these design factors describe the quality of the converters.

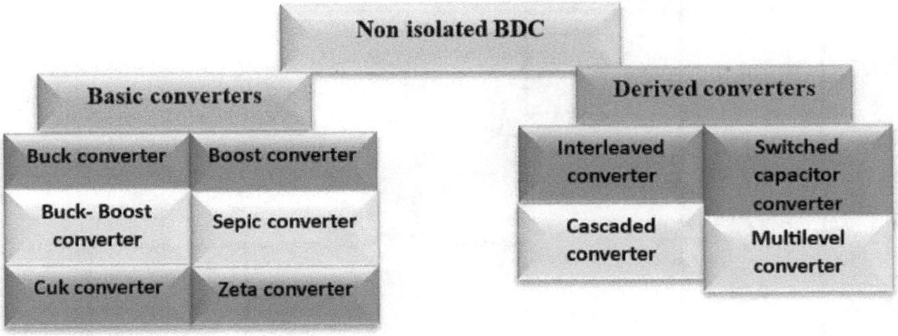

FIGURE 19.2 Classification of non-isolated BDC.

19.2 NON-ISOLATED BIDIRECTIONAL DC–DC CONVERTERS

Basically, a non-isolated BDC is made possible by adding an antiparallel diode to the switch and combining a controlled switch with a diode from the unidirectional converter architecture. Several non-isolated configurations are shown in Figure 19.2. Buck, boost, buck-boost, Cuk, and others are based on fundamental DC–DC converters. Other configurations, such as switched capacitor, interleaved multilevel, etc., are likewise largely based on voltage boosting techniques. Although isolated converters are commonly used in various applications, there are still many scenarios where galvanic isolation is not required, such as in battery-powered motor drive systems for hybrid electric vehicle (HEV) or aircrafts, and in energy storage systems within DC microgrids. In these cases, non-isolated DC–DC converters can be suitable alternatives. They typically offer higher power density and lower cost compared to their isolated counterparts.

19.2.1 Bidirectional Buck/Boost Converter

The circuit shown in Figure 19.3 works as a buck converter when the power flow occurs toward low-voltage side, V_L, and the same circuit operates as a boost converter when the power flow occurs toward high-voltage side, V_H. The buck mode operation is called the charging mode, whereas the boost operation is known as the discharging mode in this converter. In buck/charging mode shown in Figure 19.4, the output voltage of the converter is connected to the input terminal of the storage device, and it is used to step-down the output voltage to less than the input voltage of the source. In this topology, the switching device's diode is essential. It permits bidirectional current flow from the input voltage to the storage device and the other way around (Figure 19.5).

The mode of operation is illustrated in Table 19.1 for both charging and discharging modes.

The efficiency of the buck/boost converter can be calculated by performing power loss calculation, and a control scheme for the bidirectional DC/DC converter can be proposed to optimize practical engineering design [18–20].

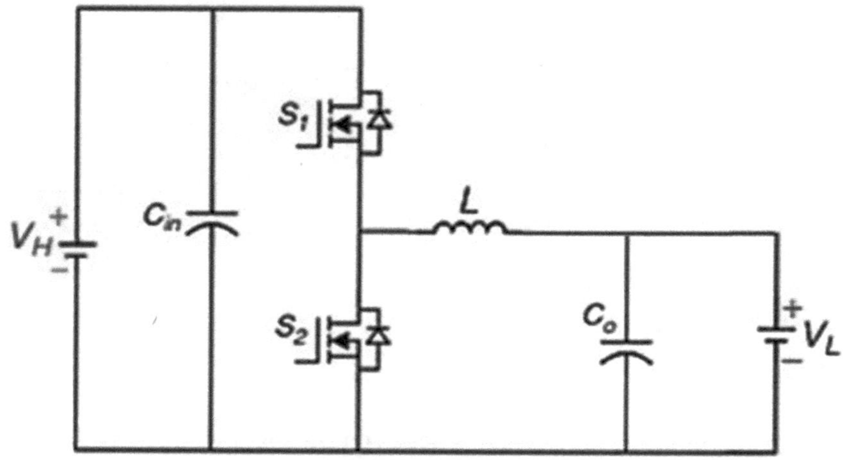

FIGURE 19.3 Bidirectional buck/boost converter.

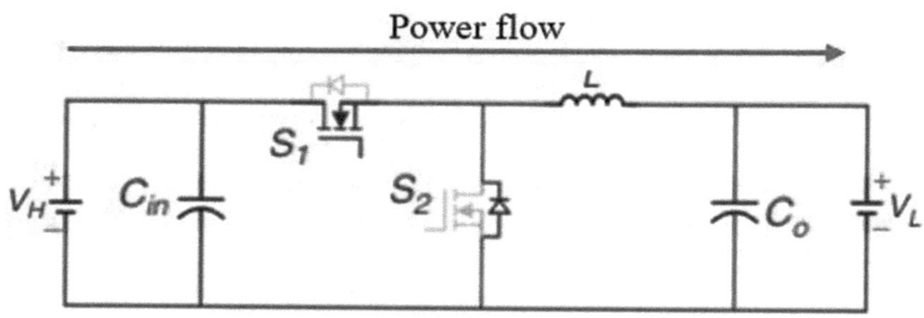

FIGURE 19.4 Buck/charging mode.

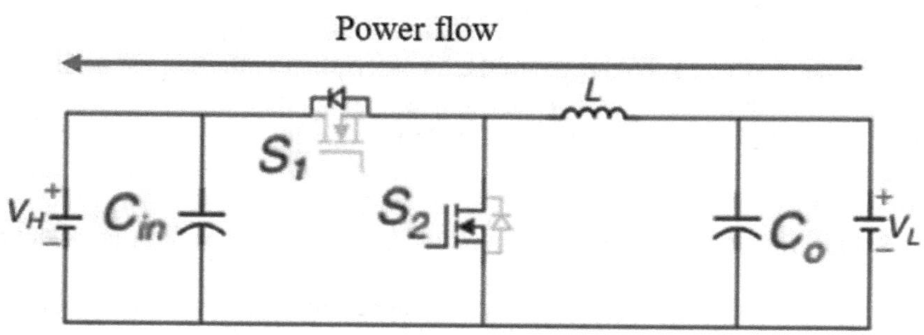

FIGURE 19.5 Boost/discharging mode.

TABLE 19.1
Mode of Operation of the Buck/Boost Converter

Mode of Operation	Switch Position				Current Path During	
	S1	D1	S2	D2	ON Mode	OFF Mode
Buck	ON/OFF	OFF	OFF	OFF/ON	$V_H - S_1 - L - V_L$	$L - V_L - D_1$
Boost	OFF	OFF/ON	ON/OFF	OFF	$V_L - L - S_2$	$V_L - L - D_1 - V_H$

19.2.2 Bidirectional Buck-Boost Converter

A bidirectional buck-boost converter is a type of power electronics device that can both step-up (boost) and step-down (buck) voltage levels, and it operates in both directions, allowing for flexible energy transfer between different components in a system. This capability is particularly useful in RES applications where the input voltage can vary and needs to be adjusted to maintain a stable output voltage. In the buck mode of operation, the higher input voltage is stepped down to the required level. In the boost mode of operation, a lower input voltage steps up the voltage to the required level. By controlling the switches and the energy storage elements (like inductors and capacitors), the converter can handle power flow in both directions, making it suitable for applications where energy needs to be stored and retrieved. The four-switch buck-boost converter, which is suitable for battery storage applications under different conditions, is detailed through the circuit diagram, which is illustrated in Figure 19.6 [21].

In both battery charge and backup modes, the four-switch power stage changes its operation mode according to the DC bus and battery voltage levels. This multi-mode control reduces switching loss and so high efficiency is achieved. However, the buck-boost BDC has some limitations, including not being suitable for high step-up or step-down due to hard switching, high duty cycle operation, voltage conversion ratio limitation, and high voltage stress on the switching device [22].

19.2.3 Bidirectional Cuk Converter

The Cuk converter is used in many applications due to its features, such as continuous input and output current. The bidirectional topology, which is illustrated in Figure 19.7, consists of two bidirectional power switches.

The coupled inductor, which is meant for eliminating the input/output current ripples [23], is adapted in the bidirectional Cuk converter, and this work becomes the ground for further development in control techniques [24].

19.2.4 Bidirectional Single-Ended Primary-Inductor (SEPIC)/ Zeta-Derived Converter

The other two types of DC–DC converters are Zeta and SEPIC, which are implemented to create an output voltage that is positive. As shown in Figure 19.8, this topology functions as a SEPIC when electricity flows from the V_L to the V_H terminals,

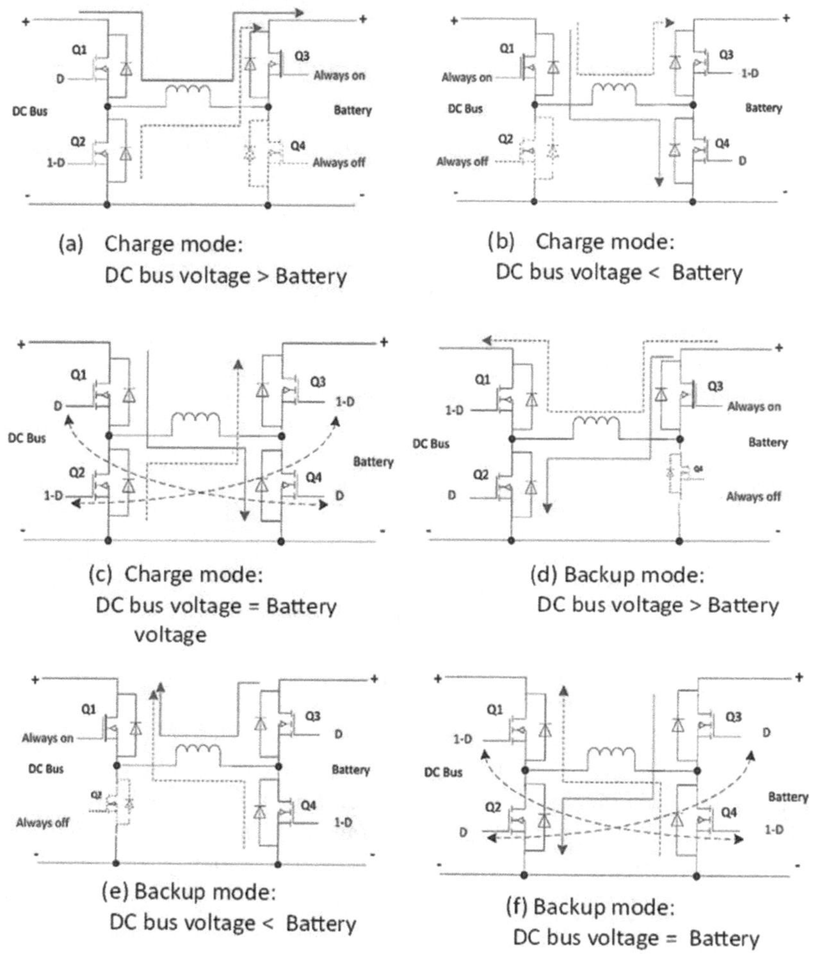

FIGURE 19.6 Four-switch buck-boost converter.

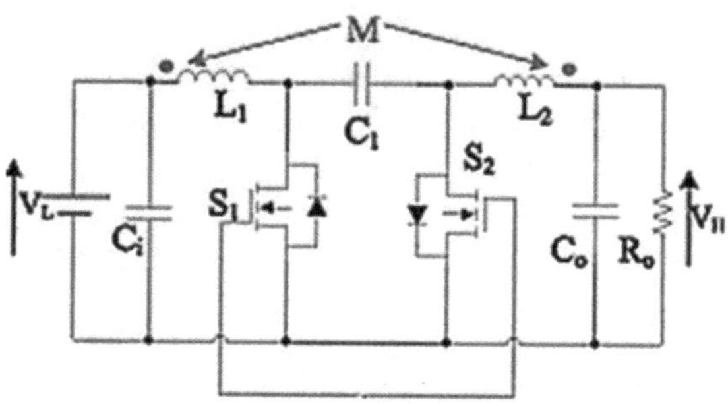

FIGURE 19.7 Bidirectional Cuk-derived converter.

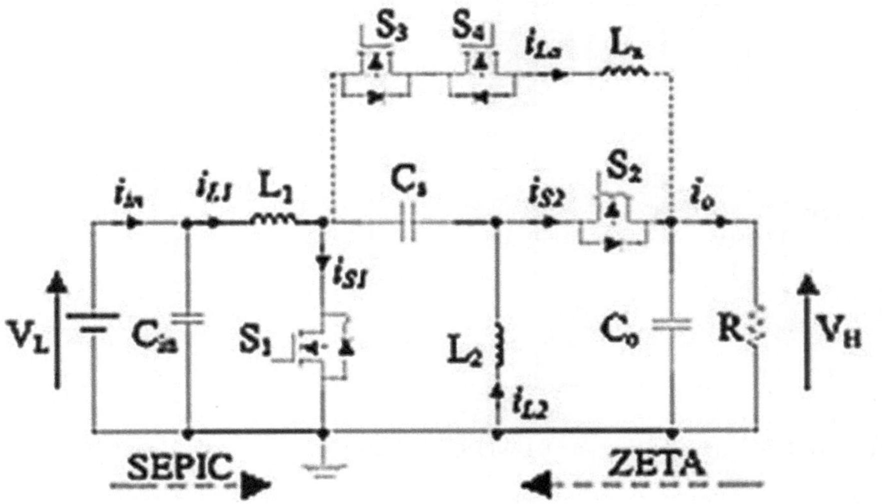

FIGURE 19.8 Bidirectional SEPIC/Zeta-derived converter.

and when transitions of electricity occur from V_H to V_L, the topology functions as a Zeta converter. In this converter, the auxiliary branch highlighted is being considered as a new direct power supply route between the input and output and lessen the ripples in the current [25].

19.2.5 Bidirectional Interleaved DC–DC Converter

Numerous studies in the literature have focused on interleaved converters designed for high-power, high current applications, including RES [25], general-purpose motor drives [26], and energy storage systems [27], among others. The output power can reach several kilowatts, as these systems often use multiple phases with semiconductor and filter components [28]. In such cases, the operating frequency becomes an integer multiple of the switching frequency, leading to smaller filter dimensions. Additionally, the drive signals for the active switches must be phase-shifted by 360°/N, which can be efficiently achieved using inexpensive microcontrollers rather than complex analog circuitry.

The basic interleaved bidirectional two-phase DC–DC converter is shown in Figure 19.9, where V_L and V_H represent the battery and the DC link, respectively. The converter operates in boost mode when transferring energy from the battery to the DC link. Conversely, it functions in buck mode when the DC link supplies energy to charge the battery. When the active switches in the same leg operate complementarily (i.e., S_1 with S_3 and S_2 with S_4), discontinuous conduction mode is not feasible. Although the direction of the inductor current may vary, it will never drop to zero during the switching cycle, this condition is referred to as forced continuous conduction mode, and the modes of operation are detailed in [29].

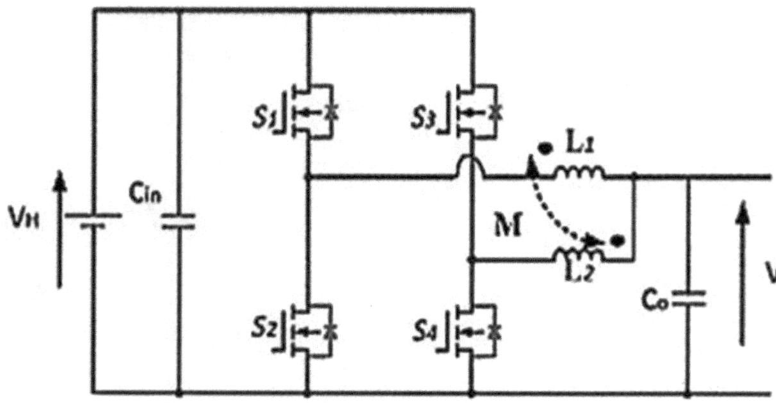

FIGURE 19.9 Bidirectional interleaved DC–DC converter

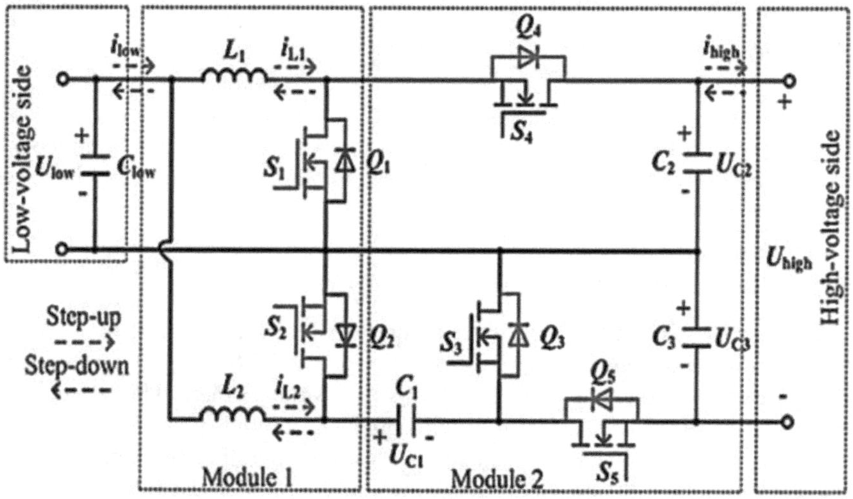

FIGURE 19.10 Bidirectional switched capacitor DC–DC converter.

19.2.6 Bidirectional Switched Capacitor DC–DC Converter

This technology uses capacitors and switches to transfer energy between different voltage levels. The converter operates by periodically connecting capacitors to different voltage sources through switches. The energy is transferred in discrete steps as the switches open and close. An interleaved switched capacitor BDC [30] is shown in Figure 19.10. The interleaved structure is connected at the low-voltage side, which reduces the ripple of the current, and the series-connected structure that appears on the high-voltage side helps to enhance the high step-up/step-down voltage gain. But the major drawback is the large number of components required to get high voltage gain. Due to many components that increase the loss, the efficiency becomes low.

The converter features an interleaved structure on the low-voltage side to minimize current ripple and a series-connected structure on the high-voltage side to achieve substantial voltage gain. Bidirectional synchronous rectification is implemented without additional hardware, enhancing the converter's efficiency. A 1 kW prototype has been developed to validate the converter's wide voltage gain range, operating between a variable low-voltage side (50V –120V) and a constant high-voltage side (400V). The converter achieves a maximum efficiency of 95.21% in step-up mode and 95.30% in step-down mode.

19.2.7 Bidirectional Cascaded DC–DC Converter

High step-up voltage gain is a desirable attribute in renewable energy applications where the RES may be integrated with the grid, thus making such DC–DC converters an attractive option [31,32].

The bidirectional cascaded DC–DC converter system combines multiple DC–DC converters in a series arrangement to achieve versatile and efficient voltage conversion. The bidirectional capability enhances flexibility, making it suitable for applications that require energy transfer in both directions. The cascaded approach provides high voltage conversion ratios and can be tailored to Figure 19.11, which illustrates such a kind of cascaded DC–DC converter, and the circuit operations under boost and buck mode are described in Figure 19.12 [33]. As in [33], the implementation of a bidirectional cascaded buck-boost converter has an average efficiency of 83% for buck mode and 81% for boost mode. Implementation of the converter used for charging a Li-Ion battery with a constant current method and current parameters 1.1 Amps and cut-off voltage 12.6V works according to the design. After charging the battery for 93 minutes. Now the battery has a State of Change (SOC) value for about 90%–95% from the previous one, which is only 40%. This value is achieved by measuring the open circuit voltage. Table 19.2 compares the different basic non-isolated DC–DC converters.

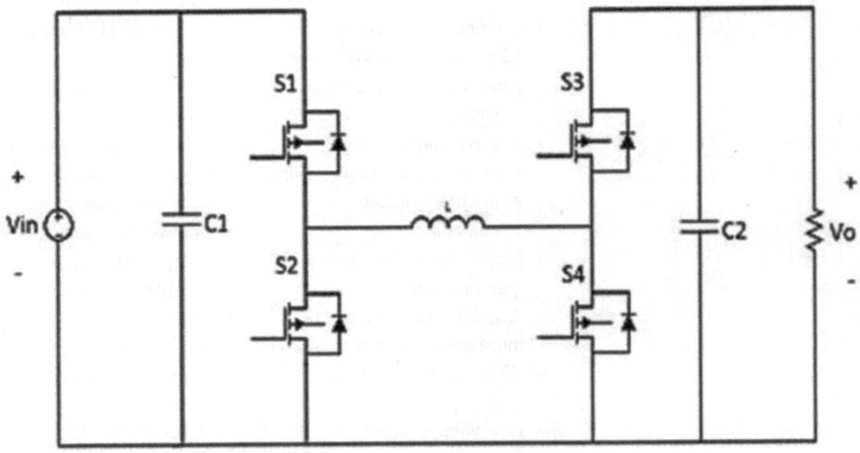

FIGURE 19.11 Bidirectional cascaded DC–DC converter.

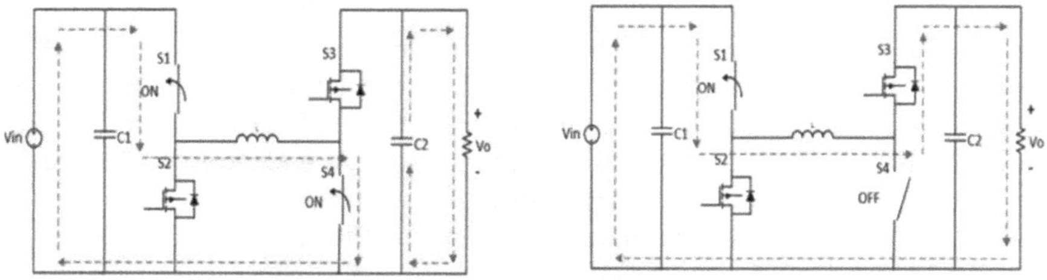

FIGURE 19.12 Circuit operation of the bidirectional cascaded DC–DC converter. (a) Boost mode. (b) Buck mode.

TABLE 19.2
Comparison of Basic Non-isolated Bidirectional DC–DC Converters

Type of the Converter	Voltage Gain	No. of Switches	Merits	Demerits
Buck	D	1	• A low number of devices • Low voltage stress • Simple structure • Easy to control • Easy to generate new topologies	• Low voltage conversion ratio • Large switching loss • Low conversion efficiency • Low-power applications
Boost	$\frac{1}{1-D}$	1	• A low number of devices • Low voltage stress • Input–output common ground	• Large switching loss • Low conversion efficiency • Low-power applications
Buck/boost	$\frac{-D}{1-D}$	2	• High efficiency • Easy to implement short-circuit protection • Less expensive	• Efficiency does not work for maximum gain • Input current and charging current are discontinuous • Output is inverted
Cuk	$\frac{D}{1-D}$	2	• Continuous current • Eliminated ripples of input current by coupling the inductors	• High number of reactive components • High current stress
SEPIC/Zeta	$\frac{1-D_2}{1-D_1}$	2/2	• Positive output voltage • Reduced current ripples using an auxiliary branch	• Pulsating output current • Capacitor requirements • Rectifier losses • Control difficulty
Interleaved inductor	$\frac{1}{1-D}$	4	• Low switching frequency current ripple • A smaller electromagnetic interference filter is required • Wide voltage conversion range • Low voltage stress	• Large number of switches • Complex structure • Input–output non-common ground • High control requirements • Pulsating current of the power source side

(Continued)

TABLE 19.2 (*Continued*)
Comparison of Basic Non-isolated Bidirectional DC–DC Converters

Type of the Converter	Voltage Gain	No. of Switches	Merits	Demerits
Switched capacitor	$\frac{2}{1-D}$	4	• Low size and weight • Continuous input current	• High current ripple in the power source side • Input–output non-common ground • Pulsating current of the power source side
Cascaded	$\frac{1}{1-D}$	4	• Higher voltage gain • Lower current stress • Wide voltage conversion range, easy to generate new topologies	• Large number of devices • Low conversion efficiency • Low power density

19.3 RECENT RESEARCH TRENDS IN NON-ISOLATED BIDIRECTIONAL DC–DC CONVERTERS

The improvements of the non-isolated bidirectional DC–DC converter are basically based on the RES and other applications. The different methods can be combined to create higher efficient topologies. The following section deals with a few of such techniques proposed by different researcher groups, which are suitable for RES-based applications.

19.3.1 Three-Phase Interleaved Boost Converter

A schematic diagram of a simple three-phase interleaved boost converter is illustrated in Figure 19.13. This topology consists of three conventional boost converters connected in parallel. Each individual converter consists of an inductor, a diode, and a controlled switch. This proposed converter [34] meets the maximum power of 5 kW. In this study, the output voltage ripple is limited to 1%, while the input current ripple is carefully controlled at 5% to reduce fluctuations and ensure stable power delivery by keeping the temperature at 25°C. Additionally, to meet the design specifications, nine modules are chosen per string, with two strings connected in parallel to achieve the 5-kW rated output. When a converter operates in discontinuous conduction mode, its efficiency can be adversely affected. Discontinuous conduction mode introduces extra switching losses due to the interruption in inductor current. This efficiency loss is typically linked to increased conduction losses and potentially higher peak currents. For a three-phase interleaved boost converter, ensuring continuous conduction mode is essential to maximize power extraction from the PV panels. Additionally, the Li-Ion battery is directly connected to the converter's output/DC link, so a separate charging circuit is unnecessary. This setup ensures that the DC link voltage stays constant. A three-phase voltage source inverter is also connected

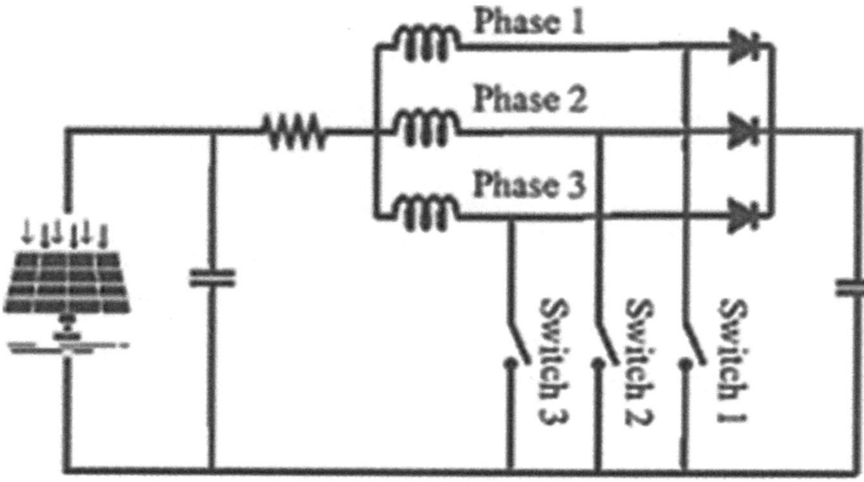

FIGURE 19.13 Three-phase interleaved boost converter [34].

to the DC link to convert the DC power to AC and deliver electricity to the AC grid. The design and loss analysis are detailed in [35] (Figure 19.13).

19.3.2 Three-Phase Interleaved Floating Output Boost Converter

The proposed converter in [36] utilizes three energy-storing capacitors to achieve a high voltage gain. In addition to its high voltage gain capability, the converter also minimizes voltage stress on the semiconductor switches. This reduction in stress allows for the use of lower-rated switching devices, which in turn helps to decrease the overall size and cost of the converter.

The proposed converter [36], which is shown in Figure 19.14, operates at a fixed switching frequency and has a constant switching period. The switches S_1, S_2, and S_3 are controlled by two pulse-width modulation signals that are 180° out of phase with each other. One pulse width modulation (PWM) signal is applied to the gate of S_2, while the other, which is 180° phase-shifted from the first one, is applied to the gates of S_1 and S_3. There is no phase shift between phases 1 and 3, both of which are 180° out of phase with phase 2. The converter is analyzed with a duty cycle D greater than 50%, resulting in four switching states within one switching period.

The analysis described in [36] reveals that traditional interleaved boost converters experience high duty cycle operation and face reverse recovery issues. In contrast, the proposed converter can achieve the same voltage gain with a more appropriate duty cycle. Additionally, the voltage stress on the switches in the proposed converter is reduced by 260% compared to that in traditional interleaved boost converters. With an efficiency exceeding 90%, the proposed converter demonstrates strong performance. These advantages make it a more suitable choice for renewable energy systems where high step-up DC–DC voltage conversion is essential.

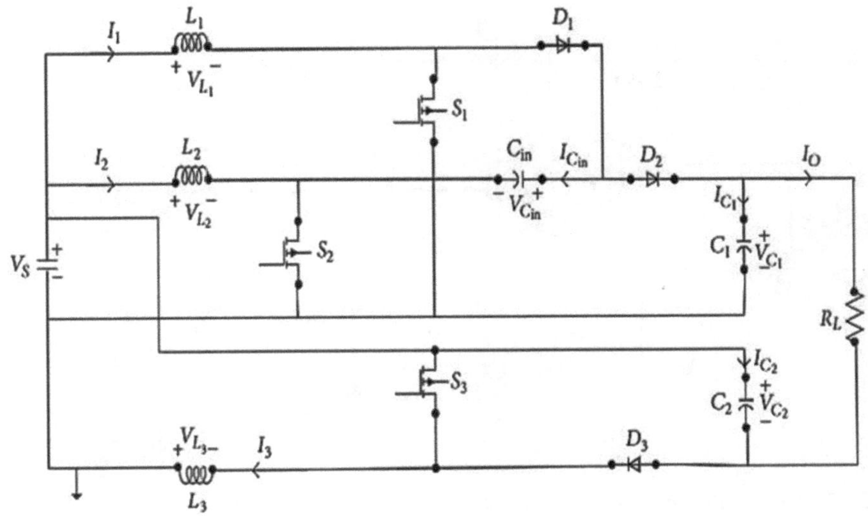

FIGURE 19.14 Three-phase interleaved floating output boost converter [36].

19.3.3 Bidirectional Two-Input Boost DC–DC Converter

A new two-input step-up DC–DC converter is introduced in [37] for use in RES. This converter is designed with few components while delivering a high step-up voltage gain and impressive efficiency. It features two input channels connected to separate energy sources through a switch-mode coupling approach. These sources can supply power to the load via the converter, and if they are rechargeable, one source can charge the other. Additionally, the bidirectional power transfer capability allows regenerative energy from the output to be fed back to the rechargeable source. The converter operates with continuous input current, making it suitable for current-sensitive applications. It includes only one main power switch on the low-voltage side, resulting in low voltage stress and allowing for the use of low-voltage rating semiconductors with minimal on-resistance. This design significantly reduces conduction losses, and the input and output terminals share a common ground. The paper outlines the theoretical background and operating principles of the proposed topology and presents experimental results from a 48/400V/1 kW prototype to validate the converter's performance (Figure 19.15).

19.3.4 High Gain Interleaved Bidirectional DC–DC Converter

A novel non-isolated interleaved bidirectional DC–DC converter that employs coupled inductors to achieve a high step voltage ratio is illustrated in Figure 19.16. The use of coupled inductors enhances the voltage conversion ratio and ensures soft switching conditions for the switches during turn-on. The interleaved structure, combined with coupled inductors on the low-voltage side, significantly reduces current ripple, which is crucial for high-power applications. This topology also facilitates

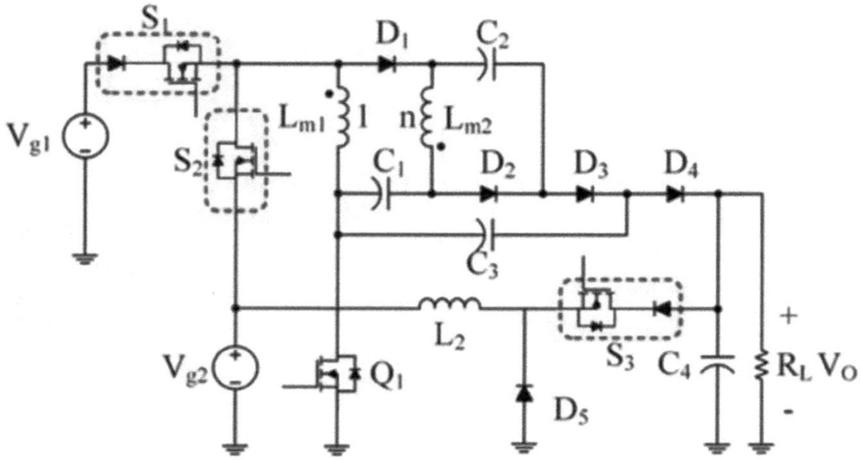

FIGURE 19.15 Bidirectional two-input boost DC–DC converter [37].

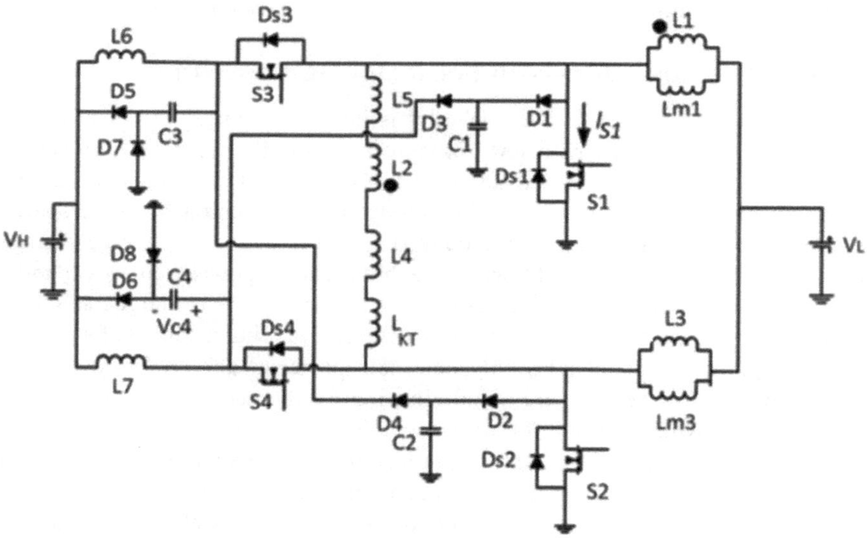

FIGURE 19.16 Interleaved bidirectional DC–DC converter with passive clamp networks and the coupled inductor model [38].

automatic current sharing between the two phases of the interleaved structure. On the high-voltage side, it provides substantial voltage gain in both buck and boost modes with appropriate duty cycles. The converter achieves zero current switching turn-off for all diodes and zero current switching turn-on for all switches without the need for additional auxiliary circuits, thereby reducing undesirable reverse recovery effects in diodes. The design features the minimum number of switches required for

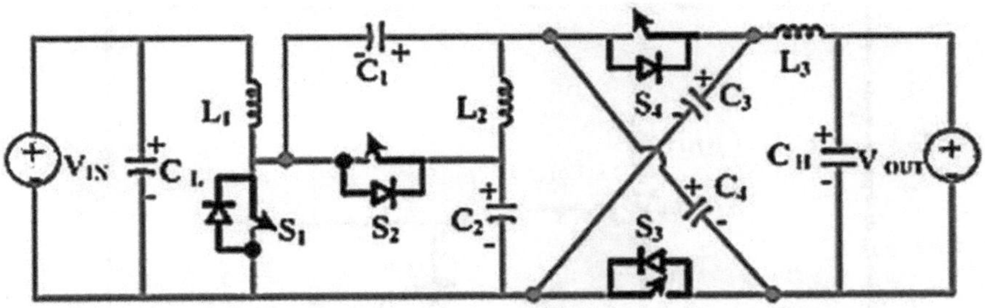

FIGURE 19.17 Switched inductor-capacitor bidirectional DC–DC converter [39].

a two-phase bidirectional structure and includes a simple passive clamp circuit to manage voltage spikes across the coupled inductors.

This structure reduces current ripple and delivers a high gain in both buck and boost modes with an appropriate duty cycle. Additionally, this topology ensures automatic current sharing among the coupled inductors on the low-voltage side.

19.3.5 Ultra-High Gain Bidirectional DC–DC Converter

The bidirectional power flow capability of the proposed converter [39] enhances its suitability for applications in microgrids. By combining switched inductor and switched capacitor cells, the converter achieves high voltage gain without the need for a bulky transformer. This design offers a wide voltage gain range and minimizes voltage stress on the power switches. The proposed switched inductor switched capacitor bidirectional DC–DC converter operates effectively in both boost and buck modes. The maximum efficiency of this converter is 95.25% and verified under both steady-state and transient conditions. Lower voltage stress and current stress across the switches have been verified from comparison with other similar topologies (Figure 19.17).

19.3.6 Capacitor Clamped Coupled Inductor Bidirectional DC–DC Converter

A novel capacitor clamped coupled inductor bidirectional DC–DC converter that delivers high voltage gain while ensuring smooth current transients and reducing capacitor stress is presented in Figure 19.18. A simplified single voltage-clamped circuit is proposed to address voltage spikes caused by the coupled inductor, effectively recovering leakage energy. The clamping capacitor further reduces output voltage ripples, which decreases switch stress and results in smoother output at the load. The converter also demonstrates a 119% reduction in input current transients and a 25.25% reduction in output voltage transients under varying line conditions. Additionally, transient reductions of 25.25% and 76.5% in input current are observed for variations in line and control parameters, respectively [40]. Moreover, the clamping capacitors efficiently reduce voltage ripples, which in turn reduces stress on the switch.

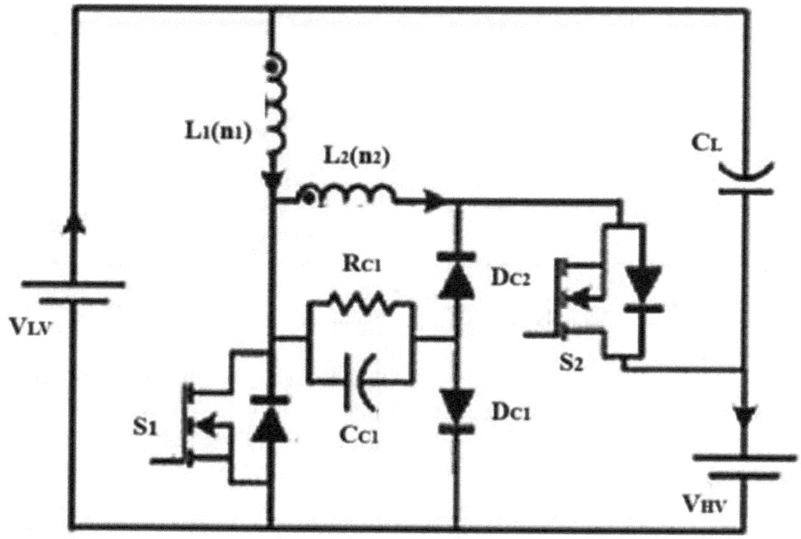

FIGURE 19.18 Bidirectional DC-DC converter based on switched Inductor-Capacitor topology [40].

19.3.7 Interleaved High Gain Bidirectional DC–DC Converter

The paper [41] proposes a high gain bidirectional DC–DC converter with an interleaved boost configuration for use in grid-integrated solar PV systems powering telecommunication base transceiver station (BTS) loads. This BDC is designed to regulate the DC bus voltage at 48V. It interfaces with a 230V AC grid and AC load through a voltage source converter, which requires maintaining a 400V DC side, thus necessitating a high gain bidirectional DC–DC converter. This converter allows for the use of components with lower ratings on the low-voltage side. The proposed converter features an interleaved boost design combined with a switched capacitor configuration, as illustrated in Figure 19.19. It employs two interleaved boost stages with a switched capacitor, C, to enhance voltage gain. Switches S_1 and S_2 are used for the interleaved step-up operation, while S_3 and S_4 handle the step-down operation of the converter.

The proposed high gain bidirectional DC–DC converter features four IGBT switches and operates under both forward and reverse power flow conditions. It has been observed that the DC current on the low-voltage side remains continuous and is evenly distributed among the interleaved inductors. In boost mode, the converter steps up the low-voltage DC bus from 48V to 400V, while in buck mode, it steps down from 400V to 48V.

19.3.8 Coupled Inductor-Based High Buck/Boost BDC

Article [42] introduces a non-isolated hybrid bidirectional DC–DC converter that offers a high step-down/step-up voltage conversion ratio. It also reduces voltage and current stresses on semiconductor devices and minimizes ripples in inductor currents (Figure 19.20).

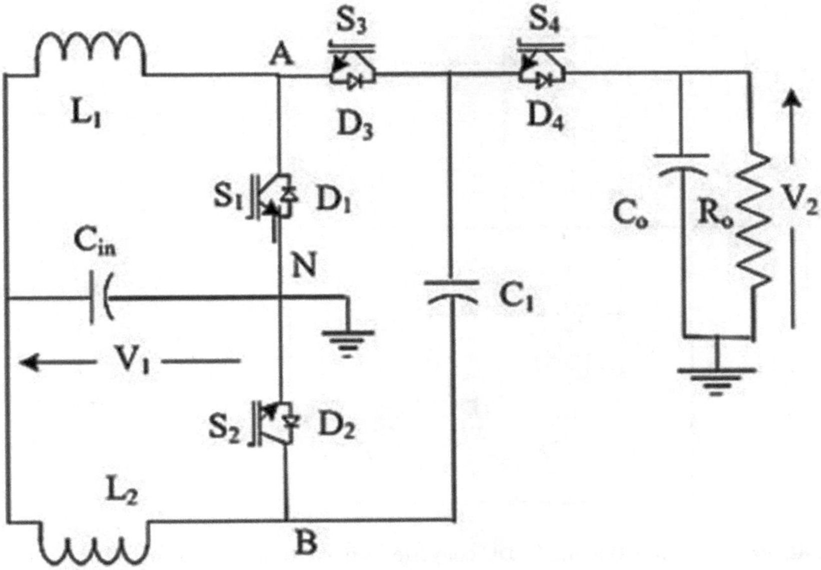

FIGURE 19.19 Interleaved high gain bidirectional DC–DC converter [41].

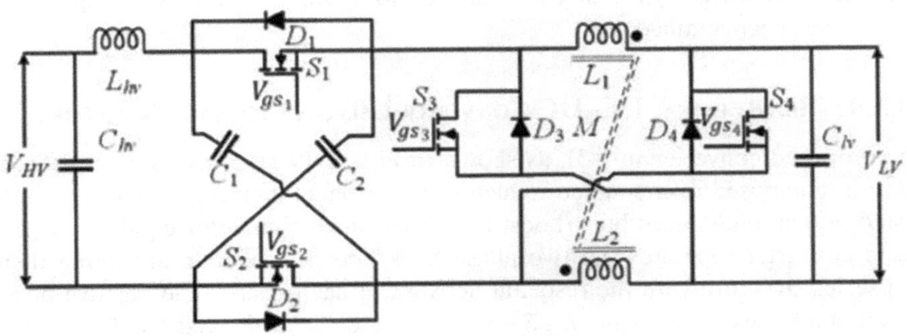

FIGURE 19.20 Coupled inductor-based buck/boost BDC [42].

A storage capacitor C_{hv} either receives or delivers high voltage. To prevent abrupt changes in current direction during charging (step-down) and discharging (step-up) operations, an inductor L_{hv} is used. The proposed converter features a bidirectional switch-diode capacitor that includes two bidirectional switches S_1 and S_2 arranged in parallel, along with two capacitors C_1 and C_2 arranged in a cross configuration. The use of inversely coupled inductors in this topology helps reduce current ripples and minimizes bulkiness.

Compared to conventional and existing hybrid BDCs, the proposed converter achieves higher output voltages with lower duty cycles. This modification in the voltage conversion ratio reduces both inductor current ripples and voltage stress on

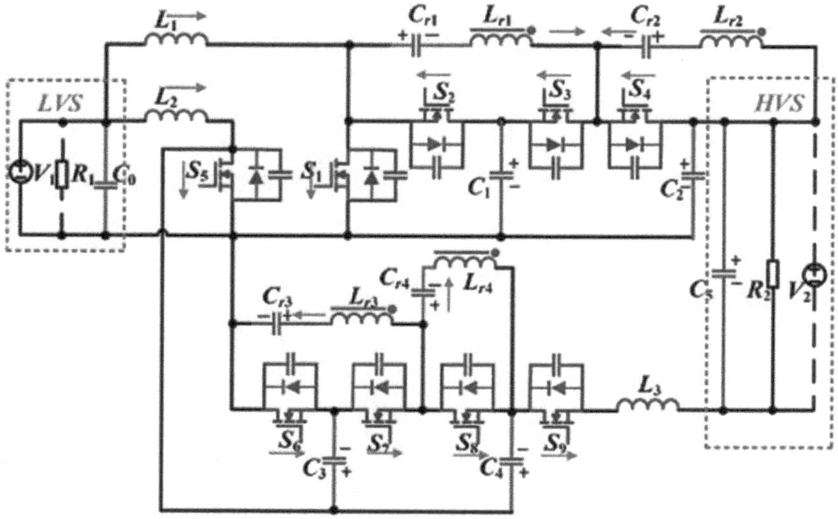

FIGURE 19.21 Bidirectional DC–DC converter using a resonant network [43].

semiconductor devices. Additionally, the converter's design has been streamlined by using single inductors, which further reduces current ripples and bulkiness. The converter demonstrates an efficiency range of 94%–98.9%, and closed-loop system stability has been attained.

19.3.9 Bidirectional DC–DC Converter Using a Resonant Network

The proposed converter in [43], as shown in Figure 19.21, consists of a resonant network composed of a coupled inductor and resonant capacitors, which is integrated into an interleaved buck/boost BDC to achieve high voltage gain. The proposed converter separates the two-phase buck/boost capacitors, arranging them in a series structure with the resonant network of each phase. The positive phase circuit comprises components L_1, S_1–S_4, L_{r1}, C_{r1}, L_{r2}, C_{r2}, C_1, and C_2. The reverse phase circuit includes L_2, S_5–S_8, L_{r3}, C_{r3}, L_{r4}, C_{r4}, C_3, and C_4. By connecting the positive phase circuit in series with the reverse phase circuit, the converter achieves a high voltage gain.

L_1 and L_2 are filter inductors at the low-voltage side, each with an inductance of L_{fL}. L_1, S_1, S_2, and C_1 together forming one conventional buck/boost DC–DC converter structure, while L_2, S_5, S_6, and C_3 constitute the other buck/boost DC–DC converter structure. Additionally, L_{r1} through L_{r4} are four windings of a coupled inductor with an equal number of turns, each connected in series with resonant capacitors C_{r1}through C_{r4}. The leakage inductors of these windings create resonant cavities with the corresponding resonant capacitors.

Additionally, the coupled inductor helps control the energy of circulating leakage inductors and suppresses variations in the resonant current. This enables soft switching across a broad voltage range, whether at light or full load. The converter also benefits from small phase currents, reduced current ripple at the low-voltage side, and lower voltage stress on the devices. This is achieved through its interleaved parallel structure at the low-voltage side and series connection at the high-voltage side. Consequently, the proposed converter is well-suited for interfacing between supercapacitor packs characterized by their widely varying terminal voltages and the high-voltage DC bus in hybrid energy source electric vehicles.

19.3.10 Bidirectional Quadratic DC–DC Converter

The main structure of the proposed converter [44], as outlined in [45], is based on the Zeta converter structure described in [46]. However, unlike the original design, the coupled inductor has been replaced with a conventional inductor to reduce current ripple. The proposed converter operates in both step-up and step-down modes. It features only two active switches per power flow direction, with the other two switches turned off and their body diodes remaining in operation. Compared to a traditional bidirectional buck-boost converter, the proposed converter offers few more benefits in both continuous conduction mode and discontinuous conduction modes. This converter offers enhanced step-up/step-down voltage gains. This topology has reduced input current ripple in step-up mode and decreased output current ripple in step-down mode. Consequently, in step-up mode, the current stress on the input power supply is minimal, and a smaller output capacitor can be used in step-down mode. To streamline the performance analysis of the proposed converter, parasitic resistances are disregarded, and only the leakage inductances of the coupled inductor are taken into account. Additionally, capacitors are assumed to be sufficiently large so their voltage remains nearly constant throughout one switching period (Figure 19.22).

19.3.11 Cascaded High Gain Bidirectional DC–DC Converter

A new multistage, non-isolated bidirectional DC–DC converter with a high voltage conversion ratio is proposed in [47]. This converter can achieve a high voltage conversion ratio (VCR) that is n times greater than that of a conventional buck or boost

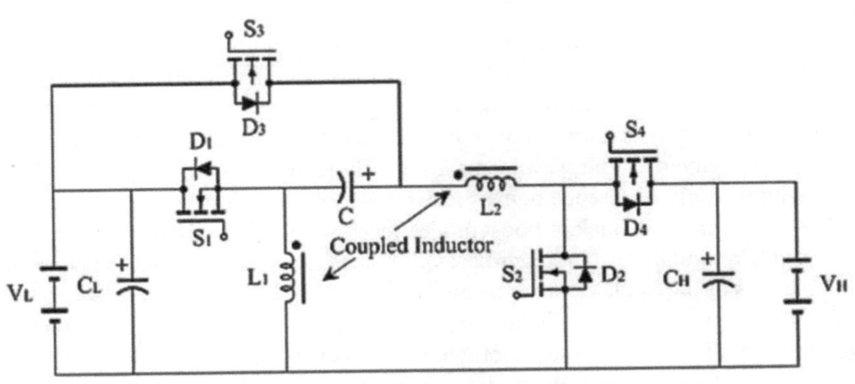

FIGURE 19.22 Bidirectional quadratic DC–DC converter [44].

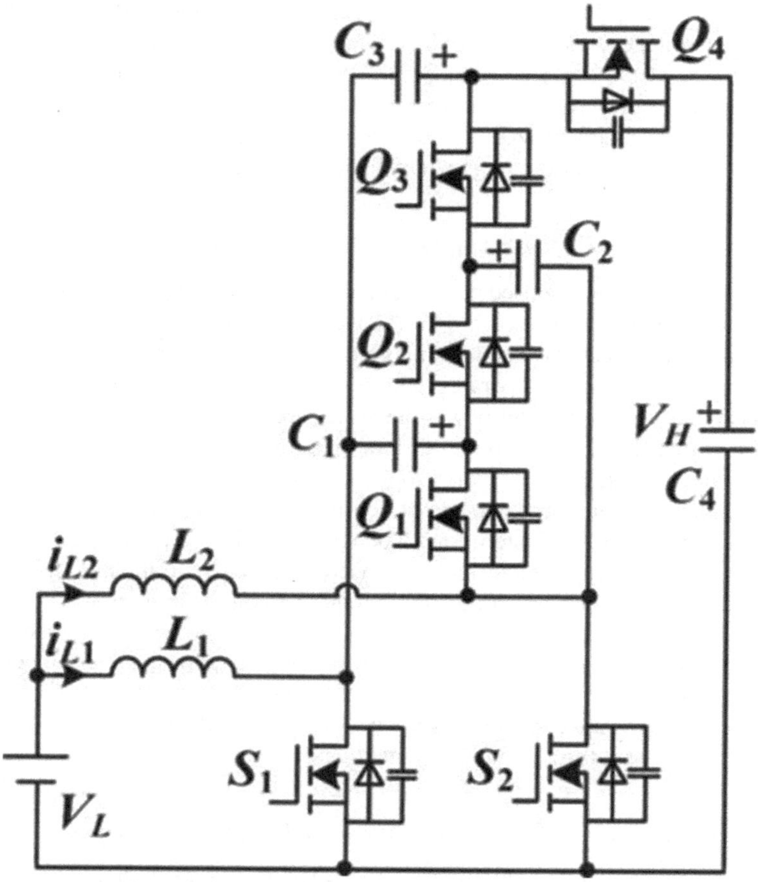

FIGURE 19.23 Cascaded high gain bidirectional DC–DC converter [47].

converter, while utilizing only $n + 2$ switches. Additionally, all switches are capable of full-range zero-voltage switching regardless of the direction of power flow.

The proposed four-stage high-VCR BDC is illustrated in Figure 19.23. In boost mode, switches S_1 and S_2 function as the main switches, while Q_1, Q_2, Q_3, and Q_4 perform synchronous rectification. In buck mode, the roles of the switches are reversed. The operating principles and waveforms are nearly identical in both boost and buck modes, with the only difference being the direction of the inductor currents.

Whether operating in buck or boost mode, once the switching frequency is set, the current ripple remains constant regardless of the load. At light loads, the positive and negative peaks of the inductor currents are sufficiently large to meet the zero-voltage switching conditions for all switches. However, a significant amount of non-active power and circulation losses reduce efficiency. At heavy loads, the average inductor currents are so high that only a small amount of reverse resonant current remains to discharge the junction capacitors, making it challenging to achieve zero-voltage switching for the main switches. To address these issues, a dynamic frequency modulation control strategy is proposed in [47]. At light loads, the switching frequency is increased to reduce the inductor current ripple and circulation losses while ensuring that zero-voltage switching conditions are met. At heavy loads, the frequency is decreased to increase the current ripple and enable soft switching.

19.3.12 Bidirectional Multiport Structure of the DC–DC Converter

The suggested topology in [48] consists of three ports in addition to one auxiliary port and is depicted in Figure 19.24. The topology that consists of three power MOSFETs Q_1, Q_2 and Q_3 and two inductors can work in buck and boost operations while reversing the directions. The multiport structure works to increase the output voltage by combining the solar panel output and the bidirectional DC–DC converter. As shown in Figure 19.24, V_{high} and V_{low} are the bidirectional ports, and V_2 acts as the second source of the converter to increase or decrease the voltage conversion ratio. During boost mode, energy is transferred to V_{low} through V_2, while in buck mode, energy is transferred to V_{high} through V_2. The merits of the proposed converter are low components, low volume, and high efficiency.

19.4 CONTROL SCHEMES IN THE BIDIRECTIONAL DC–DC CONVERTER

In addition to selecting the appropriate topology for bidirectional DC–DC converters, a high efficiency and unified control strategy is essential. However, choosing an appropriate control scheme for BDCs depends on the specific topologies and control

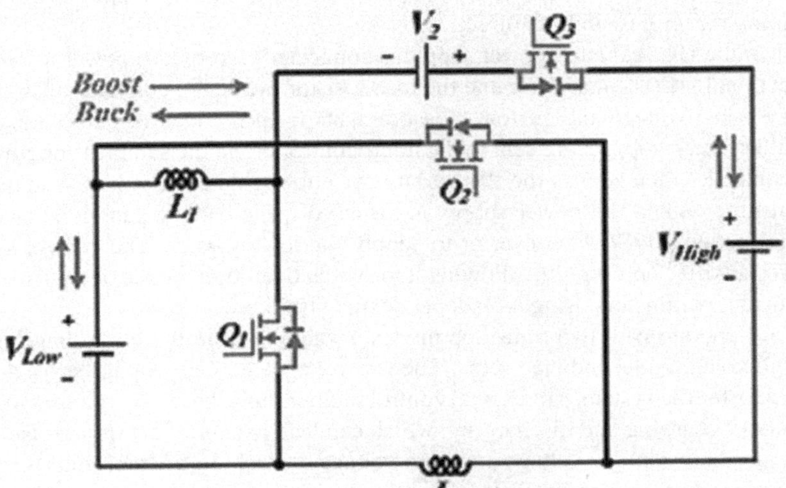

FIGURE 19.24 Multiport structure of the bidirectional DC–DC converter [48].

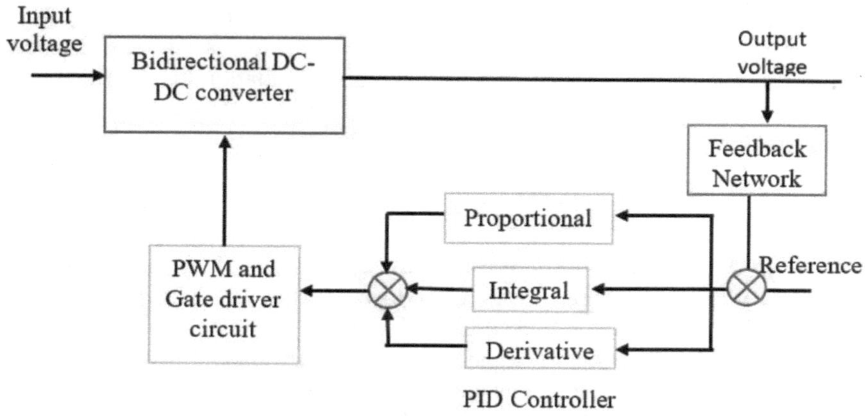

FIGURE 19.25 Structure of the general PID control scheme.

challenges encountered in practical applications. This section explores the control strategies applied to non-isolated DC–DC converter topologies, and the following section deals with the different control strategies.

19.4.1 PID Control Technique

Due to its straightforward implementation, a PID controller is often the preferred choice for designing a control strategy. The general structure of the PID controller is depicted in Figure 19.25. This control approach is widely used across various topologies and control problems and can even be integrated with other control schemes. In non-isolated bidirectional DC–DC converters, the most frequent control challenge is managing power flow in both directions.

When the DC regulated power supply is connected, it provides power to both the bidirectional DC/DC converter and the load resistor, while the converter charges the battery pack. Voltage and current measurements from the load resistor, along with the battery pack voltage, are sent to the microcontroller via the sampling circuit. The PID controller then adjusts the PWM duty cycle to regulate the voltage and current in real time. When the power supply is turned off, the battery pack is boosted by the bidirectional DC/DC converter to supply the load resistor. The control system receives auxiliary power [49], allowing it to switch operating modes or adjust voltage and current parameters using an independent control key.

BDCs operate with two transition modes: switching from the low-voltage side to the high-voltage side, and vice versa. The low voltage and high voltage serve as control inputs for the system. Traditional control methods use these two voltages to manage battery charging and discharging, which can lead to significant transients during transitions between low-voltage and high-voltage control. To address this issue, a PI regulator is employed with a PWM scheme, ensuring uninterrupted power supply to critical AC loads and reducing both the size of the DC bus capacitor and the transition time [50].

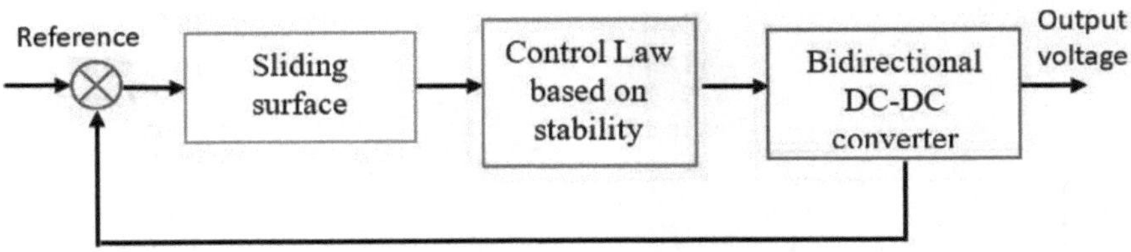

FIGURE 19.26 Structure of the general sliding mode control scheme.

Another control challenge is managing the switching time between the two directions. Fixed turn-on times for the switches can result in excessive power loss. To mitigate this, an auxiliary switch turn-on time control is used, based on a lookup table that factors in load current and the time difference between the main and auxiliary switches [51,52].

19.4.2 Sliding Mode Control Technique

Bidirectional DC–DC converters contain nonlinear elements in their topology, which result in nonlinear dynamic equations. One approach to such a nonlinear system is linearization around its equilibrium point [53,54], but linear models typically ignore perturbations and disturbances.

So, it is preferable to use nonlinear control strategies like sliding mode control, which is depicted in Figure 19.26. The finite-time convergence and low sensitivity to external perturbations of sliding mode control make it an effective control scheme for addressing the nonlinearity conditions. In [55], three distinct sliding surfaces are introduced to analyze the three specific switching states of a bidirectional Cuk converter.

When the dynamics and targeted operating points are uncertain, such as in a bidirectional DC–DC converter connected to a nonlinear load [56], using sliding mode control with a high-pass filter can be effective. This approach ensures robustness against parameter variations, regulates the DC bus voltage, and minimizes transient response under varying nonlinear loads.

However, due to challenges such as large deviations in load and line conditions, a single controller is not sufficient, and so controllers can be combined to achieve a more robust dynamic response and improved performance [57]. A fuzzy sliding mode controller has been developed for managing the energy regeneration of an ultracapacitor battery [58] to enhance the robustness in the presence of perturbations in the actual response around the desired performance.

19.4.3 Fuzzy Logic Control

The general structure of the Fuzzy control system is given in Figure 19.27. A fuzzy logic-based battery energy management system in a hybrid renewable system is proposed in [59].

The fuzzy agent can manage the duty cycle of the DC–DC converter to achieve maximum power point tracking, thereby adjusting the voltage level according to the

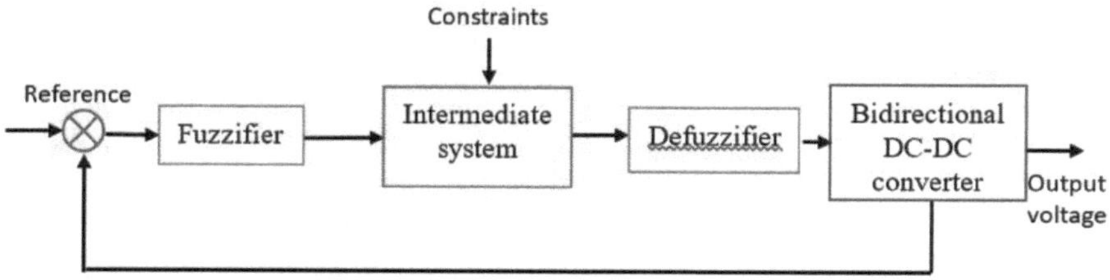

FIGURE 19.27 Structure of the fuzzy control scheme.

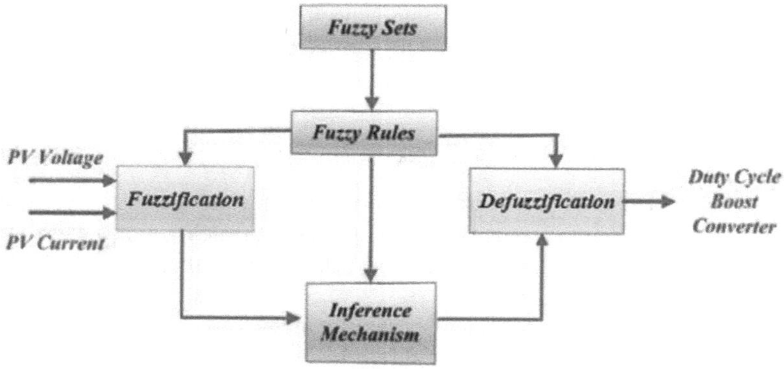

FIGURE 19.28 Structure of fuzzy control for the boost converter [59].

duty cycle. A fuzzy logic controller offers benefits such as rapid convergence, handling imprecise inputs, and managing nonlinearity. Fuzzy typically includes three stages: Fuzzification, Rule Base Lookup Table, and Defuzzification, as illustrated in Figure 19.28. The rules are established based on prior system knowledge. It is an artificial decision-making controller that functions in a closed-loop system. It uses error signals and changes in error signals as inputs. After processing these signals and determining the linguistic variables, the fuzzy logic generates the duty cycle for the buck converter based on the established rules. Compared to PI and P&O controllers, this control is considered the most efficient maximum power point track (MPPT) controller [60,61].

19.4.4 Robust Adaptive Control

An article [62] introduces a novel control strategy that integrates classical control techniques with an optimization scheme to manage the output voltage of a BDC in the presence of both matched and mismatched disturbances, as shown in Figure 19.29.

To address and counteract unknown disturbances, the paper proposes a continuous sliding mode control method based on an extended state observer, which ensures precise tracking, rapid disturbance rejection, and reduced chattering. Subsequently, an extremum seeking-based adaptive scheme is introduced to enhance system robustness

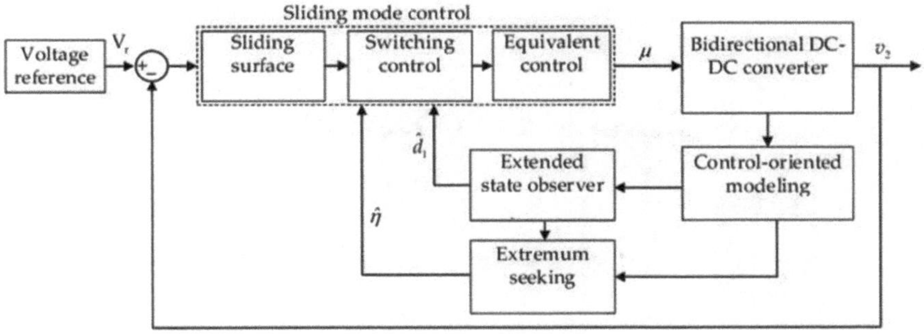

FIGURE 19.29 Structure of the control scheme proposed in [62].

and optimize control efforts across different operational conditions. In this article, different control challenges are addressed, which includes Control-Oriented Modeling, Mismatched Disturbance Estimation and Rejection, Matched Disturbance Suppression, and the adaptive scheme. With appropriate parameter selection, the main goal of tracking control amid disturbance variations and noisy measurements was largely accomplished. However, the proposed controller remained sensitive to noise and could not entirely eliminate chattering. Consequently, an extended state-based adaptive scheme was introduced to adjust the switching parameters of the Sliding Mode Controller to better adapt to varying operational conditions.

19.4.5 Model Predictive Control

A model predictive control strategy is proposed in [63], which avoids the use of proportional-integral-derivative regulators. This strategy includes two components: the Model Predictive Current and Power control scheme and the Model Predictive Voltage and Power control method. The Model Predictive Current and Power algorithm manages the bidirectional DC–DC converter in the battery energy storage system to smooth out fluctuations from RES while maintaining a stable DC bus voltage. Meanwhile, the Model Predictive Voltage and Power scheme controls the AC/DC interlinking converter to ensure a stable AC voltage supply and proper power flow between the microgrid and the utility grid. Additionally, a system-level energy management strategy is developed to ensure stable operation across various modes by taking into account fluctuating power generation, variable power demand, battery state of charge, and electricity prices. This approach, compared to traditional cascade control methods, is simpler and demonstrates improved performance.

Figure 19.30 depicts the proposed Model Predictive Current and Power strategy under grid-connected mode. This strategy involves directly managing the battery's charging or discharging current. To do this, the battery voltage and current, along with the actual DC bus voltage, are utilized to predict the battery current $I_B(k + 1)$ described in [63], and subsequently, the switching behavior that minimizes the objective function is chosen to control the buck-boost converter.

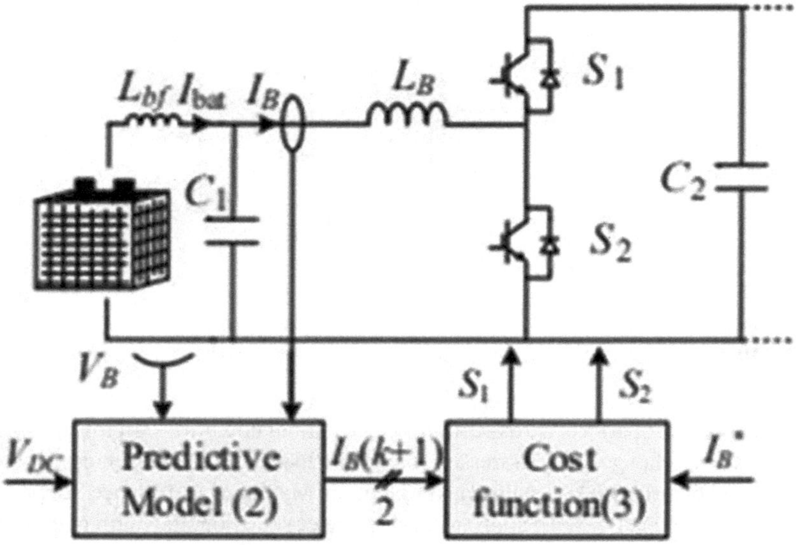

FIGURE 19.30 Structure of the control scheme proposed in [63].

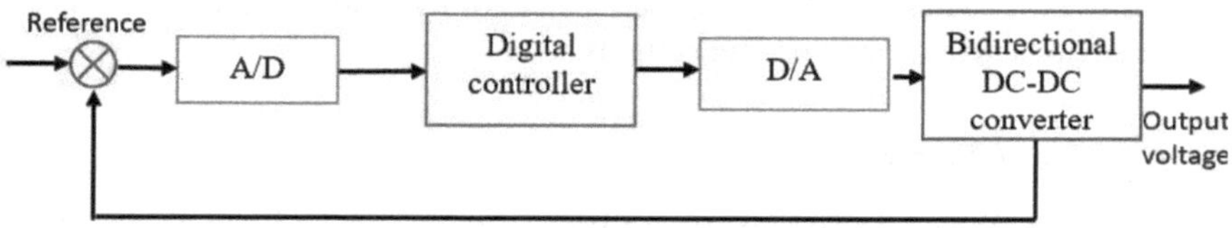

FIGURE 19.31 Structure of the digital control scheme.

19.4.6 Digital Control Scheme

The digital control system receives a continuous-time error signal, which is then converted into a form suitable via an analog-to-digital (A/D) interface. After processing, the discrete signal is sent to a controller, and the controller's output is injected into the converter. The general architecture of the digital controller is shown in Figure 19.31.

With the advent of high-speed microcontrollers, digital control strategies have become increasingly popular for use with bidirectional flyback converters [64]. The challenge in this converter is to minimize capacitive switching losses and electromagnetic interference without needing to sense high-voltage side signals in either the charging or discharging mode of the capacitive load. The proposed approach employs a valley switching technique using a digital controller. In this method, a high-speed comparator compares the input supply voltage with the drain-source voltage of a low-voltage MOSFET. The output of the comparator is then sent to the microcontroller, which detects changes in the comparator output and generates a fixed on-time pulse. This control method effectively drives a specific actuator, demonstrating high efficiency and fast charge/discharge speeds.

In many high-power converters, digital signal processors are increasingly utilized due to their high computational performance available at relatively low cost. The benefits of digital implementations include significantly greater flexibility compared to analog electronics, improved electromagnetic interference immunity, and enhanced capabilities for process and fault monitoring through external interfaces or network connections.

Intelligent digital control algorithms, such as deadband, switch, and soft-start controls, are proposed in [65] to smoothly adjust power flow directions in the converter and to protect it from inrush currents during startup. This approach enables effective bidirectional power flow control and enhances the power conversion efficiency of low-voltage DC distribution systems.

19.5 CONCLUSION

Bidirectional DC–DC converters can be analyzed from multiple perspectives. This chapter provides a thorough review of these converters, focusing on both their topological structures and control schemes. Fundamental BDCs are typically non-isolated topologies, achieved by replacing the unidirectional switches in a basic converter with bidirectional switches. Adding galvanic isolation introduces an additional degree of flexibility, such as adjusting the transformer turn ratio to enhance voltage boosting capabilities, providing isolation between the input and output sides, and enabling the use of multi-input converter designs. This chapter also examines the control schemes for non-isolated topologies of bidirectional DC–DC converters.

The study examines both the advantages and disadvantages of the modified bidirectional DC–DC converters. It compares various factors, including the control complexity, power-rating applications, efficiency, and the ability to minimize losses across all proposed converters. Additionally, each modified bidirectional DC–DC converter is evaluated from the perspective of renewable energy power generation. The comparison results aim to provide a reference for designing and developing new non-isolated bidirectional DC–DC converters based on the buck-boost architecture. Modifications involving multiport or multi-topology approaches are recommended for constructing and integrating multiple renewable energy power plants, which are clean and have long-term sustainability.

However, interleaved topologies and SiC technology have gained popularity at higher power levels due to their ability to reduce volume, despite increasing the number of components. Interleaved half-bridges are particularly favored for applications that operate without isolation. At present, proportional integral (PI) control is the most widely used method because of its simplicity. Nevertheless, to manage the nonlinearities of DC–DC converters effectively, multiple control loops need to be incorporated.

REFERENCES

1. Shahsavari A, Akbari M., "Potential of solar energy in developing countries for reducing energy-related emissions," *Renewable and Sustainable Energy Reviews*, 2018, 90:275–291.
2. Kumar J, Parhyar NR, Panjwani MK, et al., "Design and performance analysis of PV grid-tied system with energy storage system," *International Journal of Electrical and Computer Engineering*, 2021, 11:1077–1085.
3. Ponnalagarsamy S, Jothi Swaroopan NM, "Performance measures of positive output superlift luo converter using multitudinous controller," *International Journal of Power Electronics and Drive Systems*, 2018, 9:704.
4. Zarrabi N, Sandberg OJ, Zeiske S, et al., "Charge-generating mid-gap trap states define the thermodynamic limit of organic photovoltaic devices," *Nature Communications*, 2020, 11:5567.
5. Kharoubi S, El Menzhi L, "Wind turbine doubly-fed induction generator defects diagnosis using rotor currents Lissajous curves," *International Journal of Power Electronics and Drive Systems*, 2020, 11:2083–2090.
6. Li W, Wang D, Zhang Y, et al., "Defect engineering for fuel-cell electrocatalysts," *Advanced Materials*, 2020, 32:1907879.
7. Aghamolaei R, Shamsi MH, O'Donnell J, "Feasibility analysis of community-based PV systems for residential districts: A comparison of on-site centralized and distributed PV installations," *Renewable Energy*, 2020, 157:793–808.
8. Dos Santos EC, Cabral da Silva ER, *Advanced Power Electronic Converters: PWM Converters Processing AC Voltages*. John Wiley & Sons, 2014.
9. Batarseh I, Harb A., *Power Electronics: Circuit Analysis and Design*. Springer International Publication, 2017.
10. Gorji SA, Sahebi HG, Ektesabi M, Rad AB, "Topologies and control schemes of bidirectional DC–DC power converters: An overview", *IEEE Access*, 2019, 7:117997–118019.
11. Wang Y, Lucia O, Zhang Z, et al., "A review of high frequency power converters and related technologies," *IEEE Open Journal of the Industrial Electronics Society*, 2020, 1:247–260.
12. Parvez M, Pereira AT, Ertugrul N, et al., "Wide bandgap DC–DC converter topologies for power applications," *Proceedings of the IEEE*, 2021, 109:1253–1275.
13. Affam A, Buswig YM, Othman AKBH, et al., "A review of multiple input DC-DC converter topologies linked with hybrid electric vehicles and renewable energy systems," *Renewable and Sustainable Energy Reviews*, 2021, 135:110186.
14. Hossain MZ, Rahim NA, Selvaraj J, "Recent progress and development on power DC-DC converter topology, control, design and applications: A review," *Renewable and Sustainable Energy Reviews*, 2018, 81:205–230.
15. Khaligh A, Antonio MD, "Global trends in high-power onboard chargers for electric vehicles," *IEEE Transactions on Vehicular Technology*, 2019, 68:3306–3324.
16. Mumtaz F, Zaihar Yahaya N, Tanzim Meraj S, et al., "Review on non-isolated DC-DC converters and their control techniques for renewable energy applications," *Ain Shams Engineering Journal*, 2021, 12:3747–3763.
17. Alatai S, Salem M, Ishak D, et al., "A review on state-of-the-art power converters: Bidirectional, resonant, multilevel converters and their derivatives," *Applied Sciences*, 2021, 11(21).
18. Darbali-Zamora R, Quiroz JE, Hernandez-Alvidrez J, et al., Viability assessment of a real-time simulation model for a residential DC microgrid network to compensate electricity disturbances in Puerto Rico. In: *2018 IEEE ANDESCON*, Santiago de Cali, Colombia, pp. 1–6, 22–24 August 2018.

19. Sutikno T, Aprilianto RA, Purnama HS, "Application of non-isolated bidirectional DC–DC converters for renewable and sustainable energy systems: A review," *Clean Energy* 2023, 7(2):293–311.
20. Li BY, Xu C, Li C, et al., "Working principle analysis and control algorithm for bidirectional DC/DC converter," *Journal of Power Technologies*, 2018, 97(4):327–335.
21. Test Report: PMP21529, "4-switch buck-boost bi-directional DC-DC converter reference design", Texas instruments.
22. Meher J, Gosh A, Comparative study of DC/DC bidirectional SEPIC converter with different controllers. In: *2018 IEEE 8th Power India International Conference (PIICON)*, Kurukshetra, India, pp. 1–6, 10–12 December 2018.
23. Cuk S, "A new zero-ripple switching DC-to-DC converter and integrated magnetics," *IEEE Transactions on Magnetics*, 1983, 19(2):57–75.
24. Majo J, Martinez L, Poveda A, et al., "Large-signal feedback control of a bidirectional coupled-inductor Cuk converter," *IEEE Transaction on Industrial Electronics*, 1992, 39(5):429436.
25. He X, "Review of nonisolated high-step-up dc/dc converters in photovoltaic grid-connected applications," *IEEE Transaction on Industrial Electronics*, 2011, 58(4):1239–1250.
26. Huang Y, Xu Y, Zhang W, et al., "PWM frequency noise cancellation in two-segment three-phase motor using parallel interleaved inverters," *IEEE Transaction on Industrial Electronics*, 2019, 34(3):2515–2525.
27. Banaei MR, Zoleikhaei A, Sani SG, "Design and implementation of an interleaved switched-capacitor DC-DC converter for energy storage systems," *Journal of Power Technology*, 2019, 1(1):1–9.
28. Lai CM, Lin YC, Lee D, "Study and implementation of a two-phase interleaved bidirectional dc/dc converter for vehicle and dc-microgrid systems," *Energies*, 2019, 8(9):9969–9991.
29. Bastos MES, Nascimento SO, Torres VCS, et al., "Control strategy for an interleaved bidirectional DC–DC converter applied to battery management in a hybrid aircraft propulsion system," *Journal of Automation and Electrical System*, 2022, 33:965–973.
30. Zhang Y, Gao Y, Li J, et al., "Interleaved switched-capacitor bidirectional DC-DC converter with wide voltage-gain range for energy storage systems," *IEEE Transactions on Power Electronics*, 2018, 33(5):3852–3869
31. Qin Y, Yang Y, Li S, et al., "A high-efficiency DC/DC converter for high-voltage-gain, high-current applications," *IEEE Journal of Emerging and Selected Topics*, 2019, 8:2812–2823.
32. Krishnaveni S, Rajini V, "Cascaded boost converter-based high-voltage pulse generator for pulsed electric field applications," *Archives of Electrical Engineering*, 2021, 70(3):631–641.
33. Suryoatmojo H, Pratama IA, Soedibyo, "Non-inverting cascaded bidirectional buck-boost DC-DC converter with average current mode control for lithium-ion battery charger," *Journal on Advanced Research in Electrical Engineering*, 2021, 5(2):93–100
34. Pirashanthiyah L, Edirisinghe HN, De Silva WMP, et al., "Design and analysis of a three-phase interleaved DC-DC boost converter with an energy storage system for a PV system," *Energies*, 2024, 17:250.
35. Xie B, Wang J, Jin Y, et.al., "Power distribution control scheme for a three-phase interleaved DC/DC converter in the charging and discharging processes of a battery energy storage system," *Journal of Power Electronics*, 18(4):1211–1222.
36. Farooq A, Malik Z, Qu D, et al., "A three-phase interleaved floating output boost converter," *Advances in Materials Science and Engineering*, 2015, 409674.

37. Hajiabadi MK, Eshkevari AL, Mosallanejad A, "A bidirectional two-input step-up DC–DC converter with high gain, continuous input current, and common-ground for EVs and renewable energy system," *IET Power Electronics*, 2023, 16(3):458–471.
38. Shaneh M, Niroomand M, Adib E, "Non-isolated interleaved bidirectional DC–DC converter with high step voltage ratio and minimum number of switches," *IET Power Electronics*, 2019, 12:1510–1520.
39. Gaurav, NJ, Halder S, Pulavarthi SVK, et al., "An ultra-high gain compact module bidirectional DC–DC converter for energy storage system," *IEEE Access*, 2023, 11:134023–134039.
40. Kumar KSR, Pandian A, Sastry VV, et al., "Capacitor clamped coupled inductor bi-directional DC-DC converter with smooth starting," *Machines*, 2022, 10:47.
41. Sheeja V, Kalpana R, "Interleaved high gain bidirectional DC-DC converter for grid integrated solar PV fed telecommunication BTS load," In: *2018 8th IEEE India International Conference on Power Electronics (IICPE)*, Jaipur, India, pp. 1–6, 2018.
42. Biswas M, Kim HC, Park JW, "A coupled inductor-based high step-down/ step-up DC-DC nonisolated bidirectional converter with reduced ripple in current and voltage stress," *IEEE Journal of Emerging and Selected Topics in Power Electronics*, 2004, 12(4):3574–3587.
43. Yuan F, Hao R, You X, Xiang P, "A high voltage gain soft-switching bidirectional DC-DC converter with wide range using resonant network," *IEEE Transactions on Power Electronics*, 2023, 38(11):14099–14114.
44. Akhormeh ARN, Abbaszadeh K, Moradzadeh M, Shahirinia A, "High-gain bidirectional quadratic DC–DC converter based on coupled inductor with current ripple reduction capability," *IEEE Transactions on Industrial Electronics*, 2021, 68(9):7826–7837.
45. Naderi A, Abbaszadeh K, "High step-up DC–DC converter with input current ripple cancellation," *IET Power Electrotonic*, 2016, 9(12):2394–2403.
46. Rashid MH, *Power Electronics Handbook*. Elsevier, 2007.
47. Chen J, Sha D, Yan Y, et al., "Cascaded high voltage conversion ratio bidirectional nonisolated DC–DC converter with variable switching frequency," *IEEE Transactions on Power Electronics*, 2018, 33(2):1399–1409.
48. Shayeghi H, Pourjafar S, Lopez J, "A bidirectional multi-port structure of DC-DC converter for energy storage systems," *Journal of Energy Management and Technology*, 2023, 7(2):69–79.
49. Anand R, Mary PM, "Improved dynamic response of dc-to-dc converter using hybrid pso tuned fuzzy sliding mode controller," *Circuits and Systems*, 2016, 7(6):946–955.
50. Ding S, Wu H, Xing Y, Fang Y, Ma X, "Topology and control of a family of non-isolated three-port DC-DC converters with a bidirectional cell," In: *Proceedings IEEE APEC*, Long Beach, CA, USA, pp. 1089–1094, 2013.
51. Lee JH, Yu DH, Kim JG, et al., "Auxiliary switch control of a bidirectional soft-switching DC/DC converter," *IEEE Transactions on Power Electronics*, 2013, 2(8):5446–5457.
52. Engelen K, De Breucker S, Tant P, et al., "Gain scheduling control of a bidirectional DC–DC converter with large dead-time," *IET Power Electronics*, 2014, 7(3):480–488.
53. Filsoof K, Lehn PW, "A bidirectional multiple-input multiple output modular multilevel DC-DC converter and its control design," *IEEE Transactions on Power Electronics*, 2016, 31(4):2767–2779.
54. Araiza AT, Medina JLM, "Variable structure with sliding mode controls for DC motors," In: *Proceedings CIEP*, San Luis Potosi, Mexico, pp. 26–28, October 1995.
55. Martinez-Salamero L, Calvente J, Giral R, et al., "Analysis of a bidirectional coupled-inductor Cuk converter operating in sliding mode," *IEEE Transactions on Circuits System. I, Fundamental Theory Applications*, 1998, 45(4):355–363.

56. Tahim APN, Pagano DJ, Ponce E, "Nonlinear control of DC-DC bidirectional converters in stand-alone DC microgrids," In: *Proceedings IEEE CDC*, Maui, HI, USA, pp. 3068–3073, December 2012.
57. Dominguez X, Camacho O, Leica P, et al., "A fixedfrequency sliding-mode control in a cascade scheme for the half-bridge bidirectional DC-DC converter," In: *Proceedings IEEE ETCM*, Guayaquil, Ecuador, pp. 1–6, October 2016.
58. Cao J, Cao B, Bai Z, Chen W, "Energy-regenerative fuzzy sliding mode controller design for ultracapacitor-battery hybrid power of electric vehicle," In: *Proceedings of the International Conference on Mechatronics and Automation*, Harbin, China, pp. 1570–1575, 2007.
59. Nagaiah M, Chandra Sekhar K, "Analysis of fuzzy logic controller based bi directional DC-DC converter for battery energy management in hybrid solar/wind micro grid system," *International Journal of Electrical and Computer Engineering*, 2020, 10(3):2271–2284.
60. Tiwari R, Ramesh Babu N, Sanjeevikumar P, et al., "Coordinated DTC and VOC control for PMSG based grid connected wind energy conversion system," In: *Environment and Electrical Engineering and 2017 IEEE Industrial and Commercial Power Systems Europe (EEEIC/I&CPS Europe)*, pp. 1–6, 2017, Milan, Italy.
61. Kumar K, Tiwari R, Ramesh Babu N, et al., "Analysis of MISO super lift negative output luo converter with MPPT for DC grid connected hybrid PV/wind system," *Energy Procedia*, 2018, 145:345–350.
62. Trinh HA, Nguyen DG, Phan VD, et al., "Robust adaptive control strategy for a bidirectional DC-DC converter based on extremum seeking and sliding mode control," *Sensors*, 2023; 23(1):457.
63. Shan Y, Hu J, Chan KW, et al., "Model predictive control of bidirectional DC–DC converters and AC/DC interlinking converters—A new control method for PV-wind-battery microgrids," *IEEE Transactions on Sustainable Energy*, 2019, 10(4):1823–1833.
64. Thummala P, Maksimovic D, Zhang Z, et al., "Digital control of a high-voltage (2.5 kV) bidirectional DC–DC flyback converter for driving a capacitive incremental actuator," *IEEE Transaction on Power Electronics*, 2016, 31(12):8500–8516.
65. Jung JH, Kim HS, Ryu MH, et al., "Design methodology of bidirectional CLLC resonant converter for high-frequency isolation of DC distribution systems," *IEEE Transaction on Power Electronics*, 2013, 28(4):1741–1755.

Index

Kolibri 360 GmbH
Ferdinand-Jühlke-Straße 7
99095 Erfurt, Deutschland
produktsicherheit@kolibri360.de